EUGENE ZALESKI
AND HELGARD WIENERT

Technology transfer between EAST and WEST

ORGANISATION FOR ECONOMIC CO-OPERATION AND DEVELOPMENT

PARIS 1980

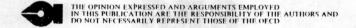

Publié en français sous le titre:

TRANSFERT DE TECHNIQUES
ENTRE L'EST ET L'OUEST

* *

TABLE OF CONTENTS

Chapter 1

EAST-WEST TRADE AND TECHNOLOGY TRANSFER
IN HISTORICAL PERSPECTIVE

Chapter 2

STATISTICAL EVALUATION OF TECHNOLOGY TRANSFER

Chapter 3

THE FORMS OF TECHNOLOGY TRANSFER

Chapter 4

EASTERN AND WESTERN POLICIES
TOWARD TECHNOLOGY TRANSFER

Chapter 5

THE INFLUENCE OF TECHNOLOGY TRANSFERS ON EASTERN ECONOMIES

Chapter 6

EFFECT OF ECONOMIC FACTORS ON EAST-WEST TECHNOLOGY TRANSFER

5

LIST OF TABLES

Chapter 2

7

ACKNOWLEDGEMENTS

The authors of this study have met on an occasional basis with the following persons, to whom they would like to express their gratitude for invaluable advice and comments:

Mr. R. AMANN
Centre for Russian and East European Studies
University of Birmingham, England

Mr. Z.M. FALLENBUCHL
University of Windsor
Canada

Mr. M. FESHBACH
Bureau of Economic Analysis
US Department of Commerce
Washington, D.C.

Mr. P. HANSON
Centre for Russian and East European Studies
University of Birmingham, England

Mr. J.P. HARDT
Congressional Research Service
Library of Congress
Washington, D.C.

Mr. P. KNIRSCH
Osteuropa Institut of the Freie Universität Berlin
Germany

Mr. A. NOVE
University of Glasgow
Scotland

Mr. H. VOGEL
Bundesinstitut für Ostwissenschaftliche und Internationale Studien
Köln, Germany

Mr. ZALESKI wishes to thank the members of the team "The Economy in and Planning Techniques of the Eastern Countries" of the Centre national de la recherche scientifique, Paris, in particular, Mr. Wilhelm JAMPEL.

INTRODUCTION

A. PURPOSE AND LIMITS OF THE STUDY

This study constitutes a review and critical assessment of the literature concerned with East-West technology transfer. Although not intended to be an original piece of research, it does seem to be the first attempt that has been made to deal with the whole spectrum of problems posed by technology transfer in the context of East-West economic relations. The purpose of the study is to provide an *overall analysis of the problem of East-West technology transfer:* to assess the current state of knowledge of East-West technology transfer in the West, and in so doing to identify, if only implicitly, areas where further study may be needed.

The scope and complexity of the subject matter has made it necessary to impose some limits on the study. First, the choice of sources to be consulted had to be restricted. It was not possible to make extensive use of East European and Soviet sources. To consult in a thorough fashion the literature of the seven CMEA (Council for Mutual Economic Assistance) countries with which this study is concerned was not possible, given the time constraints within which the study was conducted. That literature has however been referred to on occasion. Soviet sources have been used predominantly because they are most easily available and because among the CMEA countries the Soviet Union dominates the issues under consideration here.

Even for Western sources, it was not possible to consult all the available literature. US sources tend to predominate because they have been most easily accessible, and because work of an equivalent standard is to be found in only a few West European countries (notably in the United Kingdom, Germany and France).

The analysis of the subject areas covered has also been restricted in two respects. Firstly, the availability of source materials necessarily determined the direction and depth of the analysis. Secondly, greater emphasis was placed on those countries which play a predominant role in East-West technology transfer.

B. THE PERSPECTIVE OF THE STUDY

Numerous papers on technology transfer have been published in recent years, as can be seen from the selected bibliography. However, they are all confined to specific aspects of a problem or to certain countries and their conclusions tend to be restricted to the particular subject at hand. For example, a study of a particular Western country will be mainly concerned with that country's transfer problems, and a paper at the private company level

11

may fail to consider, or at least not deal fully with, national implications. Lastly, strategic and political problems may be (and frequently are) overlooked in studies of the economic benefits and profit potential of technology transfer.

Another shortcoming of this piecemeal and often short-term approach is that statistics and documentation on the terms of transfer contracts are often incomplete and obtained from indirect sources. In addition, trends of the moment are often projected too hastily and not seen in their historical, political or social contexts.

This study tries to avoid these pitfalls by approaching the subject from four different perspectives:

— the historical background of East-West technology transfer (Chapter 1);
— an attempted evaluation of technology transfer (Chapters 2 and 3);
— a survey of the political, technological and economic problems which govern and affect transfer (Chapters 4, 5 and 6);
— the effect of technology transfer on Western economies (Chapter 7).

Historical Background

Chapter 1 seeks to distinguish between the permanent or long-term features and the transient aspects of present technology transfer, so that their respective implications may be more clearly understood.

One basic feature which emerges is that technology transfer (particularly in the West-East direction) is a long standing process; at times influenced by political and economic events, but never entirely interrupted by them. Another, is that the technology flows between the Soviet Union and the Western industrialised countries differ considerably from those between the East European countries and the West. It is also clear that there are so many differences in the scale of trade, the international division of labour and policies in the various countries that any aggregation tends to be misleading.

Given that certain features of present conditions — such as the increase in bilateralism and the massive debt — are by no means new, future solutions may well be based on those of the past. This historical review also enables currently fashionable terms such as the "international division of labour", "industrial co-operation", "internationalisation of the production process" and "scientific and technical co-operation", to be seen in their true dimensions. The effects of attempts to "export" clearing practices, barter and bilateralism in trade arrangements are all too easily forgotten.

Assessment of Transfer

The first essential, of course, is to have a definition of "technology transfer". While definitions vary, the one that has been adopted here is that of a *process whereby innovations (new products or know-how) obtained in one country are then transmitted for use to another*.[1] It is essentially a process of transmitting *knowledge* and implies an active role on the part of both transferer and recipient.

Technology could be defined as systematic knowledge of applied arts, but empirically neither the science nor the product concerned can be measured. What we are focusing on is the application of science to the production of goods and services.[2] Technology transfer from one country to another signifies that the new products or know-how introduced in one

1 *Technology Transfer and Scientific Co-operation Between the United States and the Soviet Union: A Review*, prepared for the Sub-Committee on International Security and Scientific Affairs of the Committee on International Relations by the Congressional Research Service, Library of Congress, 95th Congress, 1st Session, 26th May, 1977, US Government Printing Office, Washington, D.C., 1977.
2 J. Fred Bucy, "On Strategic Technology Transfer to the Soviet Union", *Current News*, Special Edition, 11th August, 1977, p. 4.

12

country are used in another without the intervention of an independent cycle of research, project development, testing and evaluation, either directly or through the agency of a foreign producer who has followed the original inventor.[3]

Chapters 2 and 3 attempt to measure two kinds of technology transfer: the part that is directly "measurable", contributing to products entering the markets, and the part that is more difficult to quantify (information, skills, and mobility) which is identified indirectly through a study of the various forms of transfer (licensing, plant sales, co-production, etc.).

In both cases, our assessment is incomplete, because of methodological inadequacy and lack of data. The assessment of "measurable" transfer is itself difficult because of the difficulty of deciding which products to consider. Should only research-intensive products (and if so, which), or those representing high technology be included, or should the assessment cover just machinery or even only certain types of machinery? Owing to the pace of technical progress in some sectors, the statistical breakdown available is not always appropriate nor is data sufficiently uniform. In any case, the study of the value of transfers included in different contracts, in licences, and so on, does not lend itself to statistical evaluation. It requires complementary analysis of a wide body of literature, the conduct of surveys and the carrying out of case studies. While the assessment made here is both modest and incomplete, it may nevertheless give a valuable and useful idea of the flows of know-how.

Political, Technological and Economic Problems affecting Transfer

Chapters 4, 5 and 6 are the key parts of the study.

As used here, the term "political" signifies, in a broad sense, both general government policy and policy based on strategic, ideological or economic considerations. It encompasses the interaction between national and individual interests as well as individual Western and Eastern policies. In these policies it is not always easy to identify the exact role of technology transfer and there is a risk of according too much importance to certain aspects (e.g. strategic considerations).

In the case of East European policies, greater weight has been given to the USSR because of the size of the country and the fact that source material was more readily available. For Western countries, the emphasis is placed on US policy since far more work has been published on the subject in the United States than in other industrialised Western countries. Important attention is also focused on the policy of the United Kingdom, France and Germany. In any case, all Western countries do not have the same policies on technology transfer. To trace them both in the context of international organisations (United Nations, NATO, European Economic Commission, etc.) and at national level would be a task which would quite clearly be outside the scope of our study.

The question of technological level in technology transfer raises a formidable problem. Here also it was necessary to limit the analysis to study of the USSR. Using the USSR as an illustrative example we have looked at the innovative capability of planned economies from two angles: the capacity of the system to innovate itself, and the identification of factors — general and specific — which favour or impede innovation. This is followed by a review of existing studies showing the influence of Western technology on the Soviet economy in general and on innovation in particular. Finally, a summary is given of research on East-West technological lag.

Certain sectors such as armaments, agriculture, light industry and food have not been treated; either because the technological lag in those sectors has not yet been investigated or

---

3 Philip Hanson, "External Influences on the Soviet Economy Since the Mid-1950s: The Impact of Western Technology", CREES Discussion Paper, No. 7, Birmingham, United Kingdom, 1977, p. 9.

because other Western international organisations seem better placed for their study (e.g. NATO for armaments, the space industry, etc.). Furthermore, technological lead or lag is a complex, constantly shifting process, and a general survey such as this can be based only on the results of previous case studies.

The effect of economic factors on East-West technology transfer is equally difficult to assess. In particular, it is not easy to separate technology transfer from trade in general or to determine which of the economic factors is dominant. Neither is it possible to regard economic factors as *known* and merely to study their impact. We have felt it best to limit the study to four economic factors having a direct effect on technology transfer level: short-term economic trends, prices and terms of trade, debt and countertrade.

It seems that technology transfer is very sensitive to such economic factors as economic conditions in the West and the East's ability to pay. The question most frequently asked — and most imperfectly answered — is the impact of transfer on the Western economies. No overall study of this question has ever been undertaken and some specialists think that such a task would be materially impossible. Nevertheless specific issues regularly arise and Chapter 7 summarises how they have been treated.

The first issue concerns the economic benefits of technology transfer. For the West those most frequently cited are the creation of stable supply sources, job creation and profit opportunities. The discussions found in the various case studies have been far from satisfactory. The issue is still ambiguous and additional research on the subject could well provide useful background for Western policy.

Another issue is that of competition, a distinction being drawn between competition among Western firms in East European markets and the danger of competition from Western technology exported to the East. We have tried to summarise the views which have been put forward in the Western literature. Drawing the line between "normal" and "unfair" (dumping) competition has been particularly difficult in view of the differences between economic accounting practice in East and West.

The most important, though indirect, effect of Western technology transfer to the East is the trend towards bilateralism and barter in East-West trade and in the world trading system in general. This effect has not yet been adequately investigated in the West and we have only been able to collect isolated comments by representatives of Western firms. Yet this is a major problem which could well have a decisive influence on East-West technology transfer in the future.

C. TERMINOLOGY

While this study deals primarily with technology transfer from West to East, the terminology "East-West" has been used as this is the standard usage adopted in the literature. The terms "East" and "West" are used to signify two fundamentally different economic systems. The "West" comprises the system of market-economy countries and the "East" the European system of centrally planned economies. The central theme of the analysis is thus the relationship between these two fundamentally different systems.

The terms "East" and "West" are not precise, and they cannot easily be statistically defined. For this reason, in the statistical tables reference is made, in most cases, to "OECD countries" and "CMEA countries": terms which define the membership of each group.

SUMMARY

CHAPTER 1: EAST-WEST TRADE AND TECHNOLOGY TRANSFER IN HISTORICAL PERSPECTIVE

Technology transfer, particularly in the West-East direction, is a long standing process. This chapter identifies the permanent features of this process and indicates the main shifts in policy which have taken place. Five periods are discussed: 1) pre-World War I until the end of the 1920s; 2) the period immediately following the pre-World War II economic crisis; 3) the immediate post-War period — up to 1953; 4) the years 1953-1965 and, 5) the present period.

1. Pre-World War I until the End of the 1920s

Prior to the Revolution, Russian development was basically inward looking. Nevertheless, Western technology was brought in on an impressive scale to help industrialisation. Trade with the West was far from negligible and increased significantly during 1905-1913. For the Central European countries trade with the industrialised countries of Western Europe (other than the German Empire) represented only a small part of their foreign trade.

During the 1920s two broad features characterised Soviet trade policy: State monopoly of foreign trade and the orientation of Soviet trade towards co-operation with industrialised Western countries. Within this structure foreign technology and foreign trade assumed a growing importance in Soviet economic development. The first Five Year Plan (initiated in 1928) stressed the expansion of foreign trade. The plan called for an 80 per cent increase of imports, a 165 per cent increase of exports, and a doubling of the Soviet Union's share in world trade. Emphasis was placed on acquiring machinery and equipment, which grew to a large share in Western trade with the Soviet Union. In 1929 for instance, it accounted for 53.3 per cent of German-Soviet trade and 42.5 per cent of US-Soviet trade. Germany, Italy, the United States, Britain and France together accounted for 80.2 per cent of Soviet machinery and equipment imports.

Technology transfer (especially through concessions and technical assistance agreements) is believed to have been the most significant factor in Soviet economic development during the period of the 1920s.

2. Technology Transfer and Commercial Policy after the World Economic Crisis

A. *The Soviet Union*

In the 1928-1932 plan a major effort was foreseen to import technologies to support the programme of industrialisation, but by the Spring of 1930, before the plans completion,

15

Soviet economic policy was directed at achieving economic and technological self-sufficiency. That goal was written into the Soviet Constitution. By 1930 the Soviet Union had reduced its foreign trade and followed a policy which has often been called autarkic.

But the Soviet Government was still eager to develop contractual relations with Western countries to acquire modern technology. Technical assistance and trade agreements provided the vehicle to acquire that technology. Thus, while the policy of self-sufficiency led to a decline in foreign trade, technology imports continued to account for a large share of trade and Soviet imports of machinery from the West were always above average. In 1931 for instance, machinery imports still made up 53.5 per cent of the total.

B. *Central Europe*

The world economic crisis caused a drastic decline in the foreign trade of Central Europe, which had immediate repercussions on the ability of the Central European countries to pay their debts. In 1932 payments on foreign debts were suspended and exchange controls were introduced. A system of new tariffs, clearing agreements and administrative measures was erected to protect national activities. While the geographical distribution of trade changed, the main features of the technology flows which had existed prior to the First World War remained the same. Thus Germany, the industrialised countries of Western Europe and the United States remained the main exporters of machinery and equipment to Central Europe. The trade of Central Europe with the Soviet Union remained at the level of the 1920s: insignificant.

3. The Second World War and the Immediate Post-War Period (until 1953)

Established trade relations were disrupted during World War II and were never to be resumed again in the same manner. The period is marked by two phases: initially, the continued absorption of foreign technology by the Soviet Union; and later the formation of two political and economic "blocs".

Technology was made available to the Soviet Union by means of Lend Lease and the "Pipe Line Agreement", reparations and confiscation of ex-German assets. However, trade did not play a significant role during this period. Soviet trade relationships with the developed countries had been almost completely broken off just before the War. Little commercial trading took place with Central and Western Europe, due to the political uncertainty that prevailed until 1953.

The formation of the *Cominform* in 1947 and the creation of NATO in 1949 marked the political division between East and West. The economic division was illustrated by the refusal of the Soviet Union to join the Marshall Plan (1948) and the Organisation for European Economic Co-operation. The creation of the Council for Mutual Economic Assistance (CMEA) in January of 1949 marked the official resumption of the Soviet Union's policy of pursuing economic independence (within the framework of the Eastern bloc countries).

Technology transfer was also increasingly curtailed by the passage of the US Export Control Act in February, 1949, the establishment in November of that year of a Co-ordinating Committee (an informal multi-lateral co-ordinating mechanism) and enactment by the US government in 1951 of the Mutual Defense Assistance Control Act — the "Battle Act". A variety of legislation also curtailed imports from the East.

Neither the Soviet Union nor the East European countries placed much emphasis on Western trade. The share of exports of the East European countries to Western Europe, the United States and Canada, fell from 72.8 per cent in 1938 to 41.0 per cent in 1948 and 15 per cent in 1953. Soviet trade with the industrialised countries of the West shrank to a fraction of 1 per cent of GNP. Trade with the West, as a percentage of total CMEA trade fell from 42 per cent in 1948 to 14 per cent in 1953.

4. Trade with the West and Technology Transfer after Stalin: the period 1953-1965

The years 1953-1965 were a transitional period during which a climate more conducive to normal relationships between East and West developed. While trade during the period 1953-1965 showed only a modest increase, the groundwork was laid for the closer and markedly changed trading relations that followed.

The policy of economic independence began to pose serious problems for the East European countries. From 1953 the policy of autarky was increasingly questioned. A new orientation of economic policy emerged in favour of closer relations with the West. Foreign trade came to be viewed as a factor which could help counteract the slow-down in growth rates. The East viewed the import of industrial technologies as a crucial input in economic growth. Technological help from the West also enabled Eastern countries to reduce the political risks of economic reforms.

In the West, national trading interests began to respond strongly to the promise of new markets. This led to a disparity between US and West European policies towards East-West trade. Fundamental changes were less acceptable to the United States than to its partners. The countries of Western Europe took a more flexible attitude toward export and import controls than the United States. Western European nations broke with US credit policies towards the Eastern countries. West European countries initiated trade agreements with the East. These agreements led in the mid-60s to the conclusion of scientific and technical agreements.

A marked increase in trade between East and West, though still modest in terms of world trade, began during the early 1960s. The foreign trade linkage between East and West again took the pre-War form of exchanges of food, fuels and crude materials from the East, against steel, machinery and industrial equipment from the West. Machinery and transport equipment formed a very large proportion of the OECD countries' exports to the East; increasing in value by 48 per cent between 1961 and 1965.

5. Principal Recent Development in East-West Trade

Since 1965, and particularly between 1972 to 1975, East-West trade has expanded at a rapid rate. A basic condition for this rapid development is believed to be the policy of détente. For both West and East this policy provided political underpinning for "official sanctioning" of closer economic relations.

Present policy in the East shows two, not necessarily convergent, developments. These are bloc integration and support for the notion of the international division of labour. Over the past ten years steps have been taken to improve the functioning of the CMEA and to further trade integration and specialisation. However, the long-term effectiveness of these measures remains a matter of speculation and no small disagreement. Concerning the international division of labour, the Soviet Union now stresses the advantages of international economic relations and has long stressed the principle of intra-industrial specialisation; but this emphasis does not imply economic integration or assimilation with the West.

In the West, the major impetus behind the rapid expansion of trade has been economic competition among Western countries for markets in the East. "Normalisation" of economic relations is reflected in the growing number of trade agreements, long-term economic agreements and scientific and technological agreements. It is also mirrored in the easing of export and import restrictions, and by the lifting of restrictions on credit facilities. East-West economic relations developed during the early 1970s with the assistance of Western credit. The expansion of trade since 1965 has been rapid. Eastern imports from the OECD countries rose almost eight-fold, in current prices, between 1965 and 1977. Exports rose six fold. However, imports declined in 1977. The average annual growth of imports from OECD countries was 22 per cent between 1965 and 1975, but only 5.8 per cent in 1976 and

0.8 per cent in 1977. Exports to the OECD group grew on average by 16.5 per cent a year between 1965 and 1975. In 1976 the increase was 19.1 per cent and in 1977, 9.8 per cent. Machinery imports from the OECD countries occupied a large share in the total, increasing from 29.5 per cent in 1965 to 36.1 per cent in 1975 and 37.2 per cent in 1977. These trends raise a fundamental issue: namely, whether the high growth rates from 1965 to 1975 were exceptional, or whether they were representative of overall world trends and a strengthening of the international division of labour.

CHAPTER 2: STATISTICAL EVALUATION OF TECHNOLOGY TRANSFER

To measure and evaluate technology flows involves formidable methodological difficulties. The "measurable" part of technology transfer is usually only that embodied in marketed products and in the granting of licenses and registration of patents. Even that "measurable" part cannot be easily defined. Trade in technology is therefore evaluated in terms of "research-intensive" and "high technology" products, terminology which is itself problematic.

According to the United Nations, technology-intensive products formed nearly half of East European imports from the West over the period 1965-1974, but only 13-15 per cent (only about 8 per cent in the case of the Soviet Union) of exports to the West over the same period.

Aggregates indicate that machinery and transport equipment accounted for 32.3 per cent of East Europe's total exports in 1975, and 34.5 per cent of imports in the same year. During 1974 and 1975 Eastern imports of machinery and transport equipment increased at a particularly fast rate: by 35 per cent in 1974 and 55 per cent in 1975. Since 1976 the trend changed: machinery imports from the industrialised West levelled off. There was even a slight decrease in electrical machinery and transport equipment. The USSR does not seem to have been affected; in aggregate, USSR machinery imports increased again in 1977 and 1978.

During 1971-1975 there was a very steep increase in trade in machinery, in line with the general growth of trade with the OECD countries. A major change occurred in 1976; OECD exports of machinery and transport equipment to Poland, Rumania and Bulgaria decreased. They increased less steeply than previously to other East European countries. The Soviet Union appears to rely on imports of machinery and transport equipment from the OECD countries. CMEA countries' machinery and transport equipment imports from the OECD countries were very significant: 29.56 per cent of the total in 1965, 30.7 per cent in 1974, 36.1 per cent in 1975 and 38.4 per cent in 1977.

CHAPTER 3: THE FORMS OF TECHNOLOGY TRANSFER

1. Inter-Governmental Agreements on Scientific and Technical Co-operation

Since the mid-sixties the Soviet Union and the East European countries have concluded Scientific and Technical Co-operation Agreements with many countries in the West. One major purpose of these agreements is the strengthening of economic relations. Technology transfer and the exchange of information are seen as central tools for the achievement of this objective. The real scope of these agreements has yet to be defined since they are framework or enabling agreements which facilitate, but do not predicate, the conclusion of private commercial contracts mostly concluded at the request of the CMEA countries. According to experts the significance of the number of such agreements is purely relative. They demonstrate a favourable attitude and an expression of goodwill, and often depend on political factors that create the need for such a climate.

2. Industrial Co-operation Agreements between Western Firms and their CMEA Partners

Industrial co-operation contracts can be defined as arrangements whereby industrial producers agree to pool some assets and jointly coordinate their use in the mutual pursuit of complementary objectives. Definitions of industrial co-operation agreements differ. Whichever definition one adopts, what matters is the way in which know-how is transferred: the way in which knowledge can be effectively used in the importing country. Forms of transfer, i.e., the content of contracts, are of paramount importance. Relatively little is known about the number and content of agreements signed by Western firms and their East European partners, because of the secrecy maintained on both sides. The most widespread form of co-operation is co-production based on specialisation. The next most frequent form is the supply of plant or equipment and licensing in exchange for products manufactured. Joint ventures, sub-contracting — one of the original forms of co-operation — joint tendering or joint projects and tripartite agreements play a much more modest role.

According to a survey conducted by the United Nations, industrial co-operation is concentrated mainly in industries in which technological progress is vital: the chemical industry, mechanical engineering and machine tools, electrical engineering and electronics, transport equipment and metallurgy. It seems that East-West industrial co-operation agreements are a potent medium for technology transfer to Eastern Europe, particularly for applied industrial technology. Know-how transfer is particularly effective through sales of plant and joint ventures. The rising number of industrial co-operation contracts in recent years is impressive. Their attraction for the East seems to consist not only in the technology to be acquired but also in the forms of payment — credit and/or countertrade.

3. Private Enterprise Agreements on Scientific and Technical Co-operation

Little information is available about private scientific and technical agreements. According to the United Nations, agreements on scientific and technical co-operation accounted for 6.9 per cent of the 204 agreements concluded in 1972. For other years, the United Nations' surveys do not provide information concerning the share of scientific and technological agreements. It is difficult to form an idea of the number of commercial contracts confined solely to scientific and technical co-operation. Some of the industrial co-operation agreements signed by Western private firms are enabling agreements; most of them have been signed with the Soviet Union.

4. East-West Licensing Transactions

Licences take many different forms, ranging from a straightforward authorisation to exploit an individual patent, to complex agreements on industrial co-operation. The latter may include the provision of licences for the use of patents linked to the import of certain capital goods and know-how, technical assistance in building turnkey plants or other industrial installations, and licences to use trade marks. The sale of a licence often constitutes the first step towards broader co-operation. In 1972, 28 per cent of East-West co-operation agreements were based on licensing agreements; the figure in 1975 was 26 per cent, and in 1976 17 per cent. Out of 434 co-operation agreements signed by the United States with the CMEA countries, current or ended on 1st January, 1976, 183 were licence or know-how

19

sales. The biggest customers for licences were the Soviet Union and Poland. Topping the list of countries selling licences to the West was Czechoslovakia. Among Western countries, Germany was the leading licensor. While it is difficult to estimate the value of licences traded between East and West, it does seem that the balance of receipts is heavily weighted in favour of the West and that the Eastern deficit is growing. Eastern countries seek licences because they advance technological progress in the relevant field. Furthermore, the licence sold to one East European country may be used by other CMEA countries; and licences help to promote exports. Preferred fields are the technically more dynamic industries producing automatic machine tools, light chemicals, cars, aerospace electronic equipment and data processing equipment. Licensing transactions may be expected to increase in the future, but in absolute terms Eastern sales of licenses will also grow introducing more reciprocity in the flow of know-how. It can be anticipated that the East European countries will increasingly seek to tie fees paid for licences to the export of products manufactured under the licence. This would be a major change since former contracts provided for the payment of a lump sum, or for fees based on domestic production under the licences.

5. Technology Transfer and the Supply of Turnkey Plants

The supply of turnkey plants more often than not includes start-up assistance and in many cases training courses. As such it is one of the most effective channels of technology transfer. It developed strongly during the last decade, particularly in the chemicals, steel and motor industries. The value of the contracts is often considerable. For the East the main advantages of a turnkey plant are superior technology and the speed with which it can be commissioned. While no general review of the supply of turnkey plants has been carried out, available data indicate that the Soviet Union has a particular interest in this form of co-operation. For instance, out of 112 co-operation agreements between the United States and the Soviet Union, 74 are for turnkey projets. Other Eastern countries seem less interested in turnkey projects and more interested in importing applied industrial technology and know-how. According to UN data, in 1975-1976 the share of turnkey plant contracts in all industrial co-operation agreements (not including Yugoslavia) was around 21-22 per cent, although this percentage varied from country to country.

6. Co-production and Specialisation

Under co-production and specialisation agreements the partners specialise either in the production of certain parts of a finished product, or in the production of a limited number of articles in the production range, which are then exchanged so that each partner can offer the full range. The technology is usually provided by one of the partners, although in some cases it is the result of a joint R & D effort. Co-production and specialisation agreements generally also include co-operative marketing arrangements. Co-production and specialisation are the most flourishing forms of industrial co-operation. Most of the agreements concluded by the Soviet Union dealt with R & D (42.6 per cent of the total industrial co-operation agreements concluded), while the other Eastern countries, especially Hungary, Bulgaria and Poland, concentrated on co-production agreements. UN data for 1976 indicate that the chemical industry was the leading sector for such agreements (25.9 per cent), followed by mechanical engineering (23.9 per cent), transport equipment (13.5 per cent) and electronic equipment (11.1 per cent).

7. Joint Ventures

Joint ventures in Eastern Europe are different from those in the West and much less common. Yugoslavia was the first country to authorise joint ventures (in 1967). Rumania

20

has allowed them since 1971 and permits companies to engage directly in production operations. Hungary authorised joint ventures in 1972. The Western partner is required to be a minority shareholder and the only access to production is indirect: through Hungarian enterprises. In Poland joint ventures have been authorised since May 1976, but are limited to light industry, domestic trade and consumer services. The joint venture agreements which have been permitted so far by the East have been rather small in scale, though there are some indications they may find more favour in the future.

8. Sub-contracting, Joint Tendering or Joint Projects and Tripartite Co-operation

These are important technology transfer media because they imply close co-operation. The most important of the three is tripartite co-operation. Although the scale of tripartite co-operation agreements is still small, these agreements have attracted a great deal of attention because of their special economic and political implications. Through tripartite agreements, Western countries hope to lower the costs of associating with the East: who are mainly interested in obtaining foreign currency and technical knowledge.

CHAPTER 4: EASTERN AND WESTERN POLICIES TOWARD TECHNOLOGY TRANSFER

Policy makers in both East and West have been rather ambivalent about expanding economic relations with one another. This ambivalence is sharply focused with respect to the transfer of technology. The evolution of attitudes toward technology transfer has generally followed rather closely any changes in the East-West political climate.

1. The Eastern Perspective

Eastern need for Western technology — a need which seems to be growing — is a fundamental factor. This need can be demonstrated in a variety of ways. Statistical data indicate that the share of machinery and transport equipment in total CMEA imports from the OECD countries has increased considerably. Industrial co-operation agreements illustrate that transfer of disembodied technology has also greatly increased since the beginning of the 1970s. Growing interest in Western technology is related to such internal difficulties as slow growth, the need to increase factor productivity, and the lack of incentives for domestic innovation. This interest is illustrated by recent Soviet stress on the advantages of foreign technology for domestic production, and by the prominent role given to foreign technology in the Ninth and Tenth Five-Year Plans.

The notion that trade may procure substantial gains has led to a new approach to the international division of labour. Expanded economic relations with the developed countries of the West were policy objectives in the Tenth Five Year Plan. Since it is problematic whether or not East-West trade can be expanded on the basis of present export structures (Soviet exports to the Western industrialised countries consist primarily of energy products and raw materials), Soviet industrial leaders are being urged to study foreign needs and to create or expand specialised lines of production to increase the present range of export products in demand on the world market, in particular machinery and equipment. This shift is happening in the other CMEA countries whose trade with the Soviet Union must be complemented by trade with the West and their exports and production structure reoriented.

The shift in Soviet attitudes towards Western technology is, in part, ideologically based. In the 1961 Party Programme science was designated as a "direct productive force". This enhanced the role that science and technology were to play in economic development. The new and enhanced role appears in various economic reforms which have been attempted and in the succession of Plans. The 25th Party Congress in 1976 stressed the critical role of Western technology in improving Soviet performance. This may have signalled a new Soviet attitude to technology transfer: acceptance that Western technology has a direct role to play in the fulfilment of Soviet economic plans.

These ideological shifts have occasioned a good deal of dissension among Soviet leaders. The debates at the highest Party and government levels reflect disagreement about alternative ways to accelerate technological progress. Their outcome will have an important effect on the future role of Western technology in Soviet development. They will also affect Western policy towards the East — itself partly based on assumptions about how Eastern systems will evolve.

The current policy toward technology imports has been accompanied by changes in the forms of transfer. The most important change has been the shift in emphasis from non-negotiable to negotiable transfer channels. This is primarily a shift in favour of embodied transfer, exemplified by machinery imports, and the commercial purchase of licences and know-how. The new transfer forms not only transmit knowledge through the delivery of entire factories but include provisions for continuous Western technical assistance. They facilitate the procurement of whole production systems, provide the possibility of obtaining new, non-standardised technology generally not available through the market, or provide the possibility that the full technological capability will be assimilated through continuing co-operation with the firm providing the technology. Furthermore, they seem to help the East import high-technology machinery and equipment while minimising hard currency outlays.

Apart from these economic issues, technology has also been used as an instrument to alter political relations. For instance, a direct link between scientific and technological co-operation and co-existence was made at the 25th Party Congress. Such co-operation has also been linked to détente, which has eased the way for closer trade relationships and hence wider access to Western technology. However, Soviet interpretation of détente does not imply wide-reaching political concessions. In the East, détente is officially presented as a matter of mutual convenience and Soviet leaders exclude any Western assumptions that expanded economic relations will influence or lead to fundamental changes within the socialist system.

2. The Western Perspective

Western policies toward the East are strongly conditioned by assumptions about the advantages or disadvantages of technology transfer. The critical issue seems to be whether such transfers will involve selling a critical capability which the West will subsequently regret sharing with the Soviet Union.

Clear-cut policy formulation is bedevilled by disagreements about how to define and assess the military-strategic aspects of technology transfer, by differing views about what constitutes military balance, and by uncertainty about the relationship between Western technology transfers and the strengthening of Soviet military capabilities. On the national level these disagreements have led to arguments about the usefulness of export control measures. On the international level changes in approach have led to an agreed multi-lateral list which is now shorter and more flexible than originally conceived.

Unlike those of the East, where economic considerations have always seemed to dominate the policy of seeking closer relations with the West, Western trading objectives have been and remain mixed. For Western Europe, the primary reason for establishing closer trading links with the East has always been gaining access to Eastern markets. Political reasons have, by and large, been secondary. For the United States, the picture is somewhat mixed. Economic interests and the promise of commercial gains have, of course, been the key elements motivating business and industry to seek closer commercial relations with the East. For the US government however, political considerations, in addition to economic ones, have been a very important — if not primary — reason for seeking closer commercial relations with the East and, in particular, with the Soviet Union. Increased trade has frequently been viewed as a "bargaining chip" in the attempt to achieve foreign policy goals such as a moderation of the arms race or an improvement in détente. While "linkage policy" in one form or another has been supported in the United States, the policy is itself a subject of controversy, both within the United States and among its allies. The link that has been made between closer scientific and technological relations, trade and détente, raises as many problems as it was hoped to solve. However, one development is evident: technology transfer is not isolated from foreign policy considerations, but very much influenced by them.

The economic and commercial considerations which led the Western countries to seek closer trading relationships with the East are varied. All rest on assumptions about the advantages of closer commercial relations; assumptions which are also disputed and not "proven"; but which have led to the present relatively liberal trade policy. These assumptions include arguments made that increased trade helps to reduce balance of payments deficits and provides a stimulus to selected sectors of the economy, generates greater employment possibilities, stimulates industrial growth, and provides access to needed raw materials, primary products and commercially useful technology.

A different set of assumptions relates to the effects of technology transfer on East-West economic relations and on the evolution of the Eastern systems. Two central assumptions, closely related one to the other, are currently the subject of keen debate within the United States, in particular. The argument turns about the following propositions: (a) that closer economic relations will/will not lead the East to increased interdependence/interrelatedness with the West and ultimately to a reordering of Soviet economic priorities; and (b) that closer economic relations with the West will/will not bring about a change in the basic features of Eastern societies, and the Eastern system as a whole. Major arguments on both sides illustrate that both proponents and opponents of the above mentioned propositions base their reasoning on *interpretation* of past events and *estimates* made as to possible future developments. In the absence of "hard" data, policy is influenced by the general political climate.

CHAPTER 5: THE INFLUENCE OF TECHNOLOGY TRANSFERS
ON EASTERN ECONOMIES

This Chapter sums up analyses of the technological gap between the USSR and the West. Scientific and technical progress has been a fundamental goal of the Soviet Communist Party ever since it came to power. The insistence on scientific and technical progress has resulted in the introduction of a system of R & D planning and management, and in more prominence being given to science and technology in the national plans. But continuous pressure from the Eastern authorities to promote scientific and technical progress does not seem to have ensured effective innovation. There seem to be two forces pulling against each other: the government, which is adopting (in the Soviet Tenth Five-Year Plan) a new growth strategy based on scientific and technological progress and technology imports, and the inertial force of the administrative planning system itself, which discourages innovation. Western technology imports are therefore part of the new government policy, the success of which is of crucial significance.

It is difficult to make an overall assessment of technical progress. The results of any research on the growth and productivity of such factors as capital, work and "technological progress" are inevitably approximate. Attempts which have been made to assess the rate of Soviet technological progress indicate that from 1970 to 1973 it was slightly faster in the USSR than in the United States and the United Kingdom, but slower than in Germany and much slower than in Japan. An interesting point is that the improvement in technological progress (compared to the 1965-1970 period) coincided with a reduction in Soviet industrial production growth rates. Industrial growth remained higher than that of the leading industrial countries apart from Japan and the gap in industrial production narrowed. This convergence of growth rates was accompanied however by appreciable differences in the diffusion of new products and manufacturing processes. During the past 20 years, growth in traditional industries in the USSR has continued steeply whereas elsewhere it has slowed considerably. The same trends appear in per capita production of crude steel, cotton and other industrial products.

After sketching Soviet technological progress in some advanced technology industries, this Chapter sums up current knowledge about the over-all contribution of Western technology to Eastern economies as follows: the contribution is unknown; it is probably not dominant, but possibly appreciable. Comparing 1975 machinery imports from the West with domestic investment, it seems that the industries most dependent on foreign technology included chemicals, computers, shipbuilding and motor vehicles. Machinery imports from the West are likely to rise during the Tenth Five Year Plan in the chemical, oil, gas and coal industries, gold extraction and metal working. But since the diffusion of Western technologies is still sluggish in the East, even prolonged assistance from Western technicians is not sufficient to ensure technical progress equivalent to that in the West.

There are other problems in measuring the technological gap between East and West. The determination of a technological "lead" or "lag" is often derived from adventitious public information, not from representative surveys. Some comparisons have been made in specific branches of industry. These are summarised. From these branches covered (high voltage electric transmission, fuel mining and processing, machine tools, computers, turbines, process control, passenger cars, chemicals and consumer durables) it seems that the Soviet Union lags behind the West. Available research suggests that there is no evidence of a substantial diminution of the technological gap between the USSR and the West in the past 15-20 years, either at the prototype/commercial application stage or in the diffusion of advanced technology.

CHAPTER 6: EFFECT OF ECONOMIC FACTORS ON EAST-WEST TECHNOLOGY TRANSFER

Technology transfer, like all trade, is particularly sensitive to changes in the economic climate. Economic factors undoubtedly affect East-West technology transfer, but their effects are not easy to identify or quantify. When economic conditions are buoyant it is easy for Eastern countries to find markets for their products. In periods of recession they are often badly hit when demand for their exports contracts. Growth trends in the East of course also affect trade with the West. For instance, from 1971-1975, the official growth rate for Soviet national income averaged 5.7 per cent. In 1976 it was only 5.2 per cent, in 1977 3.5 per cent, in 1978 4 per cent and in 1979 2 per cent. The 1980 plan forecasts 4 per cent. Prior to 1976 Western exports had been increasing steeply. In 1977 they slowed considerably, with the exception of machinery exports which also seem to be less sensitive to price fluctuations. In 1977, in contrast to the overall downswing of 7 per cent, machinery exports to Eastern Europe only decreased by 3 per cent.

One of the biggest economic problems affecting East-West technology flows is the question of payment. The Eastern debt has increased steeply in recent years, making

24

prospects for the future somewhat doubtful. Estimates of the total net debt at the end of 1979 give a figure of $65 billion. The Soviet Union and Poland were together responsible for more than half of it. The deficit in the CMEA countries' balance of payments is mainly due to their trade deficit. Machinery and transport equipment are largely responsible for that deficit. The connection between technology transfer and the trade deficit is not simply an accounting phenomenon. It stems from the mechanism by which Western machinery is sold to the CMEA countries. Western firms who want to do business with countries suffering from a chronic strong currency deficit are forced to extend credit. The need to give credit for sales to the East is not disputed in the West. Indeed, there is some competition among Western countries for contracts with the East and in practice this has resulted in the grant of more favourable credit terms. Western technology also contributes indirectly to increase the indebtedness of the CMEA countries through the payment of interest on loans. Future prospects for the transfer of Western technology will largely depend on the credit-worthiness of the CMEA countries; that is now a debated issue. On balance, the consensus in the West seems to be that the credit-worthiness of the Soviet Union is good, that of the other CMEA countries more doubtful: but that the present level of indebtedness is not alarming. Some believe that, if Western firms are to be paid or even if their outstanding credits are merely to be kept at the present level, Western trade balance surpluses must be reduced.

One way in which this can be done is through counterpurchase and compensation agreements. With the increase in their countries' hard currency debt, Eastern partners are increasingly pressing for such arrangements in industrial co-operation contracts. Some analysts think the CMEA countries are beginning to equate compensation with co-operation: so that the Western partner has to find the credit, which is then repayable by counterpurchase or in the form of products made with imported plant or equipment.

Emphasis in counter-trade is placed on compensation agreements: their value and importance, Western participants, the share they represent of Eastern countries' exports, and the types of commodities exported and imported are discussed. The primary advantages of countertrade for the East are that such trade helps improve balances of payments, penetrate Western markets, up-date technology and create Western interest in improving Eastern products. For the West, the main advantage of countertrade is to facilitate access to the Eastern market, particularly valuable in periods of recession. Western firms sometimes concede to Eastern insistence on countertrade when the details of the buy-back agreements are tolerable. However, countertrade is not really thought to be a solution to the debt problem. In fact it is said that countertrade in its present form can only be a make-do arrangement. It does seem to enable the Eastern countries to exert pressure on the West to accept their products even if the quality of these products is not sufficiently high. This type of bilateral trade thus seems to reduce trade turnover to the level that can be afforded by the less well endowed partner.

CHAPTER 7: EFFECTS OF EAST-WEST TECHNOLOGY TRANSFER
ON WESTERN ECONOMIES

It is difficult to measure the impact of technology transfer on the Western economies. East-West trade is only a small part of total Western trade (2.9 to 4.6 per cent for OECD countries between 1970 and 1977) and its effects are over-shadowed by other economic factors. Trade in the East-West direction would have to reach a significant level for technology transfer to have an impact on Western economies. The medium-term prospects that this will happen do not seem to be very bright. The uncompetitivity of their exports on Western markets, combined with their internal domestic problems and other factors, mean that neither the Soviet Union nor the East European countries will be in a position to significantly expand exports to the West. Barring high rates of Western inflation or further large price increases in key export items the USSR, Poland and most of the other East European countries will be hard pressed to match 1972-1975 rates of growth in exports during the 1979-1985 period. Nevertheless, regardless of East European countries' export potentials, the economic situation in the Western countries will also have to be taken into account.

This Chapter briefly discusses a number of issues, very sensitive in the West, which affect policies toward West-East technology flows; all are keenly debated though few seem to have been sufficiently studied for definitive conclusions to be drawn. Among these issues, the following are mentioned: employment, the profitability of technology transfer to the East, the dangers of competition — both in Western and in Third World markets — and the problem of dumping.

An equally important issue is the nature of East-West trade. In the market-economy countries the system of trade and international payments is regulated by certain rules established by international agreements and institutions. Domestic and international prices are determined essentially by supply and demand, which trade tends to equalise. Eastern countries are not necessarily guided in their foreign trade decisions by domestic/foreign price differentials. Their domestic prices are fixed by the government. They do not reflect equilibrium between supply and demand, and products for which there is excessive demand are rationed. CMEA countries have always operated exchange controls and used multiple rates of exchange. Nevertheless, in trade with the West debts were promptly settled in hard currency. This was also the procedure used by the Soviet Union in the period between the two World Wars.

With the sizeable purchases of Western machinery, equipment and grain by the Eastern countries since the early 1970s, this policy is now changing. The CMEA countries are now constantly pressing for a larger proportion of countertrade in their agreements with the West. In other words they are "exporting" the bilateralism currently prevailing in the CMEA. The share of compensation agreements in exports to the West is likely to be about 11 per cent of the total for the CMEA countries in 1976-1980 (20 per cent for the Soviet Union), and to rise to 35 per cent, including counterpurchase, in 1980. This raises the problem of the present and future economic implications of East-West compensation trade.

The issue is whether current forms of trade are helping to fit Eastern and Western economic structures into one another. Opinions on this issue differ, but it could be maintained that present trends in East-West trade, particularly the growth of barter, bi-lateral and buy-back deals, will not be conducive to increased trade nor to the complementary development of Western and Eastern economies. Increased bilateral countertrade could mean that the East European economies will be "protected" from the necessity to adjust themselves to Western demand. Even if the East European countries were prepared, in their Economic Plans, to adapt their economies so as to facilitate East-West trade, it could be argued that their systems would still be too rigid, and influenced too much by non-economic considerations of power, to effect such adaption. In other words, there might still be a tendency for Western buyers to be offered products "not allocated" under the plan so that deficits can be made up. On the whole, it can be said that the future development of East-West trade will be greatly affected by the existing technological gap.

Chapter 1

EAST-WEST TRADE AND TECHNOLOGY TRANSFER IN HISTORICAL PERSPECTIVE

INTRODUCTION

The purpose of this survey is to set forth the evolution, present levels and future potential of technology transfers between East and West. To date, the great bulk of such transfers has been from the advanced Western nations, represented in the main by the OECD area, to the Eastern countries rather than in the reverse direction. The Eastern demand for technology has been conditioned by, but has also itself partly determined — given its importance in Eastern imports from the West — developments in the over-all trade policy of the two groups of countries.

The present Chapter surveys the main shifts in policies towards technology transfer with particular emphasis on the period which followed the accession of the Communist Party to power in the Soviet Union after the first World War. It also summarises some of the changes in East-West technology flow patterns for the six smaller East European countries, both before and after they installed centrally planned economic systems.

The Chapter seeks to illustrate the continuous ambivalence in Eastern attitudes to technology imports from ideological rivals, whose expertise is nonetheless seen to be an essential element in Eastern economic development. Eastern demands for Western technology have been consistent despite periods which can be characterised as being "autarkic". Here changes in world economic conditions which have sometimes worked against, but more recently (since 1973) in favour of, expanded Eastern trading activity on external markets are of importance.

The main difference between East-West technology flows today and in the past is one of scale. None of the choices and dilemmas confronting either the Eastern or Western countries are new. National attitudes and policies toward technology transfers have sometimes shown new features but have also sometimes reverted to earlier patterns. In consequence, some current problems have been foreshadowed by events in earlier years. This past experience should not be ignored. If nothing else, it indicates the need for a certain flexibility on the part of Western policy makers in assessing future developments in Eastern attitudes and actions.

I. TRADITIONAL APPROACHES TOWARDS TECHNOLOGY TRANSFER

1. Technology Transfer Prior to the First World War

A. *The Soviet Union*

Technology transfer is not a recent phenomenon. The transfer of advanced foreign technology to Russia occurred as early as the 18th century, particularly under the reigns of Peter the Great and Catherine, when technology from the West was brought in on an impressive scale.

Again in the 1890s, foreign capital and technology were purchased to support industrialisation. It has been estimated that by 1900 "foreign companies owned more than 70 per cent of the capital in mining, metallurgy and machine building in Russia. As a result, foreign technology was brought into Russia, both in the form of advanced capital equipment itself and in the form of human capital... Foreign investment was thus responsible for the implantation of advanced techniques in several key industries".[1]

Foreign technologists, engineers and managers not only established foreign firms in Russia, but also participated in the management of Russian firms. Foreign technology was incorporated with little or no adaptation. Economic growth proceeded fitfully and the State made efforts from time to time (primarily when military needs were pressing) in order to accelerate it.

Ambivalency about the issue of national independence was already evident at this early stage. In the 1890s for instance, the Minister of Finance (Count Witte) defended his programme of bringing in foreign capital both on the grounds that it had an "invigorating effect on the productive resources of the national economy" and because the programme "would free the country from dependence on foreign supplies of manufactured goods".[2]

Prior to the First World War, Russian development was basically inward-looking and industry was increasingly controlled by Russian bankers and industrialists (after 1903 and the removal of Witte, the State withdrew as the main initiator of economic development). The desire to maintain independence in crucial areas was still balanced, however, by advocacy of "the development of foreign markets for the products of Russian industry".[3] Thus during the "1905-13 period, Russian trade increased significantly and Russian banks and corporations began to participate in West European capital markets".[4] The Revolution terminated this process.

B. *Central Europe*

"Central Europe" includes the six East European countries which are now part of the CMEA.[5] Prior to the First World War most of this territory was a part of the three big powers: Germany, Russia and Austro-Hungary. Only Rumania (without Transylvania) and Bulgaria were independent political entities.

The present German Democratic Republic was, at that time, the central part of the German Empire. Two-fifths of present Polish territories belonged to Germany, two-fifths belonged to Russia and one-fifth was part of the Austro-Hungarian Empire. The latter, in addition to Polish territories, held some territories which are presently part of the Soviet Union, Rumania and Yugoslavia.

---

1 Herbert S. Levine, *et al., Transfer of US Technology to the Soviet Union: Impact on US Commercial Interests,* Strategic Studies Center, Stanford Research Institute, February, 1976, pp. 36-37.
2 *Ibid.,* p. 41.
3 *Ibid.,* p. 50.
4 *Ibid.*
5 Bulgaria, Czechoslovakia, German Democratic Republic, Hungary, Poland and Rumania.

Compared to the present technology transfer situation, technology flows in the period prior to World War I were among the three latter powers; that is, *internal* trade flows. The main feature of trade at that time was the vertical division of labour in which the more industrialised regions or countries sent their manufactured products to the less developed regions or countries in exchange for raw materials and food.[6] Austria, Bohemia and Hungary in part, furnished most of the industrial products to the less developed part of the Austrian Empire as well as to Rumania and Bulgaria. Those Russian territories which are now a part of Poland, furnished a large part of the industrial products to Russian markets.

In their relations with the industrialised countries of Western Europe and the United States, Germany and to a degree Austria-Hungary demonstrated a horizontal division of labour in which the exchange of industrial products, raw materials and food was carried out according to the region's specialisation. In their trade with Russia, Germany and Austria-Hungary primarily exchanged industrial products for raw materials and food. This exchange represented only a small part of their foreign trade.[7]

2. Technology Transfer and Commercial Policy in the 1920s

A. *The Soviet Union*

During the 1920s two broad features characterised Soviet trade. One was the stress placed on State monopoly of foreign trade (in order to protect State ownership of the means of production) and the second was the orientation of Soviet trade towards co-operation with industrialised Western countries. Within this structure, the role of foreign technology in Soviet economic development and in foreign trade assumed a growing importance.

1. *Foreign Trade Policy*

Following the signature of a separate peace treaty with Germany (February 1918) and the refusal of the Soviet Government to honour the debt of the former régime, a number of Western countries halted exports to Soviet Russia.[8] In 1920, however, Soviet Russia was able to establish trading offices in several Western countries.[9] However, opposition to trade with Soviet Russia continued as British, French and American banks refused to accept Russian gold in payment for Soviet imports.[10] This restriction was only abolished after the signature (16th March, 1921) of a preliminary trade agreement between the RSFSR[11] and the United Kingdom and followed Lenin's speech announcing the introduction of the new Economic Policy (NEP). Market relations with Soviet Russia were subsequently restored as a result of NEP.[12]

After the lifting of the economic blockade and the signature of peace treaties with neighbouring countries, foreign trade was resumed. However, the lack of commercial treaties

6   Bohemia was the most industrialised part of the Austro-Hungarian Empire. In the German Empire, the Western part (Ruhr) was the most industrialised. The Dresden-Berlin regions (now part of the German Democratic Republic) and Poland (Silesia) were highly industrialised.

7   Between 1901-1912, Austro-Hungarian trade with Russia represented (on average) 6.3 per cent of its imports and 3.6 per cent of its exports. Austro-Hungarian trade with Germany was roughly 40 per cent of exports and imports and with the industrialised countries of Western Europe, 16.3 per cent of imports and 19.0 per cent of exports. Eugène Zaleski, *Les courants commerciaux de l'Europe danubienne au cours de la première moitié du 20ème siècle,* Librairie générale de droit et de jurisprudence, Paris, 1952, p. 46.

8   An economic blockade was called for by the Supreme Allied Council at the beginning of 1919, but cancelled by the Allies one year later.

9   For instance in Great Britain, Sweden, Estonia, Italy and Austria.

10   Soviet Russia was obliged to send its gold through Estonia and Sweden thereby losing up to 30 per cent (of currency price) in each transaction.

11   It was only at the end of 1922 that Soviet Russia became the USSR.

12   A. Stoupnitzky, *Statut international de l'URSS Etat commerçant,* Paris, 1936, p. 470; and Serge N. Prokopovicz, *Histoire économique de l'URSS,* Paris, 1952, p. 488.

hindered the development of normal trade relations with the principal Western countries.[13]

Following the trade agreement signed with the United Kingdom, other Western countries followed suit.[14] However, the turning point in trade relations with the West was the agreement signed between the Soviets and Germany. Following the Geneva conference (February 1922), the separate peace treaty signed with Germany in Rapallo (16th April, 1922) and the renouncement by Germany of claims for confiscated German assets, Germany and Soviet Russia granted each other most favoured nation treatment. On 12th October, 1925 the Soviet government signed a commercial treaty with Germany.

The expansion of foreign trade was heavily dependent on foreign credits. Between 1922-23 and 1927-28, credits rose from 35 million rubles to 520 million rubles. Between October, 1928 and the end of 1931 credits to finance imports rose to 779 million rubles.[15]

The first Five Year Plan (initiated in 1928) stressed the expansion of foreign trade.[16] The plan called for an 80 per cent increase of imports and a 165 per cent increase of exports. The Soviet Union's share in world trade was supposed to double.[17]

2. Concessions

During the relatively free period of the NEP, with its system of market socialism, the Soviet Union attempted to introduce foreign technology through a programme of foreign concessions.[18] The exact importance of this programme is a matter of debate, yet its weight was far greater than has usually been supposed. Some authors believe that concessions were a main vehicle for technical transfers from 1920 to 1930.[19]

Among the various objectives which the concessions were to achieve, "introduction of technical progress" was key.[20] Most of the foreign capital imported through the concessions was invested in the mining and processing industries.[21] According to Soviet official figures, by 1 October, 1929, 162 concessions were in operation of which 59 were "pure" concessions.[22]

While Germany claimed the greatest number of concessions (31 of the 97 in existence by 1st June, 1926) Great Britain and the United States had a higher volume of investment.[23] By 1928, foreign capital represented between 0.6 and 1 per cent of total investment of Soviet industry.[24]

13 For instance, Soviet exports risked being confiscated through court seizure by persons whose assets had been confiscated (nationalised) in Soviet Russia. See *50 let Sovetskoy Vneshnoj torgovli* (Fifty Years of Soviet Foreign Trade), Moscow, 1967, p. 24.

14 Norway (2nd September, 1921), Austria (7th December, 1921), Italy (26th December, 1921), Czechoslovakia (5th June, 1922) and Denmark (23rd April, 1923). A. Stoupnitzky, *op. cit.,* pp. 468-472.

15 Serge N. Prokopovicz, *Histoire économique de l'URSS, op. cit.,* pp. 495-496.

16 The plan launched the industrialisation programme. Its slogan (one which pertains to all the plans) was to "catch up and surpass the developed capitalist countries".

17 In constant prices. Eugène Zaleski, *Planning for Economic Growth in the Soviet Union, 1918-1932,* Chapel Hill, 1971, pp. 251-252.

18 Foreign technology was considered to be indispensable for the construction of socialism.

19 Antony C. Sutton, *Western Technology and Soviet Economic Development, 1930 to 1945,* Vol. II, Hoover Institution Press, Stanford University, Stanford, California, 1971, p. 10.

20 W. Beitel and J. Notzold, "Les relations économiques entre l'Allemagne et l'URSS au cours de la période 1918-1932 considérées sous l'angle des transferts de technologie", *Revue d'études comparatives Est-Ouest,* Vol. VIII, No. 2, juin 1977, p. 109.

21 *Ibid.*

22 As given by Antony C. Sutton, *op. cit.,* p. 10. The different kinds of concessions are discussed in Antony C. Sutton, *Western Technology and Soviet Economic Development, 1917 to 1930,* Vol. I, Hoover Institution Press, Stanford University, Stanford, California, 1968.

23 The distribution of fixed capital of the concessions by country was as follows: Great Britain: 28 per cent, United States: 23.6 per cent, Germany: 13.5 per cent, Sweden: 12.5 per cent, other countries: 22.4 per cent. W. Beitel and J. Notzold, *art. cit.,* p. 109.

24 *Ibid.*

It should be noted that while the Soviet Union sought concessions to fill gaps in its technological structure and to speed industrialisation, the concessions had always been opposed on ideological grounds. Lenin, speaking in 1920, had stated that "concessions — these do not mean peace with capitalism, but war on a new plane".[25] Bukharin foreshadowed the ultimate fate of the concessions when he said that "on the one hand, we admit capitalist elements, we condescend to collaborate with them; on the other hand our objective is to eliminate them completely, to conquer them, to squash them economically as well as socially...".[26]

3. *Technical Assistance*

Technology transfers are believed to have been "the most significant factor in Soviet economic development" during this period.[27] Technical assistance agreements (combined with the concessions) played a very important role in this transfer process. The first technical assistance agreement was signed in 1923. The greatest number of agreements were concluded in conjunction with the initiation of the first Five Year Plan and by the end of 1929-30, 134 foreign technical assistance agreements were in existence.[28] Germany concluded the greatest number, though it is thought that United States technology was more influential in stimulating Soviet economic and industrial development.[29]

Particular emphasis was placed on assistance agreements in mechanical construction, the chemical and electrotechnical industries, automobiles, coal, steel, metallurgical industries and construction of electric power plants. The majority of the agreements (concluded during the course of the first Five Year Plan) were linked to the provision of equipment for new factories, to the supervision of their construction and to their productive operation.[30] In terms of distribution among industrial branches, most of the agreements were directed towards helping the metallurgical industry. The chemical, electrical, minerals and fuels industries followed in descending order of priority.[31]

The technical assistance agreements included a provision which enabled foreign technicians and skilled workers to work in the Soviet Union. Such personnel were engaged to help in construction of plants or in the supervision of their installation. Most worked on a contract basis.

4. *Soviet Trade with the West: The Part of Machinery in Trade*

During the Civil War, Soviet foreign trade was virtually non-existent. Between 1918-1920, Soviet exports (in 1913 weights) were 0 to 0.5 per cent of the pre-War level and imports ranged from 0.2 to 7.7 per cent.[32] During the NEP period and the first years of the first plan period, Soviet trade expanded continuously achieving almost half of its 1913 level.[33] Between 1928-1930 alone, trade volume increased by half, due to the policy of forcing exports.[34]

25 Antony C. Sutton, Vol. II, *op. cit.*, p. 30.
26 *Ibid.*, p. 29.
27 It was noted in *Pravda*, as one example, that "the attempt of the Potash Trust to carry on work without foreign technical assistance proved futile. Thus in 1927-28 several large companies were hired for technical assistance in the construction of the first potash mine...", Antony C. Sutton, *op. cit.*, p. 2.
28 W. Beitel and J. Notzold, *art. cit.*, p. 110. Sutton gives the number of technical assistance agreements signed by March 1930 as 104. See Antony C. Sutton, Vol. II, *op. cit.*, p. 10.
29 W. Beitel and J. Notzold, *art. cit.*, p. 110.
30 *Ibid.*
31 Antony C. Sutton, Vol. II, *op. cit.*, p. 11.
32 Franklyn D. Holzman, *Foreign Trade Under Central Planning*, Harvard University Press, Cambridge, Mass., 1972, p. 41.
33 44.4 and 56.6 per cent for exports in 1929 and 1930 respectively and 48.3 and 65.7 per cent for imports (1913 = 100 in 1913 prices): Franklyn D. Holzman, *op. cit.*, p. 41.
34 *Ibid.*, p. 41.

The recovery of foreign trade during the 1920s was accompanied by a shift in the geographical distribution of trade from the pattern followed by the former régime. In 1913 Russian *exports* were directed primarily to Germany (29.8 per cent of the total), Britain (17.6 per cent), the Netherlands (11.6 per cent), France (6.6 per cent) and Italy (4.9 per cent). By 1929 only 23.3 per cent of Soviet exports were sent to Germany followed by Britain (21.9 per cent), France (4.6 per cent), the United States (4.6 per cent) and Italy (3.6 per cent). The shift in the geographical distribution of Soviet *imports* between 1913 and 1929 was much more pronounced. Germany's share diminished from 47.5 (pre-War) to 22.1 per cent, Britain's from 12.6 to 6.2 per cent, and that of France fell from 4.1 to 3.6 per cent. At the same time the share of the United States which had been 5.7 per cent in 1913 increased to 20.1 per cent in 1929.[35]

From 1913 to 1929 the proportions of machinery and equipment in Soviet imports increased. These rose from 15.9 to 23.9 per cent between 1913 and 1928 and reached 29.9 per cent in 1929.[36]

The share which machinery and equipment represented in the trade of the Western industrialised countries with the Soviet Union was even higher. In 1929 it accounted for 53.3 per cent of German-Soviet, 43.3 per cent of Italian-Soviet, 42.5 per cent of United States-Soviet, 34.5 per cent of British-Soviet and 30.7 per cent of French-Soviet trade. In terms of a global figure, 80.2 per cent of Soviet machinery and equipment imports were received from the five above cited Western countries.[37]

B. Central Europe

The emergence of national states in Central Europe after the First World War had important consequences for foreign trade flows. A great part of Austro-Hungarian, German and Russian internal trade became foreign trade. War damages and the political tensions resulting from the Bolchevik takeover, greatly influenced economic relations.

The climate in which relations between Central Europe and Western industrialised countries evolved differed fundamentally from that of Soviet Union-Western relations. The climate of the latter was marked by mistrust due to the confiscation of all foreign assets in Russia by the Revolutionary Government. On the other hand, Western countries contributed, through credits given individually and under guarantees by the League of Nations, to Central European economic recovery. Recovery was also aided by the introduction of convertible currencies (Central European) and by the progressive removal of the quantitative restrictions on trade.

It is difficult to judge to what extent pre-War trade flows with the industrialised West were restored during the 1920s. During the years 1901-1912, 55.9 per cent of imports and 59.4 per cent of exports of Austria-Hungary were absorbed by the German Empire and industrialised Western countries.[38] However, trade of the Austro-Hungarian successor States with the same countries was substantially lower — despite great increases (overall) during the twenties.[39] Trade among the successor States increased, but diminished in relative

---

35   In current prices. Calculated from *Vneshnyaya Torgovlya SSSR za 1918-1940 gg. Statisti-cheskij Obzor* (Foreign Trade of the USSR for 1918-1940, Statistical Review), Moscow, 1960, pp. 21-28 and 37.

36   Franklyn D. Holzman, *op. cit.,* p. 49.

37   Calculated from *Vneshnyaya Torgovlya za SSSR 1918-1940 gg, op. cit.,* pp. 21-28, 37, 301, 471, 544, 622, 782, 1068.

38   *Ibid.*

39   In 1929, the share of Germany and of industrialised Western Europe in imports was as follows: 33.9 per cent for Austria, 38.5 per cent for Czechoslovakia, 31.8 per cent for Hungary and 29.0 per cent for Yugoslavia. For exports, the share was: 31.4 per cent for Austria, 33.6 per cent for Czechoslovakia, 22.8 per cent for Hungary and 17.4 per cent for Yugoslavia. *Ibid.,* p. 92. It is impossible to make the same comparison for Poland, but it is likely that there was a relative reduction of trade flows.

importance to total trade,[40] Central European trade with Soviet Russia fell sharply during the twenties. None of the central European countries signed commercial agreements with the Soviet Union and commercial relations between Hungary, Rumania and Bulgaria on the one hand, and the Soviet Union on the other, were almost non-existent. In 1928, Polish trade with the Soviet Union accounted for only 1.2 per cent of imports and 1.5 per cent of exports.[41] The same percentages applied roughly to trade between the Soviet Union and Czechoslovakia.[42]

To what extent were technology flows affected by the changes in Central Europe? It would seem that the most important occurrence was the interruption and sharp reduction of technology flows from Central Europe to Soviet Russia. West European technology flows to Central Europe were favoured by the extension of Western credits. Czechoslovakia and, to a lesser degree Poland and Hungary, remained the primary exporters of technology to the less developed countries of Central Europe.

II. TECHNOLOGY TRANSFER AND THE COMMERCIAL POLICY
AFTER THE WORLD ECONOMIC CRISIS

A. The Soviet Union

1. *Soviet Foreign Economic Policy and the World Crisis*

During the first Five Year Plan period (1928-1932) foreign trade played an important role in Soviet economic development.[43] A major effort was made to import technologies to support the programme of industrialisation.[44] Imports were deliberately selected (and considered to be necessary for) to stimulate economic growth. Trade was essentially aimed at building import substitution capacity and emphasis was placed on industrial capital formation. Thus, imports of machinery and equipment began to assume great importance. The Soviet Union was, therefore, eager to foster trade and other commercial relations with the West. The technical assistance, commercial and trade agreements signed by the Soviet Union illustrate this aspect of Soviet commercial policy. However, even before the completion of the first Five Year Plan, Soviet economic policy was aimed (in the Spring of 1930) at achieving economic and technological self-sufficiency. The pursuit of self sufficiency was written into the Soviet Constitution, and "according to Soviet foreign trade authorities, became the guiding objective in the formulation of Soviet foreign trade policy and planning".[45]

By 1930 the Soviet Union reduced its foreign trade and followed the policy of economic self-sufficiency, a policy which has often been characterised as being "autarkic".[46]

40   It is not possible to compare pre-War flows. However, even during the 1920s, the relative importance of intra-Central European trade diminished. *Ibid.,* p. 91.
41   *Maly Rocznik Statystyczny* (The Small Statistical Yearbook), Warsaw, 1939, p. 166.
42   In 1927-28 (fiscal year beginning 1st October, 1927 and ending 30th September, 1928), Soviet exports to Poland represented 14.9 million rubles and to Czechoslovakia, 3.7 million rubles. Soviet imports from Poland represented 7.4 million rubles and from Czechoslovakia 17.4 million rubles. *Vneshnyaya Torgovlya SSSR za 1918-1940 gg, Statisticheskij Obzor, op. cit.,* pp. 26-27.
43   In fact, the plan was officially approved in the Spring of 1929.
44   The inability of the Soviet Union to fulfil its first industrialisation plan by "going alone" was openly discussed in the Soviet press.
45   P. Chervyakov, *Organizatsiya i tekhnika vneshney torgovli SSSR,* Moscow, 1958, p. 2; and D. Mishustin, ed., *Vneshnyaya Torgovlya Sovetskogo Soyuza,* Moscow, 1938, p. 9 as given in Herbert S. Levine, *et al., op. cit.,* p. 42. Self-sufficiency was also, of course, closely linked to the policy of "catching up" with the developed countries, an aspect stressed by Stalin in his famous 1931 speech. See J. Stalin, *Selected Writings,* Moscow, 1942, p. 200 given in *Ibid.,* p. 41.
46   "Imports declined absolutely during this period of rapidly increasing Soviet GNP". Michael Dohan and Edward Hewett, "Two Studies in Soviet Terms of Trade", *Studies in East European and Soviet Planning, Development and Trade,* No. 21, International Development and Research Center, Indiana University, Bloomington, Indiana, November 1973, p. 28.

The abrogation of the concessions can be (and has been), interpreted as a move to stop the more "classical" economic involvement with market economies. Finally, following the position taken in the 1930s, Soviet planners no longer stressed the economic efficiency advantages of foreign trade and stopped arguing "that international economic relations of a socialist economy would increase with economic growth".[47]

The extent to which autarky was a deliberate economic policy resulting from careful planning to achieve import substitution and/or the extent to which withdrawal from world trade was forced upon Soviet planners by economic events (the increasing inability to compete on world markets) are important issues affecting the interpretation given to later developments.[48] The elements which illustrate Soviet policy during this period are described below.

2. *Soviet Foreign Trade Policy*

The economic crisis of 1929-31 severely hindered fulfilment of the ambitious Soviet Foreign Trade Plan. The price of some imports decreased much more slowly than the price of Soviet exports, provoking an unfavourable trend in Soviet terms of trade: 89.3 in 1930, 71.1 in 1931 and 64.9 in 1932 (1927-28 = 100).[49]

In order to obtain the equipment and machinery necessary to realise its investment plan, the Soviet Union forced exports, but at lower prices. This move enabled the Soviet Union to show a sharp increase in its foreign trade (until 1931-1932) even though World trade declined during this period. However, the policy of increasing exports by lower prices, in order to ensure imports, amplified distortions on the world markets. The Soviet Union was accused of dumping,[50] and a number of countries issued anti-dumping regulations.[51]

The unfavourable terms of trade led to deficits in the Soviet trade balance, instead of the surplus foreseen by the first Five Year Plan: 23 million rubles in 1930; 294 million in 1931 and 129 million rubles in 1932. Given the manner in which foreign trade operations were registered, the actual deficit was much higher.[52] By 1932, the Soviet foreign debt stood at 1.400 million rubles.[53] The Depression had thus caused a "major disaster for the Soviet economy and forced costly changes in economic policy".[54]

In order to control this situation, a number of measures were taken. First, imports were

47 Herbert S. Levine, *et al., op. cit.,* p. 44.
48 See Chapter 4 below. As already indicated (see also Dohan and Hewett, *art. cit.,* p. 28) the notion that autarky was deliberately planned is one explanation offered in the Soviet press. There are also other Soviet authors who deny that the goal of the Soviet Union was ever to achieve autarky. For instance, "while seeking economic independence from the capitalist countries, the Soviet Union has never tried to achieve autarky". G. Rubinshteyn, *Vneshnyaya Torgovlya,* 1960: 5, as given in Herbert S. Levine, *et al., op. cit.,* p. 42.
49 Eugène Zaleski, *Planning for Economic Growth in the Soviet Union, op. cit.,* p. 255. See also Franklyn D. Holzman, *op. cit.,* p. 59.
50 The manner in which Soviet exports were organised before being sold (goods were sent abroad according to plan but stored in warehouses owned by Soviet Trade Delegations) enabled a reduction in their price and subsequent adaptation to the unfavourable conjuncture of world markets. Alexander Baykov, *The Development of the Soviet Economic System,* Cambridge, United Kingdom, 1950, p. 265.
51 The United States, France, Belgium, Canada, Rumania and Yugoslavia. Some countries combined anti-dumping measures with human rights considerations. For instance, proof that timber exported by the Soviet Union was not obtained by use of forced labour. *50 let Sovetskoy Vneshnoj torgovli, op. cit.,* p. 40; and Alexander Baykov, *op. cit.,* p. 266.
52 Exports were registered at the time they were sent abroad. However, as noted, goods were sold much later at lower prices. The loss was not recorded in trade balance accounts. Serge N. Prokopovicz, *op. cit.,* p. 499.
53 *Ibid.,* p. 497.
54 Michael Dohan and Edward Hewett, *art. cit.,* p. 36. It has been estimated that the Depression caused three types of losses to the Soviet economy: the terms of trade which negatively effected the balance of trade; increases in the real cost of debt repayment; and the loss of import capacity. *Ibid.,* p. 48.

severely curtailed while exports continued to be pushed.[55] Secondly, new credits were negotiated: 300 million Reichsmark with Germany in April 1931[56] and 10 million Pounds with England in July 1936.[57] A credit and commercial agreement was signed with Italy in April, 1931.[58]

Notwithstanding the slowdown in Soviet foreign trade, the Soviet Government was eager to develop its contractual relations with Western industrialised countries in order to acquire modern technology.

3. Contractual Relations: Foreign Trade Policy Instruments

a) Concessions

In 1930, the policy of concessions was formally ended. Liquidation of some pure and mixed concessions had already started in 1923, but the final stage began with the adoption of a resolution in December, 1930 which repealed all concession legislation formely passed.[59]

However, even while the concessions were in the process of being liquidated, proposals for new concessions were being solicited. The last known of these concessions was granted in March, of 1930. By 1933 no manufacturing concessions remained and the few trading concessions which had still existed at the end of the 1920s were terminated by 1935.[60]

Some analysts view the liquidation of the concessions as a clear signal of Soviet intention to withdraw from active involvement with the developed West. Thus it is thought that the concessions were liquidated in favour of technical assistance agreements. The latter were more acceptable to the Soviets because the "Western operator had not even a theoretical ownership claim and the Soviets could control more effectively both the transfer of technology and the operations inside the USSR".[61]

b) Technical Assistance Agreements

As noted technical assistance agreements replaced the concessions. While the majority of the technical assistance agreements in force in 1930 were with German and US companies, by 1945, technical assistance agreements had also been concluded with Danish, Dutch, French, Italian, Swedish and Swiss firms.[62]

It has been estimated that by the end of 1932 some 6,800 foreign specialists of all kinds

---

55   This allowed the Soviet Union to register a favourable trade balance for the years 1933-1937 and to repay a good part of the foreign short and medium-term credits. In 1936, Soviet foreign debt was reduced to 86 million rubles. Since 1935-36, all trade operations were conducted in cash. Serge N. Prokopovicz, *op. cit.,* pp. 494-497.

56   W. Beitel and J. Notzold, *art. cit.,* p. 103.

57   *50 let Sovetskoy Vneshnoj torgovli, op. cit.,* p. 42.

58   A. Stoupnitzky, *op. cit.,* p. 470. France and the United States were among the countries which refused to extend credit on the grounds that the Soviet Union had not honoured the pre-1917 debts. In 1934, the United States reversed its policy not to extend credits. Thus, the US Export-Import Bank was founded in that year "specifically to finance trade with the USSR". Rogers Morton, *The United States Role in East-West Trade,* US Department of Commerce, August, 1975, p. A-1. In the late 1930s US attitudes again underwent a shift "away from the granting of credit on favourable terms and toward conditioning trade terms on political concessions". See Herbert S. Levine, *et al., op. cit.,* p. 38.

59   Antony C. Sutton, Vol. II, *op. cit.,* p. 17. As the author points out, technical assistance agreements, however, were specifically omitted from repeal.

60   The only exceptions were the Danish telegraph concessions, the Japanese fishing, coal and oil concessions, and the Standard Oil lease, *Ibid.*

61   See Antony C. Sutton, Vol. II, *op. cit.,* p. 16. In the same volume the author discusses the process by which the concessions were liquidated.

62   See Antony C. Sutton, *Western Technology and Soviet Economic Development 1930-1945, op. cit.,* p. 11 and Appendix C. Some 200 technical assistance agreements were in force between 1930 and 1945.

were working in Soviet heavy industry, of which the majority were US engineers.[63] Most of these foreign workers left after 1932 because of the currency crisis. Nevertheless between 1936 and 1941 a "number of highly important, but unpublicised, agreements were made with American companies in aviation, petroleum engineering, chemical engineering, and similar advanced technological sectors in which the Soviets had been unable to develop usable technology".[64]

c)   *Trade Agreements*

Numerous trade agreements were signed by the Soviet Union during this period. A commercial agreement was signed with France on 11 January, 1934.[65] In 1935, "the first United States-Soviet trade agreement was signed granting the Soviet Union most-favoured nation tariff treatment in exchange for a commitment to purchase a fixed amount of American products each year".[66] Other important trade agreements (or additions to existing agreements) were signed with the United States (August 1937), Turkey (October 1937), Belgium-Luxembourg (September 1935 and November 1937) and Germany (19 and 23 August, 1939).[67]

4.   *Trade with the Developed Countries: Trade as a Major Factor in the Import of Embodied Technology*

During the first Five Year Plan, relations with the industrialised West expanded, notwithstanding the severe world economic crisis which had begun in 1929.

The index of Soviet exports (in physical terms) increased from 100 in 1929 to 136 in 1930 and 146 in 1931. Soviet imports (in constant prices) increased even more quickly: index 141 in 1930 and 162 in 1931 (1929 = 100). The opposite trend was registered in world trade which was reduced by 7.0 per cent in 1930. The total reduction, in comparison to 1929, was 15.0 per cent in 1931.[68]

In 1932, both Soviet and world trade diminished, the latter by 25.5 per cent (1929 = 100). Soviet exports were reduced by 12.5 and imports by 28.3 per cent.[69]

Beginning in 1933, the pattern of Soviet and world trade again turned in opposite directions. This time, however, the volume of Soviet trade declined sharply, reaching its lowest point in 1934 for imports and in 1939 for exports; the reduction of imports was much more pronounced.[70] Following the signature of the Ribentropp-Molotov Pact in August 1939 and the commercial agreement which was made possible by the Pact, Soviet trade recovered somewhat by 1940.[71]

The 1939 agreement with Germany was very significant as far as Soviet-German trade

---

63   *Ibid.*, p. 11. The figures are from official Soviet sources. Other estimates give a figure of 5,000. See W. Beitel and J. Notzold, *art. cit.*, p. 111.

64   Antony C. Sutton, *Ibid.*, p. 2.

65   A. Stoupnitzky, *op. cit.*, p. 469.

66   Rogers Morton, *op. cit.*, p. A-1.

67   *Ekonomicheskie otnoshenija SSSR s zarubezhnymi stranami: 1917-1967; Spravochnik* (Economic Relations between the USSR and Foreign Countries, 1917-1967; Reference Book), Moscow, 1967, p. 81.

68   "Mishutin Index" quoted by Michael Dohan and Edward Hewett, *art. cit.*, pp. 24 and 27. See also Franklyn D. Holzman, *op. cit.*, p. 41.

69   *Ibid.* For World Trade, see Alexander Baykov, *op. cit.*, p. 265.

70   The index of Soviet exports (base 100 in 1929) declined from 128 in 1932, to 119 in 1933, 103 in 1934, 91 in 1935, 68 in 1936, 68 in 1937 and 62 in 1938. The index of Soviet imports (1929 = 100) declined from 116 in 1932, to 63 in 1933, 47 in 1934, and rose slowly to 52 in 1935, 59 in 1936, 52 in 1937 and 60 in 1938. The World Trade Index (1929 = 100) rose from 75.5 in 1932, to 76 in 1933, 79 in 1934, 82 in 1935, 86 in 1936 and 98 in 1937. The same sources as for note 68 above.

71   No index in constant prices is given for the years 1939 and 1940 in *Vneshnyaya Torgovlya SSSR za 1918-1940 gg, op. cit.* In 1950 ruble exchange value, Soviet exports went from 1,021 million rubles in 1938, to 426 in 1939 and 1,066 million rubles in 1940. Soviet imports went from 1,444 million in 1938, to 745 million in 1939 and to 1,446 million in 1940.

was concerned. This trade had greatly diminished since 1932, but was to increase by almost a factor of ten between 1939 and 1940.[72]

It should be noted that during the 1930s Soviet involvement in foreign trade declined continuously. In 1913 Russian exports represented 10.4 per cent of national income. By 1929, these had fallen to 3.1, rising to 3.5 in 1931 (the peak year of Soviet trade during this period), and declining steadily to 2.6 in 1932, 1.3 in 1935 and 0.5 per cent in 1937.[73]

While the policy of self-sufficiency during the 1930s was responsible for the relative and, often, absolute decline in foreign trade, this was not the case for imports of machinery and equipment. These continued to account for a large share of trade: 46.4 per cent of total imports in 1930, 53.5 per cent in 1931 and 55.2 per cent in 1932 (the peak). Declining somewhat in 1934-1935 (to 24.4 and 22.6 per cent), the share increased slowly to 34.5 per cent in 1938, 38.7 per cent in 1939, and then declining to 32.4 per cent in 1940.[74]

Soviet imports of machinery from the Western industrialised countries were always above average. In 1931 the share of machinery imports (of total imports) was 53.6 per cent. Machinery accounted for 94.3 per cent of imports from the United States, 79.5 per cent from Switzerland, 72.9 per cent from Belgium, 79.5 per cent from Italy, 65.4 per cent from Germany and 65.9 per cent from Sweden. These countries furnished to the Soviet Union in 1931, 1937 and 1940, 58, 89 and 94 per cent of total machinery imports, respectively.[75]

The main feature of changes in the geographical distribution of Soviet foreign trade in the 1930s, was the reduction of trade with Germany following the accession to power of Hitler. In 1931 Germany was the Soviet Union's second most important client for *exports* with 15.2 per cent of the total (the leading client was the United Kingdom with 32.8 per cent of the total). Other important clients were Italy (4.9 per cent), Holland (3.6 per cent), France (3.5 per cent) and the United States (2.8 per cent).

In 1937 Britain remained the first client for Soviet exports (32.1 per cent). Germany (with its 6.2 per cent) was preceded by Belgium and the United States (7.7 per cent each) and by Holland (6.4 per cent). In 1940, 52.2 per cent of Soviet exports went to Germany and 8.0, 3.3 and 2.2 per cent to the United States, Sweden and Switzerland, respectively. Trade with other industrialised countries was practically non-existent.[76]

The geographical distribution of Soviet *imports* during the 1930s followed a somewhat different pattern. In 1931, 1937 and 1940 Germany accounted for 37.2, 14.9 and 29.0 per cent of imports respectively. In the same years the United States accounted for 20.8, 18.4 and 31.0 per cent respectively. Imports from Britain were rather low in 1931 and 1937 (6.6 and 4.7 per cent respectively) and almost non-existent in 1940 (0.9 per cent). Imports from other industrialised countries were less important.[77]

B. Central Europe

The world economic crisis greatly affected the foreign trade of Central Europe. In 1933, it amounted to 29-31 per cent of the 1929 level and had recovered only slowly by 1937.[78] The

72   Index 27 in 1936, 13.3 in 1939 (1934 = 100) for exports and 21 and 4 for imports. Calculated in 1950 ruble exchange value. *Vneshnyaya Torgovlya SSSR za 1918-1940 gg, op. cit.*, p. 23.
73   Franklyn D. Holzman, *op. cit.*, p. 42. Unfortunately, the method by which these percentages were computed are not indicated.
74   *Ibid.*, p. 49, for the years up to 1936. See also *Vneshnyaya Torgovlya SSSR za 1918-1940 gg, op. cit.*, pp. 368 and 402.
75   Calculated from *Vneshnyaya Torgovlya SSSR za 1918-1940 gg, op. cit.*, pp. 301, 368, 402, 476, 481, 504, 551, 558, 586, 589, 665, 668, 786, 790, 825, 827, 847, 851, 1073 and 1078.
The part of machinery in Soviet imports from Western industrialised countries was also above average in 1937 and 1940. For those countries from which the Soviets also imported petroleum, ferrous metals, copper or other materials, this part was reduced for some years.
76   Calculated from *Vneshnyaya Torgovlya SSSR za 1918-1940 gg, op. cit.*, pp. 21-28 and 37. In current prices and actual exchange value of the ruble.
77   *Ibid.*
78   With the exception of Rumania, where the value of foreign trade fell only to 44.4 per cent (1929 = 100). *Statistical Yearbook of the League of Nations*, Years 1937 and 1938.

recovery was more rapid for the agrarian countries (index 64 for Rumania and 65 for Bulgaria — 1929 = 100) and less pronounced for Czechoslovakia (index 40), Poland (index 41) and Hungary (index 51). It should be noted that during the same period the index of world trade fell (1929 = 100) to 35 in 1933 and recovered only to 46 by 1937.[79]

The drastic decline in trade had immediate repercussions on the ability of the Central European countries to pay their debts. Servicing of foreign debts remained a heavy burden in 1931 requiring 48 per cent of Hungarian exports and 28, 24, 16 and 5 per cent of Rumanian, Polish, Bulgarian and Czech exports, respectively.[80] Attempts were made to withdraw foreign capital which had been invested in the 1920s. This led to the suspension of payments on foreign debts in 1932 and the introduction of exchange controls.[81]

The world economic crisis strongly reinforced the protectionism of the Central European nations. Austria and Czechoslovakia expanded their agricultural production while the other countries expanded their industries. This, in turn, provoked a new reduction (in value and in percentage) of intra-Central European trade. A whole system of new tariffs, clearing agreements and administrative measures was erected to protect national activities. The economies of these nations thus evolved in an autonomous fashion, often competing with one another and thus departing from the former existing division of labour.

The main shifts in the geographical distribution of foreign trade during the 1930s involved Germany and other highly industrialised Western countries. The agrarian Central European countries (Hungary, Rumania and Bulgaria) increased their trade with Germany from 1929 to 1937, even though a decline was observed in 1933.[82] For Czechoslovakia and Poland, trade with Germany declined sharply during the crisis in 1933 and continued to drop until 1937.[82] An opposite trend could be observed in the trade with the other industrialised West European countries.[83] While trade with Bulgaria declined, that with Poland, Czechoslovakia and Rumania increased, and Hungarian trade remained stationary.[84]

During this time exchange control and clearing arrangements dominated the trade flows. The increased role of Germany in trade with the agricultural countries of Central Europe resulted from the deliberate policy of the former to buy the latter's agricultural surpluses. This policy forced the Central European agricultural countries to buy German industrial goods, equipment and technology in exchange. Long-term commercial agreements entered into by Germany with these countries after 1934 reinforced this trend.

Notwithstanding these changes in the geographical distribution of trade, the main features of the technology flows which existed prior to the First World War remained in place. Thus Germany, the industrialised countries of Western Europe and the United States remained the main exporters of machinery and equipment to Central Europe. In the case of

79   *Ibid.*
80   Jan Marczewski, *Planification et croissance des démocraties populaires – Analyse historique,* Paris, 1956, p. 57.
81   Only Czechoslovakia did not suspend the payment of foreign debts. Poland suspended payment only in June 1936. See Jan Marczewski, *Ibid.,* p. 70.
82   The German portion of Hungarian *imports* was 20 per cent in 1929, 19.6 per cent in 1933 and 26.2 per cent in 1937. For Rumania, the corresponding figures were 24.1 per cent in 1929, 18.9 per cent in 1933 and 28.9 per cent in 1937; and for Bulgaria 22.2 per cent, 38.2 per cent and 54.8 per cent respectively.
The German fraction of Hungarian *exports* was 11.7 per cent in 1929, 11.4 per cent in 1933 and 24.1 per cent in 1937. For Rumania the corresponding figures were 27.6 per cent in 1929, 10.6 per cent in 1933 and 19.2 per cent in 1937. For Bulgaria the corresponding percentages were: 29.9 per cent, 17.7 per cent and 43 per cent respectively. *Statistical Yearbook of the League of Nations,* 1937 and 1938.
The German portion of Czechoslovak *imports* was 25 per cent in 1929 and 15.3 per cent in 1937, while the portion of Czechoslovak *exports* amounted to 19.3 per cent in 1929 and 13.7 per cent in 1937. For Poland the corresponding figures for *imports* were 33.2 per cent in 1929 and 19.1 per cent in 1937; for *exports* 41.7 per cent in 1929 and 19.4 per cent in 1937. *Statistical Yearbook of the League of Nations,* 1937 and 1938.
83   Britain, France, Switzerland, Belgium and the Netherlands.
84   The most pronounced increase in percentage terms was in trade between the industrialised West and Czechoslovakia and Poland.

Poland, during the years 1935-1938, these industrialised countries furnished almost all of the machinery, instruments, transport equipment and chemical products. Germany, alone, provided almost half of these imports.[85] In the case of Bulgaria, a detailed review of machinery and transport equipment imports from 1933-1937 reveals that the principal sources were mainly Germany, and in part Western Europe and Czechoslovakia.[86]

It should be noted that during the 1930s, the trade of Central Europe with the Soviet Union remained at the level of the 1920s. For Bulgaria and Rumania it was close to zero and almost non existent for Hungary.[87] The contribution of the Soviet Union to Polish imports declined from 2.1 per cent in 1933 to 1.1 per cent in 1937, while the contribution to Polish exports fell from 6.2 per cent to 0.4 per cent during the same period.[88] Only Czechoslovak trade with the Soviet Union remained significant, (even though fluctuating) with Soviet machinery imports at 18.9 million rubles in 1929, 14.5 million in 1936, 4.9 million in 1937, and 14.8 million in 1940.[89]

III. TECHNOLOGY TRANSFER DURING THE SECOND WORLD WAR AND THE IMMEDIATE POST-WAR PERIOD (UNTIL 1953)

Established trade relations were disrupted during World War II and, because of the political and demographic changes brought about by the War, were never to be resumed again in the same manner. The period is marked by two occurrences: initially, the continued absorption of foreign technology by the Soviet Union and the formation of two economic and political "blocs".

1. Technology Transfer to the Soviet Union; the Recovery Period

According to some analysts, the industrial capacity of the Soviet Union was greater at the end of the Second World War (despite war damages) than it was in 1940.[90] The mechanisms by which the technology became available to the Soviet Union were:

a) Lend-Lease and the "Pipe Line Agreement"

It has been estimated that through Lend-Lease and the associated "Supply Protocols" some $1.25 billion worth of the most modern American industrial equipment was injected into the Soviet economy.[91] About "50 per cent of the equipment supplied had reconstruction potential, equalling one-third of Soviet pre-War industrial output".[92] A rare Soviet statement acknowledging Lend-Lease substantiates the significance of the help that was extended.

85   *Maly Rocznik Statystyczny, op. cit.,* p. 177.
86   *Statisticheski Godisnik na Carstvo Bolgarija, Godina XXX,* (Statistical Yearbook of the Kingdom of Bulgaria, XXX-year), Sofia, 1938, pp. 511-517.
87   *Ibid.,* pp. 486-487, *Vneshnyaya Torgovlya SSSR za 1918-1940 gg, op. cit.,* pp. 22-26.
88   Calculated from *Maly Rocznik Statystyczny, op. cit.,* pp. 166-167.
89   *Vneshnyaya Torgovlya SSSR za 1918-1940 gg, op. cit.,* pp. 806-813.
90   Antony C. Sutton, *op. cit.,* p. 345.
91   Antony C. Sutton, *Western Technology and Soviet Economic Development 1945-1965,* Hoover Institution Press, Stanford University, Stanford, California, 1974, p. 14.
92   G. Warren Nutter, *The Growth of Industrial Production in the Soviet Union,* Princeton University Press, Princeton, N.J., 1962, p. 214 as given in Antony C. Sutton, *Western Technology and Soviet Economic Development, 1930-1945, op. cit.,* p. 345.

"Stalin paid tribute to the assistance rendered by the United States to Soviet industry before and during the War. He said that about two-thirds of all the large industrial enterprises in the Soviet Union had been built with United States help or technical assistance."[93]

b)   *Reparations*

The Soviet Union profited from massive injections of technology through reparations directed at the former "enemy" nations, Germany (East), Hungary, Rumania and Bulgaria. In addition, factories were dismantled and requisitioned arbitrarily in all the countries, enemy or "friendly", occupied by the Soviet army. Estimates of the payments made through reparations and the actual value of the commodity deliveries and dismantlement of plants differ, but they are thought to have been in the billions (US dollar equivalent).[94] The officially calculated total paid by the countries of Central Europe to the Soviet Union from 1945 up to October 1956 was estimated at 1.25 billion dollars per year (in 1938 dollars). Until 1950, the German Democratic Republic alone paid 609.6 million dollars each year (in 1938 dollars).[95]

c)   *Confiscation of Ex-German Assets*

The number of plants dismantled and shipped to the Soviet Union from Germany was sizeable. It has been estimated that a total of 25 per cent of industrial plants in the Western Allied zones was allocated to the Soviet Union and the best industry (together with key technical staff) from the East zone was moved to the Soviet Union.[96]

d)   *Trade*

Trade did not play a significant role during this period. The trade relationships of the Soviet Union with the developed countries had almost been completely broken just prior to the War. With respect to Central and Western Europe, little commercial trading took place due to the political uncertainty that prevailed at that time (until 1953) and because of the inclusion of the Central European countries (following the War) into the Soviet sphere of influence. The destruction of the European economies and the economic dislocation of what had been the Central European countries (and Germany) also contributed to the suppression of trade. The scope of trade relations, in terms of trade volume and evolution is illustrated in the subsequent section.

2.   **The Economic and Political Division Between East and West During the Cold War Period — 1945-1953: Re-orientation of Commercial Flows**

a)   *Political Division Between East and West*

The political division, one factor leading to the clear-cut bi-polarisation between East and West, was signalled by the formation of the *Cominform* in September of 1947. The

93   Conversation between Stalin and a United States official as reported by W. Averell Harriman, US Ambassador, US State Department, Decimal File 033.1161, Johnston, Eric/6-3044: Telegram 30th June, 1944 as given in Antony C. Sutton, *Western Technology and Soviet Economic Development, 1930-1945, op. cit.,* p. 3.
94   Antony C. Sutton, *Western Technology and Soviet Economic Development, 1945-1956, op. cit.,* Chapter 2. See also Franklyn D. Holzman, *International Trade Under Communism,* Basic Books Inc., Publishers, New York, 1976, Chapter 3.
95   Eugène Zaleski, *Stalinist Planning for Economic Growth, 1933-1952,* The North Carolina University Press, Chapel Hill, 1980, p. 346.
96   Antony C. Sutton, *Western Technology and Soviet Economic Development, 1945-1956, op. cit.,* pp. 26-32. See the same volume for a discussion of the assets thus acquired.

purpose of *Cominform* was to help the Soviet Union consolidate (politically and militarily) its authority over those European countries now under its sphere of influence. In the West, the formation of *Cominform* and events such as the Prague coup (February 1948) and the Berlin blockade (a few months later) were viewed as evidence of aggression on the part of the Soviet Union vis-à-vis the countries of Western Europe. The resultant response was the creation of NATO in 1949.

b)   The Economic Division Between East and West

i)   Attempts Made to Prevent the Separation

Attempts had been made by the West prior to the first embargo action taken by the United States against the Soviet Union (end 1947 and beginning of 1948) to arrest the deterioration of East-West economic relations. The Marshall Plan (1948) and the Organisation for European Economic Co-operation (founded in the same year, later to become the Organisation for Economic Co-operation and Development) were initially conceived as being open to the East European countries. However, the Soviet Union refused to join or to allow its allies to join. Even the Economic Commission for Europe (ECE), created in March of 1947 with the purpose of establishing and promoting economic co-operation between East and West, was unable to stop the growing economic division. The prevailing political conditions in the early years of the ECE's existence prevented effective operation.

ii)   Rejection of the Marshall Plan and Creation of CMEA

The creation of the Council for Mutual Economic Assistance (CMEA) in January of 1949 marked the official resumption of the Soviet Union's policy of pursuing economic independence.[97] The purpose of CMEA was to "establish wider economic co-operation between the countries of people's democracy and the USSR... with the task of exchanging economic experience, extending technical aid to one another, and rendering mutual assistance".[98]

Most Western analysts agree that CMEA was a direct response to the creation of the Marshall Plan, portrayed as "violating the sovereignty and interests of the Soviet Union and the East European countries". The economic division between East and West had been foreshadowed by the failure of the Bretton Woods Conference in 1944 and the refusal of the Soviet Union to join the International Monetary Fund. These events indicated that the Soviet Union had disassociated itself from certain international attempts to reform the world system of trade.[99]

iii)   Export Restrictions Against the East

The response of the West to the apparent political and military danger from the East was the passage of the United States Export Control Act in February 1949 and the establishment in November of that year of a Co-ordinating Committee — an informal multi-lateral co-ordinating mechanism.

The primary purpose of the Export Control Act was to prevent the export of goods which could make a significant contribution to the military potential of the Eastern bloc

---

97   In the literature the Council is also referred to as CEMA or COMECON. CMEA is the abbreviation adopted here, except in citing the title of a publication or a text where the adopted usage there is maintained.

98   Michael Kaser, *COMECON,* 2nd edition, Oxford University Press, London, 1967, p. 12 as given in Franklyn D. Holzman, *International Trade Under Communism, op. cit.,* p. 68.

99   See Peter Knirsch, "Interdependence in East-West Economic Relations", paper prepared for the Marshall Plan Commemoration Conference, OECD, Paris, 2nd-3rd June, 1977, p. 3.

countries.[100] The multi-lateral mechanism was set up in order to achieve a measure of co-ordination and co-operation among the Western allies in restricting the flow of strategic technology to the East. Discussions concerning goods which were not to be exported to the East led to the establishment of multi-lateral Lists. These Lists (periodically reviewed and revised) were to guide member governments in the execution of their national export control policies.[101]

In 1951, following the outbreak of the Korean War, the United States passed the "Battle Act" (formally known as the Mutual Defense Assistance Control Act) under which the President was empowered to terminate all military, economic and financial aid to any nation which shipped strategic products to the East.[102] In essence the purpose of the measure was to reinforce the system of international controls previously in effect and provide a link with US strategic trade controls under the Export Control Act of 1949.[103]

iv) *Import Restrictions Against the East*

In the United States the basic legislation designed to control imports from the East was the Trade Agreements Extension Act of 1951. This Act authorised the President to suspend or halt any tariff concessions accorded to the Soviet Union and to the East European countries under its control. The legislation also prohibited the importation of certain commodities.[104] Imports were also negatively affected by Proclamation 2935, of 1 August, 1951 through which the President suspended all concessions provided for in the Tariff Act of 1930, including those negotiated by GATT (General Agreement on Tariffs and Trade).[105]

GATT, founded in 1947, had as its major purposes the "increase of world trade by reducing trade barriers, particularly tariffs, and to encourage non-discriminatory trade primarily through the mechanism of most-favoured-nation agreements among members".[106] Since the Eastern countries were non-members (they adamantly opposed GATT), imports from the East were naturally not included which also served to depress economic relations between East and West.

3. Trade Relations with the West: Imports of Embodied Technology

As has been pointed out in the preceding sections, prior to World War II the Soviet Union had almost no economic ties with the former Central European countries, and trade

---

100   See *A Background Study on East-West Trade*, prepared for the Committee on Foreign Relations, US Senate, by the Legislative Reference Service. Library of Congress, 89th Congress, 1st Session, Washington, D.C., April 1965, p. 38.

101   These Lists do not replace national unilateral lists. Unanimous agreement by members of the co-ordinating mechanism is required in basic policy matters and on changes in the strategic Lists.

102   See *A Background Study on East-West Trade, op. cit.*, p. 39.

103   The President has never been requested to apply the sanctions foreseen in the Battle Act. The threat that the United States would withdraw all aid to any nation which shipped strategic goods to the East was potentially only tenable during the period of the Marshall Aid Plan. For details concerning the administrative provisions of the Act see, for instance, Nicolas Spulber, "East-West Trade and the Paradoxes of the Strategic Embargo", in Alan A. Brown and Egon Neuberger, editors, *International Trade and Central Planning*, University of California Press, 1968.

104   Under the terms of the Act MFN (Most Favoured Nation) treatment for all East European countries, except Yugoslavia, was withdrawn in 1951. The number of items prohibited from the USSR were limited and pertained mainly to furs and skins. See *A Background Study on East-West Trade, op. cit.*, p. 40.

105   Anne de Tinguy, "Les relations économiques et commerciales soviéto-américaines de 1961 à 1974", *Revue d'études comparatives Est-Ouest*, Vol. 6, No. 4, Editions du CNRS, Paris, December, 1975, p. 123.

106   Franklyn D. Holzman, *International Trade Under Communism, op. cit.*, p. 156.

with the industrialised countries of the West has been estimated to have shrunk to a fraction of 1 per cent of GNP.[107] In the immediate post-War years, however, East-West trade had begun to show an upward turn, and by "1948 trade between Eastern and Western Europe (excluding the Soviet Union and East Germany) had reached pre-War levels".[108]

During the period of the Cold War, East-West economic relations deteriorated. Thus, total exports of the East European countries (including Albania) to Western Europe, the United States and Canada, fell from 72.8 per cent in 1938, to 41 per cent and 15 per cent in 1948 and 1953 respectively.[109]

Neither the Soviet Union nor the East European countries placed much emphasis on foreign trade. By 1946, the development of an increasingly inward-looking East European "economic" bloc was becoming apparent.[110] After 1949 under Soviet pressure (politically and economically motivated) the East European countries could only develop trade relations with each other and the Soviet Union. For instance, in 1946 the "socialist countries of the area already accounted for 55 per cent of the total foreign trade of the Soviet Union. Two years later, their share rose to 60 per cent". The figures are from official Soviet sources.[111] Similar trends have been noted using Western estimates which describe intra-CMEA trade as follows: 10 per cent of area exports in 1938, 44 per cent in 1948 and 64 per cent in 1953.[112]

The changed trade pattern forced upon the East European countries by the Soviet Union (and affected by Western export-import restrictions) affected the trade patterns of the Soviet Union and the East European countries in different ways. Between 1950-1953 the East European countries were losing, in various degrees, their former markets in the West — their exports fell by 7.8 per cent. On the other hand Soviet exports rose by 91 per cent. And while the Eastern European imports of machinery and equipment fell by 34 per cent, Soviet imports increased by 23 per cent.[113]

East-West trade also did not develop at the same rate as that of the rest of world trade during the early 1950s. For instance, it is estimated that "trade with the West, as a percentage of CMEA's total trade, actually fell from 42 per cent in 1948 to 14 per cent in 1953.[114] Other estimates indicate an even lower figure and assign to Eastern trade 1.3 per cent of the share of world trade in 1953".[115]

107   US trade with the Eastern bloc was virtually non-existent by 1953. Franklyn D. Holzman, *International Trade Under Communism, op. cit.,* p. 67.
108   Christopher Cviic, "Comecon and East-West Trade Policies", in *COMECON: Progress and Prospects, Colloquium–1977,* NATO, Directorate of Economic Affairs, Brussels, 16th-17th March, 1977, p. 211.
109   Nicolas Spulber, *art. cit.,* Table 3, p. 114. (Sum of percentages given for Western Europe, United States and Canada).
110   The emphasis on intra-CMEA trade was perhaps inevitable since the Soviet Union had to forge a bloc from nations which had previously been market economies and often hostile. Thus, the "stern political and military measures required to accomplish this objective (of forging a bloc) were matched initially by dramatic economic measures". Another factor was the reaction of the Soviet Union to the policy objectives of the Western alliance, particularly of their leader, the United States. One of these objectives was to "slow up the economic and military development of the socialist nations" which was to be accomplished primarily by depriving these nations of strategic goods. See Franklyn D. Holzman, *International Trade Under Communism, op. cit.,* p. 125.
111   *Spravochnik po vneshnei torgovle,* Moscow, 1958, p. 117.
112   Nicolas Spulber, *art. cit.,* p. 113. Holzman states that "by 1950, intra-bloc trade exceeded 60 per cent of the total; by 1953, it amounted to almost 80 per cent. Trade with the USSR accounted for the largest part of the change in trading patterns". Franklyn D. Holzman, *International Trade Under Communism, op. cit.,* p. 70.
113   The percentages have been computed from millions of US dollars, f.o.b., equivalent. See Nicolas Spulber, *art. cit.,* pp. 113-114.
114   Christopher Cviic, *art. cit.,* p. 213.
115   J. Wilczynski, *The Economics and Politics of East-West Trade, A Study of Trade Between Developed Market Economies and Centrally Planned Economies in a Changing World,* London, 1969, p. 52.

In sum, the Cold-War period saw East-West commercial relations reach a nadir as political considerations dominated the world scene. In the West, measures taken to control the flow of exports to the East (as well as their imports) served to curtail trade. In the East, the creation of CMEA (despite the fact that its goal of economic integration remained largely unachieved in the early years of its existence) and the concentration on intra-CMEA trade, primarily for political reasons, also served to bring East-West trade to a virtual standstill.

IV. TRADE WITH THE WEST AND TECHNOLOGY TRANSFER AFTER STALIN: THE PERIOD 1953-1965

1. General Developments

In general the years 1953-1965 can be characterised as a transitional period during which (beginning with the death of Stalin in 1953) a climate conducive to more normal relations between East and West became evident. Many factors are held responsible for this change, among which the following may be mentioned:

First, the ideological problems, political conflict and violent disturbances within the Eastern bloc (German Democratic Republic in 1953, Poland and Hungary in 1956) indicated that Soviet power and influence were not unlimited. The awareness of this situation undoubtedly lessened the West's acute fears of the Soviet Union. Second, the introduction of "de-Stalinisation" by Khruschev in 1956, judged partially on the basis of less overt and less shrill anti-Western outbursts, was believed to have a "softening effect" on the Soviet Union's domestic and foreign policy.[116] This belief was supported in the early 1960s by Khruschev's espousal of the concept of "peaceful coexistence". In addition the commencement of the Soviet "offensive of international amicability" (following Khruschev's accession to power in 1955) undoubtedly helped improve trade relations.[117]

In the West, these developments reduced the anxiety about Soviet expansionism. The rapidly changing political climate led to considerable optimism in some Western circles regarding future development of East-West relations. This was witnessed by the efforts of some Western theorists to promote the theory of convergence. It was also supported by the Johnson Administration (United States), which initiated the policy of "building bridges". The latter policy, in many respects, can be considered to be a precursor of the current policy of détente.[118]

The importance of theories such as convergence lay not in their merit or whatever degree of realism which they may have reflected. Rather, their importance can be traced to the fact that these theories prepared public opinion and with it Western policy, to regard "closer economic relations with the East as politically acceptable and even desirable as a contribution to internal changes in Eastern Europe".[119]

These developments together with the growing commercial interests in Eastern markets, primarily on the part of Western Europe, served to decrease the pressures on the West for maintaining a unified economic policy vis-à-vis the East. Thus many members of the co-ordinating mechanism began to press for less stringent export controls. Indeed, Western Europe had never been as interested as the United States in adhering to a rigid

116   Peter Knirsch, *art. cit.,* p. 4.
117   Adam Ulam, *The Rivals,* Viking, New York, 1971, p. 227 as given in Franklyn D. Holzman, *International Trade Under Communism, op. cit.,* p. 140.
118   The theory of convergence, whose leading proponent was Jan Tinbergen, (first proposed in 1961) was never accepted by the East nor by many Westerners.
119   See for instance Zbigniew K. Brzezinski, *Alternative to Partition,* New York, 1965, as given in Peter Knirsch, *art. cit.,* p. 5. The possible influence of closer relations with the West on the evolution of Eastern policies is discussed in Chapter 4.

system of controls, and the West European unilateral embargo lists against Eastern Europe were shortened as early as 1954.[120]

Thus, while the actual trade figures during the period 1953-1965 show a relatively modest increase in trade between East and West, the period was important in that it laid the groundwork for the closer and markedly changed trading relations that followed.

2. Eastern Attitudes Towards International Economic Co-operation

As has been mentioned, economic independence was the mainstay of Soviet economic policy during the post-War and Cold War periods. The resultant policy of self-sufficiency was imposed by the Soviet Union on its Eastern partners. Soviet influence (exercised through economic exploitation of its partners and the hindrance of trade among the East European countries) was used to turn CMEA into an instrument for concentrating trade of the East European nations on the Soviet Union. This was achieved by means of long-term commercial agreements, "while holding back the measures of integration amongst themselves that had been launched by the East and Central European nations in the years directly following the War and replacing them with less far-reaching trade accords".[121]

This policy of economic independence posed serious problems for the East European countries. In the past the latter had depended on foreign trade for their economic reconstruction and growth. The policy of autarky began to be questioned by Hungary in 1953 and by the German Democratic Republic in 1954. Elements of the planning system were challenged on the theoretical level by Polish economists in the late 1950s and by Czechoslovak economists in the mid-1960s. These discussions (on the "law of value") led to a reappraisal of the advantages and disadvantages of commercial relations with the West. They were also to lead to a new orientation of economic policy in favour of closer relations with the West,[122] as foreign trade came to be viewed as a growth factor (input) which could help counteract the progressive slow-down in the previously achieved East European high economic growth rates.

A renewed interest in pursuing trade with the West became increasingly manifest in the East due to the slow-down in economic growth and the difficulties experienced in industrial exploitation of scientific and technical knowledge. In other words, it is thought that the East viewed the import of "technical progress" from the West, that is, highly developed industrial technologies in the form of machinery and equipment, as a crucial input in stimulating economic growth.[123]

The emphasis that had begun to be placed on the import of technologies and know-how by the East has also been interpreted in political terms. The internal economic reforms, attempted in the early 1960s, would have entailed the transition from extensive to intensive growth, based on rapid innovation. Successful economic reforms implied "too much political as well as economic upheaval... which is why the Party leaders opted for the alternative course of buying technological progress from abroad, especially from the West"...[124]

The diminution of interest in pursuing trade within the confines of CMEA is illustrated

---

120   If anything, the West European countries have always been more interested in working out common import quotas. Concerning exports, it was during the Marshall Plan period, in particular, that the United States was in the position to more directly exert its viewpoints concerning export controls. Thus its policies relative to such controls tended to predominate.

121   Alan Smith, "Soviet Economic Influence in COMECON", in *COMECON: Progress and Prospects, Colloquium–1977, op. cit.,* pp. 238-239.

122   These discussions also led to attempts to bring about internal economic changes — changes which were not tolerated by the Soviet Union. This aspect falls outside of the framework and purpose of the analysis and is, therefore, not taken up.

123   F. Levcik and J. Stankovsky, *Industrielle Kooperation Zwischen Ost und West,* Studien über Wirtschafts und Systemvergleiche, Springer Verlag, Wien, New York, 1977, p. 65.

124   Christopher Cviic, *art. cit.,* pp. 213-214. This aspect, particularly relevant for assessing current developments, is discussed in more detail in Chapter 4.

by trade trends. The increase in "commercial exchange within CMEA from 1961-1965 was the smallest in recent history: 55 per cent against a growth of 71 per cent for 1956-1960 and 85 per cent for 1951-1955".[125]

3. Evolution of Western Commercial Policy

We have said that a growing interest in developing commercial relations with the East was manifested by the West during this period. Western national trading interests began to assert themselves strongly in response to the promise of appreciable new markets. This interest was to lead to a disparity between US and West European policies towards East-West trade. For the United States fundamental changes towards such trade were less acceptable than they were to its partners. The following serve as illustrative examples:

a)   *Export-Import Controls*

i)   *Western Europe*

The West European countries adopted the viewpoint that the Lists "should contain only truly strategic commodities that were not available to the Socialist nations. Items which could only be justified on grounds of slowing down Communist economic growth were to be excluded".[126]

The Western countries revised their international Lists for the second time in 1957, reducing the embargo List to some 100 items and the watch list to 20.[127]

A clear shift towards relaxation of import quotas against Eastern goods in the European Free Trade Association (EFTA — founded in 1960) and the EEC (established under the Treaty of Rome in 1957) was also evident. The United Kingdom had already decided to "abolish its severe quota system in favour of a more flexible trade-inducing policy of autonomous liberation", in 1964.[128] Within the Common Market of the Six, France was the first to free imported goods from quantitative restrictions. In 1964, "around 90 per cent of imported goods were restricted — with the exception of raw materials — by 1966, over half of the 1,100 restricted items had been freed".[129]

ii)   *United States*

Contrary to the countries of Western Europe, the United States, in 1962, extended the concept of "strategic significance", as applied in the Export Control Act of 1949 to encompass goods which had *economic* as well as military importance or significance. Further, the United States continued to maintain an extensive number of items on its "positive" list: 1,303 separate entries at the end of the first quarter of 1964.[130]

125   Samuel Pisar, *Coexistence and Commerce, Guidelines for Transactions between East and West*, Allen Lane, Penguin Press, 1970, p. 16.
126   Franklyn D. Holzman, *International Trade Under Communism, op. cit.*, p. 149.
127   Nicolas Spulber, *art. cit.*, p. 109. Originally there were three different types of Lists: an embargo List (List I), a quantitative controls List (List II), and a surveillance List (List III), conforming to the classification of the "strategic" importance of the goods considered. In 1957, List I was substantially revised and Lists II and III were replaced by a new watch list consisting mainly of items formerly on List I, and still considered to be of some strategic importance. Apart from implements of war and atomic energy items, the lists contain only a selective range of the most advanced industrial materials and equipment. *Ibid.*
128   Samuel Pisar, *op. cit.*, p. 105.
129   *Ibid.*, pp. 105-106.
130   Nicolas Spulber, *art. cit.*, p. 110. This is not to say that the US did not also make some changes in its control list. For instance, between 1956-1962 roughly 700 items in 57 categories were decontrolled to permit their export to the Soviet Union and to Eastern Europe. See Zygmunt Nagorski, *The Psychology of East-West Trade*, 1974 as given in Arthur H. Hausman, "East-West Technological Co-operation: Likely Transfer Patterns (1976-1980)", in *East-West Technological Co-operation, Colloquium-1976*, NATO, Directorate of Economic Affairs, Brussels, 17th-19th March, 1976, p. 332.

Imports were regulated by the Trade Expansion Act of 1962. Under this Act the United States Congress withdrew all Presidential discretion in the matter of restoring MFN treatment to individual East European countries or to the Soviet Union. However, an amendment to the 1963 Foreign Assistance Act gave to the President the power to expand trade when he judged it to be "in the national interest" and, in 1964 the President used this provision to relax trade restrictions with Poland and Yugoslavia.[131]

US policy towards East-West trade during this period showed a difference of viewpoint between the Administration and the Congress,[132] primarily due to difficulties encountered in defining "national interest".[133] On the whole US policy towards the Soviet Union did not really change compared with the previous period, notwithstanding indications of Soviet interest in re-establishing commercial ties.[134] However, policy towards the East European countries had begun to differ as early as 1956. For instance, controls on exports to Poland were reduced in 1957 and MFN treatment was restored in 1960.[135]

b) *Financial Assistance: Credits*

i) *Western Europe*

The Western European nations broke with US credit policies towards the Eastern bloc countries in the early 1960s. International efforts to align export credit practices date back to the "gentlement's agreement" of 1934 — the so-called Berne Union.[136] In 1958, under pressure from the United States, agreement was reached to limit credits to the Eastern countries to five years. "The first actual departure of a Berne Union member from the five-year rule in East-West trade was apparently a $7.2 million ten-year insured credit by Belgium for the sale of a chemical plant to Hungary in 1963".[137] In 1964, the United Kingdom granted a "twelve-year credit to Czechoslovakia for the purchase of fertilizer plants and a fifteen-year $300 million grant to the USSR for chemical factories".[138] France also began to authorise and insure long-term credit around 1964. The Franco-Soviet Commercial Agreement of 30th October, 1964 arranged for very favourable credit conditions for factory delivery or construction.[139] It also included a French commitment to provide over the next five years government supported credits with five to seven year maturities in excess of Frs 3,5 billion (approximately $700 million) for the purchase of French capital goods.[140] Italy has likewise extended long-term credits. Most notable of the large export transactions with Italian government support was the FIAT project at Togliattigrad which included a long-term credit by IMI of over $350 million.[141]

In short, long-term credit became the rule as a mood of "every man for himself", began

131 Arthur H. Hausman, *art. cit.,* p. 332.
132 For instance the 1962 Trade Bill which would have expanded East-West trade was defeated, yet the 1963 amendment enabled the President to expand trade when he judged it to be "in the national interest". Again, the "East-West Trade Relations Bill of 1966" was defeated.
133 See Chapter 4.
134 In June, 1958 Khruschev proposed a comprehensive trade agreement for a large expansion of trade in peaceful goods with the United States. This bid was turned aside. See Franklyn D. Holzman, *International Trade Under Communism, op. cit.,* p. 140.
135 Rogers Morton, *op. cit.,* p. A-4.
136 Japan, which was not a member of the Berne Union, already extended credits of more than five years, in the late 1950s. Japan was one of the initiators of official credit support programmes (OCS) and OCS in " connection with massive projects has become an important facet in Japanese bilateral relations with the USSR": Thomas A. Wolf, "East-West Trade Credit Policy: A Comparative Analysis", in Paul Marer, ed., "US Financing of East-West Trade. The Political Economy of Government Credits and the National Interest", *Studies in East European and Soviet Planning, Development and Trade,* No. 22, Indiana University, Bloomington, Indiana, August 1975, p. 174.
137 *Ibid.,* p. 172.
138 Franklyn D. Holzman, *International Trade Under Communism, op. cit.,* p. 149.
139 Samuel Pisar, *op. cit.,* p. 113.
140 *Ibid.*
141 *Ibid.,* pp. 113-114. IMI is the Istituto Mobiliare Italiano.

to exert itself and credit was being granted almost exclusively on commercial grounds. Furthermore, while until the 1950s the "clearing system" dominated bi-lateral accounting, the 1960s saw the transition to payment in freely convertible currencies".[142]

ii) *United States*

A long-term credit (20 years) was first made available by the United States to Poland, through the Export-Import Bank, in 1957. Bulgaria, Czechoslovakia, Hungary and Poland were granted medium-term guarantees (also through the Export-Import Bank) in 1964, under Presidential determination of "national interest".[143] These guarantees authorised financing of US agricultural credit sales.

Regarding the Soviet Union, the Trade Expansion Act of 1962 (under which Presidential discretion in matters of MFN had been withdrawn) was "so reinterpreted by the Administration to allow the grant of medium-term credits linked to United States' export transactions. This Act, together with the longshoremen's agreement to have 50 per cent of all cargoes travel in non-United States' ships, paved the way for the first wheat sale to the Soviet Union in 1964".[144]

4. Commercial and Technical Agreements Between East and West

In a divergence from US practice, Western Europe did initiate trade agreements with the East. By 1960, there were 23 long-term bi-lateral trade agreements between East Europe and the industrialised Western nations.[145] These agreements were a step to regularising East-West trade relationships. In the early stages most were simply trade agreements. However, by the mid-1960s they incorporated the additional element of scientific and technological co-operation. In 1965, for example, "the USSR and France signed an agreement which involved scientific and technological as well as economic collaboration, establishing a permanent high-level commission and many joint working groups".[146]

Subsequently, the smaller East European states concluded technical agreements with the European Economic Community regarding farm produce, their "biggest and most important export item to Western Europe. Poland concluded such an agreement in 1965, and was followed by Bulgaria, Hungary and Rumania".[147]

5. Trade and Technology (Embodied) Flows Between East and West [148]

A marked increase in trade between East and West, though modest in terms of world trade, began during the early 1960s. For instance, from 1960 to 1965, the average annual growth in CMEA countries' imports from Western countries was 8.8 per cent; exports increased by 9.1 per cent.[149] During the same period, intra-CMEA trade increased substantially less — both imports and exports by only 6.8 per cent annually.[150]

---

142   F. Levcik and J. Stankovsky, *op. cit.,* p. 149. Only trade between the German Democratic Republic and the Federal Republic of Germany is still based on bi-lateral clearing. However, the former has virtually free access to Germany's market under the Berlin agreement of 1951 (still in force) and by credit from the Federal Republic of Germany — the "swing", also still in operation. See Christopher Cviic, *art. cit.,* p. 214.
143   Rogers Morton, *op. cit.,* p. A-4.
144   Christopher Cviic, *art. cit.,* p. 217.
145   F. Levcik and J. Stankovsky, *op. cit.,* p. 150.
146   Franklyn D. Holzman, *International Trade Under Communism, op. cit.,* p. 150. See also Chapter 3.
147   Christopher Cviic, *art. cit.,* p. 214. This did not represent formal recognition of the Community. The evolution of the relationship between the EEC and the CMEA is discussed in Section V.
148   Trade trends and trade by commodities is discussed in detail in Chapter 2. See also Annex tables.
149   See Annex Table A-4.
150   See Annex Table A-4.

As has been noted by Spulber, during the period 1954 to 1963, the main "Western exporters and importers to and from the East were West Germany (excluding its trade with East Germany), Italy and the United Kingdom. The main CMEA exporters and importers to and from the West were the USSR and, increasingly, the least developed countries of the area — in contrast with the post-War period, when Czechoslovakia and Poland were the main traders with the West. The foreign trade linkage between East and West again took the pre-War form of exchanges of food, fuels and crude materials from the East, against steel, machinery and industrial equipment from the West".[151]

Machinery and transport equipment formed a very large proportion of the OECD countries' exports to the East, increasing (in value, US millions of dollars) by 48 per cent between 1961 and 1965. Imports by the Soviet Union increased by only 28 per cent, those of Poland by 6 per cent. On the other hand, imports by the German Democratic Republic, Czechoslovakia and Hungary increased by 155, 83 and 86 per cent respectively.[152]

By 1965 the share of machinery and equipment in total OECD exports to the seven Eastern countries was 29.5 per cent.[153] Finland was the most important OECD source accounting for 21 per cent, followed by the Federal Republic of Germany with 17.3 per cent and Japan with 13 per cent of total exports to CMEA countries.[154]

V. PRINCIPAL RECENT DEVELOPMENTS IN EAST-WEST TRADE

1. General Considerations

Since 1965, and particularly between 1972 and 1975, East-West trade expanded at a very rapid rate. A basic condition for this rapid development of East-West economic relations is believed to be due to the policy of détente. This policy provided for both West and East (in particular the East European countries) the political underpinning of the "official sanctioning" of closer economic relations.[155]

For the West, the major impetus behind the rapid expansion of trade was the economic competition among the Western countries for markets in the East. It has been noted, for instance, that the progressive decontrolling of the multi-lateral Lists by all the Western countries (and Japan) and the "more general dramatic turnaround in US East-West trade policy from control to promotion, has undoubtedly led to a greater competition for Western Europe on the export markets of the CMEA".[156] The motivation behind rapid trade

151   See Nicolas Spulber, *art. cit.,* p. 117.
152   See Annex Tables A-11, A-13, A-16, A-18 and A-24.
153   See Table 9.
154   See Tables 9, 13 and 14.
155   Closer economic relations between East and West was one of the major topics at the Conference on Security and Cooperation in Europe (CSCE, Helsinki, 1973-1975). The main goal of the CSCE was to further the process of détente. Economic interdependence was recognised as a growing *global* need by all the signatories of the Helsinki Final Act (August 1975). Basket II of the Helsinki Act calls for "Co-operation in the Field of Economics, of Science and Technology and of the Environment". The first three subsections of Basket II affirm the signatories' desire to promote mutual trade and ease commercial endeavours as a way to "contribute to the reinforcement of peace and security". See The Commission on Security and Co-operation in Europe, *Implementation of the Final Act of the Conference on Security and Co-operation in Europe: Findings and Recommendations Two Years After Helsinki,* Report to the Congress of the United States, Washington, D.C., August 1st, 1977. Détente and its relationship to technology transfer is discussed in Chapter 4.
156   Thomas A. Wolf, "East-West European Trade Relations", in *East European Economies Post-Helsinki – A Compendium of Papers,* Joint Economic Committee, Congress of the United States, 95th Congress, 1st Session, US Government Printing Office, Washington, D.C., August 25th, 1977, p. 1043. In general countries are in agreement that items should be dropped from the Lists as they become obsolete in terms of Eastern technological capabilities and that items representing new strategic technologies should be added.

expansion in the East is related to the inability to exploit scientific and technical progress industrially. It is thought that since the Eastern systems, "in their present form, are not yet efficient enough in carrying out this process it has to be replaced by economic relations with the West and through the transfer of industrially applicable modern technologies, primarily in the form of imported plant and machinery".[157] What is believed to be a "carefully considered political decision" to expand economic relations with the West has led the CMEA countries to:

  i) "increase substantially their imports, particularly of capital goods";

 ii) incur considerable longer-term indebtedness with the West to finance these imports, and

 iii) look for new types of economic relations with the West, such as industrial co-operation, so as to ensure a more efficient transfer of modern technology".[158]

The tendencies which both characterise and explain recent developments are taken up below.

2. Bloc Integration and International Division of Labour

As opposed to the previous period where Soviet approaches to East-West economic relations were largely influenced by the old Stalinist idea of "two world markets", present Eastern policy can, by and large, be characterised by two, not necessarily convergent, developments. These are bloc integration and support of the notion of the international division of labour.

a) *Bloc Integration*

Over the past ten years, and particularly since 1969, a series of steps have been taken to improve the functioning of the CMEA. Up to that period, and despite far-reaching goals of co-ordination and co-operation in economic, technical and scientific spheres foreseen by that body's Charter and in the "Basic Principles of International Socialist Division of Labour", CMEA had little influence on matters related to guiding intra-bloc trade or economic development.[159] The CMEA's goal of achieving a highly integrated economic bloc remained a disputed and contentious "non-achievement".[160]

In 1969, a new effort was initiated to form an integrated economic unity. In April of that year the Soviet Union initiated a major discussion on reform of CMEA institutions which resulted in the "Comprehensive Programme for the Further Extension and Improvement of Co-operation and the Development of Socialist Economic Integration" of 1971. In essence, this is a complicated plan for a total reform of CMEA. It envisages a transitionary period of 15-20 years during which the union (merger) of the economies of the member countries should be more closely tied to the goal of integration.[161] Under the Comprehensive

157 Peter Knirsch, *art. cit.*, p. 9.
158 *Ibid.* For a more detailed discussion of attitudes and motivations behind trade expansion (East and West) see Chapter 4.
159 The Charter was ratified in 1959 and went into effect in 1960. The Basic Principles were adopted in 1962. See Franklyn D. Holzman, *International Trade Under Communism, op. cit.*, p. 92.
160 Dispute centred on such issues as co-ordinated planning, specialisation, multi-lateralisation and prices. These issues became focused on the broader problem of "supra-national economic power", a concept initiated by the Soviet Union in 1962 and vigorously opposed by its CMEA partners. It is a concept that continues to be opposed. See for instance *Ibid.*, Chapter 3 and Arthur J. Smith, "The Council of Mutual Economic Assistance in 1977: New Economic Power, New Political Perspectives and Some Old and New Problems", in *East European Economies Post-Helsinki – A Compendium of Papers, op. cit.*
161 "Die 32te Tagung des Rates für Gegenseitige Wirtschaftshilfe", Wiener Institut für Internationale Wirtschaftsvergleiche. Mitgliederinformation, 6/1978, Oktober, 1978, p. 6.

Programme the CMEA countries were to "more carefully co-ordinate their long-term and five-year plans, to undertake joint prognoses on major economic aggregates, to actually undertake joint planning for the production and consumption of selected products, and to conduct regular exchanges of information on the nature of economic reforms in their respective countries".[162]

In June 1975 the 29th Session of the CMEA passed the "Co-ordinated Plan of Multilateral Integration Measures for 1976-80" which among other things, provided for international specialisation and co-operation in production, joint projects for developing major primary product and fuel resources and co-operation in scientific-technical matters.[163] In essence, this programme called for the co-ordination of each of the plans of the individual member countries with the corresponding plans of all the others. Of the several sections of the plan, the one "on the joint projects appears to be by far the most tangible and carefully worked-out part of the Co-ordinated Plan".[164] Ten projects were included in this section of which eight were foreseen for the Soviet Union.[165] Of these, the Orenburg gas pipeline to the western border of the USSR is perhaps the most publicised.[166] The total value of these joint projects is in the neighbourhood of 9 billion transferable rubles.[167]

Co-ordinated planning would seem to be becoming a longer term proposition, witness the 31st session of the CMEA (June 1977) where it was agreed that "the main task in co-ordination of national economic plans for the oncoming five-year period (1981-1985) is a co-ordinated solution of the most important economic problems encountered in the elaboration of long-term specific programmes of co-operation which were of mutual interest. The CMEA member countries were advised to be guided by the above Programme in their co-ordination of national economic plans, to fulfil it and to sign relevant documents, including inter-governmental agreements, not later than the first half of 1980".[168]

Perhaps the most interesting, if not important, development was the decision also taken by the 29th CMEA session to adopt the so-called "Long-term Target Programmes". These are programmes "spanning 10-15 years designed to combine forces in CMEA for the solution of key problems associated with their further industrialisation. There are five programmes: i) fuel, energy and primary products; ii) machine-building; iii) agriculture and food supply; iv) consumption goods and, v) transport connections".[169] These Long-term Target Programmes are to be implemented by means of multi-lateral and bi-lateral agreements. Elaboration of the programmes was begun in 1976 and contracts were to be approved by 1979. These programmes were to be integrated into the five year plan for 1981-1985, into the bi-lateral long-term specialisation and co-operation programmes and into the Co-ordinated Plan of Multilateral Integration Measures for 1976-1980.[170] The

162 Edward A. Hewett, "Recent Developments in East-West European Economic Relations and their Implications for US-East European Economic Relations", in *East European Economies Post-Helsinki – A Compendium of Papers, op. cit.,* p. 187.

163 *Ibid.,* p. 189.

164 *Ibid.*

165 *Ibid.* It has been pointed out that these projects are not centrally co-ordinated by the Soviet Union, but are carried out by means of bi-lateral agreements. Furthermore, the joint projects are the "property" of that country in which they are carried out. "Die Aktionslinien des Comecon", *Neue Zürcher Zeitung,* No. 66, 21 März, 1978, p. 8.

166 See Edward A. Hewett, *art. cit.,* p. 189.

167 Council for Mutual Economic Assistance, Secretariat, *Collected Reports on Various Activities of the CMEA in 1977,* Moscow, 1977, p. 23. This figure evidently refers to eight projects where actual funds have been committed. Even if we do not know exactly the conversion rate of the transferable ruble into currencies of Eastern Europe, it has been estimated that this sum works out "somewhere between 1% and 2% of the total investments of the CMEA countries during 1976-1980 ... It is a five-year plan for a very small part of investments. Apparently the Soviet Union is paying half of the investment costs". Edward A. Hewett, *art. cit.,* p. 190. See also Arthur J. Smith, *art. cit.,* pp. 155-157.

168 Council for Mutual Economic Assistance, *op. cit.,* p. 10.

169 Edward A. Hewett, *art. cit.,* pp. 191-192.

170 "Die 32te Tagung des Rates für Gegenseitige Wirtschaftshilfe", Wiener Institut für Internationale Wirtschaftsvergleiche, *art. cit.,* p. 8.

multilateral and bilateral agreements (drawn up on the basis of the Long-term Target Programmes) "contain obligations of the parties concerned and will envisage in their national plans material, financial and labour resources for their implementation".[171] Execution of the Long-term Target Programmes is through "the conclusion of long-term agreements on specialisation and cooperation in production and economic agreements on joint construction projects".[172]

The first three Long-term Target Programmes were approved at the 32nd CMEA Session in June 1978. Their purpose is: "to meet the member countries' requirements in the basic types of energy, fuel and raw materials up to 1990; to further improve... cooperation in agriculture and the food industry... and to supply machines and equipment needed for implementing the provisions of the long-term programmes...".[173] Of these three the programmes on energy, fuel and raw materials and on machinery and equipment were accorded major importance.[174]

Since 1971 there has also been a significant increase in a new form of specialised intra-CMEA organisation as witnessed by the creation of "international economic organisations". The objective of these organisations is to establish specialised co-operation and co-ordination in industrial production at the enterprise level and across national frontiers. In some respects the international economic organisations may be roughly described as socialist multi-national corporations.[175] Legally speaking they are "juridical persons" of the host country, they are financed through the currency of the host country (in transferable rubles) and are designed to make profits.[176] According to the East European press, specialisation and cooperation in production is becoming more and more important in the context of economic relations among CMEA countries.[177] "As of January 1, 1978 the Soviet Union had signed 211 agreements on specialisation and cooperation in production with other CMEA countries (88 multilateral and 123 bilateral agreements)".[178] Furthermore, since the adoption of the Co-ordinated Plan for 1976-1980, renewed emphasis has been given to the CMEA Standardisation Programme to establish "common scientific, technical and industrial standards for industry, science, agriculture and other economic sectors throughout all Member countries".[179]

Another aspect of integration is the "increasing number of small, regional agreements, many of which might be regarded as joint ventures in the Western sense of the term".[180] Most

171   Council for Mutual Economic Assistance, *op. cit.,* p. 227.
172   V. Moiseyenko, "Specialisation and Cooperation in the Machine-Building Industry — An Important Factor in Promoting CMEA Member Countries' Mutual Trade", *Foreign Trade,* No. 2, USSR Ministry of Foreign Trade (English translation), 1979, p. 7.
173   N. V. Faddeyev, "CMEA's Role in Strengthening the Community of the Socialist Nations", *Foreign Trade,* No. 1, USSR Ministry of Foreign Trade, (English translation), 1979, p. 8.
174   The development of nuclear energy occupies the central position in the energy programme. By 1990, capacity of nuclear energy industries is to be increased 4-5 times. "Die Aktionslinien des Comecon", *Neue Zürcher Zeitung, art. cit.,* p. 8. The energy problems encountered by the CMEA countries — which undoubtedly account for the importance given to this programme — are discussed in Chapter 4. Relative to the programme on machinery and equipment, it is designed to prepare the technological basis for the fuel and agricultural programmes. "In the 1981-1985 period the CMEA countries are expected to double their output of engineering products as compared to the current five-year period, and to treble such output in the five-year period following 1981-1985". V. Moiseyenko, *art. cit.,* p. 7.
175   Arthur J. Smith, *art. cit.* pp. 159-160.
176   *Ibid.* See also "Die 'sozialistischen multinationalen Unternehmungen' der Comecon-Länder", *Neue Zürcher Zeitung.* Fernausgabe No. 276, 25ten November, 1977, p. 11.
177   See for instance Ference Nagy, "Hungary and Long-Term CMEA Cooperation", given in *US Joint Publications Research Service,* No. 72900, March 1, 1979, pp. 1-7; Gerhard Weiss, Council of Ministers of the German Democratic Republic "Resolutely on the Road to Socialist Economic Integration", given in *US Joint Publications Research Service,* No. 72301, November 24, 1978, pp. 50-57; and Ion Stoin (Rumania), "Principles Guiding Activity in CMEA Discussed", in *US Joint Publications Research Service,* No. 72204, November 8, 1978, pp. 1-7.
178   V. Moiseyenko, *art. cit.,* p. 6.
179   Arthur J. Smith, *art. cit.,* p. 161.
180   *Ibid.,* p. 162.

of these agreements "promote a relatively intensive degree of economic coordination and cooperation in highly developed industrial areas and within a small geographic area".[181]

Financial support for trade and development is provided by the International Bank for Economic Co-operation (IBEC — founded in 1964) and the International Investment Bank (IIB — established in 1970). Both banks are "increasingly active in Western hard currency markets and they have been increasingly important in the promotion of intra-CMEA trade and the development of CMEA infra-structure".[182] The IIB in particular has become a significant force in developing financial relations with the West. Both banks are becoming "valuable intermediary agencies for rounding up Western financial support — through Eurodollar and other borrowing — for CMEA multilateral projects. In this latter operation they represent the entire CMEA bloc in dealing with Western financial institutions".[183]

The long-term effectiveness of the various measures taken to further CMEA integration remains a matter of much speculation and no small amount of disagreement. For one thing, the CMEA continues to face difficulties regarding the adjustment of prices. It has been noted that the "lack of rapid growth in intra-CMEA hard currency trade indicates... that the intra-CMEA trading system is not undergoing fundamental change, in the direction of multilateral trade within CMEA and the convertibility of CMEA currencies with each other and with the currencies of noncommunist countries".[184] Furthermore, the issue of national sovereignty remains a sensitive issue and "some CMEA countries seem to be pretty self-assured in insisting on their sovereignty — at least where the fulfilment of plans and trade policy, with its importance for foreign exchange revenues, are concerned".[185] It is interesting to note in this regard that at the 31st Session of the CMEA it was stressed that the "further extension and improvement of co-operation and the development of socialist integration... will be carried out... on the basis of respect for State sovereignty, independence and national interests, non-interference in internal affairs of countries, complete equality of rights... Socialist economic integration proceeds on a fully voluntary basis, is not followed up by the establishment of supra-national bodies and does not involve matters of internal planning, financial and self-sustained activities of national agencies".[186] The issue of sovereignty was again stressed at the 32nd Session of the CMEA, particularly by Rumania, both in relation to the CMEA's competence in foreign relations and in economic planning. Concerning the latter, and in direct opposition to Mr. Kossygin, the Rumanian delegate (Prime Minister Manescu) stressed that "the Long-term Target Programmes had as their purpose only to strengthen co-ordination in economic co-operation and did not have (the programmes) the character of a document to plan the economies of the member countries".[187] Judging by the communiqué of that session it seems that no changes were initiated in the CMEA's charter and that the principle of sovereignty was maintained.[188]

181 *Ibid.*
182 *Ibid.*, p. 157.
183 *Ibid.*, p. 159. For a discussion of legal aspects and their influence on Western banks, see Axel Lebahn, "Neuentwicklungen der Geschäftstätigkeit und Rechtsgrundlagen der internationalen Comecon-Banken IBEC und IIB", *Recht der Internationalen Wirtschaft*, Januar, 1979, pp. 4-13.
184 Martin J. Kohn and Nicholas R. Lang, "The Intra-CMEA Foreign Trade System: Major Price Changes, Little Reform", in *East European Economies Post-Helsinki – A Compendium of Papers, op. cit.*, p. 146.
185 Carl A. Ehrhardt, "EEC and CMEA Tediously Nearing Each Other", *Aussen Politik* (English edition), Vol. 28,2/77, 2nd Quarter 1977, p. 173.
186 *Council for Mutual Economic Assistance, op. cit.*, pp. 25 and 222.
187 "Die 32te Tagung des Rates für Gegenseitige Wirtschaftshilfe", Wiener Institut für Internationale Wirtschaftsvergleiche, *art. cit.*, p. 6.
188 *Ibid.*, pp. 4-6. One implication of this is, as noted by a representative of CMEA, that while the 5 year plans are coordinated, "we have no internationally agreed to plans which prescribe for the member countries what they ought to produce". See "Die Aktionslinien des Comecon", *Neue Zürcher Zeitung*, Fernausgabe No. 66, *art. cit.*, p. 8. Furthermore, it has been pointed out that "economic nationalism and fear of Soviet preponderance in any joint undertaking, which have retarded Comecon integration efforts for the last three decades, continue to be major stumbling blocks, despite the lip service that is paid to intrabloc solidarity". See "Comecon Politics: Impact on Integration Plans", *Business Eastern Europe*, Vol. 7, No. 50, December 15, 1978, p. 396.

Nevertheless, it appears that the Soviet Union has begun a serious "effort to develop a planning apparatus in CMEA similar to the USSR planning apparatus".[189] The major economic problems facing the CMEA countries provide the momentum behind this effort. Furthermore, the Soviet Union probably regards CMEA (indirectly) as a tool for competition with Western Europe, particularly the EEC, and "therefore at least part of the Soviet motives towards improving the functioning of CMEA have their roots in the success of the EEC".[190] This aspect has led some observers to interpret the drive for integration as a strengthening of autarkic tendencies within the East bloc. The restrictive import policy vis-à-vis the Western industrialised countries is cited as one argument supporting this notion. Thus it is believed that "the Soviets are trying, by means of further CMEA integration, to reduce the trading engagement of the Eastern bloc countries with the West".[191] Lastly, there appears to be an incentive for Eastern Europe to use the CMEA as "a way to reach agreements with the USSR on reducing the uncertainty of supplies of products crucial to production — fuels and primary products".[192]

b)   *International Division of Labour*

The attempts which have been made to strengthen the CMEA (leaving aside the political/economic motivations and whatever repercussions integration may have on East-West economic relations) reflect one aspect of the international division of labour, albeit limited to the centrally planned economies.[193]

Increasing *international* involvement, in step with economic development, has been the general pattern for Western industrialised nations. This has not been the case for the centrally planned economies of Eastern Europe. As noted earlier, following the political decisions taken in the 1930s (with the accompanying changes in economic policy) economic efficiency advantages of foreign trade were not stressed in the Soviet literature. It was not until the beginning of the 1970s that a major emphasis had been placed on the expansion of Soviet economic relations with the industrialised West.[194]

In contrast to earlier periods where, as has been pointed out, imports were used to fill gaps or to "catch up", and in exchange raw materials and goods were exported in amounts necessary to pay for the imported goods, current discussions on the effectiveness of trade have become more sophisticated. As noted, "even aspects of the doctrine of comparative advantage, regarded in the past as an insidious capitalist tool to maintain imperialist domination over underdeveloped countries, are being treated in a positive manner.[195] For instance:

> "Mutual advantage is one of the leading principles of Soviet foreign trade associations, and they adhere to this principle in trade with their Western partners. In developing economic ties with Western Europe, our country receives an opportunity to make fuller and more rational use of its own resources and possibilities and at the same time to

189   Edward A. Hewett, *art. cit.*, p. 193.
190   *Ibid.* CMEA integration is viewed with mixed feelings in the West. Some fear has been expressed that "funds, resources and labour needed to develop their own economies would have to be allocated to meet joint Comecon projects, the majority of which to date are located in the Soviet Union". "Comecon Politics: Impact on Integration Plans", *Business Eastern Europe, art. cit.*
191   "Gebremster Westhandel und verstärkte Autarkie im Comecon", *Neue Zürcher Zeitung,* Fernausgabe No. 199, 30. August, 1978, p. 9. During a recent visit to Czechoslovakia, Secretary General Breshnev warned the Czechs against further development of trade with the West. For the bloc countries as a whole, imports from the West were reduced by 10 per cent during the first half of 1978. *Ibid.*
192   Edward A. Hewett, *art. cit.,* p. 193.
193   For instance, as early as 1956, Khruschev in his secret speech indicated that within the CMEA, "the international division of labour was to be exploited in heavy industry". P. J. D. Wiles, *Communist International Economics,* Praeger Publishers, New York, 1969, p. 316.
194   See for instance, Kosygin's speech at the November meeting of the Supreme Soviet, in *Gosplan SSSR, Gosudarstvenny pyatiletniy plan razvitiya narodnogo khozyaystva SSSR na 1971-1975 gody,* Moscow, 1972.
195   Herbert S. Levine, *et. al., op. cit.,* p. 44.

acquire, by way of commercial exchange, goods of other countries that are not produced in our country or whose production would cost more than it does to import them. Thus, foreign economic ties offer a more efficient solution to a number of problems arising in the course of economic construction".[196]

The advantages of foreign trade have been interpreted by the East as leading to mutual economic advantages and also furthering détente and co-operation among all the European countries.[197]

Nevertheless it is important to note that while the Soviet Union stresses the advantages of international economic relations and has long stressed the principle of intra-industrial specialisation, this emphasis does not imply a shift acknowledging economic integration or assimilation.[198] Indeed, even discussion on intra-CMEA integration and trade "stresses the maintenance of the sovereignty of the individual nations within the bloc, rather than stressing the growth of their inter-dependence".[199] In the Soviet literature it is likewise consistently stressed that "expanded economic relations with the West do not involve capitalist countries in the basic core of the Soviet economy, but operates at the margin to improve the performance of the Soviet economy".[200]

Support for the notion of international division of labour and subsequent acceptance of the economic advantages (to the East) of foreign trade do certainly indicate a much more open policy towards East-West economic relations. At the same time, and as is discussed in subsequent chapters, the interpretation that can be given to these developments is itself a "policy issue".[201]

3. Development of East-West Contractual Relationships

a) *Affiliation with International Organisations*

As has been noted, during the post-War years the East European countries followed the Soviet Union's lead in refusing to co-operate in attempts to regulate the system of international trade. It was not until the end of the 1950s that the East European countries began to join certain economic international organisations. Membership in GATT was usually the first step.

Czechoslovakia was a charter member of GATT (1947) and Poland became a member as early as 1967. Rumania and Hungary followed in 1971 and 1973, respectively. Bulgaria has had observer status since 1967. The USSR and the German Democratic Republic have not joined the organisation and have shown little interest in direct cooperation with or membership in GATT.[202]

196 Minister of Foreign Trade Patolichev, *Pravda*, 27th December, 1973, p. 4 as given in Herbert S. Levine, *et al., op. cit.*, p. 45.

197 W. Iskra, "RGW und EWG: Möglichkeiten der Zusammenarbeit", in *Probleme des Friedens und des Sozialismus*, Vol. 19, No. 6, 1976, p. 784, as given in Max Baumer and Hanns-Dieter Jacobsen, "Die Entwicklung der Beziehungen Zwischen EG und RGW, vor der KSZE-Folgekonferenz in Belgrad", Stiftung Wissenschaft und Politik, Ebenhausen, March 1977, p. 26.

198 This aspect, which brings into consideration the notions of inter-dependence and inter-relatedness, concepts related to assumptions made as to future patterns of East-West economic relationships is discussed in Chapters 4-6.

199 Herbert S. Levine, *et al., op. cit.*, p. 46.

200 *Ibid.*, p. 47.

201 For a general discussion of comparative international integration, see P. J. D. Wiles, *Communist International Economics, op. cit.*, Chapter XII.

202 Rumania was also the first and only Eastern country to join the IMF and the IBRD (International Bank for Reconstruction and Development) — in 1973. Yugoslavia is here treated as a "special case" since she is an Associated Country of the OECD, an Associate Member of the CMEA and a member of the IMF, GATT and IBRD. For a general discussion of GATT, see for instance, Karen C. Taylor, "Import Protection and East-West Trade: A Survey of Industrialised Country Practices", in *East European Economies Post-Helsinki – A Compendium of Papers, op. cit.* It is interesting to note that Czechoslovakia's membership in GATT was unaffected (as was her trade with the West) by the events of 1968. See Franklyn D. Holzman, *International Trade Under Communism, op. cit.*, Chapter 4.

The membership of these countries in GATT marks a change in attitude on the parts of East and West. The major purpose of GATT was to increase world trade by reducing trade barriers, particularly tariffs, and to encourage non-discriminatory trade primarily through the mechanism of MFN agreements among its Members. The main arguments against extending membership to Eastern countries were the "centrally planned and discriminating character of their economies".[203]

Opposition by the East was generally linked to the policy of non-recognition of international economic organisations. During the period of the 1950s and early 1960s, the East attempted to bypass GATT membership while still obtaining the benefits of MFN and a reduction of quota restrictions.[204] The membership in GATT signalled the formal willingness and desire of the CMEA countries to intensify economic relations with the West and to adjust trade relationships. It also signalled a reciprocal readiness on the part of the West to take into account Eastern constraints and objectives.

b) *Trade Agreements*

i) *Western Europe*

As has been pointed out, the "normalisation" of political and economic relations has shaped East-West trade since 1965. Thus, the number of trade agreements concluded — and which have served to regularise East-West trade relations — jumped from 23 in 1960, to 55 in 1966, and to 90 in 1974.[205] A report of the Economic Commission for Europe noted that as of November 1976, 94 trade agreements were in existence.[206] The list of agreements in force by September 1975 is given in Table A-41. With the exception of the German Democratic Republic, all East European countries and the Soviet Union have a well-developed structure of long-term trade agreements with Western nations. The main provisions of these agreements deal with the expansion and conditions of trade, the application of most-favoured nation treatment, the planned lifting of quantitative restrictions, terms of settlement and prices.[207] The provisions which have the greatest practical influence deal with quantitative import restrictions and financial settlement (payment in convertible currency).

ii) *United States*

The evolution of trade arrangements between the United States and the Eastern bloc countries followed a different course. At the time as the countries of Western Europe were embarking on a policy designed to construct a common framework for their relations with Eastern Europe, the United States was still in the process of adjusting trade relations with the East.

Negotiations for a US-USSR commercial agreement began in November 1971 when

203   Max Baumer and Hanns-Dieter Jacobsen, "CMEA and the World Economy: Institutional Concepts", in *East European Economies Post-Helsinki – A Compendium of Papers, op. cit.,* p. 1011.

204   As has been indicated, these attempts were to a great extent achieved as MFN was extended to some of the bloc countries and import quota restrictions were reduced.

205   F. Levcik and J. Stankovsky, *op. cit.,* p. 150. Holzman gives a different figure. He states that by 1968, "the socialist nations of Europe as a group had 149 trade agreements with advanced Western nations; 17 involving the USSR. The only Western nations with no such bi-lateral agreements were Ireland and the United States". See Franklyn D. Holzman, *International Trade Under Communism, op. cit.,* p. 150. The author is quoting J. Wilczynski, *Economics and Politics of East-West Trade, op. cit.,* pp. 106-109.

206   As given by Susanne S. Lotarski, "Institutional Developments and the Joint Commissions in East-West Commercial Relations", in *East European Economies Post-Helsinki – A Compendium of Papers, op. cit.,* p. 1020. The author cites United Nations Economic Commission for Europe (ECE), *Long Term Agreements on Economic Co-operation and Trade,* Secretariat Note, TRADE/302, 25th October, 1974.

207   MFN is compulsory for GATT members. For non-members it is applied to customs tariffs. For EEC members it does not override the Community's common customs tariffs.

Maurice Stans, then Secretary of Commerce, was visiting Moscow. Further progress was made in May of 1972 with the signing of the "Basic Principles of the Relations Between the United States and the Union of Soviet Socialist Republics" by President Nixon and Secretary General Brezhnev. This action gave rise to the 1972 Trade Agreement (October 1972) as well as to a number of scientific and technological agreements.[208]

The 1972 agreement was to have removed the main obstacles to trade between the two countries. It settled the thorny question of the Soviet lend-lease debt, made provisions for the tripling of the volume of trade in three years (compared with the three years 1969-1971), offered facilities to governmental and private organisations for dealing with trade formalities, and granted reciprocal most favoured nation treatment. A special article gave each government the right to take the necessary steps to protect itself against market distortion.[209]

The key point of the 1972 trade agreement with the Soviet Union, and on which its entry into force depended, was the grant of MFN treatment. In June, 1974 a long term agreement to facilitate "Economic Industrial and Technical Cooperation" was signed by the United States and the Soviet Union.[210] In December, 1974, the United States Congress passed the Trade Act.[211] It included the provision (the so-called Jackson-Vanik amendment) whereby the granting of MFN to the Soviet Union, Soviet eligibility for government-sponsored credits, and conclusion of a trade agreement were made legally dependent on freer emigration from the Soviet Union.[212]

In January 1975, (and prior to its ratification by the US) the Soviet Union renounced the 1972 Trade Agreement on the grounds that the Jackson-Vanik amendment represented interference in the internal affairs of the Soviet Union. In fact, it is thought that the Soviet Union was most upset by the financial aspect. "The Soviet Union was engaged in negotiations for a number of deals at the end of 1974 reportedly worth about $3 billion of United States' exports. When it became clear that the credits would not be available, the Trade Agreement lost much of its value to the Soviet Union".[213]

Attempts have been made to rehabilitate the 1972 Trade Agreement. Although these attempts have not been successful to date, the number of commercial agreements between private firms and the Soviet Union has been increasing. It has even been noted that the drive for imports on the part of the Soviet Union has not been seriously damaged.[214] In fact, a bill introduced for hearings (in 1979) by the US Congress could greatly ease commer-

208   Rogers Morton, *op. cit.*, p. A-7.
209   Franklyn D. Holzman, *International Trade Under Communism, op. cit.*, pp. 163-168; *Background Materials Relating to United States-Soviet Union Commercial Agreements*, Committee on Finance, United States Senate, 93rd Congress, 2nd Session, Washington, D.C., 1974.
210   See *Treaties and Other International Acts Series 7910, Economic, Industrial and Technical Cooperation*, Agreement between the United States of America and the Union of Soviet Socialist Republics. The Agreement is designed to "facilitate economic, industrial and technical cooperation in keeping with established practices and applicable laws and regulations in the respective countries". This is not a trade agreement in the sense of those mentioned in this section, rather as is stated in the text, it is designed to facilitate cooperation between the two countries.
211   Franklyn D. Holzman, *International Trade Under Communism, op. cit.*, p. 168 and Christopher Cviic, *art. cit.*, p. 218.
212   Christopher Cviic, *Ibid*. See also *Public Law 93-618*, 93rd Congress, H.R. 10710, January 3, 1975.
213   Christopher Cviic, *art. cit.*, p. 218. Supporters of the amendment had argued that the "US should not provide trade and credit benefits which would strengthen the military-industrial capabilities of long-standing adversaries before a permanent and tangible improvement in relations is achieved. The issue of freer emigration was considered an important test of such relations". Paul Marer, ed., *US Financing of East-West Trade, op. cit.*, pp. 3-4.
214   Other US-USSR agreements signed during this period include those on maritime (14th October, 1972), transport (19th June, 1973) and tax matters (20th June, 1973). Joint Commissions (or Joint Councils) were formed with several East European countries: Poland, Rumania, Hungary and Bulgaria. The commercial agreement with Rumania signed in April 1975, was approved by Congress in July of that year. Rogers Morton, *op. cit.*, pp. A-15 to A-17. The Rumanian agreement is not included in Table A-41.

cial relations between US firms and the Soviet Union. The so-called "Stevenson Bill" proposes to establish new methods for granting trade benefits to the Soviet Union (and China). The bill would, it is said, in effect replace the Jackson-Vanik amendment and give the Soviet Union de facto MFN status. In particular, it would allow the President to grant trade benefits "if he determines that doing so would lead substantially to the achievement of the free-emigration objective called for in the Jackson-Vanik amendment of the Trade Act of 1974".[215] Support for extending MFN trade status to the Soviet Union (pending Congressional approval) has been expressed at the Presidential level.[216] Since extension of MFN status not only needs Congressional approval but most likely also renegotiation of the 1972 Trade Agreement, it is fair to surmise that any legislation designed to alter the present trading arrangements would elicit a great deal of controversy and debate. Nevertheless, the above-mentioned bill is an interesting illustrative example of the continuing fluctuations in policy toward the East.

c) *Long-term Economic Agreements*

Another development which marks this period is the attempt by both East and West to encourage new forms of co-operation. Thus the more "classical" trade agreements are being replaced by long-term economic agreements which call for economic, industrial, and scientific and technical co-operation as well as co-operation in the area of production.[217] These agreements have brought about a "net of overlapping government to government agreements which seek to regulate the economic relationships of the concerned parties".[218] In 1966-67 there were 17 such agreements and in 1974, 85 were in force.[219]

The transition from trade agreements to economic co-operative agreements involves more than a change of name. While the signatories assumed few obligations, they undertook to exert efforts to fulfil the objectives of the agreement. Accordingly, the function of the joint commission has changed from that of hearing mutual "complaints" to "framing measures to facilitate co-operation, defining the forms of co-operation which might be pursued, and determining the branches of production which might be of interest".[220] Joint commercial commissions have become identified primarily as an institution of East-West trade.

215 "US MFN and Exim Credits Proposed for the USSR", *Business Eastern Europe,* Vol. 8, No. 7, February 16, 1979, p. 49. Specifically, "the proposed Stevenson amendments (s. 339) to the Export-Import Bank Act and the Trade Act would: 1) delete provisions in the Export-Import Bank Act and the Trade Act which single out the USSR for discriminatory treatment with respect to credits: 2) establish a new limitation on Bank support for US exports to any single Communist country; and 3) revise the "waiver" provisions concerning emigration practices and eligibility for MFN treatment and Eximbank credits". See Ronda A. Bresnick, "The Setting: The Congress and East-West Commercial Relations", in *Issues in East-West Commercial Relations – A Compendium of Papers,* submitted to the Joint Economic Committee Congress of the United States, 95th Congress, 2nd Session, US Government Printing Office, Washington, January 12, 1979, p. 2.

216 An article in the *International Herald Tribune* quotes President Carter as follows: "I would guess the Soviet Union is upgrading (emigration) to the point where it would qualify for most-favoured-nation status... I would hope, in the next few months, we would have most-favoured-nation status granted to... the Soviet Union". "Carter: Upgrade Soviet Trade Tie", *International Herald Tribune,* Paris, March 1, 1979, p. 1. The United States has extended MFN treatment to Poland, Rumania and (as of 1978) to Hungary.

217 See F. Levcik and J. Stankovsky, *op. cit.,* p. 151. The economic agreements are either long-term (at least 10 years) or with duration unspecified. The same time frame applies to the periods of notice that must be given for termination.

218 *Ibid.,* pp. 151-152. It has been noted that the replacement of the classical trade agreements by long-term economic agreements is due to the ability of the EEC countries to discuss economic co-operation, but their inability to discuss trade which is covered by the common commercial policy.

219 A list of agreements in force at the end of September 1975 is given in Table A-41. A different source indicates that as of November 1976, there were 128 long-term commercial agreements. See Susanne S. Lotarski, "Institutional Development and the Joint Commissions in East-West Commercial Relations", in *East European Economies Post-Helsinki, op. cit.,* p. 1020.

220 Susanne S. Lotarski, *art. cit.,* p. 1021.

As has been noted, the type of co-operation foreseen under the long-term economic agreements is not only a mechanism to reconcile the asymmetries in structure and development between market and centrally planned economies but also one designed to provide a political link. Thus the creation of the Franco-Soviet *Grande Commission* in 1966 was believed to be one element of the policy of "Europe to the Urals". Even greater political importance was attached to commercial relations and the joint commission in the development of the United States' policy of détente with the Soviet Union.[221]

It is instructive to indicate the broad spectrum of industrial co-operation that has been worked out under the umbrella of the economic co-operation agreements. According to an EEC estimate, industrial co-operation agreements include inter-firm agreements, framework agreements between Eastern administrative organs and Western firms, and mixed East-West companies (joint equity ventures) established in both East and West.[222] It has been estimated that most likely "now more than 1,000 such East-West arrangements are in existence, involving virtually the entire spectrum of Western industry".[223]

The United States was initially reluctant to follow this path. The United States has three joint Commissions: the Joint United States-USSR Commercial Commission, the Joint American-Polish Trade Commission (both established in 1972) and the American-Rumanian Commission (1973). Each is the product of the "summit meeting" of the early 1970s. Unlike their West-East European counterparts, the joint commissions in which the United States participates were not created under the terms of trade or long-term co-operation agreements.[224]

d) *Scientific and Technological Co-operation Agreements*

One purpose of S & T Co-operation Agreements is to strengthen economic relations between East and West. Technology transfer and the exchange of knowledge and information are seen as central tools for the achievement of this objective.

During the 1960s scientific and technological co-operation agreements were being concluded between East and West. Inter-governmental scientific and technology agreements have been signed between the Soviet Union and Australia, Austria, Belgium, Canada, Denmark, France, Germany, Greece, Italy, Japan, the Netherlands, Norway, Sweden, the United Kingdom and the United States.

Between the United States and the USSR alone, eleven co-operative agreements on science and technology have been signed.[225] Under these, some 150 authorised projects are

221   *Ibid.,* pp. 1034 and 1038.
222   Carl H. McMillan, "East-West Industrial Co-operation", in *East European Economies Post-Helsinki – A Compendium of Papers, op. cit.,* p. 1184.
223   Rogers Morton, *op. cit.,* pp. 12-13. Industrial Cooperation Agreements are generally entered into by private Western firms for commercial benefit. It is doubtful whether the majority of them are under the umbrella of government-to-government Long-Term Economic Agreements. It is believed that even if they are not, the Industrial Cooperation Agreements would have been concluded in any event. It is interesting to recall that the final Act of the Helsinki Conference did stipulate the following definition of industrial co-operation: "... Industrial Co-operation encompasses a series of forms of economic co-operation which go beyond the framework of present trade arrangements". While establishing a matter of principle, it is generally thought that the Act has had little effect on actually furthering industrial co-operation.
224   Susanne S. Lotarski, *art. cit.,* p. 1027. It should be noted that not all of the Western countries distinguish between Long Term Economic Agreements and Scientific and Technological Cooperation Agreements. For countries such as Sweden, to use one example, they are one and the same.
225   These agreements are based on a provision contained in the "Basic Principles of Relations Between the United States and the USSR" agreed to in 1972. This provision called for the development of mutual contacts and co-operation in the fields of S & T. See Lawrence H. Theriot, "United States Governmental and Private Industry Co-operation with the Soviet Union in the Fields of Science and Technology", in *Soviet Economy in a New Perspective – A Compendium of Papers,* Joint Economic Committee, Congress of the United States, 94th Congress, 2nd Session, United States Government Printing Office, Washington, D.C., 14th October, 1976, p. 740.

either functioning, soon to be initiated or are planned.[226] These agreements cover the following fields: science and technology, environmental protection, medical sciences and public health, space co-operation, agriculture, transportation, studies of the oceans, atomic energy (peaceful uses of), artificial heart research and development and housing and other construction.[227]

One of the eleven agreements, the Agreement for Co-operation in the Fields of Science and Technology, established a basic format for the ten subsequent inter-governmental accords. Eight of the eleven agreements contain similar articles (usually number 4) which "have become particularly significant, because they have been interpreted by the Soviets as a juridical basis for joint co-operation directly between Soviet agencies and private United States companies".[228] Under the terms of article 4 of the basic Science and Technology Agreement, fifty-three "co-operation agreements" have been concluded between US corporations and the Soviet Union's State Committee for Science and Technology. The distribution of agreements is as follows: radio, television and electronic equipment: 9; engineering: 6; data processing: 5; aircraft and parts: 5; machine tools: 4; and food product machinery: 4.[229] These areas are believed to reflect Soviet priorities and their interest in "commercially usable, applied science and technology". They are also thought to reflect a modest concern with basic scientific research — at least as a subject for co-operation.[230]

4. Relationship Between the European Economic Community and the CMEA Countries

Mutual readiness to establish closer trading relations is also clearly illustrated in the changing nature of EEC-CMEA relationships. When the EEC was first established, it was greeted with a great deal of hostility, in particular by the Soviet Union. Apparently, the Soviet Union feared that the EEC would reduce CMEA sales to the West and that it "might become a powerful magnet of attraction for its smaller CMEA partners".[231]

During the 1960s declining growth rates and a number of economic reforms, designed among others to establish more organic links with the world market, led to a more "positive" view of the EEC on the part of the CMEA. In any event, since the EEC as of 1973 represents its member countries in all trade agreements with other nations and EEC members may not negotiate new trade agreements on a bi-lateral basis, recognition of the EEC by the CMEA became a *sine qua non*.[232]

226 *Review of United States-USSR Co-operative Agreements on Science and Technology, Special Oversight Report No. 6,* Committee on Science and Technology, United States House of Representatives 94th Congress, 2nd Session, November 1976, p. 9. See also Chapter 3.
227 Lawrence H. Theriot, *art. cit.,* p. 740. See also Chapters 3 and 4 for a general discussion of the scientific and technological co-operation agreements and their significance for the technology transfer issue.
228 *Ibid.,* p. 741.
229 *Ibid.,* p. 750. See also Chapter 3 and Annex Table A-21.
230 *Ibid.,* p. 751.
231 See Franklyn D. Holzman, *International Trade Under Communism, op. cit.,* p. 155; and Christopher Cviic, *art. cit.,* p. 124. While first refusing to recognise the EEC, a number of CMEA members implicitly recognised the organisation by "signing agreements which the EEC had signed and by negotiating agreements with an agency of the EEC on trade in specific agricultural commodities". Franklyn D. Holzman, *International Trade Under Communism, op. cit.,* p. 156.
232 Nearly all the long-term trade agreements negotiated by EEC Member states with the Eastern European countries expired on 31st December, 1974, the remainder expiring by 31st December 1975. "The European Community and the Eastern European Countries", *Information,* 163/77, Commission of the European Communities, External Relations, Brussels, p. 6.

EEC-CMEA relations entered a new phase in the autumn of 1974 when the Community offered to each of the CMEA countries to conclude long-term non-preferential trade agreements based on reciprocity and including: MFN treatment in tariff matters (i.e. EEC-MFN in return for reciprocal concessions by the CMEA countries); creation of a framework (Joint Committee) within which to search for possibilities of liberalising import arrangements; discussion on a case-by-case basis of problems of agricultural imports; creation of safeguard machinery which takes account of the differences in economic systems and provisions on payments problems and export credits.[233]

The Community's offer was very largely ignored by the CMEA countries.[234] In February 1976, following contacts at working level between the EEC Commission and the CMEA Secretariat, the CMEA proposed an agreement between the CMEA and its member countries on the one hand and the EEC and its member states on the other.[235] The draft agreement contained far-reaching provisions on trade, credits, preferences and so forth.[236] The EEC replied in November 1976 with a draft framework agreement providing for the establishment of working relations between the two organisations. However, the establishment of working relations was limited to a few areas.[237]

Following an exchange of letters in June 1977 and following an initial meeting in Brussels in September 1977, negotiations began in Moscow in May 1978 and continued with meetings in Brussels in July and November 1978. The November meeting broke off without reaching significant agreement on anything. In this regard the *Financial Times* stated that while the talks in Brussels had failed to produce a mutually agreed-to basis for negotiation of a formal accord between the two organisations, "Comecon's secretary, Mr. N. Fadeev of the Soviet Union, has agreed that Comecon will consider a compromise proposal for an agreement put forward by the EEC Commission... The agreement envisaged by the Commission would be limited in content to the exchange of information on industrial and trade statistics, economic planning, and environmental affairs. Its text would also refer to trade relations between the two organisations. As an added incentive, the EEC has offered to meet Comecon's quest for greater international recognition by suggesting that the eventual agreement should be formally concluded on the European side by the Council of Ministers.

233 Information furnished by a representative of the EEC. See also Max Baumer and Hanns-Dieter Jacobsen, *art. cit.*, p. 1006.
234 It has been pointed out that the "assessment of the EC offer of negotiations to countries with state-controlled economies differed from country to country and was therefore anything but uniform within the CMEA. None of the partner nations of the Soviet Union wanted, or were, in a position to take the decisive step en route to Brussels before the Soviet Union — no matter how obvious such a step would have been in view of the trade interests involved". Carl A. Ehrhardt, *art. cit.*, p. 169.
235 The wording of the agreement reflects the basic conflict that has plagued the attempt to formulate an EEC-CMEA agreement. Namely, the position taken by the EEC that the "CMEA is not an equal partner (jurisdictionally competent) with which the Community can resolve the crucial internationally legal rules concerning the development of East-West trade". Axel Lebahn, "RGW und EG – Faktoren des Ost-West-Handels", *Aussenpolitik,* No. 2, 1978, p. 126. The CMEA proposal that the two parties to the agreement are not to be designated as CMEA and EEC but as "member nations" of each body "gave rise to considerable criticism in the EEC". Article 11 of the CMEA draft, first paragraph, "provides that individual questions of trade and economic relations in connection with the framework treaty be regulated by bilateral and multilateral agreements between the member states of the two parties". But according to the second paragraph, "individual concrete questions can also be decided upon on "the basis of the principles of this agreement" by direct contacts, arrangements and agreements "between the member nations of the CMEA and the authorities of the EEC, between the member nations of the EEC and the authorities of the CMEA as well as between their relevant economic organisations". Depending on the interpretation, this formula can help to eliminate either all or none of the obstacles". Carl A. Ehrhardt, *art. cit.*, pp. 174-175.
236 The list of topics proposed by the CMEA in its draft is based on Basket II, Part two of the Final Act of Helsinki. *Ibid.,* p. 174. See F. Levcik and J. Stankovsky, *op. cit.*, pp. 268-271, which gives the text of the draft proposal.
237 Axel Lebahn, "RGW und EG-Faktoren des Ost-West-Handels", *art. cit.*, p. 130.

The Commission would, however, remain in charge of negotiations".[238] Whatever may be the outcome of the EEC-CMEA negotiations, it is clear that the CMEA countries are gradually overcoming their initial unwillingness to deal with the EEC on matters of trade.

Meanwhile, the EEC has concluded a number of agreements with member countries of the CMEA in particular sectors; textiles, steel products and technical arrangements on agricultural products. For instance, certain technical arrangements concerning imports of agricultural products have been concluded with Poland, Hungary, Rumania and Bulgaria. An agreement on textiles with Rumania was initialled on 10th November, 1976.[239] Furthermore, at the beginning of February, 1979, the Common Market Ministers gave the EEC Commission "the green light to negotiate a trade agreement with Rumania".[240]

Parenthetically. MFN treatment continues to be a highly controversial subject, witness the Final Act of the Helsinki Conference which gave little satisfaction to the Eastern countries.[241] Linking the concept of reciprocity to MFN treatment is an important development, since reciprocity is defined in the Final Act as "permitting, as a whole, an equitable distribution of advantages and obligations of comparable scale". It can be interpreted as follows: because of the differences in economic systems, each side must look at concrete advantages it is giving and receiving rather than at the formal identity of the concessions made on each side.[242]

5. Relations Between the United States and the Eastern Countries

The evolution of liberalised trade relations is very evident in the development of US policy. Regarding export controls, for instance, the United States in the second half of the 1960s adjusted its criteria of the strategic and economic embargo in order to achieve closer conformity with the criteria already adopted by the other Western countries. In the Autumn of 1966, hundreds of items were consequently removed from the US unilateral lists thereby re-establishing a "de facto consensus on a limited but highly flexible range of embargoed

238   Guy de Jonquières, "EEC and Comecon Fail to Agree on Basis for Talks", *Financial Times,* November 27, 1978. According to accounts given of the November meeting, the 1976 CMEA proposal will be abandoned. Furthermore, the CMEA delegation recognised that commercial questions are the responsibility of national governments. "The EEC has consistently rejected the Eastern bloc's demands for a full trade agreement on the grounds that Comecon is not equipped to deal with trade matters, having no common rules on tariffs, quotas or free movement of goods. It has insisted that any trade agreements must be concluded bilaterally between Brussels and the individual East European countries". *Ibid.* It is thought that "behind this somewhat legalistic stance lies the EEC's fear that such an agreement would be used by the Soviet Union to tighten its control over the independent trading activities of its East European satellites". "Hitch in EEC Trade Talks with Comecon", *The Times,* November 27, 1978.

239   This was the first trade agreement concluded by the Community with an Eastern European country. *Information,* 163/77, *art. cit.,* p. 8. In fact, this involved "the technical and administrative implementation of the multilateral 'All Fibres Agreement' within the framework of GATT of which Rumania, as a member of GATT, was a signatory". Carl A. Ehrhardt, *art. cit.,* p. 172.

240   "EEC to Offer Trade Pacts to Rumania, Yugoslavia", *International Herald Tribune,* Paris, February 7, 1979, p. 1. The same article notes that Rumania had asked for an agreement "thereby breaking ranks with the rest of Comecon. Talks between Comecon and the EEC commission are stalled on the question of whether the communist trade organisation alone or its member countries individually have the right to conclude trade agreements with the Common Market".

241   At the CSCE Conference, the "East Bloc countries failed to succeed in pushing through their demands in Basket II to declare most favoured nation treatment a general principle". Carl A. Ehrhardt, *art. cit.,* p. 165.

242   The definition of reciprocity can also be a major restriction since it is generally agreed that reciprocity, as normally defined, can only be applied in a competitive situation. The EEC does extend *de facto* MFN treatment in tariff matters to all the East European countries, and *de jure* MFN treatment to those which are members of GATT.

goods".[243] The passage of the US Export Administration Act of 1969 was a major turning point in trade relations between the United States and the CMEA countries and represented the beginnings of economic détente. The evolution of US attitudes towards trade with the Eastern countries, as mirrored in this Act, reflects the general policy of "bridge-building" (later this was called détente) which became current in the mid-1960s. Thus, the Act "included a specific endorsement of trade in peaceful goods between US firms and all countries with which the United States had commercial relations".[244] Reference to the economic significance of an export was dropped with passage of the 1969 Act. Furthermore, it provided for the abolition of controls in those instances where it could be shown that the proscribed goods could be purchased by Communist buyers from other Western countries or Japan.[245] In many respects this action brought the United States conceptually closer to the "working concept" of controls which had been previously adopted by the other members of the co-ordinating mechanism. The Export Administration Amendments of 1977 determined that US export policy towards individual countries was not to be made exclusively on the basis of "a country's Communist or non-Communist status" but was to be based on the present and potential relationship to the United States.[246] These amendments also extended the Export Administration Act until September 30, 1979. As may have been anticipated, the administration of export controls in East-West trade has been a focal point in Congressional consideration of bills to extend and amend the Act. The debates which surrounded the passage of and amendments to the Export Administration Act underlined two major themes: "the threat to US national security posed by the sale of dual-use technologies to the communist world, and the importance to the US national interest of a positive trade balance and therefore of a healthy export sector". Both themes are reflect in the current Export Administration Act of 1979. This Act expires on September 30, 1983.

243   Nicolas Spulber, *art. cit.*, p. 110.
244   Rogers Morton, *op. cit.*, p. A-7.
245   Marshall I. Goldman, *Détente and Dollars, Doing Business with the Soviets,* Basic Books Inc., New York, 1975, p. 50. See also Thomas W. Hoya, "The Changing US Regulations of East-West Trade", *Columbia Journal of Transnational Law,* Vol. 12, No. 1, 1973. The Export Administration Amendments of 1974 amended and extended the authority of the President and the Secretary of Commerce to regulate US exports under the original Act. Other US legislation which provides for control over US exports includes the Munitions Control Act, the Trading with the Enemy Act and the Equal Export Opportunity Act of 1972. See *A Factbook Concerning the Relationship between Technology and Trade,* Vol. II, Legal/Institutional Data, Centre for Policy Alternatives, MIT, Cambridge, Massachusetts, August 1976.
246   *Public Law 95-52,* 22nd June, 1977, 95th Congress, 91 Stat. 235, "Export Administration Amendments of 1977", 22nd June, 1977, H.R. 5840. See also *Export Administration Regulations, June 1, 1977,* US Department of Commerce, Domestic and International Business Administration, Bureau of East-West Trade, Office of Export Administration, Washington, D.C.; *Export Administration Act of 1969, As Amended,* Public Law 95-223 (H.R. 7738), 91 Stat. 1625, approved December 28, 1977; and *Export Administration Act of 1969 as Amended and Extended,* Public Law 95-435, October 10, 1978. The latter is given in *Export Administration Bulletin,* Supplement to Export Administration Regulations, No. 189, US Department of Commerce, Washington, D.C., October 26, 1978. It has been pointed out that the "1977 Amendments represent a significant effort by the Congress to strengthen the framework for East-West trade by facilitating the export of US goods and technology while clarifying and simplifying the export licensing process". See Ronda A. Bresnick, "The Setting: The Congress and East-West Commercial Relations", *art. cit.,* p. 2. *Technology and East-West Trade,* Office of Technology Assessment, Congress of the United States, Washington, D.C., November, 1979, p. 124. The Export Administration Act of 1979 (Public Law 96-72) "closely follows H.R. 4034 and S. 737", the primary legislation considered by Congress. Both bills aimed, in essence, to strengthen controls on exports of technology to the Soviet Union. It is of course too early to venture an opinion as to whether these and other developments indicate a turning away from the "liberalisation trend" which emerged in the mid-1960s. It should be noted that items eligible for export are authorised by means of either a "general license" or a "validated license". Distinctions are also made according to the criterion of "country destination". The Commodity Control List (CCL) is the key to determining whether a specific shipment is exportable under an established general license authorisation, or whether a validated export license (for which an application must be filed), is needed. For all the commodities licensed by the Office of Export Administration, the CCL shows the destinations for which each commodity requires a license document.

During the fourth quarter of 1976 license applications for items valued at $123.6 million were approved for export to the USSR, Eastern Europe (and the People's Republic of China). Of this Rumania accounted for $27.3 million. Most of the approvals covered electronic computing equipment, magnetic recorders and parts, fixed wing aircraft and accessories and communications equipment.[247] New export control regulations were announced by the Department of Commerce on August 1, 1978 which provide "for a case-by-case review of proposed exports to the USSR of petroleum and natural gas exploration and production commodities, and related technical data... A considerable number of applications for export licenses have already been reviewed and approved under this regulations. None have been denied".[248] Among the items approved were "twenty-two licenses ranging from an offshore drilling station to the production of paraffin wax. The "backlog" of pending oil and gas cases has been eliminated. The dollar value of the cases approved under the August guidelines now totals $276 million".[249]

6. Import Restrictions

With regard to imports, a traditional concern to the Western European nations, the trend towards liberalisation which began in the early sixties continued unabated. As has been noted, most "Western European nations have for some time either *de jure* or *de facto* extended non-discriminatory, or MFN tariff status to imports from Eastern Europe".[250] For instance, France maintained only some 200 articles on its restrictive list by 1968.[251] This number was reduced to 97 in 1969 and to 10 by the end of 1974.[252] Italy instituted a major

247   See *Export Administration Report, 11th Report on United States Export Controls to the President and the Congress,* Semi-annual: October 1976-March 1977, US Department of Commerce, Domestic and International Business Administration, Bureau of East-West Trade, Washington, D.C.

248   Excerpts from an address by US Secretary of the Treasury W. Michael Blumenthal in Moscow, December 6, 1978, at the opening meeting of the USSR Trade and Economic Council, following the conclusion on December 5, 1978 of the seventh session of the Soviet-American Commission on Trade and Economic Relations. See also *Export Administration Bulletin,* Supplement to Export Administration Regulations, No. 185, US Department of Commerce, August 1, 1978.

249   W. Michael Blumenthal, *op. cit.* See also "US-USSR Trade Relations at a Turning Point?", *Business Eastern Europe,* Vol. 7, Nos. 51 & 52, December 22, 1978. Export licenses for the so-called "Verity Projects" — a list of 28 projects drawn up jointly by C. William Verity, Chairman of Armco Steel and the Soviet Deputy Minister of Foreign Trade, Vladimir N. Sushkov — have also been favourably regarded by the Carter Administration. As noted by the US Secretary of the Treasury, the "Administration welcomes the participation of US firms in projects which are of benefit to both countries... The license for one such project—the Akron Standard Plant for the construction of larger tires—has already been approved by the Administration, and we have approved and completed export licensing for portions of three others dealing with gas lift equipment, technical data for production of television tubes, and technical data for an aluminium smelter". Two projects on the list have been awarded to non-US firms: "Technip S.A. of France is to supply gas lift equipment for the Samotlor oil field, and a Japanese firm has the main contract for a colour TV plant". See W. Michael Blumenthal, *op. cit.,* and "US-USSR Trade Relations at a Turning Point?", *Business Eastern Europe, art. cit.* Whether these (US) licenses will be affected by the US suspension of all exports of sophisticated technology and machinery to the Soviet Union, following the Soviet Union's invasion of Afghanistan (December 1979), is not known by the authors. The Commerce Department "will review already approved export licenses for the sale of high-technology products to the USSR to determine whether some or all of them should be revoked or suspended in the wake of President Carter's action to constrain such sales following the Soviet invasion of Afghanistan". *Aviation Week & Space Technology,* January 14, 1980, p. 16. See also Chapter 4.

250   Thomas A. Wolf "East-West European Trade Relations", *art. cit.,* p. 1044.

251   Samuel Pisar, *op. cit.,* p. 106.

252   "Les relations commerciales de la France avec les pays de l'Europe de l'Est à commerce d'Etat", *Journal officiel,* Avis et rapports du Conseil économique et social, Paris, 22 octobre 1976, p. 1005.

revision in 1967 of its import quotas removing about 80 per cent of items from restrictions.[253] All in all, every Western European country has liberalised its import policy. As already mentioned, with the exception of the German Democratic Republic and the Federal Republic of Germany, bi-lateral clearing agreements have been replaced by settlement in convertible currency.

7. Credit Policy

Many restrictions on credit facilities have been lifted. Thus, East-West economic relations developed during the early seventies on the basis of Western credit.[254] Again, US policy was adjusted more slowly than that of its West European partners who had already eased credit conditions in the early sixties.

Concerning the United States, the 1964 Congressional amendment to the Export-Import Act of 1945 had allowed Eximbank credit support for sales to the Eastern countries only when deemed by the President to be in the national interest. In 1968, even more restrictive legislation was passed; this had the effect of removing all Presidential discretion in this matter, and amounted to an absolute prohibition on Eximbank credits to the East. Waiver authority along the lines of the 1964 legislation was not restored until 1971.[255]

National interest determinations allowing Eximbank official credit support have been made available to Rumania (November 1971), to the USSR (October 1972) and to Poland (November 1972).[256] The Trade Act of 1974 had, as has been noted, tied US Government supported credit to freer emigration. Exceptions were made for Poland and Yugoslavia and

---

253 Samuel Pisar, *op. cit.,* p. 106. The United States maintains only minor restrictions against imports from the East. An interesting recent development is the request by a 'group of US ammonia producers' that the US government put a tariff on ammonia imports from the Soviet Union. The group has claimed that the "low-priced imports are putting US producers out of business and threatening the security of US agriculture... The complaint is the first of its kind against a Soviet product, and only the third action filed under the Trade Act of 1974 seeking protection against imports from a Communist country". "US Ammonia Makers Seek Protection from USSR", *International Herald Tribune,* Paris, July 13, 1979.

254 See Peter Knirsch, *art. cit.,* p. 15. It has been pointed out that while economists and government officials may argue among themselves whether Eastern Europe can manage the level of its debt to the West, the banks seem to have no such qualms. "The volume of new loans went up in 1977 and even Poland, whose overall external debt is considered to be the most worrisome, had no difficulty in arranging a recent credit of 2 billion DM". "Western Banks Not Worried About East Bloc Borrowing", *International Herald Tribune,* Part I, Paris, December, 1977, p. 1. According to the same article, "the level of borrowing has not decreased despite a decline in Comecon's trade deficit with the West... The overall lending by US banks is modest due to the limits placed on US banks". *Ibid.,* pp. 1 and 3. How credit policy towards the Soviet Union will evolve in the long run, following the invasion by that country of Afghanistan remains to be seen. According to an article in the *International Herald Tribune,* the United States "is asking Japan and West European countries to halve official and government-guaranteed credits to Moscow and to end preferential export-credit terms to protest the Soviet invasion of Afghanistan". "US Said Seeking Soviet Credit Cut", *International Herald Tribune,* Paris, January 9, 1980, p. 7. According to a more recent report, the nine Common Market countries have decided to no longer accord favourable credit terms to the Soviet Union. As noted in the article, France, the United Kingdom and Italy could apply waivers under the consensus which existed within the OECD framework which foresees a minimum rate of 7.75 per cent interest on the credits their countries accord to the USSR. The Nine agreed to no longer apply these waivers. *Le Figaro,* Paris, February 6, 1980. It should be noted that there are Western bankers and company managers who believe that, in effect, the USSR is the least susceptible of all the East European countries to any potential credit restrictions because the Soviets have been able to repay a substantial amount of their debt ahead of schedule. See for instance "Credit Squeeze: No Threat for Suppliers to the USSR", *Business Eastern Europe,* Vol. 9, No. 7, February 15, 1980. See also Chapter 6.

255 See Thomas A. Wolf, "East-West Trade Credit Policy: A Comparative Analysis", *art. cit.,* p. 175; and Rogers Morton *op. cit.,* p. A-5.

256 John P. Hardt and George D. Holliday, "East-West Financing by Eximbank and National Interest Criteria", in Paul Marer, ed., *US Financing of East-West Trade, op. cit.,* pp. 291-295.

subsequent waivers were obtained for Rumania and Hungary.[257] While the Eximbank has played a central role in recent efforts to promote and facilitate trade with the bloc countries — and continues to do so — the Eximbank amendments of 1974 (P.L. 93-646) signed into law by President Ford on January 4, 1975, made some important changes.[258] Namely, "it is still US policy to foster exports by Government financing, but in the future the impact of loans must be explicitly assessed on the basis of short material supplies and employment. Moreover, Eximbank's reporting requirements to Congress have been greatly expanded".[259]

8. Expansion of Trade and Technology (Embodied) Flows

As has been noted, there has been a rapid expansion of trade since 1965. Eastern European imports from the OECD countries rose almost 8 fold, in current prices, between 1965 up to 1977. Exports rose 6 fold. Poland and the Soviet Union registered the highest increase in imports (10 times). The share of the Soviet Union's imports among the CMEA countries from OECD countries rose from 38.8 per cent in 1965 to 49.7 per cent in 1977. Poland's share rose from 14.0 per cent in 1965 to 18.3 per cent in 1977.

Imports remained almost unchanged in 1977. While the average growth of imports from the OECD countries was 22 per cent between 1965 and 1975, it was only 5.8 per cent in 1976 and 0.8 per cent in 1977. CMEA countries' exports to the OECD group registered an average annual growth of 16.5 per cent between 1965 and 1975. In 1976 the increase was 19.1 per cent and in 1977, 9.8 per cent.

Machinery imports from the OECD countries occupied a large share in total imports, increasing from 29.5 per cent in 1965 to 36.1 per cent in 1975 and to 37.2 per cent in 1977.[260]

It would be important to know whether the high growth rates in East-West trade from 1965 to 1975 were exceptional or whether they were representative of the overall world trends and strengthening of the international division of labour. The following chapters attempt to shed some light on this issue.

257   "The role of Eximbank is to facilitate US exports through direct loans, as well as guarantees and insurance for commercial bank lending... Poland has been eligible for Eximbank credits since 1972; the USSR qualified from 1972 until the passage of the Trade Act; Rumania satisfied the Trade Act provisions in 1975; Hungary did so in 1978. Of the several Eximbank programs, direct loans have been most significant in financing East-West trade, accounting for 85 per cent of the value of Eximbank's total operations to date with Poland, Rumania and the Soviet Union". Allen J. Lenz and Lawrence H. Theriot, "The Potential Role of Eximbank Credits in Financing US-Soviet Trade", in *Issues in East-West Commercial Relations – A Compendium of Papers, op. cit.,* p. 220.

258   The legislative requirement that Eximbank loans, guarantees and insurance be used only after Presidential determination of national interest was "based on the general belief that trade with Communist countries requires special attention to the divergence between national and private commercial interest". John P. Hardt and George D. Holliday, "East-West Financing by Eximbank and National Interest Criteria", *art. cit.,* p. 291. (The central issue as to what should be the criteria for determining national interest is taken up in Chapter 4).

259   *Ibid.* "Under the 1974 Eximbank Act Amendments, total new authorisations for direct loans and guarantees [for the Soviet Union] cannot exceed $300 million without prior approval of the Congress... The $300 million ceiling could be raised by a Presidential national interest determination, subject to approval by concurrent Congressional resolution". Allen J. Lenz and Lawrence H. Theriot, "The Potential Role of Eximbank Credits and Financing US-Soviet Trade", *art. cit.,* pp. 220-221. It is believed that the Soviet Union sees the removal of Eximbank restrictions as a major policy objective. It has been reported for instance, that at the time Mr. Breshnev endorsed the previously mentioned "Verity list", he "insisted that the only way to ensure significant growth in trade is to remove the restrictive MFN and Eximbank credit provisions in the Jackson-Vanik and Stevenson amendments... The Soviets repeatedly stated that the major industrial projects such as energy extraction in Siberia will require government-backed credits such as those offered by Germany, the United Kingdom and Canada". "US-USSR Trade Relations at a Turning Point?", *Business Eastern Europe, art. cit.,* p. 402. A proposal was made (see H.R. 1835) to "establish a new limitation on Eximbank support for US exports to all non-market nations of $2 billion". See Ronda A. Bresnick, "The Setting: The Congress and East-West Commercial Relations", *art. cit.,* p. 2.

260   The figures have been taken or calculated from Tables 9, 10 and 11 of the report. A detailed discussion of changes in trade between East and West is given in Chapter 2.

Chapter 2

STATISTICAL EVALUATION OF TECHNOLOGY TRANSFER

1. MEASUREMENT DIFFICULTIES

It has already been stressed in the introduction that technology is not just like any other commodity. It is a *science* or an *art* and it is transferred only if the "art" used for production purposes by one organisation is used by another. Thus, the receiving enterprise must possess the rights of ownership or use and also be in a position to apply the technology. A distinction therefore has to be made between the purchase of a product or an item of technical information, simply injected into the production process, and a transfer of knowledge enabling the new product to be produced.[1]

The first difficulty in measurement therefore stems from the fact that only part of technology transfer is measurable and even then, subject to the reservations made above. The measurable part consists of the purchase or hire of physical goods. The following forms of transfer cannot be measured:

— *Information transferred* (via direct observation of technological processes, consultation of experts or the acquisition of documentation);

— *Improvement in skills* (via normal vocational training or apprenticeship);

— *Transfer through staff movements* (staff movements within a given firm or between different firms, recruitment of new permanent or temporary staff).

These phenomena, however, do not wholly elude statistical investigation — the number of exchanges or training courses can be evaluated and so can the number of specialists trained, the cost of training and staff movements — but even if data is available on the number of persons involved in these processes and the cost of the operation, the quality of the knowledge that is transferred cannot be evaluated. In practice, allowance for this is made in the residual part of the computation which takes into account not only the transfers in question but also the margin of error in calculating other, more or less definable, factors.

Monographs concentrating on certain sectors (case studies), however, can be very useful provided their object and scope are clearly defined. The "measurable" part of technology transfer is therefore confined to that embodied in marketed products and in the granting of

1 As Philip Hanson says, a specialised machine-tool may be installed in the Togliatti works and used effectively in the production of Zhiguli cars, without the Russians being able to produce the drawings and tooling it embodies. Philip Hanson: "External Influences on the Soviet Economy Since the Mid-1950s; the Impact of Western Technology", CREES Discussion Paper, No. 7, *art. cit.,* p. 9.

licences and registration of patents.[2] However, the "measurable" part of technology transfer is itself not easily defined and presents complex methodological and statistical problems. Attempts are therefore made to evaluate trade in "research-intensive" products and in "high technology" products.

2. THE SHARE OF RESEARCH-INTENSIVE PRODUCTS IN EAST EUROPEAN COUNTRIES' FOREIGN TRADE

The evaluation of trade in research-intensive products comes up against both methodological and statistical difficulties.

The available literature shows that opinions vary as to how it should be calculated. In 1970, OECD experts defined fifty research-intensive products among the various SITC categories.[3] The largest group is machinery (18 product groups) but there are many other products as well, such as chemicals, instruments, iron and steel products, etc.

Since the publication of its study in 1970, the OECD has continued research on methods of evaluating R and D expenditure. One result has led to the proposal of three main indicators:

— R & D expenditure in terms of gross production;
— R & D expenditure in terms of value added;
— R & D manpower in terms of total payroll.[4]

Other possible indicators include self financing of R & D as a percentage of actual profits; purchase of patents, licences and know-how as a percentage of intra-mural expenditure on R & D, and sales as a percentage of purchases of patents, licences and know-how.

Thus far there have been no studies of East-West trade on the basis of these proposed indicators. Even the existing indicators based on the fifty product groups selected by the OECD have not been fully treated in the studies undertaken to date.

The method nearest to the OECD classification of technology-intensive products is that used in the United Nations study published in 1976.[5] It covers 204 categories represented by the four digit headings in the *Standard International Trade Classification* (SITC) and, allowing for the disaggregation of a certain number of these headings, the total number of product groups used reaches 244.[6] The study gives no clue on the extent to which these groups correspond to the research-intensive products referred to in the OECD reports.[7] A rough comparison between the technology-(or research) intensive products in the United Nations study (see Table 1) and the OECD classification (see Annex A-1) shows that the OECD list includes more industrial sectors (in addition to chemicals and the engineering industries) but that the number of engineering industry products is smaller.

The results of the United Nations computation given in Table 1 show that technology-intensive products formed nearly half of East European imports from the West over the

2   The most advanced study on technology transfer based on an evaluation of trade in machinery is the study by Werner Beitel "Technological Co-operation with the Soviet Union", in *East-West Technological Cooperation,* Colloquium 1976, *op. cit.,* pp. 275-313. A study on the granting of licences and registration of patents is included in the same publication, pp. 197-239: Jiri Slama and Heinrich Vogel, "Technology Advances in Comecon Countries: An Assessment". In view of the particular problems of evaluating technology transfer through licensing, this question is dealt with separately in Chapter 3.
3   See Annex Table A-1.
4   Provisional calculations so far have revealed marked differences between the indicators. R & D expenditure is relatively highest in the electrical, chemical and aerospace industries and lowest in services and agriculture.
5   *Economic Bulletin for Europe,* United Nations, New York, 1976, Vol. 28, p. 117 (prepared by the Secretariat of the Economic Commission for Europe, Geneva).
6   The United Nations Study uses the term "technology intensive" (see Table 1) and the OECD definition (Annex Table A-1) "research intensive". It is possible that these two definitions are not identical.
7   *Ibid.,* pp. 117-118.

Table 1

**Western Exports to and Imports from Eastern Europe and the Soviet Union:
Shares by Factor Intensity**

Sample Ten Countries in the West: Total Exports and Imports = 100

	Labour intensive		Resource intensive		Technology intensive	
	Exports	Imports	Exports	Imports	Exports	Imports
All products						
1965-1968	27	27	22	60	51	13
1971-1974	30	26	24	59	46	15
1974	33	24	21	62	46	14
Specialised items [a]						
1965-1968	17	14	6	45	37	2
1971-1974	19	12	10	44	33	4
1974	22	14	6	47	33	4
Specialised items [b]						
1971-1974	14	12	9	18	16	6
The Soviet Union						
All products						
1965-1968	32	8	14	85	54	7
1971-1974	33	11	24	81	43	8
1974	40	7	17	85	43	8

a) *Shares higher than average for 204 product groups:*
Imports:
Labour intensive: meat, preserved fruit, sawn lumber, vegetable oils, blooms, slubs and coils, bars and rods, universals and heavy plates and sheets (all from iron and steel), clothing (knitted and non-knitted), footwear.
Resource intensive: live animals, fresh vegetables, fur skins (undressed), pulpwood, sawn logs, raw cotton, crude fertilizers, sulphur, iron ore and concentrates, iron and steel scrap, coal, coke, crude petroleum, motor spirit, distilled fuels, residual fuel oil, pig iron, copper, nickel, aluminium.
Technology intensive: hydrocarbons, potassic, fertilizers, machine tools for working metals, cars, ships and boats.
Exports:
Labour intensive: feeding stuffs, synthetic fibres, synthetic fabrics and knitted fabrics, bars and rods, angles, shapes, universals and heavy plates and sheets, other plates and sheets, tinned plates and sheets, boats and ships, tubes and pipes (all made of iron and steel), finished structural parts, manufactures of metals, clothing knitted, footwear.
Resource intensive: wheat, maize, worked copper, some types of yarn.
Technology intensive: paints, enamels, varnishes, products of condensation, polymerization, insecticides, fungicides and disinfectants, machine tools for working metals and metalworking machinery, textile machinery, paper mill and pulp mill machinery, construction and mining machinery, mineral crushing, sorting and moulding machinery, heating and cooling equipment, pumps and centrifuges, mechanical handling equipment, powered tools, other non-electric machines, machinery and mechanical appliances, electric apparatus for electric current, insulated wire and cables, electric machinery and apparatus, bodies, chassis and frames for vehicles, ships, scientific instruments.
b) *Rate of volume growth faster than average.*
For individual commodities see the section on the volume growth. *Economic Bulletin for Europe, op. cit.*
Note: Labour intensive products: textiles, processed foodstuffs, other light industry, metal and wood products;
Resource intensive products: raw agricultural forestry products, industrial materials, fuels, non-metallic mineral products, ferrous and non-ferrous metals;
Technology intensive products: chemicals and machinery.
Source: UN Trade Statistics, as quoted in the *Economic Bulletin for Europe.* Vol. 28, *op. cit.,* pp. 117-118.

period 1965-1974 but only 13-15 per cent (only about 8 per cent in the case of the Soviet Union) of that area's exports to the West over the same period.

As pointed out by the authors of the United Nations' study this distribution of trade by factor intensities is not sharply defined.[8] Although, overall, the production of machinery is technology-intensive, a number of sub-items are resource-and labour-intensive. Eastern machinery and equipment exports are dominated by labour and material-intensive products while Western exports consist predominantly of more advanced machinery embodying a larger degree of processing, substantial research and a generally higher degree of sophistication.[9] These differences are reflected in unit values, and in almost all groups, Western

8   *Ibid.,* p. 119.
9   *Ibid.,* p. 119.

Table 2

**Commodity Composition by Factor Intensity of Western Exports and Imports
from Eastern Europe and the Soviet Union**

Total exports and imports = 100

	Capital-intensive products		Labour-intensive products		Natural resource-intensive products		Technologically advanced products	
	Exp.	Imp.	Exp.	Imp.	Exp.	Imp.	Exp.	Imp.
	In per cent							
Eastern Europe and USSR								
1965-1969	12	15	17	12	23	64	48	9
1973-1977	19	20	16	15	19	53	46	12
of which: 1976-1977	18	21	15	17	18	50	49	12
Soviet Union								
1965-1969	11	19	22	7	18	69	49	5
1973-1977	23	27	15	6	18	60	44	7
of which: 1976-1977	21	25	15	8	17	60	47	7

		Trade Balance (f.o.b. - f.o.b.) (in $ million)						
Eastern Europe and USSR	Total							
1965-1969	39	—93		217		—1 728		1 643
1973-1977	5 967	1 092		1 023		—4 374		8 226
of which: 1976-1977	6 879	814		852		—4 990		10 203
Soviet Union								
1965-1969	—315	—198		232		—1 078		729
1973-1977	2 237	200		1 016		—2 928		3 949
of which: 1976-1977	2 938	251		1 095		—3 780		5 372

Source: Economic Bulletin for Europe. Vol. 30, No. 1, pre-publication text, Table B.8. The layout of this table has been rearranged to conform with that of Table 1. The definition of the groups is not the same in the two tables. The definitions for this table are as follows:

Capital-intensive products: beverages, tobacco manufactures, iron and steel, road vehicles, petroleum and coal refining, soap and related goods, paints and allied products; rubber tubes and other articles, cement.

Labour-intensive products: stone, leather goods, wood manufactures, plywood and furniture, paper articles and printed matter, textiles (yarn, thread, fabrics and clothing), metal manufactures, non-metallic minerals and their products (concrete, structural clay products, glass and pottery products), heating, lighting and plumbing equipment, household appliances, tyres, railway vehicles, ships and boats, radio, television, office supplies, plastic articles, toys, sporting goods, jewellery and silverware, and miscellaneous consumer goods.

Natural resource-intensive products: agricultural products (excluding tobacco manufactures), natural rubber, wood, pulp, waste paper, paper, mineral fertilizers, sulphur, iron ore and concentrates, non-ferrous metal ores, coal, coke, briquettes, crude petroleum, natural gas, precious stones, pearls, non ferrous metals.

Technologically advanced products: organic and inorganic chemicals, medicinal products, plastics and synthetic materials, miscellaneous chemicals, power machinery, agricultural machinery, office machines, metalworking machines, special industrial machines, electric power, distributing machinery, miscellaneous electrical apparatus, communication equipment, aircraft, instruments and related apparatus, photographic supplies, watches, clocks, ordinance and ammunition.

The basic classification is taken from H. Giersch, *The International Division of Labour. Problems and Perspectives,* J. C. B. Mohr (Paul Siebeck) Tübingen. Some help in breaking down traditional goods was obtained from P.B.W. Rayment, "The Homogeneity of Manufacturing Industries with Respect to Factor Intensity: The Case of the UK", in *Oxford Bulletin of Economics and Statistics,* Vol. 58, No. 3, August 1976 and H. B. Lary, *Imports of Manufactures from Less Developed Countries,* National Bureau of Economic Research, New York, 1968.

countries received a considerably higher price per ton for exports than was paid for imports, and the difference increased over time.[10] In 1977, the United Nations diversified its product classification by introducing a new category — capital-intensive products (see Table 2). So far statistics are available up to 1977. The relative importance of technologically advanced products[11] in Western exports to the East is still as high — 49 per cent in 1976-1977

10 *Ibid.,* p. 119. Regarding prices, see the section on the economic aspects of transfer.

11 According to the United Nations definition, technologically advanced products are "characterised by a share of professional, technical and scientific personnel which exceeds 9 per cent of the labour force in producing industries. Until they mature, information about production processes involved is not freely available, technology is unstable and output is highly skill-intensive. Once in the mature stage, technologies become stable, the need for skill in production diminishes". Source: *Economic Bulletin for Europe,* Vol. 30, No. 1, 1978, pre-publication text, p. 69.

compared with only 12 per cent for Eastern exports to the West (7 per cent in the case of the USSR). With an average trade balance of $10.2 billion in favour of the West for 1976-1977, technologically advanced products are largely responsible for the East's trade deficit, which averaged $6.9 billion in 1976-1977 (see Table 2).

East-West High Technology Transfer

An interesting point is that in some studies, the "technology-intensive" (or research intensive) concept is replaced by "high technology".[12]

The list of "high technology" products proposed by the Bureau of East-West Trade is given in Annex Table A-2. A straight comparison with the OECD list published in 1970 shows that it is much narrower and more precise as it uses four — and even five — digit SITC headings. It covers mainly machinery and instruments but excludes chemicals, special steels, etc.

There seems to be a considerable difference between "technology-intensive (or research-intensive) and "high technology" (R & D expenditure may be high without actually resulting in high technology), and it is interesting to look at some of the figures given by the Bureau.

Table 3 shows that the share of high technology in the total exports of fifteen industrialised countries to the Eastern countries varied very little between 1972 and 1977. It was highest (though with a downward tendency) for exports to the Soviet Union (14 to 17.6 per cent). It was stable for exports to Eastern countries as a whole (11.7 to 13.5 per cent) and slightly higher than the average for the whole world (9.9 to 10.9 per cent).

An interesting point is that five product groups (see Table 4) alone represented approximately 70 per cent of high technology exports in 1976 and 1977. Germany supplied the biggest share of high technology to the Soviet Union (34 to 36 per cent of the total between 1972 and 1977) (see Table 5), followed by Japan (14.4 per cent in 1972 and 16.9 per cent in 1977) and then France. The United States provided only 6.7 per cent in 1972, rising to 12.7 per cent in 1976 and falling to 9.1 per cent in 1977. Table 6 shows that high technology accounted for 12.1 per cent of US exports to the communist countries but 17.3 per cent of that country's total exports in 1977.

The authors of this study are fully aware of the snags in quantifying "high technology". The SITC breakdown does not go far enough in many cases and the figures arrived at may be too high.[13] The proposed list does not have universal agreement and in any case, it is bound to be a rapidly changing one. The SITC headings that are not included also contain varying shares of high technology, and this tends to give figures that are too low. These over- and under-estimates, obviously mean that the overall evaluation is far from perfect but it does present the advantage of highlighting advanced technology exports. Even so, it is important to compare these results with those on trade in machinery.

12 *Quantification of Western Exports of High Technology Products to the Communist Countries,* Office of East-West Policy and Planning, Bureau of East-West Trade, 17th October, 1977, Washington, D.C. This report, same title, was presented in 1978 and 1979 under the name of John P. Young.

13 For instance, heading SITC 7142 "calculating machines, accounting machines and similar machines incorporating a calculating device" covers both computers of varying complexity and simple calculating and accounting machines, which means that the high technology share tends to be overestimated. *Quantification of Western Exports of High Technology Products to the Communist Countries, op. cit.,* p. 1.

Table 3

Comparison of High Technology Exports with Total Exports of Fifteen I. W. Countries to the World and to the Eastern Countries in 1972, 1974, 1976 and 1977

Millions of US Dollars

Destination	1972 Total	1972 High Techn.	1972 %	1974 Total	1974 High Techn.	1974 %	1976 Total	1976 High Techn.	1976 %	1977 Total	1977 High Techn.	1977 %
USSR	3 317	582	17.5	6 250	1 036	16.6	11 653	1 627	14.0	11 412	2 003	17.6
Eastern Europe	5 098	619	12.1	11 322	1 223	10.8	12 757	1 525	12.0	12 866	1 741	13.5
Yugoslavia	2 117	270	12.8	4 503	482	10.7	4 034	561	13.9	5 407	801	14.8
Cuba	257	27	10.5	817	42	5.1	942	83	8.8	993	93	9.4
China	144	64	4.5	4 369	414	9.5	3 423	343	10.0	3 585	248	6.9
Total Communist Countries	12 234	1 562	12.8	27 261	3 197	11.7	32 808	4 140	12.6	34 263	4 886	14.3
Total — World	273 045	29 092	10.7	498 470	49 314	9.9	590 833	64 366	10.9	669 393	71 576	10.7

I. W. Countries: United States. Canada. Japan. Belgium-Luxembourg. France. Federal Republic of Germany. Italy. Netherlands. Austria. Norway. Sweden. Switzerland. United Kingdom. Denmark.
Eastern Europe: Bulgaria. Czechoslovakia. German Democratic Republic. Hungary. Poland. Rumania. USSR.

Source: John P. Young. *Quantification of Western Exports of High Technology: Products to the Communist Countries*, *op. cit.*, p. 10.

Table 4

Top Five High Technology I. W. Exports to the Eastern Countries, 1976 and 1977

SITC		Value ($ millions)	% of total High Tech. exports	% of total exports	Value ($ millions)	% of total High Tech. exports	% of total exports
		1976			1977		
7151	Machine tools for working metal	1 110	26.8	3.4	1 257	25.7	3.7
7192	Pumps and centrifuges	648	15.7	2.0	775	15.9	2.3
7299	Electrical machinery and apparatus, n.e.s.	451	10.9	1.4	560	11.5	1.6
71992	Taps, cocks, valves, n.e.s.	438	10.6	1.3	495	10.1	1.4
72952	Electrical measuring and controlling instruments, n.e.s.	255	6.2	0.8	307	6.3	0.9
	Top Five Total	2 902	70.2	8.8	3 394	69.5	9.9

Source: John P. Young. *Quantification of Western Exports of High Technology Products to the Communist Countries, op. cit.*, p. 12 ; and 2nd draft 1979, p. 12.

Table 5

Soviet Imports of Advanced Technology from the Western Industrialised Countries[a]

	$000	%	$000	%	$000	%	$ 000	%
	1972		1974		1976		1977	
United States	39 008	6.7	137 581	13.3	207 109	12.7	182 748	9.1
Canada	1 330	0.2	729	0.1	8 587	0.5	7 267	0.4
Japan	83 916	14.4	88 559	8.6	225 506	13.9	338 700	16.9
Belgium/Luxembourg	2 690	0.5	13 685	1.3	14 390	0.9	14 849	0.7
France	66 077	11.3	155 501	15.0	177 187	10.9	228 703	11.4
Germany	209 484	36.0	352 446	34.0	560 777	34.5	683 962	34.1
Italy	62 463	10.7	78 713	7.6	136 713	8.4	223 540	11.2
Netherlands	2 315	0.4	41 214	4.0	39 148	2.4	19 007	1.0
Austria	9 036	1.6	20 446	2.0	38 469	2.4	72 880	3.6
Norway	417	0.1	1 316	0.1	7 231	0.4	1 364	0.1
Sweden	11 769	2.0	41 197	4.0	49 265	3.0	75 196	3.8
Switzerland	32 211	5.5	63 511	6.1	100 503	6.2	98 311	4.9
United Kingdom	56 062	9.6	36 391	3.5	53 594	3.3	43 434	2.2
Denmark	5 662	1.0	4 919	0.5	8 627	0.5	13 234	0.7
Total	582 440	100.0	1 036 208	100.0	1 627 106	100.0	2 003 195	100.0

a) For the definition of Western Industrialised Countries, see note for Table A 1.

Source: John P. Young. *Quantification of Western Exports of High Technology Products to the Communist Countries, op. cit.*, p. 15 ; and 2nd draft 1979, p. 12.

Table 6

**United States High Technology Exports to the Communist Countries
and to the World, 1976 and 1977**

$ 000

Exports to	1976			1977		
	High Tech. Exports	Total Exports	% High Tech.	High Tech. Exports	Total Exports	% High Tech.
Cuba	3	89	3.4	126	600	21.0
People's Republic of China	22 907	135 390	16.9	15 147	171 300	8.8
Yugoslavia	60 307	296 882	20.3	86 680	356 300	24.3
Bulgaria	2 872	43 320	6.6	4 713	23 900	19.7
Czechoslovakia	8 716	147 470	5.9	7 073	74 000	9.6
German Democratic Republic	1 462	64 770	2.3	1 192	36 100	3.3
Hungary	4 068	62 960	6.5	12 876	79 700	16.2
Poland	38 703	621 040	6.2	37 014	436 500	8.5
Rumania	20 158	249 030	8.1	23 569	259 400	9.1
USSR	207 109	2 305 930	9.0	182 748	1 623 500	11.3
Total Communist Countries	366 305	3 926 871	9.3	371 144	3 061 300	12.1
World	19 445 897	113 323 145	17.2	20 443 800	117 962 800	17.3

Source : John P. Young. *Quantification of Western Exports of High Technology Products to the Communist Countries, op. cit.,* p. 16 ; and 2nd draft 1979, p. 12.

3. THE SHARE OF MACHINERY AND TRANSPORT EQUIPMENT IN EAST EUROPE'S FOREIGN TRADE

The value of using statistics on trade in machinery to study technology transfer has been pointed out by Werner Beitel:

"The question remains whether an analysis of the transfer of machinery and transport equipment can be regarded as representative of the transfer of technical progress to the USSR. Since the latter cannot be measured directly, one would have to determine the productivity progress achieved through the use of imported machinery and equipment as opposed to other types of transfer, but also in comparison to other investment goods. As long as this question cannot be resolved, the value of individual forms of technology transfer is uncertain. One indicator could be the orders of magnitude reached with regard to investment goods imports and the acquisition of licenses".[14]

Machinery and transport equipment figure largely in East Europe's foreign trade. Aggregates, as shown in the Annex (Table A-3a) indicate that these commodities accounted for 31.8 per cent of these countries' total exports in 1975 and 36.7 per cent of imports in the same year. Trade in machinery and transport equipment has expanded considerably between 1969-1977, growth rates being highest in 1975.[15]

This balanced picture with its continuous expansion of trade in general, and of trade in machinery and transport equipment in particular, does not hold if we disaggregate the figures by major areas: developed market economy, developing and East European countries.

14   Werner Beitel, *art. cit.,* p. 278.
15   The statistics reproduced in Tables 7, 8, 12, 15, 18, 19 and 20 and those given in several tables in the Annex (A-3, A-4 and others) are taken from East European countries' national statistics. Because the recording periods for the FOB and CIF calculations differ from country to country, the figures lead to different results from those based on the statistics of the countries of the West given in OECD tables.

74

Table A-4 in the Annex lists the main figures showing trends in the foreign trade of East European countries by area from 1960 to 1975. The flows showing the most rapid expansion are those with the developed market economy countries (a nine-fold increase), followed by those with the developing countries (x 8.4) and lastly those among the countries of East Europe (x 4.6). On the other hand, while imports and exports between the East European countries were in balance, this was not true of trade with the developed market economy countries. Exports to these countries increased by a factor of 7.3 whereas imports from them advanced by 10.5 times. As a result of this the overall trade balance showed a shortfall of $5 billion in 1974 and $11 billion in 1975.

In this context, the distribution of product groups in the different trade areas is particularly important. Table 7 reflects the situation as it was in 1970. West Europe's exports of machinery and transport equipment to East Europe were equivalent to less than 5 per cent of Western machinery exports, although they accounted for 37.5 per cent of those countries' total exports to East Europe. Conversely machinery exports to Western Europe accounted for only 6.1 per cent of East Europe's total machinery exports and only 9.6 per cent of total exports of all kinds to the West. There was also a big trade gap in chemicals which accounted

Table 7

Commodity Composition of Selected West and East European Trade Flows, 1970

	Western Europe			Eastern Europe		
	Total Exports	Intra-Trade	Exports to Eastern Europe	Exports to Western Europe	Intra-Trade	Total Exports
Millions of US dollars (f.o.b.)						
Machinery and transport equipment	47 190	28 390	2 180	590	7 010	9 560
Chemicals	13 400	8 380	720	340	870	1 500
Iron and steel	9 300	7 440	580	475	1 470	2 300
Non-ferrous metals	4 450	3 480	195	260	390	750
Crude materials, oils and fats	7 960	6 660	365	1 010	1 610	3 120
Textiles	7 110	5 050	275	130	200	520
Clothing	3 250	2 570	130	125	570	730
Other manufactured products	24 450	15 010	810	530	2 190	3 410
Food, beverages, tobacco	13 970	10 280	490	1 010	1 570	3 260
Mineral fuels and related materials	4 610	3 780	54	1 100	1 530	3 010
Miscellaneous transactions and commodities	1 810	780	41	550	1 040	2 370
Total	137 500	91 850	5 840	6 120	18 480	30 530
Percentage distribution						
Machinery and transport equipment	34.3	30.9	37.3	9.6	38.1	31.3
Chemicals	9.7	9.1	12.3	5.6	4.7	4.9
Iron and steel	6.8	8.1	9.9	7.8	7.9	7.5
Non-ferrous metals	3.2	3.8	3.4	4.2	2.1	2.5
Crude materials, oils and fats	5.8	7.3	6.3	16.5	8.7	10.2
Textiles	5.2	5.5	4.7	2.1	1.1	1.7
Clothing	2.4	2.8	2.2	2.0	3.1	2.4
Other manufactured products	17.8	16.4	13.9	8.7	11.9	11.1
Food, beverages, tobacco	10.2	11.2	8.4	16.5	8.5	10.7
Mineral fuels and related materials	3.3	4.1	0.9	18.0	8.3	9.9
Miscellaneous transactions and commodities	1.3	0.8	0.7	9.0	5.6	7.8

Source: United Nations. *Monthly Bulletin of Statistics.* April and July 1972. Special Table C. "Commodity Composition to the Revised SITC" as quoted in *Analytical Report on Industrial Co-operation among ECE Countries.* United Nations, Geneva, 1973, p. 72.

Table 8
East-West Trade in Machinery

	Western imports from the East (c.i.f.)						Western exports to the East (f.o.b.)					
	Value in millions of dollars	Percentage change over same period of preceding year					Value in millions of dollars	Percentage change over same period of preceding year				
	1978	1974	1975	1976	1977	1978	1978	1974	1975	1976	1977	1978
Total Eastern Europe	2 395	12	32	7	18	18	10 988	35	55	0	10	16
of which: USSR	543			16			5 819			7		
of which:												
Non-electrical machinery (total East)	1 008	27	38	1	1	15	7 674	26	48	4	13	14
of which: USSR	215											
Electrical machinery (total East)	435	21	16	11	18	17	1 834	46	38	−5	24	17
of which: USSR	57											
Transport equipment (total East)	952	−10	34	12	43	22	1 480	87	109	−10	−16	24
of which: USSR	271											

Eastern Europe: USSR, Bulgaria, Hungary, Poland, German Democratic Republic, Rumania, Czechoslovakia. Trade between the Federal Republic of Germany and the German Democratic Republic has not been included in these figures.
Western Countries: OECD countries (Western Europe, United States, Canada, Japan).

Sources: Economic Bulletin for Europe, Vol. 28, *op. cit.,* p. 95; *Economic Bulletin for Europe,* United Nations, Vol. 30, (Pre-publication), Tables 3.4 and 3.9; Vol. 31, (Pre-publication), Table 3.9.

for 12.3 per cent of Western exports to East Europe but only 5.6 per cent of East Europe's exports to the West.

A point to note is that, during 1974 and 1975 Eastern imports of machinery and transport equipment from Western Europe increased at a particularly fast rate (by 35 per cent in 1974 and 55 per cent in 1975) mainly because of the particularly large imports of transport equipment (see Table 8). In addition to machinery, imports of apparatus and instruments from Western Europe also increased steeply — by 34 per cent in 1974 and by 23 per cent in 1975.[16]

In 1976 the trend changed: machinery imports from the industrialised West levelled off and there was even a decrease of 5 and 10 per cent in the case of electrical machinery and transport equipment respectively (see Table 8). The USSR does not seem to have been affected by this drop in 1976, for, in aggregate, her machinery imports went up in that year by a further 7 per cent. In 1977 and 1978 there was a renewed relative increase (10 per cent in 1977 and 16 per cent in 1978) in machinery imports from the industrialised West (see Table 8) but this coincided with a general slowdown in the increase of imports to the East from the West (see Table 9). Except for Hungary and Rumania, the CMEA countries showed lower total import figures in 1977 although machinery imports from the OECD countries increased.[17] As a result machinery's share of total imports increased to an average of 37.2 per cent, the highest level since 1965 (see Table 9).

The effects of this trend reversal on East-West trade are not yet known but certain changes are noticeable in Soviet foreign trade in 1977 and 1978.[18]

4. THE SHARE OF MACHINERY AND TRANSPORT EQUIPMENT IN EAST EUROPE'S FOREIGN TRADE WITH OECD COUNTRIES

The composition of trade between the OECD countries and East Europe by commodity groups is given in the Annex, Tables A-5, A-6, A-7.

It can be stated at the outset that machinery and transport equipment have never played a major role in OECD country imports; on the other hand exports of these commodities are relatively important (some 30-37 per cent). It is interesting to note, in this connection, that the commodity balances for raw materials and fuel (see Table A-8) are always negative for the OECD countries, and have increased between 1974 and 1977 because of higher prices for heating and motor fuels.[19] The balance for manufactured goods, on the other hand, shows a

16   *Economic Bulletin for Europe*, Vol. 28, *op. cit.*, p. 96.
17   Except for the German Democratic Republic and Bulgaria. There was a slight drop in machinery imports by Poland.
18   In particular, the Soviet Union has tried to strengthen its trade links with CMEA countries. Its imports from these countries increased by 13.7 in 1977 and 22.2 per cent in 1978. Exports increased by 16.9 and 13.3 per cent in 1977 and 1978 respectively. In trade with the non-Socialist countries, the Soviet Union has mainly aimed at increasing its exports — the figure went up by 24.6 per cent between 1976 and 1978 — and keeping imports at as low a level as possible — they increased by only 1.3 per cent between 1976 and 1978. *Vneshnyaya Torgovlya SSSR v 1977 godu* (USSR Foreign Trade in 1977), Moscow, 1978, p. 8; *Commerce extérieur de l'URSS*, Supplément No. 3, 1979. In interpreting these figures it is necessary to bear in mind the changes in Soviet grain imports (higher in 1978) and income from Soviet oil sales; which for the main OECD importing countries, (17 out of 23: Portugal, Turkey, Canada, United States, Australia and New Zealand excluded) fell from 4.36 billion rubles in 1977 to 4.22 billion rubles in 1978. *Vneshnyaya Torgovlya SSSR v 1978*, Moscow, 1979, p. 61.
19   From 1973 to 1976, total oil and oil product exports from the USSR went up by 26 per cent in weight and by 219 per cent in value, but the Soviet Union delayed applying the price increases to fuels exported to the CMEA countries. At the same time the Western countries were made to pay higher increases than the USSR average. Oil and oil product exports to France went up by only 7.5 per cent in volume between 1973 and 1976 but the prices France paid went up by a factor of 4.1 over the same period. These figures are based on *Vneshnyaya Torgovlya SSSR* (USSR Foreign Trade), Statistical data for 1974 and 1976, pp. 29 and 201, and 26 and 189.

Table 9

Total Exports of OECD Countries to the Seven East European Countries and Share of Exports of Machinery and Transport Equipment from 1965 to 1977

(SITC, Section 7)[a]

At current prices in million US dollars

Seven East European Countries	1965 Total	1965 SITC Section 7	1965 %	1970 Total	1970 SITC Section 7	1970 %	1974 Total	1974 SITC Section 7	1974 %	1975 Total	1975 SITC Section 7	1975 %	1976 Total	1976 SITC Section 7	1976 %	1977 Total	1977 SITC Section 7	1977 %
USSR	1 346	462	34.3	2 643	1 028	38.9	7 501	2 309	30.8	12 510	4 576	36.6	13 755	4 909	35.7	13 736	5 375	39.1
German Democratic Republic[b]	295	56	19.0	438	159	36.6	1 000	220	22.1	1 131	336	29.6	1 302	372	28.6	1 196	306	25.6
Poland	484	113	23.3	882	241	27.3	4 558	1 528	33.5	5 483	2 084	38.0	5 509	1 990	36.1	5 071	1 865	36.8
Czechoslovakia	445	110	24.7	784	284	36.2	1 727	566	32.8	1 874	689	36.8	2 080	707	34.0	2 084	789	37.9
Hungary	293	67	22.9	625	140	22.4	1 784	408	22.9	1 830	471	25.7	1 822	535	29.4	2 321	725	31.2
Rumania	319	142	44.3	703	266	38.0	2 042	719	35.2	1 990	696	35.0	2 014	534	26.5	2 342	883	37.7
Bulgaria	221	72	32.4	327	102	31.2	844	230	27.2	1 099	498	45.3	941	405	43.0	905	340	37.6
Total	3 468	1 022	29.5	6 404	2 219	34.6	19 450	5 980	30.7	25 917	9 351	36.1	27 427	9 451	34.5	27 656	10 283	37.2

a) SITC: Standard International Trade Classification.
b) Trade in goods between the Federal Republic of Germany and the German Democratic Republic and East Berlin are not covered by these statistics.

Source: Calculated from *Statistics of Foreign Trade*, "Trade by Product Summarised by Markets". OECD, Paris.

surplus. This surplus increased steeply from $0.9 billion in 1973 to $4.8 billion in 1975 before falling to $3.1 billion in 1977. However, the reductions in the surplus for 1977 were not due to machinery and transport equipment for which the OECD countries' surplus remained at the level of $8.4 billion in 1977 (see Table A-8).

It is worthwhile taking a closer look at the trends affecting trade in machinery and transport equipment between the OECD countries and East Europe using statistics which take into account the respective situations of all East European countries.

Table 9 gives figures for exports of machinery and transport equipment from the OECD countries to East Europe. We have already said that these exports represented about 30-38 per cent of the total. The question now is to analyse how this breaks down among the various countries of the East.

The most vigorous growth of exports of machinery and transport equipment was registered in Poland whose share of total exports from OECD countries increased from 23.3 per cent in 1965 to 38.0 per cent in 1975 and to 36.8 per cent in 1977. The smallest share, in relative terms, (22-31 per cent) was accounted for by Hungary despite the existence of an open-door policy. Hungary's high degree of dependence on imported raw materials and semi-finished products may be the explanation of this exception to the rule (see Table A-3 in the Annex). The figures for the German Democratic Republic are somewhat skewed (trade between the German Democratic Republic and the Federal Republic of Germany is not foreign trade and, consequently, the figures about intra-German trade are not published in OECD statistics) although trade in technology is significant. A very steep increase in exports from the OECD countries other than Germany is apparent. For the other countries of Eastern Europe, machinery exports from OECD countries account for a very large share of the total, often reaching 35-45 per cent. The percentages vary, however, from one year to the next and the very high figures reached in 1970 for certain countries, like the USSR and Rumania, have not been repeated in 1977. It also looks as though these machinery exports are affected by such unforeseeable factors as credit, grants, loans and the signing of contracts. The "internalisation of production" theories do not appear to be confirmed by these trade flows.

Table 10 shows that the OECD countries' machinery imports from East European countries accounts for a small share of total imports (from 5.3 to 8.6 per cent) and that the recent increase is relatively slight. However, there are considerable differences in the figures for the various East European countries. Machinery and transport equipment account for a large share of imports from the German Democratic Republic and Czechoslovakia but, whereas the trend is up for the former, it is down for Czechoslovakia. Perhaps the reason for this is that the German Democratic Republic benefits from the types of trading arrangements between it and the Federal Republic of Germany that focus on technological progress. Czechoslovakia, on the other hand, may find it difficult to keep pace with technical progress in the West or perhaps is too busy with technological exports to the other East European countries.

It is interesting to note that the OECD country imports of machinery from Poland, Rumania and Bulgaria, practically insignificant in 1965, had increased considerably between 1974 and 1977 reaching approximately 8-16 per cent of total OECD imports from those countries. The poorest performer in this field is the Soviet Union which is still supplying very little machinery and transport equipment to the OECD countries. The latter accounted for only 2.7 per cent of imports from the USSR in 1965, 4.3 per cent in 1976, falling to 3.1 per cent in 1977. In absolute figures, in 1977, the USSR exported to the OECD countries less machinery and transport equipment than Poland and slightly more than Czechoslovakia and the German Democratic Republic despite the existence of a greater economic potential.

Table 11 sums up the main trends in machinery and transport equipment trade between the OECD countries and the East European countries.

Table 10

Total Imports of OECD Countries from the Seven East European Countries and Share of Imports of Machinery and Transport Equipment from 1965 to 1977
(SITC Section 7)ᵃ⁾

At current prices in million US dollars

Imports

Seven East European Countries	1965 Total	1965 SITC Section 7	1965 %	1970 Total	1970 SITC Section 7	1970 %	1974 Total	1974 SITC Section 7	1974 %	1975 Total	1975 SITC Section 7	1975 %	1976 Total	1976 SITC Section 7	1976 %	1977 Total	1977 SITC Section 7	1977 %
USSR	1 749	48	2.7	2 554	93	3.6	7 915	238	3.0	8 490	372	4.4	10 761	462	4.3	12 128	371	3.1
German Democratic Republicᵇ	257	56	21.8	410	93	22.7	952	218	22.9	1 039	229	22.0	1 078	257	23.8	1 144	278	24.3
Poland	633	16	2.5	1 061	62	5.8	2 870	241	8.4	3 172	364	11.5	3 591	481	13.4	3 829	604	15.8
Czechoslovakia	422	69	16.4	723	135	18.7	1 529	234	15.3	1 638	287	17.5	1 696	274	16.2	1 896	303	16.1
Hungary	279	12	4.3	536	34	6.3	1 343	95	7.1	1 250	116	9.3	1 436	141	9.8	1 675	169	10.1
Rumania	284	2	0.7	554	30	5.4	1 578	89	5.6	1 662	114	6.9	1 972	152	7.7	1 911	158	8.3
Bulgaria	158	3	1.9	240	17	7.1	415	31	7.5	392	34	8.7	488	42	8.6	517	44	8.5
Total	3 848	205	5.3	6 078	464	7.6	16 602	1 147	6.9	17 643	1 516	8.6	21 022	1 809	8.6	23 090	1 927	8.3

a) SITC: Standard International Trade Classification.
b) Trade in goods between the Federal Republic of Germany and the German Democratic Republic and East Berlin are not covered by these statistics.

Source: Calculated from *Statistics of Foreign Trade,* "Trade by Product Summarised by Markets". OECD. Paris.

Table 11

Growth Rates of Total Foreign Trade between OECD Countries and the Seven East European Countries and of Share of Machinery and Transport Equipment

(SITC Section 7)[a]

Seven East European Countries		Growth rates in %							
		1970/1965 (1965 = 100)		1975/1970 (1970 = 100)		1976/1975 (1975 = 100)		1977/1976 (1976 = 100)	
		Total	SITC Section 7	Total	SITC Section 7	Total	SITC Section 7	Total	SITC Section 7
USSR	A	196	222	473	445	110	107	99.9	109.5
	B	146	193	333	400	127	124	112.7	80.3
German Democratic Republic	A	149	284	258	211	115	111	91.9	82.2
	B	160	166	254	246	104	112	106.1	108.2
Poland	A	182	213	621	865	100	95	92.0	93.7
	B	168	388	299	214	113	132	106.6	126.6
Czechoslovakia	A	176	258	239	243	111	103	100.2	111.6
	B	171	196	226	213	104	95	111.2	110.6
Hungary	A	213	209	293	336	100	114	127.4	135.5
	B	192	283	233	341	115	122	116.6	119.9
Rumania	A	220	187	283	262	101	77	116.3	165.4
	B	195	159	300	380	119	133	96.9	103.9
Bulgaria	A	148	142	335	488	86	81	96.2	84.0
	B	152	567	163	1 134	124	124	105.9	104.8
Total Seven countries	A	185	217	405	421	106	101	100.8	108.8
	B	158	226	290	327	119	119	109.8	106.5

a) SITC – Standard International Trade Classification.
A = Exports of OECD countries to the seven East European countries.
B = Imports of OECD countries from the seven East European countries.

Source: Calculated from Tables 5 and 6.

During the years 1965-1970, in practically all cases,[20] trade in machinery increased impressively and more steeply than trade in general. During the period 1971-1975 there was also a very steep increase in trade in machinery, a trend that applied to the growth of trade in general with the OECD countries. This increase applied to the Soviet Union, the German Democratic Republic and Czechoslovakia and to some extent Rumania and Bulgaria. Poland is the one exception, showing an extraordinary increase in machinery purchases and a somewhat less steep increase in other imports from the OECD countries.

However, a major change occurred in 1976 and in 1977. OECD exports of machinery and transport equipment to CMEA countries decreased in the case of Poland, the German Democratic Republic and Bulgaria and increased less steeply than previously in the case of the Soviet Union. Only Rumania and Hungary increased their imports of machinery. At the same time the Eastern countries tried to sell as much as possible to the industrialised countries of the West. Success was achieved particularly in the case of the USSR.

5. THE SHARE OF MACHINERY IN TRADE BETWEEN EAST EUROPE AND THE DEVELOPING COUNTRIES

We have already referred to the considerable expansion in trade between East Europe and the developing countries. Between 1960 and 1975 trade increased by 10.8 times for exports to and 8.8 times for imports from the developing countries, giving East Europe an

20 Except for machinery and transport equipment exports from the OECD countries to Rumania and Bulgaria which show a slightly lower growth rate than for total OECD exports to these countries.

overall surplus in the trade balance.[21] To what extent are machinery and transport equipment involved in this trend?

It is clear from Table 12 that machinery and transport equipment play a very important part in exports to the developing countries. In this field the Eastern countries occupy a comparable position to that of the OECD countries. In the case of the Eastern countries, it is the USSR which plays the part of the "most developed" country by supplying the highest proportion of machinery. However, in trade between East European and OECD countries the percentage of machinery in total USSR *imports* is one of the highest for East European countries.[22] Eastern European imports from the developing countries equal Eastern European exports to the OECD countries.[23]

Here again it is interesting to observe a certain loss in momentum in East European exports of machinery and transport equipment to the developing countries; the shares of total exports accounted for in 1965 and 1970 by machinery and transport equipment have not been repeated and the trend was sharply down in 1974. It would be interesting to know the reasons for this: some offered include credit policy and competition from the OECD countries supplying more advanced technology.

Table 12

Share of Machinery and Transport Equipment in Trade between East Europe and the Developing Countries

(SITC Section 7)

As a percentage of total exports/imports

	1965	1970	1973	1974	1975
Exports to Developing Countries					
East Europe	42.6	45.2	41.3	30.7	..
of which from the USSR:	..	..	47.3	33.7	36.9
Imports from the Developing Countries					
East Europe	6.8	5.2	5.4	4.5	..
of which to the USSR:	..	..	3.7	3.9	4.7

Source: See Annex. Tables A-9 and A-10.

6. SPECIFIC FEATURES OF TRADE IN MACHINERY AND TRANSPORT EQUIPMENT WITH THE SOVIET UNION

It has already been pointed out that the Soviet Union is very dependent on imports of machinery and transport equipment from the OECD countries. Tables 13, 14 and 15 give figures on the geographical distribution of trade in machinery between the USSR and the OECD countries and the "non-socialist" countries respectively. Tables 13 and 14 have been derived from OECD *Foreign Trade Statistics* and Table 15 from official Soviet statistics. We thus have two evaluations — that of the exporting country and that of the importing country — which do not always coincide.[24] It is, however, possible to detect the main trends.

21 See Annex, Table A-4.
22 See Table 9 above.
23 See Table 10 above.
24 On the comparison between American and Soviet statistics for trade between the United States and the USSR see: Anne de Tinguy, "Les relations économiques et commerciales soviéto-américaines de 1961 à 1974", *art. cit.*, pp. 132-139.

Table 13

Share of Main OECD Countries Exporting Machinery and Transport Equipment to the USSR

(SITC Section 7)

	1961	1965	1970	1971	1972	1973	1974	1975	1976	1977
Total exports of machinery and transport equipment to the USSR	100	100	100	100	100	100	100	100	100	100
of which:										
United States	4.7	1.1	4.4	7.0	5.1	11.8	9.7	12.0	12.3	7.0
Japan	—	13.0	10.0	13.0	16.2	9.7	10.0	12.2	13.9	14.3
Finland	—	21.0	12.1	7.0	7.0	9.3	9.4	8.4	13.1	14.7
France	15.8	5.8	13.8	15.3	9.7	9.7	12.6	12.5	8.4	10.7
Germany	31.9	17.3	16.0	18.7	30.4	33.3	32.9	29.5	28.5	28.0
Italy	8.0	6.9	16.9	15.5	10.6	8.9	7.4	8.5	7.3	10.4
United Kingdom	21.1	10.8	9.8	9.3	8.0	6.0	3.0	4.6	4.2	3.6

Source: Foreign Trade Statistics, OECD, Paris.

Table 14

Share of Main OECD Countries Importing Machinery and Transport Equipment from the USSR

(SITC Section 7)

	1961	1965	1970	1971	1972	1973	1974	1975	1976	1977
Total	100	100	100	100	100	100	100	100	100	100
of which:										
United States	—	—	1.1	0.7	2.7	0.7	0.8	1.4	0.9	0.9
Japan	—	4.6	5.9	6.5	3.4	2.5	3.6	1.5	1.6	4.0
Finland	—	36.9	15.3	15.1	14.2	17.3	20.3	24.9	21.3	20.9
France	4.0	4.0	8.0	6.2	5.2	6.0	8.9	9.7	8.6	10.8
Germany	9.5	6.0	18.6	9.2	7.4	11.0	2.5	10.0	12.3	9.6
Italy	5.0	1.5	5.7	2.2	3.3	4.9	6.6	4.1	3.3	6.8
United Kingdom	13.5	4.0	5.9	6.6	8.0	9.6	12.3	9.5	6.9	11.5

Source: Foreign Trade Statistics, OECD, Paris.

Soviet imports of machinery and transport equipment (i.e. OECD countries' exports to the USSR — see Tables 13 and 15) are largely obtained from six highly industrialised countries and Finland. At the top of the list is Germany which, in 1977, alone accounted for 28 per cent of imports from OECD countries (Table 13) or 20.8 per cent of those from the non-socialist countries (Table 15). Important, however, are the changes that have taken place over 16 years in the relative positions of the main Western supplier countries. The greatest progress has been made by the United States which increased its share of Soviet imports of machinery and transport equipment from the OECD countries from 1.1 per cent in 1965 to 12.0 per cent in 1975, but only to 7 per cent in 1977.

Table 13 also shows the frequent changes in the various industrialised countries' shares of machinery and transport equipment exports to the USSR. Germany's share, extremely high in 1961 (31.9 per cent) fell to 16.0 per cent in 1970. After recovering to 33.3 per cent in 1973 it declined to 28.0 per cent in 1977. A slight upward trend seems to be apparent in 1978.[25]

25   Total Soviet imports from Germany increased from 1 745 to 1 942 million rubles between 1977 and 1978. *Vneshnyaya Torgovlya SSSR v 1978 godu, op. cit.,* p. 10.

Table 15

Pattern of Trade in Machinery and Transport Equipment between the USSR and the Non-Socialist Countries

Section "Machinery and Transport Equipment" = 100

In million Roubles

A. Exports	1970	1975	1976	1977	1978
Total exports from the USSR to the Non-Socialist Countries	3 990	9 446	11 575	14 154	14 414
of which:					
Machinery and Transport Equipment	735	1 198	1 391	1 305	1 507
Machinery as a percentage of total	18.4	12.7	12.0	9.2	10.5
Percentage share of main trade partners:					
— United Kingdom	0.1	1.9	1.5	1.9	2.1
— Italy	0.6	1.2	0.8	1.2	1.6
— France	0.8	1.6	2.4	2.2	2.1
— Germany	3.5	2.1	3.0	2.0	2.3
— Finland	1.8	7.3	4.9	4.0	4.3
— Egypt	20.2	7.1	5.8	6.2	4.7
— Iran	19.5	16.1	12.2	13.1	10.2
— Syria	4.1	5.4	5.7	3.4	4.8
B. Imports					
Total imports to the USSR from the Non-Socialist Countries	3 686	12 702	13 627	12 926	13 812
of which:					
Machinery and Transport Equipment	1 045	3 852	4 709	4 461	4 826
Machinery as a percentage of total	28.4	30.3	34.6	34.5	34.9
Percentage share of main trade partners:					
— Italy	16.9	8.0	6.8	10.9	9.9
— United States	2.1	11.8	13.2	7.9	5.7
— Germany	11.7	26.2	23.2	23.2	20.8
— Finland	11.5	8.1	11.0	15.0	15.2
— France	14.9	11.0	10.8	12.7	14.3

Source: Vneshnyaya Torglovlya SSSR v 1973 godu (USSR Foreign Trade in 1973) pp. 10, 19, 53-54, 94 ; in 1975, pp. 8, 17, 51-52, 91 ; in 1976, pp. 8, 46, 47, 83 ; in 1977, pp. 8, 18, 45, 46, 80 and in 1978, pp. 8, 18, 45, 46, 80.

France's share of Soviet imports decreased sharply between 1961 and 1965, increased fairly steeply in 1971, but has since fluctuated. The United Kingdom's share has a distinctly downward trend whilst Finland, often with a considerable share, shows a very high figure again (14.7 per cent) in 1977.[26]

Soviet exports of machinery and transport equipment to the OECD countries (or to non-socialist countries) (see Tables 14 and 15) have changed in many different ways in terms of the position of each country. Finland accounts for a particularly high share, though Soviet and OECD figures differ considerably. The impression is even gained that part of the trade simply passes through Finland. However, the matter needs clarification. More generally too, explanations are wanting for the fluctuations in Soviet machinery exports to the Western countries. The bargaining that goes on when trade agreements are negotiated could possibly be an important factor in this connection.

The *internal structure* of the Soviet Union's imports of machinery and transport equipment has a vital bearing on the study of technology transfer to East Europe. In this respect, the Soviet statistics are far less detailed than those of the OECD countries but some comparison of the data may be attempted. For the OECD's statistics it means analysing the

26 See Table 13. It is something of a surprise to find Finland as one of the USSR's main machinery suppliers. The question deserves more detailed study.

Table 16

Pattern of Machinery and Transport Equipment Exports from the OECD Countries to the USSR, 1961-1977

Percentages

		1961	1965	1970	1971	1972	1973	1974	1975	1976	1977
	Total Section 7	100.0	100.0	100.0	100.0	100.0	100.0	100.0	100.0	100.0	100.0
71	Machinery, other than electric	78.4	48.5	63.0	68.3	77.7	81.0	75.4	68.2	70.8	72.7
711	Power generating machinery	—	2.6	1.3	1.9	2.2	1.8	1.9	2.8	2.8	..
712	Agricultural machinery and implements	—	—	2.9	2.0	0.4	1.9	0.7	3.1	3.7	3.4
714	Office machines	—	5.0	2.3	3.3	2.7	2.0	1.2	1.2	1.4	..
715	Metalworking machinery	7.2	5.2	18.7	15.8	21.1	21.7	23.1	13.9	13.4	13.8
717	Textile and leather machinery	—	6.3	2.8	2.0	0.7	3.6	4.2	4.6	3.6	..
718	Machines for special industries	19.1	6.1	7.5	7.4	6.0	6.6	7.5	6.4	6.7	5.3
719	Machinery and appliances, other than electrical, and machine parts, n.e.s.	38.8	29.0	27.5	35.9	41.6	43.4	37.9	36.2	39.3	44.0
72	Electrical machinery, apparatus and appliances	14.1	10.2	15.9	15.0	13.4	11.1	14.1	12.1	10.0	13.7
722	Electric power machinery and switchgear	5.0	1.9	1.8	2.5	1.9	1.4	1.9	1.6	2.0	3.1
723	Equipment for distributing electricity	—	1.2	2.4	2.5	1.9	1.6	2.6	2.0	1.6	..
724	Telecommunications apparatus	—	0.9	1.7	0.9	1.0	1.0	1.0	0.5	0.6	..
726	Electric apparatus for medical purpose and radiological apparatus	—	0.6	0.4	0.7	0.6	0.5	0.5	0.5	0.5	..
729	Other electrical machinery and apparatus	6.0	3.7	9.4	8.1	7.9	6.5	7.9	7.3	5.1	7.1
73	Transport equipment	7.5	41.6	21.2	16.7	8.9	7.9	10.5	19.7	19.2	13.6
731	Railway vehicles	—	—	0.6	1.3	1.0	1.1	1.1	1.0	0.7	..
732	Road motor vehicles	0.6	0.9	2.9	4.1	3.6	1.5	2.1	8.5	7.6	2.8
733	Road vehicles other than motor vehicles	—	—	—	—	0.1	0.2	0.2	0.3	0.4	..
735	Ships and boats	5.0	40.6	17.6	11.3	4.2	5.0	7.0	9.8	10.4	10.0

Source: Statistics of Foreign Trade, OECD, Paris. Calculated from figures in Annex Table A-13.

Table 17

Pattern of Machinery and Transport Equipment Imports into the OECD Countries from the USSR, 1961-1977

Percentages

	1961	1965	1970	1971	1972	1973	1974	1975	1976	1977
Total Section 7	100.0	100.0	100.0	100.0	100.0	100.0	100.0	100.0	100.0	100.0
71 Machinery, other than electric	55.0	41.7	54.8	55.6	51.9	47.2	47.9	44.6	42.0	33.1
712 Agricultural machinery and implements	5.0	4.2	3.2	2.2	3.3	4.1	6.3	6.5	6.1	6.2
715 Metalworking machinery	15.0	10.4	20.4	20.7	18.0	12.7	15.1	15.6	11.3	10.8
718 Machines for special industries	10.0	10.4	6.5	10.4	6.0	2.6	3.4	3.8	4.3	2.3
719 Machinery and appliances - non-electrical - parts	10.0	14.6	17.2	16.3	18.6	22.5	16.4	11.5	9.7	5.7
72 Electrical machinery, apparatus and appliances	15.0	10.4	12.9	8.1	15.8	12.7	15.1	11.0	10.2	11.9
722 Electric power machinery and switchgear	—	2.0	3.2	3.0	3.8	2.6	2.9	3.5	3.9	2.5
729 Other electrical machinery and apparatus	10.0	2.0	4.3	2.2	6.6	4.9	5.9	3.5	3.0	5.1
73 Transport equipment	30.0	45.8	32.3	36.3	31.7	40.0	37.4	44.4	47.8	55.0
732 Road motor vehicles	15.0	22.9	9.7	9.6	16.4	21.7	26.5	30.6	30.7	45.3
735 Ships and boats	15.2	14.6	20.4	24.4	8.7	15.4	6.3	10.2	15.2	5.9

Source: Statistics of Foreign Trade. OECD. Paris. Calculated from figures in Annex Table A-14.

five-digit categories which entails that a considerable amount of work of trends over several years is required.

Table 16 shows the percentage calculated for the main sub-headings of the machinery and transport equipment Group in OECD countries' exports to the USSR.

The largest category among Soviet machinery and transport equipment imports corresponds to SITC Group 71 (machinery other than electric). The share of this Group in total machinery and transport equipment imports from the OECD countries was 78.4 per cent in 1961, 48.5 per cent in 1965, 81 per cent in 1973 and 72.7 per cent in 1977. The largest sub-category in this Group is that headed machinery and appliances — non-electrical — parts (SITC 719). As a percentage of total USSR machinery and transport equipment imports from the OECD countries it has fluctuated considerably: 38.8 in 1961, 27.5 in 1970, 43.4 in 1973, 36.2 in 1975 and increasing to 44 per cent in 1977. Imports of transport equipment (SITC Group 73) fluctuated in a similar manner: 7.5 in 1961, 41.6 in 1965, 7.9 in 1973, 19.2 in 1976 and 13.6 per cent in 1977.[27]

These fluctuations in Soviet imports of the various categories of machinery would probably be even greater if the analysis were made in terms of sub-categories. The varying level of imports of machinery and appliances — non-electrical — parts (SITC 719) seems to suggest the absence of any clearcut policy in this field. More generally, much research can be done in this area. More detailed figures need to be obtained for the categories of imported machinery and compared with data relating to agreements and commercial policy on the one hand and the Soviet internal economic situation and annual and five-year plans on the other.

Research is also lacking into Soviet machinery and transport equipment exports. A few figures for the main machinery groups and sub-groups are set out in Table 17. A striking feature is the large share that transport equipment accounts for, motor vehicles in particular, from 1974 onwards (43.1 per cent in 1977). This is largely due to the construction of the Fiat works in the USSR. Fluctuations in the ships and boats Sub-category (SITC 735) also need explanation.

7. THE SHARE OF MACHINERY IN THE FOREIGN TRADE OF THE INDUSTRIALISED EAST EUROPEAN COUNTRIES

OECD foreign trade statistics for Germany do not include figures for trade between the German Democratic Republic and the Federal Republic of Germany. Further work would have to be done to incorporate them but some comparisons regarding trade in machinery are possible on the basis of national statistics.

Table 18 gives information about the German Democratic Republic's trade in machinery and transport equipment with various types of country groups.

The striking feature of the German Democratic Republic's exports is the relatively minor importance of machinery in exports to the industrialised countries and the very considerable share — well above the average for the East European countries — of exports to the developing countries. Another interesting point is the considerable difference in the shares accounted for by machinery in exports to the OECD countries (not including trade between the German Democratic Republic and the Federal Republic of Germany) and to the Federal Republic of Germany. Whereas the German Democratic Republic manages to sell a large proportion of its machinery to the OECD countries (not including Germany), the exports to the Federal Republic of Germany are only one half that of the OECD countries.

27 See Table 16 and Annex, Table A-13.

87

Table 18

Share of Machinery and Transport Equipment in the German Democratic Republic's Foreign Trade

Annual average

	1965	1970	1971	1974	1975	1976	1965-1970	1971-1975	1977	1978
Exports:										
To the industrialised countries^a	—	—	18.7	13.4	—	—	—	—	—	—
To the OECD countries^b	20.2	22.5	—	22.9	—	—	—	—	—	—
To Germany^c	—	—	—	—	9.9	10.5	13.7	10.7	11.0	10.5
To the developing countries	—	—	51.7	56.9	—	—	—	—	—	—
Imports:										
From the OECD countries^b	18.7	36.6	—	22.1	29.6	—	—	—	—	—
From Germany^c	—	—	—	—	22.8	28.9	22.9	23.8	30.3	31.0

a) Official German Democratic Republic statistics. See *Statistisches Jahrbuch der DDR*. 1973, p. 301 ; 1974, p. 299 ; 1975, pp. 263, 279 ; 1976, p. 281.
b) OECD statistics, see Tables 9 and 10 (not including trade between the German Democratic Republic and the Federal Republic of Germany).
c) Statistics for the Federal Republic of Germany. See Annex Table A-15.

A comparison between the share of the German Democratic Republic's machinery imports from the OECD countries — excluding Germany — and in intra-German trade does not allow a conclusion to be drawn. The differences do not seem large but research is needed to calculate the actual share of machinery in the German Democratic Republic's trade with Western industrialized countries.

The disaggregation of Czechoslovakia's trade in machinery and transport equipment by destination is given in Table 19. It is interesting to note that half of the imports stem from Germany and Austria, the proportion steadily increasing over several years. It is difficult to trace any clearcut pattern regarding the other Western countries that are Czechoslovakia's traditional suppliers although imports from France seem to be gaining some ground at the expense of those from Sweden and the United Kingdom. For the latter a decline was visible in 1977. Czechoslovakian exports seem to be fairly well distributed among the various countries, Egypt and India accounting for large but decreasing shares and Iraq's share being rather important. However, whereas Germany's share of Czechoslovak machinery imports has been climbing steeply to reach more than one-third in 1977, Czechoslovak machinery exports to Germany have been much less strong.

The figures in Table 20 are an attempt to divide the East European industrialized countries' trade in machinery and transport equipment into three sub-groups.

Hungary seems to be a special case with respect to *exports* to the OECD countries. Over half of Hungarian exports of machinery and transport and equipment are in the electrical machinery, apparatus and appliances sub-group. The share of this sub-group is increasing. For the other countries, machinery other than electrical, takes precedence, although declining slightly. Within this sub-group metalworking machinery is first (particularly in Czechoslovakia) but its share is declining in every case.[28]

Transport equipment also figures largely in exports to the OECD. Its share is particularly high in Poland's case (30-54 per cent of all machinery between 1961 and 1977) the main constituent being ships and boats. It is also high in Czechoslovakia and in the

28 See Annex, Tables A-17, A-19, A-22, A-25.

Table 19

**The Share and Structure of Exports and Imports of Machinery and Transport Equipment
in Czechoslovakian Foreign Trade with Non-Socialist Countries**
(SITC Section 7)

In million Kcs

A. Exports from Czechoslovakia	1970	1975	1976	1977
Total Exports	8 017	13 274	13 436	15 486
of which (*):				
Machinery and Transport Equipment	2 370	3 757	3 980	4 604
Percentage share	29.6	28.3	29.6	29.7
of which:				
France	3.1	3.8	4.0	2.7
Germany	10.0	6.9	6.7	7.9
United Kingdom	2.5	5.6	5.4	5.0
Iraq	. .	. .	. .	. .
India	8.9	5.1	5.1	2.8
Egypt	7.4	5.7	5.5	5.8
Italy	3.2	4.0	4.0	3.8

B. Imports to Czechoslovakia				
Total Imports	8 143	15 318	16 890	19 020
of which:				
Machinery and Transport Equipment	2 338	4 500	4 952	5 839
Percentage share	28.7	29.4	29.3	30.6
of which (*):				
France	12.1	8.3	11.8	5.2
Italy	12.7	7.2	6.6	3.9
Germany	27.3	34.4	32.8	33.6
Austria	10.9	16.4	15.6	28.1
Sweden	7.5	4.3	3.3	3.3
United Kingdom	6.9	5.5	4.1	3.6

* Exports (Imports) of Machinery and Transport Equipment = 100.

Sources: Czechoslovakian Foreign Trade Yearbook, 1977, p. 81, and 1978, p. 77, Czechoslovakian Chamber of Commerce Publications.
Statisticka Rocenka CSSR (Czechoslovakian Statistical Yearbook) 1977, pp. 472-476, and 1978, pp. 474-480.

German Democratic Republic but a downward trend, particularly for motor vehicles, is apparent for Czechoslovakia.[29]

Imports from the OECD countries are headed, in every case, by machinery, other than electric. The figure is relatively stable at about 70 per cent [except in the case of the German Democratic Republic].[30] Machinery and appliances — non-electrical — accounts for roughly half of this group.[31] This share holds at a fairly stable level, unlike the situation in the USSR where it is higher and more variable.

The share of aggregate Group 7 (machinery and transport equipment) imports into each of the East European industrialized countries that transport equipment accounts for varies considerably from country to country. In the German Democratic Republic it has been very

29 *Ibid.*
30 Not including trade between the German Democratic Republic and the Federal Republic of Germany which is given in Table A-15.
31 See Annex, Tables A-16, A-18, A-21, A-24.

Table 20
Breakdown of East European Countries' Trade in Machinery and Some Transport Equipment by the Three Main Groups
As a percentage of each country's total Exports/Imports of Machinery and Transport Equipment

	1961	1965	1970	1974	1975	1976	1977
Exports to OECD Countries (Section 7 = 100)							
Division 71. Machinery							
German Democratic Republic:							
— OECD statistics [a]	70.0	41.7	58.1	46.3	41.9	42.0	41.4
— national statistics [b]	. .	. .	50.3	65.7	. .	. .	. .
Czechoslovakia	53.8	53.6	67.4	62.8	62.4	61.5	57.4
Poland	44.4	56.3	43.6	47.3	51.9	37.2	34.4
Hungary	38.0	33.3	38.2	31.6	34.5	29.8	30.8
Division 72. Electrical Machinery, Apparatus and Appliances							
German Democratic Republic:							
— OECD statistics [a]	16.7	23.2	30.1	28.9	32.8	30.0	34.2
— national statistics [b]	. .	. .	15.1 [c]	29.6	. .	. .	. .
Czechoslovakia	9.6	7.2	10.4	16.2	13.6	16.5	18.8
Poland	11.1	12.5	14.5	16.2	13.5	12.3	11.4
Hungary	50.0	50.0	50.0	56.8	56.0	56.7	55.0
Division 73. Transport Equipment							
German Democratic Republic:							
— OECD statistics [a]	13.3	32.1	11.8	25.2	25.3	30.1	24.5
— national statistics [b]	. .	. .	34.6	4.7	. .	. .	. .
Czechoslovakia	34.6	39.1	23.0	20.9	23.7	22.3	23.7
Poland	44.4	31.2	41.9	36.5	34.6	50.5	54.1
Hungary	12.5	16.7	11.8	11.6	10.3	13.5	13.9
Imports from OECD Countries (Section 7 = 100)							
Division 71. Machinery							
German Democratic Republic							
OECD statistics	45.5	55.4	74.2	68.2	57.4	51.1	60.0
Czechoslovakia	71.7	73.6	78.2	73.9	74.7	73.1	75.2
Poland	74.5	71.7	67.2	69.4	66.1	69.3	70.0
Hungary	63.9	62.7	63.6	70.8	69.1	69.2	69.7
Division 72. Electrical Machinery, Apparatus and Appliances							
German Democratic Republic							
OECD statistics	45.5	16.1	11.3	12.7	10.4	9.7	13.4
Czechoslovakia	21.7	19.0	14.8	19.3	18.9	20.7	18.9
Poland	21.7	22.1	19.5	14.5	14.6	17.6	18.2
Hungary	27.8	23.9	23.6	21.3	23.8	21.5	21.7
Division 73. Transport Equipment							
German Democratic Republic							
OECD statistics	9.1	28.6	13.8	18.6	32.4	39.0	26.6
Czechoslovakia	6.7	9.1	7.0	6.9	6.5	6.2	6.0
Poland	3.8	6.8	13.3	16.1	19.2	13.1	11.9
Hungary	8.3	13.4	13.6	8.1	7.4	9.5	8.8

a) Not including trade between the German Democratic Republic and the Federal Republic of Germany.
b) Trade with non-Socialist countries (it is not possible to establish whether this includes or excludes trade between the German Democratic Republic and the Federal Republic of Germany).
c) 1971.
Source: Annex. Tables A-16. A-17. A-18. A-19. A-21. A-22. A-24. A-25.

high since 1965 [mainly due to her imports of railway rolling stock][32] but it is fairly low in Czechoslovakia reaching only 6-7 per cent of total machinery imports. In Poland and Hungary the reverse is true. In Poland, the share of transport equipment has increased considerably (from 3.8 per cent in 1961 to 19.2 per cent in 1975, though falling to 11.9 per cent in 1977). In Hungary, where motor vehicles are included in the figures, the trend is down.[33] Points to note are that the Central European industrialized countries' imports of transport equipment are far less unstable than those of the Soviet Union and also that, in relative terms, more vehicles and railway rolling stock are imported.[34] Overall, however, the seven CMEA countries' transport equipment imports from the OECD countries are subject to severe fluctuation: 6.8 per cent of the total in 1961, 26.6 per cent in 1965, 9.4 per cent in 1973, 18.2 per cent in 1975 and 12.5 per cent in 1977.[35]

32   $72 million in 1975 or 21.4 per cent of all machinery imports from the OECD countries (not including trade between the German Democratic Republic and the Federal Republic of Germany); *Statistics of Foreign Trade,* "Trade by Commodities, Market Summaries", Series C, Vol. III, January-December, 1975, OECD, Paris, p. 439 and Annex, Table A-16.
33   See Annex, Table A-25.
34   See Table 16; and Annex, Tables A-16, A-18, A-21, A-24.
35   See Annex, Table A-11.

Chapter 3

THE FORMS OF TECHNOLOGY TRANSFER

1. INTER-GOVERNMENTAL AGREEMENTS ON SCIENTIFIC AND TECHNICAL CO-OPERATION

Inter-governmental scientific and technical agreements between East and West constitute a very wide field of study and clearly call for specific research. Many Western countries have entered into such agreements and those signed with the Soviet Union include the following:[1]

France:
Agreement of 30th June, 1966 replaced by the 10 year agreement of 27th October, 1971, supplemented by the agreements of 10th and 27th July, 1973.

United States:
Eleven agreements signed on 23rd and 24th May, 1972, 19th June, 1973, and 21st and 28th June, 1974.

United Kingdom:
Agreement of January 1968 supplemented in April 1974.

Italy:
Agreement of April 1966 amended in February 1974.

Sweden:
Agreement of 1970.

Germany:
Agreement negotiated in 1973 now signed. May 1978, 25 year general agreement for economic co-operation signed with the USSR.

Finland:
Agreement of 16th August, 1955. Ten-year programmes of 16th October, 1974 and 28th October, 1975. Fifteen-year programme of 18th May, 1977.

Canada:
Agreement of 1971.

Japan:
Agreement of October 1973.

Australia:
Agreement of January 1975.

East European countries apart from the Soviet Union have also entered into agreements on scientific and technical co-operation with the industrialized countries of the West.[2] In 1973 over 80 such agreements were in force.[3] These agreements are usually signed for a period of five to ten years and in most cases supplemented by annual protocols. The

---

1 "Les relations commerciales de la France avec les pays de l'Europe de l'Est à commerce d'Etat", *Journal officiel, op. cit.,* Paris, 22 octobre 1976, p. 1031; Lawrence H. Theriot, "Governmental and Private Industry Co-operation with the Soviet Union in the Fields of Science and Technology", in *Soviet Economy in a New Perspective, op. cit.,* pp. 740, 752-755. See also Chapter 1.
2 J. Wilczynski, *Technology in Comecon, op. cit.,* p. 298.
3 *Ibid.,* p. 298.

agreements are often administered by a joint Commission with a Minister presiding on either side. In many cases, too, provision is made for working parties to be set up and in some cases, co-operation in the industrial, economic or even cultural as well as scientific and technical co-operation fields is included.[4]

Since 1973, the number of East/West long-term bilateral commercial, economic, scientific and technical co-operation agreements has continued to increase. In 1978 they totalled 250, 83 for commercial co-operation and 155 for economic, industrial, scientific and technical co-operation.[5] The agreements are becoming more complex, an example being that signed between the Soviet Union and Germany in 1978. To illustrate the fields covered by these agreements we shall now look at those which France and the United States have with the Soviet Union. The first French agreement[6] on *scientific, technical and economic co-operation,* signed on 30th June, 1966, was amended on 27th October, 1971, by a ten-year agreement on *economic, technical and industrial co-operation,* and then supplemented by two other agreements — one dated 10th July, 1973 and incorporating a *programme for Franco-Soviet development and co-operation in the economic and industrial field for a period of ten years.* Another, dated 27th July, 1973, sets out a *programme for furthering Franco-Soviet co-operation in the scientific and technical field* for a period of ten years.

Thus, co-operation between France and the Soviet Union in the scientific and technical field has, since 1973, been the subject of specific agreements, separate from those of industrial co-operation. The agencies implementing both kinds of agreements, however, are the same, being those established in the 1971 agreement.

All Franco-Soviet economic, scientific and technical co-operation is governed by the joint *Franco-Soviet Commission,* known as the "Grande commission" which meets alternately in Paris and in Moscow at least once each year. The French Delegation is led by the Minister for Economic and Financial Affairs and includes the Ministers for Foreign Trade and for Industry and Research together with senior government officials. The Soviet Delegation is led by the President of the USSR Committee for Science and Technology (who is also Vice-President of the USSR Council of Ministers) and includes the Minister for Foreign Trade, the Vice-President of the Commission on the State Plan, the President of the State Committee for Atomic Energy, etc.

The "Grande commission" formulates general guidelines for co-operation but leaves specific dealings to the "Petite commission" (the joint Franco-Soviet Commission for Scientific, Technical and Economic Co-operation) which also meets alternately in Paris and Moscow once each year. The French Delegation is led by the Director for Economic and Financial Affairs in the Ministry for Foreign Affairs and the Soviet Delegation by one of the Vice-Presidents of the Committee of State for Science and Technology. The "Petite commission" leads and co-ordinates the activity of the "sectoral co-operation groups" (14 joint groups, each of which deals with one industrial sector and three groups with general responsibilities) which meet regularly and decide on co-operation programmes including the arrangements of specialised sub-groups, meetings, exchanges of experts, missions, etc. The sectoral groups also deal with scientific and technical co-operation in their respective fields.

The Franco-Soviet agreement on scientific and technical co-operation also designates the main fields of application. The field of basic and applied science primarily relates to nuclear energy, space research, environmental protection and joint projects on colour television, theoretical low and high temperature physics, semi-conductors, medicine and public health, the development of new types of plant and new animal breeds, etc. Technical co-operation in the agreement extends to various industrial sectors and specifies products and new processes.

4   *Ibid.,* pp. 297-298.
5   *Economic Bulletin for Europe,* United Nations, Vol. 30, No. 1, Pre-publication, p. 46.
6   «Les relations commerciales de la France avec les pays de l'Europe de l'Est à commerce d'Etat», *op. cit.,* pp. 1031-1035 and 1066-1070.

The work of the Franco-Soviet "Grande" and "Petite" commissions has had a fairly wide impact in France and newspapers and periodicals have devoted a large number of articles and much comment to it. However, no general study of the results of this co-operation has yet been published.

Far more information is available on governmental agreements on scientific and technical co-operation between the USSR and the United States. The United States Congress has some excellent publications which reproduce the content of these agreements and analyse the results.[7]

Thus far, the United States and the USSR have entered into eleven separate agreements on scientific and technical co-operation in the following fields:

— scientific and technical co-operation (24th May, 1972);
— co-operation in environmental protection (23rd May, 1972);
— co-operation in medical sciences and public health (23rd May, 1972);
— co-operation in space (19th June, 1973);
— co-operation in agriculture (19th June, 1973);
— co-operation in transportation (19th June, 1973);
— co-operation in oceanographic research (19th June, 1973);
— scientific and technical co-operation in the peaceful uses of nuclear energy (21st June, 1973);
— co-operation in energy (28th June, 1974);
— co-operation in research on the artificial heart (28th June, 1974);
— co-operation in housing and other construction (28th June, 1974).

The main agreement between the two countries is that on scientific and technical co-operation and constitutes the formal framework and model for ten other agreements. As noted in Chapter 1, Article 4 of the agreement is the most important commercially. Section 2 of this article states that all exports of American technology or equipment must comply with the legislation of the two countries or, in other words with the United States *Export Administration Act* (and Soviet legislation as well, of course).

Nine of the eleven US-USSR agreements on scientific and technical co-operation established Joint Commissions to implement them.[8] These Commissions meet at least once a year; in the interval the relevant responsibilities are exercised by the governmental bodies at ministerial or departmental level. In the United States, three government bodies (National Science Foundation, Department of State and Department of Commerce) have special responsibilities in the implementation of these agreements. On the Soviet side, responsibility lies mainly with the State Committee for Science and Technology, the foreign trade agency *Vneshtechnika* and the specialised agency that deals with licences, *Licencintorg*.

The Joint Commissions have handled the general administration of the agreements and the preparation of joint research projects. Near the end of 1975, 206 joint projects had already been approved and 89 were under preparation. The large range of scientific and technical problems will be noted, but obviously only the specialists concerned are in a position to assess the true scope of their respective areas. It may be noted that the Joint Commissions have also organised exchanges of people. It is because of this arrangement that

7 See *inter alia: Background Materials on US-USSR Co-operative Agreements in Science and Technology*, Committee on Science and Technology, US House of Representatives, 94th Congress, 1st Session, Washington, November 1975; *Review of US-USSR Co-operative Agreements on Science and Technology, op. cit.*; Lawrence H. Theriot, *art. cit.*, pp. 739-766; *Technology Transfer and Scientific Co-operation between the United States and the Soviet Union: A Review*, prepared for the Subcommittee on International Security and Scientific Affairs, Committee on International Relations, Congressional Research Service, Library of Congress, 95th Congress, 1st Session, US Government Printing Office, Washington, D.C., 26th May, 1977.

8 Lawrence H. Theriot, *art. cit.*, p. 740.

the number of Americans visiting the USSR increased from 262 in 1972 to 910 in 1975, and the number of Russians visiting the United States from 246 in 1972 to 900 in 1975.[9]

US Government authorities have carried out or commissioned surveys among the American participants in order to assess the results of these exchanges and joint projects. Several meetings of the House of Representatives' Committee on Science and Technology have been devoted to this subject with the object of evaluating the pros and cons of these agreements. Recommendations have been made as a result of the meetings, and these recommendations together with some replies to questionnaires are quoted in published documents.

What strikes the reader of these documents — which represent only a small part of the work of assessment carried out in the United States — is the amount of thought, unequalled in the other countries of the West, given to these agreements on scientific and technical co-operation with the USSR. The explanation, of course, lies in the political importance of the agreements. The published documents include some highly critical comments, e.g., those on the bureaucratic nature of the arrangements, the lack of interest in the agreements, the inertia and lack of response on the part of the Soviet participants and also the infiltration of "idiot" police disguised as translators.[10] However, there are also some penetrating replies regarding the evaluation of specific projects.

The real scope of inter-governmental agreements on scientific and technical co-operation has yet to be defined since they are framework or enabling agreements, (agreements which facilitate, but not predicate the conclusion of private commercial contracts) mostly concluded at the request of CMEA countries and, very occasionally, of Western countries. They smooth the workings of the administration in CMEA countries and thus help Western firms to make contacts. According to the experts, the significance of the number of such agreements is purely relative. They constitute a favourable attitude and an expression of goodwill, and often depend on the political factors that create the need for such a climate. Nevertheless, when it comes to putting the agreements into effect, in terms of "practical business", bargaining is just as hard as if they did not exist. This explains the relatively small number of contracts which are signed within this framework.

2. INDUSTRIAL CO-OPERATION AGREEMENTS BETWEEN WESTERN FIRMS AND THEIR CMEA PARTNERS – GENERAL OUTLINE

a) Nature of the Contracts

In Chapter 2 it was mentioned that several types of technology transfer could not be measured: information transfer, improvement in skills, etc. The dimensions of transfers of this kind can be partly gauged by studying *industrial co-operation* contracts.

Generally speaking, industrial co-operation contracts can be defined as "constituting arrangements whereby industrial producers agree to pool some assets and jointly to coordinate their use in the mutual pursuit of complementary objectives".[11] "As a mechanism for the international co-ordination of relations among producers, co-operation is interme-

9 *Technology Transfer and Scientific Co-operation between the United States and the Soviet Union: A Review, op. cit.,* p. 13.
10 *Ibid.*
11 Carl H. McMillan, "East West Industrial Co-operation", *art. cit.,* p. 1178.

diate to the market and the firm. It is in a sense a hybrid, combining attributes of both of the other two mechanisms, to which it can be regarded as a basic institutional alternative".[12]

Definitions of what is meant by (or covered by) *industrial co-operation* agreements differ. Carl H. McMillan distinguishes three types of classification:

i)   *The Broad Definition* covers all the contracts listed in Annex, Table A-26. It is preferred by the USSR and has also been used by Paul Marer and Joseph C. Miller in their study for the United States Department of Commerce (see Annex, Table A-27). The broader categories used by the CMEA countries include all but sales of equipment for complete production systems or "turnkey" plant sales (category 1 of Table A-26).

ii) *The Intermediate Definition* used by the United Nations covers categories 6 to 14 of the Annex, Table A-26 and is also used by Carl H. McMillan. This definition covers the following contracts:[13]

—   licensing with or without payment in resultant products;[14]
—   supply of complete plants or production lines with or without payment in resultant products;
—   co-production and specialisation, including
     —   co-production of a final product by specialisation in components,
     —   specialisation within a range of final products,
     —   co-production and specialisation involving R & D only;
—   short and long-term sub-contracting;
—   joint ventures
     —   involving marketing only,
     —   involving production, marketing and R & D;
—   joint tendering or joint projects;
—   tripartite co-operation.

iii) *The Narrow Definition* covers categories 7 to 13 of Table A-26 only, excluding sub-contracting, sales of plants and full equipment with payment in the form of compensation and certain joint ventures.

b)   Relationship between the Form of Contracts and Transfer Effectiveness

The above definitions of industrial co-operation are mainly of statistical interest. They enable the scope and significance of various surveys to be compared. For technology transfer studies, however, another type of classification is required in order to group contracts according to transfer effectiveness. *The way in which know-how is transferred* rather than its content is what matters; in other words the way in which knowledge can be *effectively used* in the importing country. Forms of transfer, i.e. the content of contracts, are of paramount importance in this connection.

In assessing the scale of these transfers, a first distinction has been drawn between one-time and long-term enterprise relationships between East and West. Table 21 reproduces the simple model constructed by Eric W. Hayden and Henry R. Nau. The essential distinction proposed by Hayden and Nau between "one-time" and "long-term" transfers is

12   *Ibid.*, p. 1178.
13   *Analytical Report on Industrial Co-operation among ECE Countries*, United Nations, Geneva, 1973, pp. 7-15. See also Levcik and J. Stankovsky, *Industrielle Kooperation zwischen Ost und West, op. cit.*, p. 38.
14   The definition used by Carl H. McMillan is not generally accepted. The most accepted definitions of industrial co-operation involving licensing agreements are those which are connected with payments in resultant products. Other forms are usually considered as trade. This definition was accepted by the Economic Commission for Europe and by such countries as Sweden.

not the duration of the transaction but the attempt to continue the relationship beyond the production phase. According to these authors, a long-term relationship must include some combination of the following elements:

— a technical assistance agreement calling for the transfer of current and new technology developed during the lifetime of the agreement;
— the commitment of trademark rights;
— the buy-back of resultant products from the same industrial venture.

Table 21

Models of Enterprise Relationships in East-West Technology Transfer

	One-Shot	Long-term
Definition	Sale of equipment, licence and/or plant for cash or unrelated products	Provision of technical assistance and/or trademarks for cash and/or purchase of resultant products
Duration	Termination upon completion	Usually 5 to 15 years
Content	Contract for design, manufacture and delivery of equipment	Model A Technical assistance only Model B Technical assistance plus trademark rights Model C Technical assistance plus trademark rights plus resultant product purchase
Examples	Swindell-Dressler (United States) and the Kama River Truck Plant in the Soviet Union	Model A Contract with General Tire in Rumania Model B Honeywell contract in Poland Model C International Harvester and Singer contracts in Poland

Source: Eric W. Hayden and Henry R. Nau, "East-West Technology Transfer: Theoretical Models and Practical Experiences", *Columbia Journal of World Business,* Fall 1975.

Another distinction between forms of technology transfer has been made in terms of degree of activity in the seller's attitude.[15] The following forms of transfer are regarded as active and highly effective:

— supply of turnkey plants;
— licences with intensive apprenticeship;
— joint ventures;
— technical exchange with regular contact;
— training in advanced technologies;
— manufacture of equipment with know-how.

Such transfers as the provision of documentation and technical data, consultation and licences (with know-how) are regarded as *effective* and yet others, such as sales calls or propositions, are only *moderately effective.*

15 *An Analysis of Export Control of United States Technology – A DOD Perspective, op. cit.,* p. 6. This analysis is also reproduced in *Technology Transfer and Scientific Co-operation between the United States and the Soviet Union: A Review, op. cit.,* p. 68.

Non-active and *ineffective* transfers are those implying a passive attitude on the part of the seller such as:

— the sale of licences without know-how;
— the sale of products without provision for maintenance or operation;
— non-documented proposals;
— sales literature;
— fairs and exhibitions.

Other Western authors generally confirm these views. One example is the research by Jürgen Nötzold and Jiri Slama which investigates the various forms of technology transfer and their influence on the use, dissemination and development of the technology concerned.[16] The answers Nötzold and Slama give are qualified and point out the advantages and drawbacks of each type of transfer. Further in depth research on this subject would seem to be worthwhile.

c) General Outline of Industrial Co-operation Agreements

Very little is widely known about the number and contents of agreements signed by Western firms and their East European partners because of secrecy maintained on both sides. Carl H. McMillan has attempted to estimate those concluded in 1976, as shown in Table 22.

Some of the agreements, often concluded with the Soviet Union through the State Committee for Science and Technology, are also outline or enabling agreements. The highest number of substantive agreements were concluded by Yugoslavia, Hungary and Poland. While the USSR has concluded only 196 agreements — or 14.7 per cent of the total for Category II in Table 22, the specific value of these agreements is very high. Western observers estimate that their total value is much higher than that of all the other East European countries' co-operation agreements.[17]

The percentages shown in Table 22 lose much of their significance through the lack of any information about the value of industrial co-operation contracts. The same applies to the statistics that show a substantial increase in the number of agreements during the past few years. All that can be said is that the number of such agreements, based on the United Nations intermediate definition (see above), rose from 600 in early 1973 to 1,000 by the end of 1975[18] and that their value, according to United Nations estimates, is still about 3-4 per cent of total trade flows though probably within the range of 10-15 per cent of trade in manufactures.[19]

16   Jürgen Nötzold and Jiri Slama, "Der Transfer von Technologie zwischen den beteiligten Volkswirtschaften", *Osteuropa Wirtschaft*, No. 1, 1975, pp. 39-48.
17   Paul Marer and Joseph C. Miller, "US Participation in East-West Industrial Co-operation Agreements", *Journal of International Business Studies*, Fall-Winter 1977, p. 21.
18   Paul Marer and Joseph C. Miller, *art. cit.*, p. 24. The authors do not make a direct estimate of the corresponding figure for the end of 1976 on the basis of the statistics in Table 22 (which are known to them). Applying their own criteria to Carl H. McMillan's figure for the end of 1976 they arrive at 861 (not including Yugoslavia or joint marketing ventures in the West or scientific or technical agreements) and to the United Nations figure for the end of 1975 at 750 (not including Yugoslavia). It should be noted that Maureen Smith ("Industrial Co-operation Agreements: Soviet Experience and Practice" in: *Soviet Economy in a New Perspective, op. cit.*, p. 771) estimates the total number of industrial co-operation agreements concluded by the CMEA countries with the Western countries at about 1,000. According to the *Economic Bulletin for Europe, op. cit.*, p. 88, "Presently, the number of agreements [on economic co-operation] overstepped 1,000".
19   *Promotion of Trade Through Industrial Co-operation, Recent Trends in East-West Industrial Co-operation*, Note by the Secretariat, United Nations, Economic and Social Council, Economic Commission for Europe, Committee on the Development of Trade, TRADE/R. 373, 31 August 1978, p. 33.

Table 22
Estimated Universe of EWIC Agreements Concluded by Eastern Countries (as of end of 1976)[a]

Category of agrements[b]	Estimated number of agreement[c]	As percent of subtotals
I. General framework agreements with Western firms:		
Bulgaria	15	4.4
CSSR	4	1.2
German Democratic Republic	4	1.2
Hungary	5	1.5
Poland	14	4.1
Rumania	22	6.5
USSR	275	81.1
Yugoslavia	d	—
Total	339	100.0
II. Substantive nonequity cooperation agreements with Western firms:		
Bulgaria	56	4.2
CSSR	53	4.0
German Democratic Republic	26	2.0
Hungary	228	17.1
Poland	216	16.2
Rumania	77	5.8
USSR	196	14.7
Yugoslavia	478	35.9
Total	1 330	100.0
III. Joint equity ventures (by location):		
1. In those Eastern countries permitting foreign equity investment:		
Hungary	3	2.0
Poland	0	0
Rumania	6	3.9
Yugoslavia	144	94.1
Total	153	100.0
2. In the industrialised West:		
Bulgaria	18	3.8
CSSR	12	2.5
Germany Democratic Republic	12	2.5
Hungary	49	10.2
Poland	39	8.1
Rumania	27	5.6
USSR	58	12.1
Yugoslavia	265	55.2
Total	480	100.0
Grand Total	2 302	—

a) All agreements are between Eastern economic organisations and Western firms; no intergovernmental EWIC agreements are included. No information is available on Albania's participation in EWIC and it is presumed to be nil. Partners to the agreements included in this table are located in the following Western countries; Austria, Belgium, Canada, Denmark, Finland, France, Germany, Italy, Japan, Netherlands, Norway, Spain, Switzerland, United Kingdom and United States.

b) Definition of categories:
I. General agreements: Agreements or protocols with Western firms, establishing intent to cooperate and broad conditions for cooperation, but not containing specific terms of cooperation. Thus there are "umbrella" agreements under which substantive, follow-up agreements are concluded between the partners.
II. Specific agreements: Agreements establishing relationships covered by type 7 through 13 of Table A-26.
III. Agreements of type 14.

c) With a few exceptions, the figures given are the total numbers of cases individually documented by the "East-West Project", Institute of Soviet and East European Studies, Carleton Universtiy, Ottawa. Documentation of these cases is based on lists of agreements provided by official and non-official organisations (in East and West) and on published reports, adjusted to comply with the intermediate definition (types 6 through 14 of Table A-26). Exceptions are: (a) Estimates of framework agreements concluded by the USSR which are based in part on figures published by the USSR, State Committee for Science and Technology, (b) those portions of the estimates representing agreements with US partners, which are based on the findings of an Indiana University research project, adjusted to the ECE definition (Marer, et al., 1975, Ch. 4). (c) the figures for Yugoslavia under categories II and IIIA are based on an official Yugoslav submission to the United Nations and are not fully comparable with the other country figures in this table, since they do not conform exactly to the ECE definition and represent agreements with foreign partners generally, as of mid-1976. (UN/ECE, Trade/AC.3/R.8 and 2, Sept. 24, 1976). The estimate for Yugoslavia under category IIIB is based on several secondary Western and Eastern sources. (See McMillan, "Direct Soviet and Eastern European Investment in the Industrialized Economies", Working Paper No. 7, Institute of Soviet and East European Studies, Carleton University, Ottawa, February, 1977, p. 6).

d) Not available.

Source: Carl H. McMillan, "East-West Industrial Cooperation", in *East European Economies Post-Helsinki, op. cit.*, p. 1186.

These gaps in information are filled to some extent by surveys of East-West co-operation agreements based on varying sizes of sample and including those made by the United Nations Economic Commission for Europe, Carleton University of Ottawa, the joint survey made by the Bureau d'informations et de prévisions économiques (BIPE) and the Ecole des hautes études en sciences sociales (EHESS), in Paris.

The United Nations has been making surveys since the 1970s and the size of the sample (not including Yugoslavia) has increased from 202 contracts in 1972 to 658 on 1st June, 1976 and to 314 at the end of September 1978 (see Table 23).

The UN, itself, points out the limitations of these surveys. Namely, that the value of the contracts selected is not known and that the various countries are not sufficiently well represented. The number of contracts known to have been concluded by the German Democratic Republic, or, in some degree, by Bulgaria and Czechoslovakia, is so small that the information can hardly be regarded as representative.[20]

It is clear from Table 23 that the most widespread form of industrial co-operation is specialisation on either side in co-production ventures.[21] In 1978, this type of co-operation accounted for 45.2 per cent of all East-West co-operation agreements (37.6 per cent if Yugoslavia is included), 63.7 per cent relating to Bulgaria, 57.4 per cent to the USSR, 44.7 per cent to Hungary and 37 per cent to Poland.

Next comes the supply of turnkey plants, whose share tended to diminish between 1976 and 1978 when it amounted to only 17.4 per cent for the seven CMEA countries (not including Yugoslavia): 71.4 per cent involving the German Democratic Republic, 27.2 per cent the USSR and 24.1 per cent Poland.

It would be incorrect however, to conclude from this that specialisation and co-production are relatively more important than the supply of plants. In 1976, according to United Nations estimate, specialisation and co-operation agreements represented over 80 per cent of the value of industrial co-operation contracts of certain countries.[22] Their number, on average, for the seven countries (not including Yugoslavia) represents however only 20.5 per cent of the total (Table 23).

A further point is that the statistics on joint ventures given in Table 22 do not accurately reflect their relative importance because they include contracts located in both East and West and agreements relating to production as well as those concerned primarily with marketing.

The decline suggested by Table 23 in the supply of licence contracts in return for products manufactured in the plants with the equipment or under the licences concerned (28.2 per cent in 1972, 17.1 per cent in 1976 and 6.1 per cent in 1978 for the seven CMEA countries less Yugoslavia) may also be more apparent than real. The point is that licences are often supplied as a part of large-scale plant supply or joint venture contracts and are not recorded separately.

20   In the sample for the 1st June, 1976 (not including Yugoslavia) the Soviet Union accounted for 29.8 per cent, Hungary 25.2 per cent, Poland 18.8 per cent and Rumania 14.9 per cent of the total number of contracts, making a total of 88.7 per cent for the four countries. On the Western side, Germany accounts for 21.4 per cent of the total (not including Yugoslavia), France 14.4 per cent, Italy 9.1 per cent, Japan 7.4 per cent, Austria and the United Kingdom 7 per cent each and Sweden 6.2 per cent, or 86.2 per cent for the seven countries. *Statistical Outline of Recent Trends in Industrial Co-operation* – Note by the Secretariat, ECE, Geneva, 14th November, 1977, *op. cit.,* p. 2.

In the sample for late September 1978, Bulgaria accounted for 3.5, Czechoslovakia for 2.9 and the German Democratic Republic for 2.2 per cent of contracts. The USSR accounted for 41.1, Hungary for 24.2, Poland for 17.2 and Rumania for 8.9 per cent. *Promotion of Trade Through Industrial Co-operation – Statistical Outline of Recent Trends in Industrial Co-operation, op. cit.,* Figure 1.

21   According to some Swedish experts consulted, real specialisation agreements between the partners, in the true sense of the term, in East-West industrial cooperation are rare. This seems to be especially the case for the Soviet Union which over-estimates this particular form of industrial cooperation.

22   *Promotion of Trade Through Industrial Co-operation – Recent Trends in East-West Industrial Co-operation, op. cit.,* p. 6.

Table 23
Classification of East-West Industrial Co-operation Agreements
Percentages

Country	Total	Supply of licence[a]	Delivery of plant	Specialisation	Subcontracting	Joint venturing	Joint tendering or joint projects	Tripartite co-operation
Total CMEA Countries (not including Yugoslavia)								
1972	100.0	28.2	11.9	37.1	7.9	14.9	—	—
1975	100.0	26.1	21.7	33.3	6.8	2.9	9.2	—
1st June, 1976	100.0	17.1	20.5	39.3	7.4	10.5	5.1	1.1
End September, 1978	100.0	6.1	17.4	45.2	3.8	16.9	4.2	6.4
Total CMEA Countries (including Yugoslavia)								
1975	100.0	24.2	18.7	30.2	5.9	13.6	7.4	—
1st June, 1976	100.0	15.6	17.6	32.8	6.4	22.4	4.4	0.8
End September, 1978	100.0	6.1	13.3	37.6	2.7	31.3	3.1	5.9
By Country								
Bulgaria, 1975	100.0	62.5	25.0	12.5	—	—	—	—
1st June, 1976	100.0	17.1	25.7	31.4	11.4	3.0	11.4	—
End September, 1978	100.0	—	—	63.7	9.1	27.3	—	—
Czechoslovakia, 1975	100.0	35.7	—	28.6	7.1	—	28.6	—
1st June, 1976	100.0	27.3	—	22.7	9.1	18.2	22.7	—
End September, 1978	100.0	22.2	—	11.1	—	66.7	—	—
German Democratic Republic, 1975	100.0	—	—	66.7	—	—	33.3	—
1st June, 1976	100.0	—	23.5	14.2	7.1	24.6	9.2	—
End September, 1978	100.0	—	71.4	28.6	—	—	—	—
Hungary, 1975	100.0	32.3	13.2	44.1	7.4	—	3.0	—
1st June, 1976	100.0	29.5	16.3	32.6	9.6	10.2	1.2	0.6
End September, 1978	100.0	2.6	1.3	44.7	9.2	17.1	5.2	19.7
Poland, 1975	100.0	25.0	16.1	37.5	3.6	1.8	16.0	—
	100.0	21.7	21.2	23.2	6.4	6.4	8.0	1.0

1st June, 1976	100.0	19.4	25.5	14.2	7.1	24.6	9.2	—
End September, 1978	100.0	—	3.6	14.3	7.2	67.9	—	7.1
USSR, 1975	100.0	—[b]	56.6	34.8	4.3	—	4.4	—
1st June, 1976	100.0	3.2	20.4	61.5	4.7	7.1	1.6	1.5
End September, 1978	100.0	2.4	27.2	57.4	—	8.6	3.9	0.8
Yugoslavia, 1975	100.0	18.5	9.2	20.0	3.0	47.7	1.5	—
1st June, 1976	100.0	9.5	6.5	11.9	2.4	68.6	1.1	—
End September, 1978	100.0	6.1	3.1	19.2	—	66.2	0.8	4.6

a) In exchange (in part at least) for products or components.

b) It is clear from Tables A-29 and A-30 that the Soviet Union has signed a large number of agreements covering the supply of licences and it would therefore appear that these agreements do not include repayment in the form of resultant products. Such contracts are therefore not included in the sample studies by the Economic Commission for Europe.

Sources:
1972: Analytical Report on Industrial Co-operation among ECE Countries, op. cit., p. 6 (French edition). Based on 202 cases of industrial co-operation.
1975: Proceedings of the UN/ECE Seminar on the Management of the Transfer of Technology with Industrial Co-operation. Geneva. 14th-17th July. 1975. Economic Commission for Europe. ECE/SC TECH./10. 16th February. 1976. p. 20. Based on 275 cases of industrial co-operation (not including Yugoslavia).
1st June 1976: Aperçu statistique de l'évolution récente de la coopération industrielle. Note by the Secretariat. Geneva. 14th November. 1977. Economic Commission for Europe Committee on the Development of Trade. TRADE/R.355/Add. 2. Annex, Table 3. Based on 826 contracts (658 not including Yugoslavia).
End September. 1978: Promotion of Trade Through Industrial Co-operation. Statistical Outline of Recent Trends in Industrial Co-operation. Note by the Secretariat. United Nations Economic and Social Council. Economic Commission for Europe. Committee on the Development of Trade. TRADE/R.375/Add. 5. 19th October. 1970. p. 8.

Perce

Country	Total	Chemical industry	Metallurgy	Transport equipment	Machine tools	Mechanica engineering
		2	3	4	5	6
Bulgaria						
1/6/1976	100.0	25.0	5.7	14.3	2.9	28.5
End September, 1978	100.0	—	—	9.1	—	45.5
Hungary						
1/6/1976	100.0	17.5	3.6	13.3	4.8	20.5
End September, 1978	100.0	14.5	2.6	11.8	3.9	13.2
Poland						
1/6/1976	100.0	23.4	11.3	10.5	4.8	23.4
End September, 1978	100.0	16.7	7.4	9.3	1.9	25.9
German Democratic Republic						
1/6/1976	100.0	23.5	23.5	6.0	17.6	17.6
End September, 1978	100.0	14.3	—	14.3	28.6	28.6
Rumania						
1/6/1976	100.0	20.5	7.1	22.5	—	30.6
End September, 1978	100.0	39.3	10.7	21.4	—	10.7
Czechoslovakia						
1/6/1976	100.0	22.7	—	22.7	9.1	36.4
End September, 1978	100.0	33.3	—	11.1	22.2	22.2
USSR						
1/6/1976	100.0	12.1	9.7	7.7	1.5	15.8
End September, 1978	100.0	36.4	13.2	5.4	3.9	16.3
Total						
1/6/1976	100.0	23.8	7.9	12.6	3.5	22.0
End September, 1978	100.0	26.1	8.3	9.6	4.1	18.2
Yugoslavia						
1/6/1976	100.0	22.5	7.0	20.1	2.3	11.3
End September, 1978	100.0	20.8	6.2	15.4	2.3	10.8
Total						
1/6/1976	100.0	23.6	7.7	14.1	3.3	19.8
End September, 1978	100.0	24.5	7.7	11.3	3.6	16.0

Sources:
1st June 1976: *Statistical Outline of Recent Trends in Industrial Co-operation.* Note by the Secretariat. Economic Commission f
Europe. Geneva. 14th November. 1977. *op. cit.,* Annex. p. 2.
End September 1978: *Promotion of Trade Through Industrial Co-operation. Statistical Outline of Recent Trends in Industri
Co-operation.* Note by the Secretariat. Geneva. 19th October. 1978. *op. cit.,* p. 7.
1 Food and agriculture industry (including beverages):
2 Chemicals industry (including pharmaceuticals):
3 Metallurgy (including mining):

| | Comprising: | | | | | | |
| otal | Electronics | Electrical equipment | Total | Food and agriculture | Light industry | Total | Other branches |
+6	7	8	7+8	1	9	1+9	10
1.4	2.9	5.7	8.6	17.3	—	17.1	2.0
5.5	9.1	9.1	18.2	27.3	—	27.3	—
5.3	7.2	10.8	18.0	4.2	12.1	16.3	6.0
7.1	11.8	21.1	32.9	5.3	11.8	17.1	3.9
8.3	9.0	5.6	14.6	1.6	4.0	5.6	6.4
7.8	13.0	1.9	14.9	5.6	13.0	18.6	5.6
5.2	—	5.9	5.9	5.9	—	5.9	—
7.2	—	—	—	14.3	—	14.3	—
0.6	6.1	2.0	8.1	4.1	3.0	7.1	4.1
0.7	10.7	—	10.7	—	7.1	7.1	—
5.5	—	9.1	9.1	—	—	—	—
4.4	—	11.1	11.1	—	—	—	—
7.3	9.7	3.6	13.3	7.2	7.6	14.8	5.1
0.2	9.3	3.1	12.4	2.3	6.2	8.5	3.9
5.5	7.5	6.0	13.5	5.2	6.5	11.7	5.0
2.3	10.2	7.3	17.5	4.5	8.3	12.8	3.5
3.6	3.0	8.3	11.3	8.9	13.6	22.5	3.0
3.1	5.4	13.8	19.2	6.9	13.8	20.7	4.6
3.1	6.5	6.5	13.0	6.0	7.9	14.0	4.6
.6	8.8	9.2	18.0	5.2	9.9	15.1	3.8

4 Transport equipment: includes aircraft. automobiles. lorries. tractors (even for agriculture). rolling stock. earth-moving pment. diesel engines (even stationary):
5 Machine-tools:
6 Mechanical engineering (all other non-electrical engineering):
7 Electronics (computers and other office equipment. radio and television sets. communication equipment):
8 Electrical equipment (all other including electric locomotives and household appliances):
9 Light industry (textiles. footwear. rubber. glass. furniture. consumer goods):
10 Other. such as construction. hotel management. tourism. etc.

Other forms of East-West industrial co-operation are of only secondary importance, although the recent increase in the share of tripartite co-operation agreements should be noted (see Table 23).

Table 24 shows the breakdown of co-operation contracts by industry based on the United Nations surveys of 1st June, 1976 and end September 1978. As the authors point out, industrial co-operation is centred mainly on those industries where technological progress is of prime importance. In 1978, the chemical industry accounted for 26.1 per cent of signed agreements (not including Yugoslavia), mechanical engineering and machine-tools 22.3 per cent, electrical and electronics industries 17.5 per cent, transport equipment 9.6 per cent and metallurgy 8.3 per cent totalling 83.8 per cent of the sample (see Table 24).

The survey conducted by Carl H. McMillan of Carleton University, covers 218 industrial co-operation agreements in force in early 1975. It uses the same definition as the United Nations survey (mainly category II of Table 21) and its results are therefore comparable.[23] But it is original in its approach to the extent that the agreements are not classified in only one but rather in several ways.[24] The aspects of the agreements most closely linked with technology transfer are identified in detail: plant or equipment design, staff training, technical assistance, licensing, and R & D co-operation. According to Carl H. McMillan, East-West industrial co-operation agreements are potent media for technology transfer to Eastern Europe. Nearly two-thirds of the agreements surveyed include provisions for technical assistance and about half of them provisions for licensing and staff training. These three items are found in 30 per cent of the agreements comprising the sample. Other calculations show that 75 per cent of the agreements in the sample contained at least one of the major technology transfer media (items 4-7 and 16 of Table 26).[25]

Carl H. McMillan estimates that industrial co-operation agreements are less important as a way of obtaining capital goods[26] than as a means of acquiring Western technology, especially applied industrial technology. Industrial co-operation agreements go beyond procurement; they set up or consolidate a system of technical and commercial links which provide continuous access to the other party's technology through which new non-standardized technology, not generally accessible on the market, can be obtained. The reason for this is that the Western partners in such agreements continue to have an interest in the operational and commercial application of the technology transferred.

A more recent general survey was made in 1978 by Jean Cheval, François Gèze and Patrick Gutman of the BIPE and the EHESS in Paris. The authors' main object was to see whether the activity of the leading Western countries in East-West industrial co-operation mirrored the hierarchy of production systems in the West. In the survey, 474 industrial or pre-industrial co-operation contracts were picked, from the initial sample of 800, corresponding to the terms of contract under study. Unfortunately, the authors took no account of contract value, simply stating that this was known for 90 per cent of the French contracts but unknown in the case of most foreign contracts.[27]

The main value of this study is that it outlines the part played recently in industrial co-operation by six of the leading Western countries: France, Germany, United States, Japan, Italy and United Kingdom (Table 25). The relative importance of turnkey plants is very clear (53 per cent of the sample) which is closer to their approximate relative

23   The United Nations nonetheless include Category II.1. of Table 22, which is studied separately by Carl H. McMillan. However, there are very few joint ventures in Hungary and Rumania so this difference does not affect the results. Carl H. McMillan, *art. cit.*, p. 1187.

24   For instance, in specialisation agreements McMillan makes a distinction between technical assistance, licences, staff training, etc.

25   See Carl H. McMillan, *art. cit.*, p. 1189.

26   *Ibid.*, pp. 1192-1193. According to this source, 20 to 30 per cent of the agreements in the UN and Carleton University samples include equipment or plant transfer.

27   Jean Cheval, François Gèze and Patrick Gutman, *Les accords de coopération industrielle Est-Ouest,* Report presented at the colloquium of the DGRST, "Division internationale du travail", Saint-Maximin, 19th-20th October 1978, p. 18.

Table 25

Shares of the Leading Western Countries in the Various Forms of Co-operation and the Supply of Capital Goods

	Grants of licence	Grants of licence or know-how with the relevant equipment	Total	Sub-contracting	Turnkey plants	Co-production	Joint Ventures	Long-term co-operation agreements	Co-operative manufacture	Total in per cent	Total number of contracts	Contracts relating to		
												Capital goods	Inter-mediates	Consumer goods
	%	%	%	%	%	%	%	%	%	%				
	1	2	1 + 2	3	4	5	6	7	8	9	10	11	12	13
France	15	14	15	44	31	14	5	19	(40)	25	117	34	58	25
Germany	38	29	35	44	20	36	37	36	(60)	28	131	57	47	27
United States	18	36	23	—	6	23	21	19	—	12	59	27	14	18
Japan	4	11	6	—	24	9	16	6	—	16	76	21	39	16
Italy	8	4	7	—	12	5	5	19	—	11	53	12	23	18
United Kingdom	15	7	13	11	7	14	16	—	—	8	37	12	17	8
Total	100	100	100	100	100	100	100	100	100	100				
Number of contracts in sample	71	28	99	9	253	22	19	67	5	474	473[a]	163	198	112

a) The source bases the breakdown by forms of co-operation on 474 contracts and the breakdown by type of goods supplied (capital, intermediate or consumer goods) on 473.

Source: Jean Cheval, François Geze, Patrick Gutmann. *Les accords de coopération industrielle Est-Ouest*. Report presented at the DGRST Colloquium: "Division internationale du travail". Saint-Maximin. 19th-20th October. 1978. pp. 21-25.

Table 26

Component Elements of EWIC Agreements[a]

Per cent of country's agreements surveyed containing designated element

	Country							Total all countries
	Bulgaria	CSSR	German Democratic Republic[b]	Hungary	Poland	Rumania	USSR	
Number of agreements surveyed	17	18	6	75	51	30	21	218
Element:								
1. Managerial services	11.8	5.6	0	10.7	7.8	10.0	4.8	8.7
2. Capital equipment sale	29.4	22.2	0	29.3	23.5	50.0	19.0	28.4
3. Complete plant sale	41.2	11.1	33.3	5.3	19.6	33.3	42.9	20.2
4. Custom design of plant equipment	23.5	27.8	16.7	13.3	23.5	40.0	28.6	22.9
5. Training of East personnel	58.8	61.1	16.7	45.3	47.1	56.7	23.8	40.8
6. Technical assistance (know-how)	58.8	66.7	33.3	60.0	62.7	60.0	57.1	60.1
7. Licence	47.1	50.0	16.7	44.0	54.9	45.7	47.6	47.2
8. Supply parts components to East partner	35.3	66.7	16.7	61.3	49.0	63.3	28.6	52.6
9. Provision by Eastern partner of components produced according to Western specifications and incorporated in Western product	35.3	66.7	0	44.0	54.9	60.0	23.8	46.8
10. Provision by Eastern partner of products according to Western specifications, to be marketed by Western partner	41.2	27.8	16.7	56.0	33.3	33.3	23.8	39.9
11. Production specialisation and exchange of parts/components so each partner produces same end product	23.5	33.3	0	20.0	15.7	23.3	9.5	19.3
12. Production specialisation and exchange so each partner disposes of full line of final goods	11.8	11.1	16.7	2.7	5.9	6.7	0	5.5
13. Quality control	23.5	22.2	0	38.7	19.6	26.7	0	25.2
14. Co-ordination of marketing/servicing	35.3	27.8	33.3	34.7	35.3	30.0	9.5	31.2
15. Joint project in third country	35.3	22.2	66.7	17.3	31.4	26.7	9.5	24.3
16. Joint R & D	23.5	5.6	33.3	26.7	17.6	20.0	47.6	23.9

a) The data presented are based on questionnaires completed in 1975 by Western firms partner to the agreements. Details of the survey and methodology are described in Carl H. McMillan, "Forms and Dimensions of East-West Inter-Firm Cooperation", in: *East-West Cooperation in Business: Inter-Firm Studies*, edited by G. T. Saunders, Wien-New York, 1977. This is a revised version of Appendix Table V of this article based on a slightly extended sample. Agreements included conform to the ECE definition.

b) In interpreting the percentages for the German Democratic Republic, the small absolute number of agreements included in the sample should be borne in mind.

Source: Carl H. McMillan, "East-West Industrial Cooperation", *art. cit.* p. 1190.

Table 27

Operating Industrial Co-operation Agreements of United States Firms with the USSR and East Europe by Country and Type of Agreements
Agreement or Projects Completed or Contracts in Force as of 1st January, 1976

Type of Agreement	USSR	East Europe							Country Unknown	Total
		Bulgaria	Czech.	German Democratic Republic	Hungary	Poland	Rumania	East Europe Sub. Tot.		
Scientific-Technical	62						1	1		63
Know-how	8	4	4		3	7	7	25		33
Licence, direct	17	9	15	6	16	26	16	88	25	130
Licence, indirect	3	1	2	1	1	9	3	17		20
Turnkey	74	6	10	5	5	22	23	71	3	148
Subcontracting	4			1	2	4		7		11
Co-production	6	2	2		5	6	2	17		23
Joint Venture					3		2	5	1	6
Total Number of Agreements	174	22	33	13	35	74	54	231	29	434[a]
Reverse Licence	21	2						2		23
Prospective agreements	102	33	12	5	30	54	47	181		283

a) Of these 70 are "indirect" agreements. 20 of which are licences.

Source: Paul Marer and Joseph C. Miller. "United States Participation in East-West Industrial Agreements". *Journal of International Business Studies*, Fall-Winter 1977. pp. 19-20.

importance in value (80 per cent) as estimated by the United Nations. The grant of licences also accounts for a higher percentage (21 per cent) and this, too, seems closer to the truth. Particular interest attaches to the survey's findings as regards contracts relating primarily to capital goods. These account for 34.5 per cent for the sample as a whole but for 46 per cent in the case of the United States, 44 per cent in that of Germany and 29 per cent in that of France. The authors conclude that their survey confirms the existence of a hierarchy of production systems in the West. The United States and Germany are in the dominant position and the four other countries in a subordinate position. The level of Eastern Europe countries is not specified. This kind of classification of production systems would seem to be interesting in studying East-West technology transfer but would seem to need more thorough research.

d) **Industrial Co-operation Agreements between East European Countries and Certain Specific Western Countries**

Apart from these general surveys which, in principle, cover all Eastern and Western countries, some authors in the West have made a more particular study of relations between the CMEA countries and their own. The following paragraphs refer briefly to three such surveys produced in the United States, Sweden and France.[28]

The American survey was carried out at Indiana University by Paul Marer and Joseph C. Miller in co-operation with John B. Holt on behalf of the US Department of Commerce. Its main findings are given in Table 27. The sample on which it is based is narrower in some ways and broader in others. Narrower, because it only concerns US relations with Eastern Europe. Broader because it includes scientific and technical co-operation agreements with the USSR Committee for Science and Technology, know-how transfer and licensing, and licences granted by East European countries (see Table A-27). The Indiana University survey covers not only contracts in force on 1st January, 1976, but also prospective agreements.

Although based on a very large sample of 434 current and 283 prospective agreements, the survey is confined to agreements concluded by the United States. The authors point out that the United States has signed many more agreements since the early 1970s. As of 1st January, 1976 these represented about 25 per cent of all industrial co-operation agreements concluded with Eastern Europe.[29]

Scrutiny of the United States' co-operation agreements reveals a difference between the attitude of the Soviet Union and that of the other East European countries. In fifty per cent of the contracts between the USSR and the United States the object is capital investment, in 30 per cent the production of intermediates and in only 12 per cent consumer goods.[30] Table 27 shows that out of 112 co-operation agreements between the United States and the USSR, 74 are for turnkey projects.[31] In the other East European countries the share of turnkey projects is not more than one third. These countries are more interested in importing applied industrial technology and know-how. A possible explanation is the Soviet Union's wish to keep contacts between its citizens and Western experts to a minimum for political reasons: it is also easier to isolate plants geographically from one another and contacts can be kept short-lived.[32]

---

28   The German survey by Klaus Bolz and Peter Plötz is referred to later.
29   Paul Marer and Joseph C. Miller, "US Participation in Industrial Co-operation Agreements". *Journal of International Business Studies,* Fall-Winter 1977, p. 25.
30   *Ibid.,* p. 24.
31   The 112 do not include the 62 scientific and technical co-operation agreements, which are mostly enabling agreements.
32   Paul Marer and Joseph C. Miller, *art. cit.,* p. 25.

About 75 per cent of the co-operation agreements between East European countries and the United States are concluded with large, mostly multi-national firms.[33] The technology transferred often enables these countries to buy products and replacement parts on a continuous basis. American firms are preferred less because they have the monopoly of high technology than because they provide access to sources of finance (outside the United States as well), opportunities to continue Research and Development and production and marketing potential at world level.

Table 28

Difficulties Arising in East-West Industrial Co-operation Based on a Survey Among 53 Swedish Firms

		Number of Cases	Percentage of Cases:	
			Where the Difficulty Arises	Where the Difficulty Persists
1.	Technology Transfer			
	Documentation	42	31	23
	Training	33	45	26
	Production (assimilation)	36	38	38
2.	Quality	41	69	51
3.	Supply Problem			
	Planning of East-West deliveries	41	29	29
	Planning of West-East deliveries	32	19	19
	Performance of East-West deliveries	41	44	43
	Performance of West-East deliveries	32	10	7
	Performance of East-West purchases	41	28	16
	Performance of West-East purchases	32	6	2
	Delivery delays in the East-West direction	41	66	65
4.	Prices	41	42	30
5.	Marketing problems			
	Market-sharing and compliance with agreements in this field	28	26	22
	Problem of trade marks	28	4	4
6.	Management and Co-ordination			
	Co-ordination between partners	46	48	36
	Management and co-ordination in the East	46	55	51
	Management and co-ordination in the West	46	16	11

Source: B. Högberg, Andreas Adahl, *East-West Industrial Co-operation: A Survey of Swedish Firms, op. cit.*

The Swedish survey conducted by B. Högberg and Andreas Adahl covered 53 co-operation agreements, 47 of which were current in 1977. The breakdown was 24 agreements with Hungary, 16 with Poland, five each with the USSR and Czechoslovakia, two with the German Democratic Republic and one with Rumania. Of these, 45 per cent related to the mechanical engineering industry, 11 per cent to the electrical engineering industry, 9 per cent to metallurgy and 8 per cent to the transport equipment industry.

33 *Ibid.,* p. 21.

The value of the contracts is not shown but an excellent analysis is given of the difficulties arising in concluding and carrying out contracts. The main difficulties referred to by Swedish firms are summed up in Table 28. It is interesting to see the same difficulties reported whether persistent or otherwise (column 3 of Table 28).

The main problems seem to relate to product quality and slow delivery in the East-West direction. These difficulties seem to be of the persistent kind. In 26 per cent of cases, Swedish firms were able to improve quality but at a relatively high cost.[34] Difficulties with regard to deliveries from the East are a matter of inflexibility as regards quantities, dates and so on and also the simple failure to deliver. The reason is often the conflict between the various organisations in the East and between industry and foreign trade agencies.

The management and co-ordination problem is also acute both as regards co-ordination between West and East and on the Eastern side as such. Eastern partners often find it difficult to organise effective management and have no way of controlling their own sub-contractors.

Conflicts on price may be summed up as follows. To begin with, the Eastern organisations refused to include price index clauses but changed their mind after the oil crisis. Partners in the East have a rather vague notion of distribution costs and the prices of component parts. They also tend to underestimate the cost of documentation and have difficulty in training their staff in the West. This is not only because of the language problem but because persons other than those foreseen in the contracts are sent abroad. That is to say, instead of technicians, high level managers seize the chance of a holiday abroad. The considerable mobility of skilled personnel in the East also raises problems. In general, however, some improvement in technology transfer is perceivable.[35]

The French survey made in December 1978 by the "Centre d'études sur la coopération économique avec les Pays de l'Est" of the Aix-Marseille Applied Economics Faculty concerns chemical technology transfers from France to East Europe.[36] There are two aspects to the survey: an analysis of contracts and a survey of French exporters.

The former discusses the various contracts in the sample, the obligations of the parties to them, terms of payment, guarantees and penalties. A detailed analysis is made of exporters' replies giving particular attention to economic aspects, negotiation procedures, technical progress clauses, disputes, payment and compensation.

The study of this sample fills a gap in the United Nations surveys by evaluating the representativeness of the information quoted and weighting the contracts in the sample in relation to their value. The 192 agreements signed between 1970 and 1977 are listed and their value given — totalling F. 15,118 million. Ten contracts, amounting in all to F. 5,422 million, account for only 5.2 per cent of the total number of known chemical contracts — but 35.9 per cent of their total value.[37] Two contracts relating to gas drying plant and a natural gas purification plant, account for 69 per cent of the total value of the ten. Figures like these show the kind of mistake that can be made by treating each agreement as a unit in industrial co-operation statistics.

An in-depth analysis estimating the French and foreign share of the chemical plants supplied puts it at an average of 72 per cent although it is only 52 per cent for the biggest contract (F. 2,700 million) and somewhere between 82 and 99 per cent for all the others.[38]

34   B. Högberg, Andreas Adahl, *East-West Industrial Co-operation: A Survey of Swedish Firms,* n. d., 1978. To be published by the Swedish Board in Technical Development, Stockholm.
35   *Ibid.*
36   Henri Dunajewski, assisted by Michèle Brami and Gérard Double, *Etude des transferts de technologie chimique. Contrats France-URSS. Enquête France-Europe de l'Est.* Centre d'études sur la coopération économique avec les pays de l'Est, Aix-Marseille Applied Economics Faculty, Aix en Provence, December, 1978.
37   *Ibid.,* pp. 3-4 and 82-88.
38   *Ibid.,* p. 90.

e) Industrial Co-operation Case studies Concerning Specific Firms and Industries

The study by Henri Dunajewski at Aix-en-Provence shows clearly that detailed case studies are the only way to arrive at precise conclusions.

It is, unfortunately, not possible to refer to them all in this report. In this context, two recent United Nations surveys are of interest.[39]

This is not the place for a detailed survey of the various types of industrial co-operation, but the main forms of technology transfer are briefly reviewed below:[40]

— private enterprise agreements on scientific and technical co-operation;
— licensing transactions;
— turnkey plants;
— co-production and specialisation;
— joint ventures;
— tri-partite co-operation;
— sub-contracting, joint tendering or joint projects.

3. PRIVATE ENTERPRISE AGREEMENTS ON SCIENTIFIC AND TECHNICAL CO-OPERATION

As already stated, some of the co-operation agreements signed by Western private firms are enabling agreements.[41] Most are with the Soviet Union. Generally speaking these agreements (see Annex Table A-28) specify the field of co-operation, the form of co-operation (exchanges of specialists, joint seminars and so forth), financial costs and secrecy obligations. It is usually specified that all relevant legal, financial and commercial questions are to be covered by separate contracts.[42]

Unfortunately we have not found any comment in the sources consulted on the nature of these agreements entered into with non-American firms. An annex to the German survey by Klaus Bolz and Peter Plötz[43] does contain a special questionnaire sent to German firms

---

39 *Promotion of Trade Through Industrial Co-operation. The Experience of Selected Western Enterprises Engaging in East-West Industrial Co-operation: Results of a Survey of Fifteen Forms in the Machine-Tool Sector.* United Nations, Economic and Social Council, Economic Commission for Europe, Committee on the Development of Trade, TRADE/R 375, Add. 4., 14 August, 1978. *Promotion of Trade Through Industrial Co-operation. Case Studies in Industrial Co-operation: Results of a Survey of Five Western Enterprises,* United Nations, Economic and Social Council, Economic Commission for Europe, Committee on the Development of Trade, TRADE/R. 373, Add. 3, 18 September, 1978.

40 Co-operation agreements among research institutions are of course also an important form of technology transfer. This is particularly true for research institutions engaged in applied industrial research. The authors are aware that the forms discussed here are not exhaustive.

41 See Category I, Table 22.

42 It should be noted that Table A-28 mentioned above includes 53 US firms which have signed so-called Article 4 Agreements which are equivalent to letters of intent to cooperate. However, only a fraction of these agreements have led to contracts and in all but a very small number of cases these contracts have been for commercial ventures in manufacturing and sales and not for scientific and technological cooperation. All exports of information or products in the framework of agreements on scientific and technical co-operation automatically imply the intervention of the Soviet organisation responsible for foreign trade, the responsible government ministry and the USSR State Committee for Science and Technology. On the American side, the usual export licences are also necessary. Lawrence H. Theriot, *art. cit.,* p. 750.

43 Klaus Bolz and Peter Plötz, *Kooperationserfahrungen der Bundesrepublik Deutschland mit den Sozialistischen Ländern Osteuropas,* Hamburg, October 1973, pp. 30-31. These authors estimate the number of agreements on industrial co-operation between Germany and all East European countries at about 300 (100 with Hungary, 90 with Poland and 60 with Rumania but "few with Czechoslovakia and the Soviet Union"). But the survey produced only 62 replies to the questionnaire plus 11 detailed letters and the authors used press reports in the case of about 200 co-operative projects. This source therefore fails to provide any particulars about scientific and technical co-operation although it estimates the share of such agreements at 19 per cent.

asking detailed questions about the nature of co-operation and progress. However the replies received from 62 firms are not analysed in detail with respect to these concerns.

The only comments obtained come from American firms via the Congressional publications previously referred to. Most of these firms regard this kind of co-operation as a way of penetrating the Soviet market. However, in the Symposium held on this subject in 1974 it was learned that firms entering into co-operation agreements are not the same as those that usually sign contracts. One of the recognised practical advantages of these agreements was their ability to facilitate the execution of travel formalities and provide opportunities to assess Soviet technological potential and sales prospects. In any case, the United States' authorities carefully consider the pros and cons of such exchanges.[44]

It is difficult to form an idea of the number of commercial contracts confined solely to scientific and technical co-operation. In late 1974 *Eastern Europe Report* gave a figure of over 200 such agreements signed by the USSR alone with Western firms.[45] This figure is much higher than that given by Wilczynski (80) for 1973[46] for the whole of East Europe, based on Soviet sources. It is probable that the total is now very much higher.

However, the best source on this subject — the United Nations — is not very informative. After stating in 1972 that agreements on scientific and technical co-operation accounted for 6.9 per cent of the 204 agreements concluded for that year[47] the United Nations 1975 survey covering 275 cases of industrial co-operation pays no further attention to this category.[48] The same comment applies to the United Nations survey of 1st June, 1976 (Table 23). All we know is that the share of these agreements in German industrial co-operation in 1973 was 19 per cent and that the corresponding figure for the United States in 1975 was 11.6 per cent.[49] The elimination of scientific and technical co-operation agreements from United Nations statistics in 1975 is somewhat surprising. Out of the 220 agreements on industrial co-operation recently entered into by the Soviet Union and set out in Annex Table A-31, at least 73 relate to scientific and technical co-operation and R & D.

The reason for the United Nations' decision would seem to be the difficulty of singling out transactions on scientific and technical co-operation — which form part of practically every form of industrial co-operation.

The relative importance (19 per cent) of agreements on scientific and technical co-operation entered into by German firms makes that country's experience particularly interesting. The fields covered are closely related to production and practical applications. In most cases the results of research are applied on a joint basis. German firms generally provide more in the way of equipment whereas their partners in East Europe provide more in the way of manpower.[50]

44 An inventory of 230 American firms made in 1973 by the National Science Foundation and the Bureau of the East-West Trade attempted to quantify the degree of reciprocity of agreements on scientific and technical co-operation (Lawrence H. Theriot, *art. cit.*, p. 751). It may also be noted that the House of Representatives' Sub-Committee on National and International Planning in the Field of Science (*Review of US-USSR Co-operative Agreements on Science and Technology, op. cit.*, pp. 2-3) recommends that projects in progress should be critically examined periodically to assess their effectiveness and reciprocal advantages, and that projects which have become inoperative due to failure in performance on the Soviet side should be promptly terminated as in normal business relations.
45 *Eastern Europe Report,* Business International, Geneva, 13 December, 1974.
46 See above.
47 F. Levcik and J. Stankovsky, *op. cit.*, p. 42.
48 See table 15.
49 F. Levcik and J. Stankovsky, *op. cit.*, p. 42 and p. 58, footnote (1).
50 F. Levcik and J. Stankovsky, *op. cit.*, p. 55-56.

4. EAST-WEST LICENSING TRANSACTIONS

a) Forms of Transfer

Licences may come in many different types of arrangements. They range from a straightforward authorisation to exploit an individual patent to complex agreements on industrial co-operation. The latter may include the provision of licences for the use of patents linked to the import of certain capital goods and to use know-how, technical assistance in building turnkey plants or other industrial installations and licences to use trademarks.[51]

The sale of a licence often constitutes the first step towards broader co-operation. If the purchaser of a licence supplies the seller with products obtained by using the new process, the signing of the first secondary contract is already in sight.[52]

In 1972, 28 per cent of East-West industrial co-operation agreements were based on licensing agreements; the figure in 1975 was 26 per cent, in 1976 17 per cent and only 6.1 per cent in 1978 (Table 23).

Some particulars about the various forms of agreement involving the sale of licences between the United States and the countries of East Europe are given in Table 27. Out of 434 co-operation agreements current or ended on 1st January, 1976, 183 were licence or know-how sales. Another survey on co-operation agreements between the United States and CMEA countries shows that out of the 188 agreements involving the sale of licences, 87 were payable in products and 32 in cash.[53] In 1975, American firms were the sellers of licences in 92 per cent of cases and buyers in only 8 per cent of cases. In most cases, the American firms operated independently. There are only 20 cases (out of 141) of co-operation with firms in other countries.[54]

Licensing agreements constituted 40 per cent of all co-operation agreements between Germany and Eastern Europe. In 70 per cent of cases, the East European participant received know-how in the form of documentation, and in some cases the German seller undertook to provide the buyer with the results of technological progress achieved after the date of sale. In 80 per cent of cases licences were to be paid for in cash and in only 20 per cent in the form of products. In 70 per cent of cases the East European buyer was given the right to export the licenced product but mainly to the Socialist countries.[55]

b) Value of Transactions

The number of transactions involving the sale of licences is not accurately known. According to Josef Wilczynski, the total number of licences sold by Western countries to East Europe up to 1976 was over 2,400, producing an average annual income of $300 million. Sales by Eastern Europe to the West over the same period totalled 700 licences (almost half of these from Czechoslovakia), producing an annual income of some $40 million.[56] An

---

51   *Licensing and Leasing,* TRADE/INF. 2, Committee on the Development of Trade, United Nations Economic Commission for Europe, Geneva, 1976, p. 17.

52   F. Levcik and J. Stankovsky, *op. cit.,* p. 51.

53   *The US Perspective on East-West Industrial Co-operation.* International Development Centre of Indiana University, Bloomington, 1975, as quoted by F. Levcik and J. Stankovsky, *op. cit.,* p. 48. This is the same survey as the one quoted in Table 27 (preliminary results).

54   F. Levcik and J. Stankovsky, *op. cit.,* p. 48.

55   K. Bolz and P. Plötz, *Erfahrungen aus der Ost-West-Kooperation,* Hamburg, 1974, quoted by F. Levcik and J. Stankovsky, *op. cit.,* p. 52.

56   Josef Wilczynski, "Licences in the West-East-West Transfer of Technology", *Journal of World Trade Law,* March-April, 1977, pp. 126 and 131. These figures include Yugoslavia (500 licences bought, but no particulars given on the number sold). See also Wschód-Zachód: Patenty i Licencje (East-West: Patents and Licences), *Polityka,* 6th March, 1978, p. 13 and *Business Eastern Europe,* 2nd June, 1978, p. 173.

Table 29

Receipts and Payments for Licensing Transactions between East and West

	Period	Receipts accruing to East European countries	Payments to countries of the West	Balance
		US $ million		
East European Countries				
Czechoslovakia[1]	1966-1973 total	64	273	—209
Czechoslovakia[14]	1970-1976 total	92.9	349.3	—256.4
Czechoslovakia[2]	1968-1974		320	
Hungary[3]	1968-1973 total	5.6	18.6	— 13
Hungary[15]	1970-1976 total	25.1	90.3	— 65.2
Poland[4]	1970	—	about 3 times as much as receipts[4]	
Poland[5]	1975	4.0	81	
Poland[6]	1971-1975		approximately 500	
Poland[16]	1971-1975		665	
Total for CMEA countries	1960-1969 total	—	11.2 times as much as receipts[7]	—
Total for CMEA countries	1970	—	4.3 times as much as receipts[7]	
Total for CMEA countries	early 1970. Approximate yearly average	30	(360)[8]	
Total for CMEA countries	around 1974		500[9]	
Total for CMEA countries	around 1975	40[10]	200-300 per annum[11]	
Total for CMEA countries[17]	1976	40	300	
Countries of the West				
United Kingdom[12]	1972	0.2	5.9	+ 5.7
		DM million		
Germany[13]	1960-1974	148	34	+ 114
Germany[13]	1973	18.1	1.4	+ 16.7
of which:				
Chemicals and oil refining	1973	3.7	0.3	+ 3.4
Metal and metal-working industries	1973	8.9	0.9	+ 8.0
Electrical engineering industry	1973	4.8	—	+ 4.8
		US$ million		
United States[14]	1973	5	1	+ 4

Sources and notes:
1 *Licensing and leasing,* TRADE/INF.2. Committee on the Development of Trade, United Nations Economic Commission for Europe, Geneva, 1976, p. 19.
2 Moscow Narodny Bank, *Press Bulletin,* 23rd November, 1977, p. 10.
3 *Licensing and leasing,* TRADE/INF.2. Committee on the Development of Trade, United Nations Economic Commission for Europe, Geneva, 1976, p. 20.
4 Economic Commission for Europe, quoted by Philip Hanson, "International Technology Transfer from the West to the USSR", *Soviet Economy in a New Perspective, op. cit.,* p. 804.
5 "Wschód-Zachód: Patenty i Licenje" (East-West: Licences and Patents), *Polityka* (Poland), 6th March, 1978.
6 Moscow Narodny Bank, *Press Bulletin,* 23rd November, 1977, p. 10.
7 J. Wilczynski, *Technology in Comecon,* London and Basingstoke, 1974, p. 308. quoted in *Die Wirtschaft,* East Berlin, 13th January, 1972, p. 27.
8 Yu. Naido, S. Simanovskij, "Uchastie stran SEV v mirovoi torgovle litsenzii", *Voprosy ekonomiki,* 1975, No. 3, pp. 67-77, "Current" (no date) annual revenue figure from p. 67, where it is stated to be about one per cent of total world licence trade excluding (p. 68) intra-CMEA trade. Total East-West licence trade put (p. 68) at about 10 per cent world total including the national value of intra-CMEA licence exchanges (period not specified), which is put at 24 per cent of world total including intra-CMEA. Hence (apparently) payments to West = about $\frac{(10-74)}{100}$ times $\frac{74}{100}$ times receipts, or about 12 times receipts. Philip Hanson estimates this amount at 24 per cent of world total including intra-CMEA and payments to the West for licences of about 12 receipts. Philip Hanson, "Forms and Dimensions of Technology Transfer between East and West", in *Industrial Policies and Technology Transfers between East and West,* edited by C. T. Saunders, Springer-Verlag, Wien-New York, 1977, p. 157.

(cont'd on next page)

interesting point is that only about 10 per cent of these licences went to multinational firms.[57]

Among the East European countries, the biggest customers for licences were the Soviet Union and Poland[58] with 450 licences each, followed by Czechoslovakia and Hungary (250-300), and lastly Rumania, Bulgaria and the German Democratic Republic. Topping the list of countries selling licences to the West was Czechoslovakia with some 350 licences, followed by the USSR, Hungary, the German Democratic Republic and Poland. The Soviet Union began selling licences only recently. At the Frankfurt Fair in June 1970 the USSR offered 400 licences and since then has sold licences to 35 countries.[59]

Among the Western countries, Germany was the leading licensor, followed by the United Kingdom, the United States, France, Japan, Italy, Sweden, Switzerland, the Netherlands and Belgium. Over 75 per cent of the licences were granted by large corporations or their subsidiaries.[60]

The fields to which these licences apply and the quality of the new processes that are sold are not known in any detail. Lists of licensing agreements are given in the Annex (Tables A-29 and A-30) for both purchases and sales. They cover heavy engineering, the chemical industry, transport, electrical engineering and electronics, the metal and building industries, and light engineering.[61] The striking feature is the large number of licences bought and sold by the USSR. The quality of the licences sold is still a complete unknown. The report of the Economic Commission for Europe refers to certain complaints of Eastern countries to the effect that Western firms often tried to sell licences for technologies that were already obsolete.[62] J. Wilczynski seems to share this viewpoint.[63] This question would certainly seem to merit more thorough study since the only basis for assessment at the moment consists of a few complaints.

57 *Polityka*, 6th March, 1978, *art. cit.*, p. 13.
58 Excluding Yugoslavia, J. Wilczynski, *art. cit.*, p. 126.
59 J. Wilczynski, *art. cit.*, pp. 130-131. The statistics on grants of licence between East and West published by the Economic Commission for Europe for 1975, give the following surprising percentages for the shares of the various East European countries: Hungary 40.7 per cent, Poland 25.9 per cent, Bulgaria 9.3 per cent, Czechoslovakia 9.3 per cent, and the USSR and the German Democratic Republic nil. Quoted by F. Levcik and J. Stankovsky, *op. cit.*, p. 44.
60 *Business Eastern Europe*, 2nd June, 1978, p. 173.
61 *Ibid.*, p. 173.
62 *Licensing and Leasing, op. cit.*, p. 7.
63 J. Wilczynski, *art. cit.*, p. 130. The same author also points out that licences sold by East European countries include minor and unsophisticated inventions, lower in standard than those sold by the countries of the West, *Ibid.*

(Notes cont'd from Table 29)

9 Alexsandr N. Bykov, Perspectives of East-West Relations in Technology Transfer and Related Problems of Dependence, *Industrial Policies and Technology Transfers between East and West*, edited by C.T. Saunders, Springer-Verlag, Wien-New York, 1977, p. 167.
10 "Wschód-Zachód, Patenty i Licencje" (East-West: Licences and Patents), *art. cit.*
11 Moscow Narodny Bank, Press Bulletin, 23rd November, 1977, p. 10.
12 Philip Hanson, "International Technology Transfer from the West to the USSR", *art. cit.*, p. 804.
13 Jiri Slama and Heinrich Vogel, "Technology Advances in Comecon Countries: An Assessment", *East-West Technological Co-operation*, NATO, Brussels, 1976, pp. 231, 234, 235.
14 *Promotion of Trade Through Industrial Co-operation. Recent Trends in East-West Industrial Co-operation*, TRADE/R.373, 31st August, 1978, *op. cit.*, Annex II, p. 1 (1971-1975) from *Statisticka Rocenka USSR*.
15 *Promotion of Trade Through Industrial Co-operation. Recent Trends in East-West Industrial Co-operation, op. cit.*, Annex II, p. 2. For 1975 and 1976, from *Külkereskedelmi Statisticka Evkönyv* (Statistical Yearbook of Foreign Trade).
16 Exchange rate: 3.46 Zloty = $1. 23.8% paid to Germany, 17.8% paid to Great Britain, 11.9% paid to France, 8.0% paid to the United States, 7.3% paid to Italy and 6.8% paid to Sweden. W. Brzost, "Wspópraca naukowotechniczna" (Scientific and technical co-operation), *Handel Zagraniczny*, No. 8, 1977, p. 12 quoted by *Promotion of Trade, op. cit.*, Annex II, p. 2.
17 *Promotion of Trade Through Industrial Co-operation. Recent Trends in East-West Industrial Co-operation, op. cit.*, Annex II, p. 2.

The problem of evaluating the value of licences that are sold and bought in East-West trade has yet to be pursued. As the Economic Commission for Europe's report points out, data on receipts and payments in connection with licences are no more than the tip of the iceberg, the iceberg itself being the sum of all the transactions involved in technology transfer. The complexity of certain industrial co-operation agreements, sometimes involving two-way technology transfers for which payment is made in the form of co-production of related parts (in other words two or three years after the licence is granted) implies that statistics confined to payments made in cash in a given year do not always take these transfers into account.[64] International concentration in the West, in the form of holding companies, subsidiaries, etc., also results in technology transfer appearing only partly in national statistics on the sale of licences.[65] In many cases the sale of licences and know-how is included in package deals comprising the sale of machinery (e.g. turnkey plants). The particulars of these package deals, and particularly the prices set for know-how, may be embodied in the general framework of a negotiation in such a way that the figures which appear in the accounts of the Soviet trading organisations may be misleading.[66] There are several countries in West and Eastern Europe that do not systematically publish information on the value of sold licences disaggregated by country of destination.[67]

The estimated values of licences traded between East and West set out in Table 29 are therefore inevitably highly approximate. Even so, it is quite clear that the balance of receipts is heavily weighted in favour of the West. In the years 1970-1973, cash payments to the countries of the West were several times higher than during the same period for those obtained by Eastern nations (3 times for Poland and 4 times for Hungary, Czechoslovakia and for Eastern Europe as a whole). Although this difference diminished in the early 1970s, it showed signs of widening again in 1974-1976 (Table 29). Thus it appears that the deficit of the East European countries is growing.

Trade in licences with certain industrialised countries such as the United States or Germany produced even larger deficits although there is some evidence, particularly for Hungary, that the gap was closing in the early 1970s. The reason for this narrowing is that the Hungarian pharmaceuticals industry, in which expenditure and receipts are more or less balanced, is one of the largest exporters of licences to the West. The gradual decline in Hungarian expenditure on licences during 1971-1973 and the country's continuing sales of pharmaceutical product licences are responsible for the effect described.[68] Hungary's purchases of licences, however, jumped to $10.5 million in 1974, $16.0 million in 1975 and $47.6 million in 1976. Its sales of licences however kept to the low level of $3 million in 1975 and went up to only $16.2 million in 1976.[69] The Hungarian deficit has thus widened dramatically. Czechoslovakian payments for licences has been relatively stable since 1949[70] but went up to $22 million in 1976, increasing the deficit by that amount to $47.4 million in that year.[71]

64   *Licensing and Leasing, op. cit.,* p. 18.
65   Jiri Slama and Heinrich Vogel, "Technology Advances in Comecon Countries: An Assessment", *art. cit.,* p. 209.
66   Philip Hanson. "International Technology Transfer from the West to the USSR", *Soviet Economy in a New Perspective - A Compendium of Papers, op. cit.,* p. 802.
67   *Ibid.,* p. 802. Of the East European countries, only Czechoslovakia and Hungary systematically publish data on receipts and payments for licences sold or bought.
68   *Licensing and Leasing, op. cit.,* pp. 19-20.
69   *Külkereskedelmi Statistikai Evkönyv* (Statistical Yearbook on Foreign Trade), Budapest, quoted by *Promotion of Trade Through Industrial Co-operation. Recent Trends in East-West Industrial Co-operation,* 1978, *op. cit.,* Annex II, p. 2.
70   *Licensing and Leasing, op. cit.,* p. 19.
71   *Statisticka Rocenka CSSR* (Statistical Yearbook of Czechoslovakia) quoted by *Promotion of Trade Through Industrial Co-operation. Recent Trends in East-West Industrial Co-operation,* TRADE/R. 373, 31st August, 1978, Annex II, p. 1.

c) Prospects for the Development of Trade in Licences between East and West

The acquisition of knowledge through licences offers certain undoubted advantages for East European countries. Firstly, it helps to accelerate technological progress. In East European countries, according to estimates, licences bring forward technological progress in the relevant field by 7-8 years, compared to only 3-5 years for the purchase of know-how and 1-2 years for co-production.[72]

The technological progress brought about in this way by the purchase of licences creates requirements for other improvements, further imports and the purchase of further licences. Estimates suggest that, in 98 per cent of cases the purchase of Western licences in Eastern Europe is followed by further transactions. In Poland for instance, it calls for imports worth six to eight times the expenditure on licences.[73] There are many concrete examples of such transactions.[74]

Another possibility is that a licence sold to one East European country may be used by other CMEA countries. The licences supplied by Renault to Poland and by Fiat to USSR, for example, were used to manufacture cars in Bulgaria and Rumania.[75]

The purchase of licences also helps to promote exports. In Hungary, 22-25 per cent of the machinery exported to the West is produced with the assistance of Western licences. In Poland, nearly one-tenth of industrial production is based on licences bought in the West and it has been estimated that expenditure on licences has enabled exports to be increased by 4-6 times.[76, 77]

In the West this increase in exports through the use of licences has not yet generated any anxiety about competition from East Europe.[78] Firms in the West feel that sales of licences help them to finance the next R & D stage and bank on the time it takes the importing countries to innovate.[79]

The USSR seems to have seen the value of having licensing transactions regulated internationally since it joined the Paris Convention on the Protection of Industrial Property in 1965.[80] This does not mean, as Philip Hanson has pointed out, that the Soviet Union then stopped copying, but Western sellers of licences can now institute proceedings in Soviet courts.[81]

However, the purchase of licences presents certain drawbacks for those buying them. It has already been said that the licences bought are not necessarily up-to-date. In addition, the time it takes to introduce a new technology may often be extremely long. Generally it takes

---

72   A. Bodnar and B. Zahn, *Rewolucja naukowo techniczna a socializm* (Scientific-technical Revolution and Socialism), Warsaw, 1971, p. 158, quoted by J. Wilczynski, *art. cit.,* p. 128.

73   *Polityka, art. cit.,* 6th March, 1978, p. 13.

74   For example, a contract worth $29 million for the building of a man-made fibre plant in 1973 led to another contract worth $8 million for acro-nitrile production technology with Standard Oil of Ohio. J. Wilczynski, *art. cit.,* p. 127.

75   *Ibid.,* p. 127.

76   *Ibid.,* p. 128.

77   M. L. Gorodisskij, *Lizencii po vneshnej torgovle SSSR* (Licences in USSR's Foreign Trade), Moscow, 1972, p. 20, puts the total gain from foreign licences to the USSR's national economy at over 10 times their cost.

78   J. Wilczynski, *art. cit.,* p. 130.

79   J. Wilczynski, *Technology in Comecon, Acceleration of Technological Progress through Economic Planning and the Market,* MacMillan, 1974, p. 305.

80   Philip Hanson, *art. cit.,* p. 803 and J. Wilczynski, *art. cit.,* p. 124. In 1973 the Soviet Union joined the International Union for the Protection of Literary and Artistic Works (Berne Convention) as well.

81   However, a decree on patents passed in 1973 allows Soviet firms to disregard protests by the owner of a patent. See Philip Hanson, *Ibid.*

an average of two years to negotiate a licensing contract with an East European country.[82] Once the licence is bought, a considerable period of time may elapse before it can be exploited. In Hungary, the period is from one to six years; in Poland, of 87 licences bought in 1967, only 14 were put to use within one year, 11 within two years, 23 within three to five years and 3 within six or more years.[83] Similar lead times have been reported for Bulgaria and Rumania.

In view of the fact that the average life of a licence agreement is five to ten years, these delays considerably reduce the benefits of purchasing licences and thus prevent East European countries from narrowing the technology gap.[84]

Some economists in East European countries have warned the authorities against excessive imports of foreign technology. This kind of "horizontal" transfer, unlike the "vertical" transfer stemming from a national R & D effort, could well have the effect of perpetuating the technology gap. The same authors claim that dependence on foreign licences translates to a 5-10 year retardation in the technological level of the importing country.[85]

The recent increase of licence imports into East Europe has been accompanied by some change in their distribution by industrial sector. Originally, Eastern Europe's imports mainly focused on heavy industry such as the non-organic chemicals and heavy transport equipment. Later, however, a shift to the technically more dynamic industries producing automatic machine-tools, light chemicals (pharmaceuticals, plastics and man-made fibres), cars and aerospace electronic equipment, e.g. control systems, computers and electronic data processing equipment, occurred. Another change affecting the former exclusive preoccupation with the production of capital goods has been the slightly moderated increase in consumer goods.

From the history of licensing transactions it is possible to forecast some possible future trends. These transactions may be expected to increase and to feature more reciprocity (the *quid pro quo* basis referred to in the United States).[86] As scientific and technological progress gathers speed, production and technological processes will become more and more diversified, and trade in manufactured goods (especially technology-intensive products) will increasingly take the lead over trade in raw materials and semi-finished products. Licensing is the quickest and cheapest form of technology transfer, and may therefore be expected to grow vigorously.[87]

Of prime importance, however, is the way in which the terms of trade develop. The countries of Eastern Europe may be expected to want to buy the most modern technology at lowest cost. In order to oblige licensors to continue improving the technology transferred and to make it immediately available to CMEA licensees, East European countries will tend to include clauses in the contracts tieing the fees paid to the export *of products* manufactured under the licences. This would be a major change since formerly, contracts provided for the payment of a lump sum or fees based on domestic production under the licences.[88]

The West does not seem to have any very clear-cut attitude in this field. We shall return to the matter when studying the impact of East-West technology transfer on the economy of Western countries.[89]

82   *Eastern Europe Report,* Business International, Geneva, 28th December, 1973, p. 286.
83   Z. Madej, *Nauka i rozwoj gospodarczy* (Science and Economic Development), Warsaw, 1970, p. 180.
84   *Polityka, art. cit.,* 6th March, 1978, p. 13.
85   A. Bodnar and B. Zahn, *op. cit.,* pp. 186-187.
86   *Technology Transfer and Scientific Co-operation between the United States and the Soviet Union: A Review, op. cit.,* p. 34. The Committee of the American Congress has not yet found any practical opportunities for such reciprocity.
87   Provided there is no serious political crisis of course.
88   J. Wilczynski, *art. cit.,* p. 136.
89   See Chapter 6 below.

The supply of turnkey plants, more often than not includes start-up assistance and in many cases training courses, is regarded as one of the most effective channels of technology transfer.[90]

This form of co-operation developed strongly during the last decade, particularly in the chemicals, steel and motor industries. The size of the contract is often considerable. Some examples are shown in the Annex, Table A-32.

In the early 1970s contracts averaged $30 to $50 million in value, the biggest being the one concluded by Swindell-Dressler (United States) for the Kama plant (USSR). The value of contracts has sharply risen since that time, especially those concluded with the Soviet Union. Table 30 shows that out of 31 current contracts for the supply of plant or essential equipment in the USSR in 1976, five were worth between $900 and 1,500 million each, and each of another five between $400 and 670 million. Contracts negotiated in 1977-1978 with the USSR or which have been proposed seem to be for even larger sums. For instance a $1,000 to $2,000 million contract for pulp and paper plants with the United States and Japan competing against each other in the negotiations; a proposed $3,000 million contract for the development of gas fields in Western Siberia (North Star project) involving three American firms;[91] a $2,800 million project for the construction of a petrochemical complex under negotiation with three firms (American, Italian and German), and a $1 000 to 2,000 million project for a German consortium to construct an iron and steel complex.[92] The Soviet Union's preference for turnkey plant purchases is clear from the fact that 90 out of 160 industrial co-operation agreements concluded with the West until 1975 were for the supply of turnkey plants.[93]

The value of turnkey plants supplied to the other CMEA countries is much lower.[94] However, an $800 million agreement has been concluded between Poland and a consortium led by Krupp (Germany) for the construction of coal gasification plants.[95]

For the East European countries, the main advantages of a turnkey plant are the superior quality of the technology and the speed with which it can be installed and commissioned. There are, however, other advantages. According to experience in the West, it is difficult to introduce radical innovation in big old plants. The new firms perform a pioneering role and are regarded as "invaders". It may be assumed that a similar function is performed by the turnkey plants supplied to the CMEA countries.[96] The purchase of turnkey plants also helps to avoid the conflicts that may arise when the Western partner is responsible for only part of the construction work. An illustration is Swindell-Dressler's part-responsibility for the Kama truck factory. By 1974, start-up at this plant was two years late.[97]

---

90   See above. The same viewpoint is taken by Maureen R. Smith, "Industrial Co-operation Agreements: Soviet Experience and Practice". *art. cit.*, p. 774.

91   This contract was not concluded.

92   See Table A-33 which shows the compensation agreements concluded with the USSR since 1969. Since all agreements for the supply of plants above a certain size now include compensation payments, the Table can also be used to study sales of turnkey plants.

93   Maureen Smith, *art. cit.*, p. 773.

94   See Tables A-34 to 39.

95   See Table A-34 and footnote 92.

96   T. Podolski, *art. cit.*, pp. 129-130.

97   *Ibid.*, p. 130.

Table 30

USSR: Major Turnkey Projects in Operation in 1976: Nationality of Western Participant and Description of Project/Compensating Project Flows

1976

Partner/country	Project	Value (million $)	Soviet product payment
France	Gas field equipment	250.0	Natural gas
Austria	Large diameter pipe	400.0	Natural gas
Italy	Large diameter pipe	190.0	Natural gas
Finland	Pipe	n.a.	Natural gas
Germany	Large diameter pipe	1 500.0	Natural gas
France, Austria, Germany	Large diameter pipe and equipment	900.0	Natural gas, Natural gas
Japan	Forestry handling equipment	163.0	Timber products
Japan	Wood chip plant	45.0	Wood chips and pulp
Japan	Forestry handling equipment	500.0	Timber products
France	Pulp paper complex	60.0	Wood pulp
United Kingdom	Shoes	3.2	Food products, toys
Germany	Polyethylene plant	39.0	Polyethylene
Germany	Polyethylene plant	61.0	Polyethylene
France	Styrene/polystyrene	100.0	Polystyrene
Italy	Chemical plants	600.0	Ammonia
United Kingdom, United States	Polyethylene plant	50.0	Polyethylene
France	Ammonia plant	220.0	Ammonia
Italy	Chemical plants	670.0	Chemical products
United States	Ammonia plants	200.0 +	Ammonia
United States	Fertilizer storage and handling facilities	100.0	Ammonia
France	Ammonia pipeline	200.0	Ammonia
United States	Ammonia pipeline	100.0	Ammonia
Italy	Surface active detergent plant	n.a.	Organic chemicals, surface-active detergents
Italy	Polypropylene	100-130.0	Chemical intermediates
United States	Equipment, cola concentrates	n.a.	Vodka
Japan	Oil exploration	150-250.0	Oil and gas
Italy	Large diameter pipe	1 500.0	Scrap metal, coal, iron ore
Japan	Coal development equipment	450.0	Coal
Germany	Steel complex	1 200.0	Pellets, steel products
France	Aluminium refinery	1 000.0	Aluminium
Germany	Ethylene, oxide/glycol plant	80.0	Related products

Source: As quoted by Maureen R. Smith. "Industrial Co-operation Agreements: Soviet Experience and Practice". *Soviet Economy in a New Perspective, op. cit.,* p. 773.

No general review of the supply of turnkey plants has yet been produced. All we have are the findings of various sample surveys.[98] Those of the United Nations surveys are set out above in Table 23. In September 1978, the share of turnkey plant contracts in all industrial co-operation agreements (not including Yugoslavia) was 17.4 per cent although it varied from country to country: 71.1 per cent for the German Democratic Republic,[99] 27.2 per cent

98   See above.
99   It may be wondered to what extent the contracts in the survey sample are representative of this country. The share of turnkey plants in contracts signed by the German Democratic Republic was only 23.5 per cent in June 1976 and no contracts are shown in the United Nation's 1975 survey (see Table 23).

for the USSR and 24.1 per cent for Poland, but far less for the other East European countries.

Of the total plant supply contracts taken into account by the United Nations' sample of end September 1978, 63.6 per cent were for the USSR, 23.6 per cent for Poland and 9.1 per cent for the German Democratic Republic. The percentages for the other countries, are insignificant.[100] The value of individual contracts with the USSR is often particularly high, making the USSR's share all the greater (Table 30).

The United States seem to have a leading role in the supply of turnkey plants. According to the Indiana University survey, which covered 434 US agreements, the share of turnkey plants is 34.1 per cent in the case of CMEA countries as a whole and 55.2 per cent in that of the USSR (see Table 27). Another survey carried out in France in 1978, however, and covering 253 turnkey plant contracts finds that only 6 per cent of this type of contract involved the United States, compared with 31 per cent for France, 24 per cent for Japan and 20 per cent for Germany (see Table 25). The authors admit that France's share could be overestimated[101] but there is obviously a big difference between the results obtained from different surveys.

Because of the differences in the definitions used in industrial co-operation surveys[102] and in the extent to which the samples are representative (the latter often relate to agreements of *widely differing monetary values)* a numerical total would have little meaning. The concept of plant supply is equally ambiguous: for instance, the United Nations surveys interpret plant supply as comprising capital equipment for the exploitation of mineral resources and projects relating to the application of various technologies for such purposes.[103]

These differences in definition no doubt explain the differences in the statistics provided by different countries. Other major differences may arise if East European country statistics are considered. This situation underscores the absence of any comprehensive study on the supply of turnkey plants, fundamental to the pursuit of more detailed research into technology transfer via this channel.

6. CO-PRODUCTION AND SPECIALISATION

Under co-production and specialisation agreements each partner specialises either in the production of certain parts of a finished product — which are then assembled by one or both partners — or in the production of a limited number of articles in the production range which are then exchanged so that each partner can offer the full range. The technology usually comes from one of the partners and in some cases is the culmination of a joint R & D effort. Generally speaking, co-production and specialisation agreements also include

100   1.8 per cent each for Hungary and Rumania and none at all for Bulgaria or Czechoslovakia. *Promotion of Trade through Industrial Co-operation. Statistical Outline of Recent Trends in Industrial Co-operation, op. cit.,* Table 4, p. 9.
101   Jean Cheval, François Geze, Patrick Gutman, *Les accords de coopération industrielle Est-Ouest, op. cit.,* p. 19.
102   See above.
103   F. Levcik and J. Stankovsky, *op. cit.,* p. 54.

co-operative marketing arrangements. Usually the product bears the trademark of both partners, each of whom has exclusivity for the market in his own area but shares the market in other countries.[104]

Co-production and specialisation are the most frequent form of industrial co-operation (45.2 per cent of the United Nations sample of end September 1978 — see Table 23). In practice they consist in moving a production activity to a different location, usually to Eastern Europe.[105]

The main subjects of inter-firm co-operation are shown in Table 31. They relate to co-operation in R & D, communication of scientific and technological data, construction, production and marketing activities.

The United Nations surveys identify various types of co-production and specialisation agreements as follows:

— Co-operation, including or excluding sales, in which each party makes compo-
    nents or parts of a finished product, the technology being supplied by one or both
    parties (21.0 per cent of the United Nations sample of end September 1978 — not
    including Yugoslavia).[106]

This form of co-operation, found frequently in the engineering and electronics industries, resulted in many contracts being signed in the late 1960s and early 1970s but their unit value was low and their life short.[107] It seems to have actually been replaced by more elaborate forms of co-operation which go so far as to transfer entire production capacity to sub-contracting enterprises.[108]

— Co-operation in which each partner specialises in part of the production
    programme, then exchanges units in order to complete the other partner's product
    range (3.8 per cent of the United Nations sample of end September 1978, not
    including Yugoslavia).[109]

This is a more elaborate form of co-operation in which the partners can also apply a policy of deliberate specialisation and share production of the component parts of finished products to organise mass production, facilitate supplies of materials and acquire the relevant production technologies. Relations between partners may be so close as to give rise to active links between laboratories and design offices in the development of new products.[110]

---

104 *Analytical Report on Industrial Co-operation among ECE Countries, op. cit.,* p. 11. However, according to some Western experts most agreements which are called co-production and specialisation agreements do not imply a real specialisation in the sense that the Western firm gives up its own production of a certain part. Usually the Western firm continues to produce the particular part, but uses its Eastern partner as a complementary supplier. Eastern deliveries are often substitutes for payment in money for the Western technology. Specialisation often occurs when the Western firm allocates to an East European country products which are "late" in their "life cycle" and which can be sold in the West provided that production costs are reduced.

105 Carl H. McMillan, *art. cit.,* p. 1194.

106 *Promotion of Trade Through Industrial Co-operation. Statistical Outline of Recent Trends in Industrial Co-operation, op. cit.,* Table 3, p. 8.

108 *Ibid.*

107 According to a survey made among Polish enterprises in 1973, the average value of exports arising out of co-operation agreements was under US $100,000 (310,000 Zlotys used for foreign trade operations) and the average life of a contract one year and a half. *Promotion of Trade Through Industrial Co-operation. Recent Trends in East-West Industrial Co-operation, op. cit.,* p. 16.

109 *Promotion of Trade Through Industrial Co-operation. Statistical Outline of Recent Trends in Industrial Co-operation, op. cit.,* Table 3, p. 8.

110 *Promotion of Trade Through Industrial Co-operation. Recent Trends in East-West Industrial Co-operation, op. cit.,* p. 17.

Although there are reports of an increasing number of contracts of this type often associated with R & D clauses, the majority of contracts now current are still based on simpler agreements for the exchange of parts and accessories over a given period of time. Sometimes joint distribution networks are agreed to but in most cases a market-sharing arrangement is made giving the Western partner the right to sell jointly-made products in the West whereas the Eastern enterprise has exclusive rights to sell them on its own national market and in the CMEA countries.[111]

Table 31

Main Elements of a System of Co-operation between Enterprises

Activity \ Method of co-operation	Co-operation by sharing or co-ordination of tasks between the partners	Co-operation through joint use or operation of equipment or fixed installations
1. Research and development	Co-ordination of projects undertaken by the partners, in the areas of co-operation, including communication of relevant information	Use of the partner's research facilities; establishment of joint laboratories or test stations
2. Communication of scientific and technical data and transfer of property rights in these data	Exchange of technical information, exchange of patents, possibly cross-licensing, provision of technical assistance	Establishment of joint information services (e.g. data bank); access to existing systems; establishment of joint consulting companies
3. Industrial construction and works	Joint tendering, e.g. by participating in a consortium of enterprises	Establishment of joint engineering and/or building companies
4. Manufacture	Joint purchases; sub-contracting; contract manufacturing	Use of the partner's surplus production capacity; establishment of joint production plants or subsidiaries
5. Marketing	Mutual agency agreement; agreement on provision of after-sales services; market-sharing arrangements	Establishment of joint storage facilities; use of the partner's surplus storage (or transport) capacities; establishment of a joint distribution company

Source: Promotion of Trade Through Industrial Co-operation. Recent Trends in East-West Industrial Co-operation, United Nations Economic and Social Council, Economic Commission for Europe, Committee on the Development of Trade, TRADE/R. 373, 31st August, 1978, p. 19.

— Co-production and specialisation in R & D only (20.4 per cent of the United Nations sample of end September 1978 — not including Yugoslavia).[112]

It is interesting to note that the breakdown of co-production and specialisation agreements by categories varies from country to country in Eastern Europe. The Soviet Union has signed agreements mainly in the R & D field (42.6 per cent of the total in both United Nations's surveys — that for 1st June, 1976 and that for end September 1978). The

111 Ibid., p. 17.
112 Promotion of Trade Through Industrial Co-operation. Statistical Outline of Recent Trends in Industrial Co-operation, op. cit., Table 3, p. 8.

other Eastern countries, and in particular Hungary, Bulgaria, the German Democratic Republic and Poland, seem to take a greater interest in co-production.[113]

The United Nations study of the 1st June, 1976 sample also gives a distribution of co-production and specialisation agreements by economic sector. In terms of total co-production agreements, the chemical industry was the leader (25.9 per cent), followed by mechanical engineering (23.9 per cent), transport equipment (13.5 per cent) and electronic equipment (11.1 per cent).[114] Bulgaria, the USSR and Hungary concentrated on the chemical industry (the percentage of co-production agreements concluded by these countries being 33.2, 29.8 and 26.0 respectively), while Czechoslovakia, Poland and the German Democratic Republic showed a preference for mechanical engineering (40.0, 37.5 and 33.3 per cent respectively) and Rumania for transport equipment (35.7 per cent of the co-production and specialisation agreements entered into by this country).[115]

Co-production and specialisation agreements are of greatest value in fields where the two partners complement one another. They often involve vertical integration and the manufacture of fewer different products as the product range is completed by supplies from the other partner. Very often, it is difficult to differentiate among the various types of co-production and specialisation agreements as they evolve. For instance, the difference between a licensing agreement involving partial payment in the form of products manufactured under the licence and a co-production agreement based on licensed technology often lies more in *ex ante* intentions than in the *ex post* terms of the arrangements.[116]

In any case, co-production and specialisation agreements are rarely restricted to the exchange of semi-finished products, parts and finished products. They often include ancillary services in the form of know-how, production processes, designs, drawings, etc., and culminate in wide-ranging R & D co-operation.[117] Some authors even consider these supporting activities to be the essential advantages while co-production and the transfer of capital are merely secondary. In their view, the chief advantages of this form of co-operation are.

— imports of technological advances are not limited by the shortage of foreign currency;
— technological progress in individual enterprises is gradual, thus facilitating dissemination throughout a country;
— co-operation of this kind promotes personal contact, often more valuable than formal information;
— such agreements confer the advantages of long-run production;
— the pooling of R & D facilities is more conducive to success.[118]

---

113   *Ibid.*, p. 8 and *Statistical Outline of Recent Trends in Industrial Co-operation, op. cit.*, Table 3.
114   *Statistical Outline of Recent Trends in Industrial Co-operation, op. cit.*, Table 5. The same specifications are not published in the UN sample of end September, 1978.
115   *Ibid.*, Table 5.
116   Carl H. McMillan, *art. cit.*, p. 1195.
117   Norbert Leise, *Transfer Mechanismen des technischen Fortschritts in und zwischen verschiedenen Wirtschaftssystemen,* Institut für Aussenhandel und Überseewirtschaft der Universität Hamburg, Forschungsbericht No. 4, p. 33.
118   Jürgen Nötzold et Jiri Slama, *Der Transfer von Technologie zwischen den beteiligten Volkswirtschaften, op. cit.*, p. 46.

7. JOINT VENTURES

The particularly interesting feature of joint ventures is the ability to institutionalise industrial co-operation by constituting new legal entities and integrating pooled activities. In a joint venture:[119]

— assets are pooled in the form of capital, equipment, intellectual property (licences, patents, etc.) and technical and organisational facilities;
— participation is given a value in the form of shares;
— joint objectives are set and joint services established;
— risks and profits are shared;
— management is shared by participants;
— in East European countries, joint ventures usually include technology transfer and the provision of finance.

The legal aspects relative to joint ventures are given in Table 32.

Legally, the joint venture is a pattern of business organisation that can be adopted by every type of industrial co-operation. Certain specific features, however, may be noted depending on whether the ventures are set up in the East, in the West or in the developing countries.

Joint ventures in the East

Joint ventures are not very common in Eastern Europe. Yugoslavia has signed a fairly large number (144 up to 1976 and 150-160 up to 1978)[120] but the numbers in other countries have been minimal. By mid-1979 there were nine in Rumania, five in Hungary[121] and four in Poland where this type of co-operation was authorised in 1976.

Yugoslavia is the first country to have authorised the establishment of joint ventures since 1967. The United States as the most important partner provided 23 per cent of foreign investment in value terms between 1st January, 1967 and 1st July, 1976, followed by Germany (21 per cent) and Italy (17 per cent). Most of these projects are small, over three-quarters of them accounting for less than $1 million each. More than 80 per cent of the projects negotiate payment in terms of technology and know-how rather than cash. The relatively minor scale of joint ventures may be due to the reluctance of Western investors to work within Yugoslav legal requirements (under the Act of 13th April, 1973).[122] Around 1970, 15 joint venture contracts were reported for Yugoslavia for a total value of $50 million. In early 1974 the number had gone up to about 100, totalling $200 million and by mid-1978 to 150-160, the West's share being around $400 million.[123] In spite of this growth, foreign investment in Yugoslavia is still at a relatively low level accounting for something less than 1 per cent of the total capital investment in the economy.[124] Recently (1976), Dow Chemical, an American firm, concluded a joint venture contract for a

119   F. Levcik and J. Stankovsky, *op. cit.*, pp. 84-85. See also *Promotion of Trade Through Industrial Co-operation. Recent Trends in East-West Industrial Co-operation, op. cit.*, p. 25.
120   See Table 22 and *Promotion of Trade Through Industrial Co-operation. Recent Trends in East-West Industrial Co-operation, op. cit.*, p. 27.
121   One joint venture between Hiradas Technika (Hungary) and Bowmar (Canada) signed in 1975 is no longer reported for mid-1978 (see Table 33).
122   Carl H. McMillan, *art. cit.*, p. 1217.
123   *Promotion of Trade Through Industrial Co-operation. Recent Trends in East-West Industrial Co-operation, op. cit.*, p. 27.
   It should be noted (see Table 33) that Western shareholdings are never more than 49 per cent but in Yugoslavia exceptions are possible, in theory, subject to Parliamentary approval. However, some decisions have to be taken by unanimous vote, which restricts the scope of this possibility. Hungary has also recently allowed the possibility of majority shareholding in the operation of joint ventures.
124   *Ibid.*, p. 28.

Table 32
Joint Ventures in Socialist Countries

	Intra-socialist Ventures			East-West Ventures		
	CMEA Model	"Halex" (Poland and Hungary)	"Intransmash" (Bulgaria and Hungary)	Rumania	Hungary	Yugoslavia
	The joint enterprise in its relation to the economic system of the situs-state					
Legal form	Juridical person of the situs-state	Joint-stock company of the situs-state (= Poland)	Juridical person of the situs-state (= Bulgaria; branch in Hungary)	Joint-stock company or limited liability company ("joint company")	Joint-stock company or limited liability company, unlimited liability partnership or joint enterprise ("economic association")	"Basic organisation of associated labor" or "Work organisation"
Sphere of activity	Any economic activity	Coal hoisting, (utilization of coal waste)	Mechanization of internal transport systems	Permitted in industry, agriculture, construction, tourism and transport	Permitted for developing technological and economic levels as well as in trade and services	Any economic activity, except banking, insurance, domestic transport, trade and services
Property in the enterprise's assets	Right of "operative administration" (question of property remains open)	Property of the joint-stock company	Property of the juridical person	Right of "operative administration" (question of property remains open)	Property of the economic association; partner may reserve property rights	"Social property"; investor may reserve property rights
Currency system	Incorporation in the currency system of the situs-state; settlement in transferable rubles	Incorporation in the currency system of the situs-state; settlement in transferable rubles ("calculation system")	Incorporation in the currency system of the situs-state; settlement in transferable rubles ("calculation system")	According to agreement; as a rule, convertible currency ("enclave system")	Application of law of situs-state; conversion into convertible currency ("calculation system")	Incorporation in the currency system of the situs-state ("interaction system")
Planning system	Integration in the plans of the situs-state	Planning questions are decided upon jointly by both countries	Incorporation in the planning system of both countries	Enterprise draws up economic and financial programmes requiring approval	Enterprise adopts its economic plans	Basic organisation adopts its working plans
Foreign Trade System	Right to engage in foreign trade in accordance with the law of the situs-state	Right of supplying domestic enterprises; sale in Hungary through Hungarian	Foreign trade through foreign trade organisations or right to engage directly in	Foreign trade through foreign trade organisations or right to engage directly in	Right to engage directly in foreign trade	Right to engage directly in foreign trade

international treaty, law of the situs-state or foundation contract			(24% in case of reinvestments)	to a certain level; from there 50% (partial refund possible in case of reinvestments)	regulations of the member-republic so far usually 35% the profit (lower rates in case of reinvestments)

The foreign partner in his relation to the joint venture

Property relations	In the form of shares. They are indivisible and, in case of doubt, inalienable	In the form of stock issued on the names of the holders	In the form of deposits into a statute fund	In the form of shares; maximum of 49%	In the form of shares; maximum of 49%	In the form of deposits; transferable; not higher than domestic deposit
Participation in management	One voice per share (departure by contract permitted); in certain questions unanimity is required	Participation on basis of equality	One voice per country; unanimity is required	Can take over management functions; partners may agree on unanimity requirement	In accordance with the foundation; contract requiring ministerial approval	In accordance with investment contract; certain rights remain with "Workers' Councils"
Transfer of invested assets	After withdrawal or liquidation there is a right to claim a refund of the investment	Unknown	Upon liquidation settlement to be agreed upon	Upon withdrawal or liquidation transfer guaranteed	Upon withdrawal or liquidation transfer permitted	Upon termination of contract, transfer permitted (minus depreciation); beforehand only if contract so provides
Transfer of profit	Permitted without limitation	Permitted in kind or in transferable rubles	Permitted in transferable rubles	Permitted; 10% to be paid as tax	Permitted	Permitted within quotas (maximum amounts) depending on foreign exchange earned through exports

Practice and references

Distribution (number of enterprises)	Approx. 5	1	1	7	3	91
Legal basis	CMEA Uniform Statutes of 1976	Agreement between Poland and Hungary of 1959; Polish Commercial Code of 1934	Agreement between Bulgaria and Hungary of 1964	Decree of 1972; Commercial Code of 1887	Decree of 1972; Commercial Law of 1875; Limited Liability Company Law of 1930	Law of 1973

Source : Dietrich A. Loeber. "Capital Investment in Soviet Enterprises? Possibilities and Limits of East-West Trade". *Adelaide Law Review*. September. 1978. pp. 340-341.

Table 33

Western Participation in Joint Ventures in East Europe

Name, registered address and year of formation	Partner in East Europe	Partner in the West	Capital in $'000	Western company's shareholding	Industry
Yugoslavia					
Belinka, Ljubliana (1973)	Belinka	Solvay (Belgium)	6 685	49%	Chemicals
Fadip, Belgrade (1971)	Fadip	Dunlop (United States)	2 490	43%	Water pipe
Yugoslavia Commerce Belgrade (1972)	Yugoslavia Commerce				
Iskra, Kranj (1972)	Iskra	Gillette (United States)	1 470	20%	Razor blades
		Bell Telephone (Belgium)	44 120	6%	Electrical precision instruments
Kovaska Industria Zrece (1972)	Kovaska Industria	Renault (France)	4 520	22%	Forging
Kromos, Umag (1970)	Kromos	Hempel (Denmark)	7 300	38%	Sea products
Lek, Ljubliana (1970)	Lek	Bayer (Germany)	3 150	49%	Pharmaceuticals
Pliva, Agram (1972)	Pliva	Ciba-Geigy (Switzerland)	880	49%	Pharmaceuticals
Sava, Kranj (1971)	Sava	Semperit (Austria)	31 910	20%	Radial tyres
Tomos, Koper (1973)	Tomos (and others)	Citroën (France)	11 910	49%	Cars
Unioninvest Sarajevo (1973)	Unioninvest	Marlo Italiana (Italy)	3 200	20%	Air conditioning installations
Krk, near Rijeka (1976)	INA	Dow Chemical	120 000[b]	49%	Oil complex
Rumania					
Rifil, Bucharest (1973)	Fibrex	Romalfa (Italy)	2 100	48%	Acrylic yarn
Romcontrol Data, Bucharest (1973)	CIETV	Control Data (United States)	4 000	45%	Computer components
Resita Rank AG, Bucharest (1973)	Resita Uzinexport	Rank AG (Germany)	16 600	49%	Marine engines
Roniprot (1974)	Union for dyestuffs	Dainippon Inc. (Japan)	11 300	43%	Dyestuffs
Elarom	Union for electronics	L'électronique appliquée (France)	2 030	49%	Electronic appliances
Romelite (1975)	Union for heavy engineering	F. Kohmeier KG (Austria)	6 600	48%	Heavy engineering
Rom Avia SRL (1977)	Grupul Aeronautic Bucuresti	Vereinigte Flug-Werke-VPW-Fokker (Netherlands and Germany)	425 000 to 855 000	…	Jet aircraft
BAC 1-11	Grupul Aeronautic Bucuresti	British Aerospace	170 000	…	Jet aircraft
Oltcit (1977)	FTO Auto-Dacia	Citroën (France)		36%	Cars (130,000 to be produced end 1979)
Hungary					
Sincontact, Budapest (1974)	Inter-co-operation	Siemens (Germany)	530	49%	Technical consultation
Volcom, Budapest (1974)	Csepel[a]	Volvo (Sweden)	1 740	48%	Assembling of lorries
			1 …	49%	Industrial and

			(United States)	..	Publicity about American products in Hungary
... Budapest[d] (1978?)	FTO Metimpex Mahir Hungarian Publicity Company	..	Young and Rubicam Int. (United States)	49%	

Poland

... Warsaw	.. ..		Kobelinski (United States)	100%	Construction of a Hotel of 600 rooms and several motels in Warsaw and vicinity
... Szczecin	PZM (Polish Maritime Company)	..	Shipbuilding Industry (United Kingdom)	..	Construction of 22 bulk carriers and 2 floating cranes
...	..	..	Anna Falkner (Austria)	100%	Production of adhesive labels
...	..	..	S. Szewczyk (Canada)	100%	Jeans factory

a) With other partners.
b) Total credit evaluated by Carl H. McMillan at $700 million. op. cit., p. 1217 and $500-600 million by Promotion of Trade Through Industrial Co-operation. Recent Trends in East-West Industrial Co-operation, op. cit., p. 28.
c) This joint venture is not given in the list quoted by the United Nations for mid-1978. Ibid., p. 26.
d) This firm has a similar society in Frankfurt (51% for USA and 49% for Hungary). The Budapest firm did not start to work until April 1979.

Sources: J. Wilczynski, Joint Ventures and Rights of Ownership, Institute of Soviet and East European Studies, Carleton University: Ottawa, Working Paper No. 8, 1975. Based on press reports. Quoted from F. Levcik and J. Stankovsky, op. cit., pp. 100-101. Carl H. McMillan, "East-West Industrial Co-operation", art. cit., pp. 1215-1217. See also Table A-38. Promotion of Trade Through Industrial Co-operation. Recent Trends in East-West Industrial Co-operation, op. cit., p. 26. Anita Tiraspolsky. Les investissements occidentaux dans les pays de l'Est. (Western Investments in Eastern Countries): Le courrier des pays de l'Est, avril 1979, pp. 14-15.

petro-chemical complex at Krk near Rijeka, investing $500-600 million of which $120 million was the American shareholding (17 per cent).[125] This contract could mark the beginning of a new phase in co-operation between Yugoslavia and the West.

In Rumania, Act No. 1 of 17th March, 1971, allows joint ventures permitting, among others, companies to engage directly in production operations.[126] These companies have a special status allowing operation outside of the Rumanian Plan system. They are often referred to as *"enclaves,"* their transactions with Rumanian firms being conducted through Rumanian foreign trade enterprises.[127]

In spite of these advantages, the number of joint venture contracts concluded with Rumania is very small (nine by mid-1978) (see Table 33). Recently however, two major agreements have been reported (see Table A-38), one with VEW-FOKKER for a jet aircraft factory (in 1977) and another with Citroën in 1977 for $170 million.[128]

Hungary first authorised joint ventures in October 1972. The possibility of Western majority shareholding was recently accepted. The only possible access to production is indirect, through Hungarian enterprises. Co-operation in this case relates mainly to R & D, consultancy and marketing. A typical contract of this type is that signed with Siemens (Sincontact — see Table 33). Joint ventures therefore play only a limited role and no more than three were current in mid-1978. The new regulations promulgated on 6th May, 1977 have eased some restrictive rules and allow joint ventures in the production sector. For the moment, however, foreign investors have made no use of this possibility although a number of projects are reported as being under negotiation.[129]

In Poland, joint ventures have been authorised since May 1976, but are limited to such categories as light industry, domestic trade and consumer services. Their main purpose seems to be to induce Polish nationals residing abroad to repatriate funds.[130] In 1976 and 1977 there were reports of negotiations for joint ventures with General Motors for a truck manufacturing plant[131], with an Anglo-Polish shipping firm and with an Australian-Polish food firm.[132] Negotiations have also been reported regarding the establishment in Poland of firms belonging wholly to foreign investors.[133]

Joint ventures offer definite advantages to both East and West. The East European partner can stimulate the interest of the Western firm in the practical application of its technology through participation in the profits or losses of the technology through participation in the profits or losses of the venture. Without spending any foreign currency, the East European partner can obtain modern technology and learn about Western management and marketing methods.[134] In some cases he can obtain a technology not yet available for sale or on lease or when the owner is unwilling to part with all (or most) of the ownership rights. In this way he can acquire a technology that is not available on the market. The main interest, however, is the prospect of access to Western markets through association with the foreign partner.[135] A further advantage is that investments of this kind take less time than those which call on internal resources and facilitate the communication of new technologies within the country.

125   *Ibid.,* p. 28.
126   Carl H. McMillan, *art. cit.,* pp. 1213-1214.
127   *Ibid.,* p. 1215.
128   *Ibid.,* p. 1215.
129   *Ibid.,* p. 1209.
130   *Ibid.,* pp. 1212-1213.
131   T. Podolski, *art. cit.,* p. 125.
132   Carl H. McMillan, *art. cit.,* p. 1213.
133   *Promotion of Trade Through Industrial Co-operation. Recent Trends in East-West Industrial Co-operation, op. cit.,* p. 26.
134   Norbert Leise, *op. cit.,* p. 26 and F. Levcik and J. Stankovsky, *op. cit.,* pp. 88-89.
135   *Promotion of Trade Through Industrial Co-operation. Recent Trends in East-West Industrial Co-operation, op. cit.,* p. 25.

For the Western participant, the main advantage of this form of co-operation is access to East-European markets. The Western partner is also guaranteed participation on a continuous basis in the management of the plant. Thus there is a possibility of continuous, reciprocal technology transfer and profits can be guaranteed more easily than in the case of licensing. Costs may often be another advantage. A jointly owned and operated plant using Western technology but taking advantage of building, labour and manpower costs in the host country can often produce at very competitive prices.[136] A further psychological advantage is that joint ventures often create a climate of confidence that is more difficult to obtain with other forms of co-operation.[137]

The agreements concluded thus far have been rather small in scale, chiefly because of the East European ideological opposition (this would seem normal as the concept of shareholdings does not fit in with communist tenets and the belief in the superiority of the socialist system). This is reflected in legal restrictions which deter potential Western investors. However, attitudes are reported to be softening in Rumania, Hungary and Poland, and even in the Soviet Union, where opposition to Western shareholdings is at its highest. Thus the Deputy Minister for Foreign Trade, V. N. Sushkov, recently proposed the joint formation of "special enterprises".[138] However, the proposal's actual significance is not yet known.

Joint ventures located in the West

The primary purpose of joint ventures located in the West is to support East-West trade especially in the marketing field. They also relate to services, occasionally the processing of raw materials and also transport. Very few relate to production, and specialisation plays only a secondary part.

The main advantage of joint ventures to the Western partner is that of helping the Eastern partner to earn foreign currency which may subsequently serve as a source of finance to further exports to the East.[139]

Joint ventures located in the West differ from those in the East in many ways. Their duration is longer, they are more diversified in their legal form (the Eastern partner is often the majority shareholder), they are far more numerous and are often set up in the form of subsidiaries belonging completely to the East European country.

Their number and their capital value are difficult to assess. Carl H. McMillan's investigations (see Table 22) indicate that, in late 1976, there were 480 joint ventures with the CMEA countries (but only 215 if Yugoslavia is excluded) located in the industrialised West. Another estimate (for mid-1978) which excludes banks and other financial institutions gives 239 joint ventures with seven CMEA countries (i.e. not including Yugoslavia) located in 16 industrialised Western countries. The capital invested in these establishments was put at something short of $600 million.[140]

Although this form of co-operation is very widespread, there have been complaints recently on both sides, the East Europeans accusing their Western partners of failing to fulfil

---

136 *Ibid.,* pp. 25-26 and *Promotion of Trade Through Industrial Co-operation: Result of a Survey of Five Western Enterprises,* Note by the Secretariat, United Nations Economic and Social Council, Economic Commission for Europe, Committee on the Development of Trade, TRADE/R. 373/Add. 3, 18th September, 1978, pp. 32-33.

137 F. Levcik and J. Stankovsky, *op. cit.,* p. 89.

138 T. Podolski, *art. cit.,* p. 125.

139 *Promotion of Trade Through Industrial Co-operation. Recent Trends in East-West Industrial Co-operation, op. cit.,* pp. 29-30.

140 This estimate is confined to equity investment. Over 100 such equity investments are reported for Yugoslavia although the capital amount involved is thought to be low. *Promotion of Trade Through Industrial Co-operation. Recent Trends in East-West Industrial Co-operation, op. cit.,* p. 29.

their commitments to purchase and the latter complaining about poor product quality, slow delivery and exchange control restrictions in the East.[141]

Joint ventures located in the developing countries

Joint ventures located in Third World countries are generally an extension of tripartite and multipartite co-operation (see below). In other words the partners in a joint venture on a purely contractual basis later find they would like to develop the co-operation between them by making it of longer duration and sharing profits and risks.

Generally speaking, these joint ventures are financed by participation and credits in cash or kind. A number of countries provide special facilities to attract foreign capital. Nothing is known about their number or the amount of capital involved. They would appear to be greater in number than those located in the East but fewer than those located in the West and are still at an early stage of development.[142]

8.   TRIPARTITE CO-OPERATION

The purpose of tripartite industrial co-operation is to set up industrial complexes, build plants, prospect natural resources, assemble plants and provide marketing services in developing or semi-industrialised countries on a joint basis, the partners being Western firms and socialist import-export centres, with local firms and industries involved to varying extents depending on the case. The term "tripartite" merely means that there is at least one Western, one East European and one Third World country involved. It also implies active participation on the part of the member on whose territory the venture is located. The contribution of the Western and Eastern partners generally consists in the grant of patents and other industrial property rights, the supply of specialised plant and technical assistance. The main contribution from the developing country is the civil engineering and the supply of certain types of equipment.[143]

Tripartite industrial co-operation is not yet very widespread but its frequency is increasing considerably. It accounted for only 1.1 per cent (not including Yugoslavia — 0.8 per cent with Yugoslavia) of all the industrial co-operation agreements in the United Nations sample of 1st June, 1976 but the figure had increased to 6.4 per cent (5.9 per cent including Yugoslavia) in the September 1978 sample (see Table 23).

The total number of such agreements is difficult to determine. For the period 1965-1975, UNCTAD counted 132 but Patrick Gutman and Jean-Christophe Romer traced 65 in only two years (mid-1976 to mid-1978) by going through the leading specialised periodicals.[144] It would also appear that the number of tripartite outline agreements entered into by East European countries — which had been only 10 during the period 1965-1975 — went up to 38 over the period mid-1976 to mid-1978.[145]

141   Carl H. McMillan, *art. cit.*, pp. 1196-1197.
142   *Promotion of Trade Through Industrial Co-operation. Recent Trends in East-West Industrial Co-operation, op. cit.*, pp. 30-31.
143   *Ibid.*, p. 23.
144   Patrick Gutman and Jean-Christophe Romer, "Coopération industrielle tripartite Est-Ouest-Sud et dynamique des systèmes", report to the Colloquium on triangular East-West-South technology transfers. Research centre on USSR and East European countries, Strasbourg III, University, 8th December, 1978, p. 25. The periodicals concerned were: *East-West Markets, Business Eastern Europe, Moscow Narodny Ban Press Bulletin, The Reuter East-West Trade News, Ecotass, Polish Economic Survey, Marketing in Hungary.*
145   *Ibid.*, p. 27.

During this recent period (mid-1976 to mid-1978) the Western countries with the highest proportion of tripartite industrial co-operation agreements were: Germany (33.8 per cent), France (20 per cent), the United Kingdom (9.2 per cent), Austria (6.5 per cent) and the United States and Finland (6.2 per cent each). During the same period CMEA countries' shares were as follows: Poland 28.3 per cent, Hungary 18.3 per cent, USSR 13.3 per cent and Czechoslovakia and the German Democratic Republic 10 per cent each.[146]

The number of contracts, incidentally, is not, always representative of their significance and the figures need to be studied with care. It has been estimated that, between 1964 and 1973, the total value of known tripartite industrial co-operation agreements accounted for slightly more than 1/8th of the developing countries' total imports of capital goods.[147] But the relative importance of this business for each country is difficult to ascertain. In a sample of tripartite co-operation agreements (34 cases), for example, studied by Patrick Gutman, the six contracts signed by the Soviet Union were worth F. 1.8 billion whereas the 17 signed by Poland and Yugoslavia totalled only F. 0.4 billion.[148, 149]

There is no overall study on tripartite industrial co-operation. Only a few surveys covering a varying number of agreements have so far been carried out.

The first survey seems to be that produced by Patrick Gutman and Francis Arkwright relating to 26 French firms and totalling 40 cases.[150] The figures for the comparative participation of French, Socialist and Third World countries in a selected series of agreements[151] are as follows:

	Per cent
French firms:	
Design engineering	82.5
Sub-contracting	17.5
Socialist firms:	
Erection, civil engineering	30
Sub-contracting, erection and civil engineering	35
Detailed engineering and sub-contracting	17.5
Design engineering	17.5
Third World firms:	
No services	65
Sub-contracting and erection	5
Erection and civil engineering	30

Through tripartite agreements, Western countries hope to lower their costs of association with East European countries; the latter are mainly interested in obtaining foreign currency and technical knowledge.[152]

146   *Ibid.*, pp. 25 and 27.

147   I. Sronec, *The Experience of Socialist Countries of Eastern Europe in the Transfer of Technology to Developing Countries.* UNCTAD, United Nations, TD/B/C. 6/25, 1978, p. 10.

148   Patrick Gutman, *Etat présent et perspectives de la coopération industrielle tripartite Est-Ouest-Sud,* report to the colloquium on the future of North-South-East economic relations, GERPI, Paris, 20th and 21st January, 1978, p. 27.

149   Maureen R. Smith, *art. cit.,* p. 774, refers to the Soviet Union's interest in tripartite industrial co-operation.

150   Patrick Gutman and Francis Arkwright, "La coopération industrielle tripartite entre pays à systèmes économiques et sociaux différents de l'Ouest et du Sud", *Politique étrangère*, No. 6, 1975, p. 643; and (same authors) "Coopération industrielle tripartite 'Est-Ouest-Sud': Evaluation financière et analyse des modalités de paiement et de financement. Application au cas français", *Politique étrangère*, No. 6, 1976, p. 619.

151   Patrick Gutman, *Etat présent et perspective de la coopération industrielle tripartite Est-Ouest-Sud, op. cit.,* p. 27.

152   *Ibid.,* p. 28.

Because of the low costs for construction work offered by East European countries, Third World countries often prefer to optimise the overall cost of an industrial complex immediately, even if it means that their local industries and firms cannot participate effectively, although they are often technically qualified to undertake such work. Tripartite co-operation is thus reduced to East-West co-operation in Third World countries.

In East-West relations, this type of co-operation often forces the socialist trading centre, acting as sub-contractor, to accept the Western partner's ideas on production and technological and practical application, and the international technical standards adopted by the latter. Patrick Gutman concludes that this amounts at least to implicit readiness to accept the capitalist approach to production and to see the capitalist work system reproduced.[153] Although this conclusion may be unjustified (why restrict it to international technical standards used in tripartite co-operation and not to the domestic production of Eastern Europe, for instance?), this form of co-operation, like the joint venture, constitutes a powerful medium for transferring Western technology to the East (and the South). However, its future development would seem to depend far more on the political climate than on technical advantages.

Another more recent survey by the United Nations covered 51 tripartite industrial co-operation contracts involving Hungary, Rumania and Yugoslavia.[154] For the firms interviewed, tripartite co-operation accounted for 4-20 per cent of their foreign trade. 43 of the 51 contracts included the supply of plant and 3 the supervision of the plant supplied. Practically all the contracts included technical assistance. The informant firms claimed that in these tripartite industrial co-operation contracts, clients in the developing countries were able to take part in choosing and acquiring advanced technology and gaining knowledge and experience in its use.[155]

Another interesting study, recently carried out by the Wiener Institut für Internationale Wirtschaftsvergleiche covered 58 tripartite co-operation agreements negotiated by Austrian firms and in particular Vöest-Alpine and Simmering-Graz-Pauker.[156] The survey lists the main advantages and disadvantages as seen by these Austrian firms.

The advantages include:

— lower prices from sub-contractors in East Europe than those usually asked in the West;

— ease of obtaining contracts with Eastern partners when included in tripartite co-operation;

— the Eastern partner may be willing to take countertrade raw materials and crude oil goods for which the Western country would have no use;

— the Western partner may fulfil part of his own purchasing obligations;

— the Western partner may hope to improve his image in the East and to negotiate new contracts with his Eastern partner;

— Eastern technology, being less sophisticated, can be applied more easily and at lower cost by the developing countries.

153   *Ibid.,* p. 39.
154   *Promotion of Trade Through Industrial Co-operation. Tripartite Industrial Co-operation Contracts: Results of an Inquiry,* United Nations Economic and Social Council, Economic Commission for Europe, Committee on the Development of Trade, TRADE/R. 373, Add. 1, 12th October, 1978.
155   *Ibid.,* pp. 4-5, 9-10, 26-28, Annex II, p. 5.
156   "Incentives and Problems in Tripartite Co-operation", *Business Eastern Europe,* April 20, 1979, pp. 121-122. Out of these 58 contracts, 24 were with Africa, 11 with the Middle East, 8 with the Far East and 6 with South America. In the CMEA the leading countries were Hungary (12), the German Democratic Republic (11) and Poland (7). Rumania signed 19 contracts but most have remained only on paper.

The disadvantage on which most stress is laid is the highly complicated decision-making process in the East and the inability of Eastern officials to take immediate decisions. Other disadvantages[157] are:

— the reluctance on the part of the Eastern partners to provide financing for their share of the co-operation agreement even for that part which could be financed in their currency;
— difficulties arising out of the fact that technical specifications and promised parameters are not met;
— low productivity and performance of some construction and assembly workers in the Third World;
— poor after-sales service and lack of spare parts and the impact which these shortcomings have on the image of the Western partner which could affect future sales in the developing countries.

9. SUB-CONTRACTING, JOINT TENDERING OR JOINT PROJECTS

In the case of *sub-contracting* (short or long-term) the East European enterprise undertakes to produce and deliver an agreed quantity of finished or semi-finished articles manufactured in accordance with the information and know-how (and sometimes with the parts, machinery and equipment) provided by the main contractor. *Joint tendering* or *joint projects* differ from other industrial co-operation contracts in that they involve a customer who may be located in either country or, more often than not, a third.[158] It includes the supply of production lines or complete plants and the construction of infrastructure projects. It usually includes the grant of licences, co-production and contract manufacturing.[159]

The number of cases of sub-contracting and joint tendering is relatively low and it can be seen from Table 23 that proportionally sub-contracting is tending to fall (3.8 per cent of the United Nations sample of end September 1978) whereas the number of joint projects is tending to increase (16.9 per cent).

The main advantages of joint tendering relate to costs and the possibility of supplementing production capacity. The Western party may also engage an East European enterprise as contractor or sub-supplier (and vice versa) or do the design work. Another advantage is that it also helps to find the sources of credit that customers may request. Several different institutional arrangements are used in joint tendering and joint projects.[160]

---

157 "Incentives and Problems in Tripartite Co-operation", *art. cit.*, pp. 121-122.
158 *Analytical Report on Industrial Co-operation Among ECE Countries*, United Nations, Geneva, 1973, pp. 14 and 16.
159 *Promotion of Trade Through Industrial Co-operation. Recent Trends in East-West Industrial Co-operation, op. cit.*, p. 20.
160 *Ibid.*, pp. 20-21.

Chapter 4

EASTERN AND WESTERN POLICIES TOWARD TECHNOLOGY TRANSFER

INTRODUCTION

Policy makers in both East and West have demonstrated a certain ambivalence toward the expansion of economic relations with one another. This ambivalence is sharply focused with respect to the transfer of technology. On the one hand, such transfers are believed to be mutually advantageous. On the other hand, the more conservative parties among the leadership in both East and West — particularly in the Soviet Union and the United States — take the view that the "political factors associated with trade and the transfer of technology present a mixed blessing. It is not surprising then, that in this context, policies toward East-West trade should swing between 'encouragement' and 'restraint', heavily dependent upon the political climate".[1]

The division of opinion about the relative merits of the transfer of technology has assumed all the hallmarks of a major controversy. While the general attitudes toward technology transfer have changed as political postures have swung between the extreme of "cold war" and the proclivity toward détente, the main factors in the controversy have, by and large, remained the same. In many respects, the evolution of attitudes toward technology transfer has closely followed changes in the general political climate of East-West relations.

In practice, it is difficult to separate developments which have affected the evolution of trade between East and West from those which have affected policies directly related to technology transfer. However, changes in policies toward technology transfer have probably been more affected by political considerations.

This Chapter examines those factors — political, commercial and strategic — which have influenced policies in both East and West toward technology transfer in the past and are likely to do so in the future. These factors not only constitute the controversy surrounding technology transfer but affect the changes in attitudes toward such transfers. The following discussion seeks to set out those issues which complicate the technology transfer problem and to elucidate the type of thinking which leads — or has led — to the adoption of a particular policy stance.

For the *East*, greater emphasis will be placed on those considerations which explain the need for Western technology. Here the discussion will highlight the evolution of Eastern

---

1 Arthur H. Hausman, *art. cit.*, p. 333.

interest in Western technology and hence in improving trade relations with the West. Two general issues provide the focus of the discussions; the Eastern need for Western technology and know-how to help solve internal economic problems, and the use of technology as an instrument to alter political and economic relations with the West.[2]

For the *West,* emphasis is placed on the difference of opinion regarding the potential merits of the transfer of technologies and on the assumptions on which policies towards such transfers are formulated. The discussion seeks to throw light on what is regarded to be the crucial issue in the controversy, namely whether such transfers "involve the West in selling cheaply a critical capability, which we shall subsequently regret sharing with the Soviet Union".[3] This issue can be separated into a number of discrete concerns, among which are the following:

a) will greater East-West technology flows create an interdependence favourable to peace?;

b) will these flows promote, or on the contrary, provide a substitute for Soviet economic reforms? and

c) will these flows tend to draw Soviet policy-makers into unplanned complementary resource commitments at the expense of military expenditure?[4]

Equally important Western concerns are the potential consequence of continuing transfers and whether such technology transfers produce long-term advantages or disadvantages for Western economies.[5]

2 A number of issues treated here are closely related to the question of the East/West technological gap and to the impact of Western technologies on the Eastern economies — see Chapters 5 and 6 respectively. To minimise duplication and respect the framework of this chapter they are not treated in depth.
3 Philip Hanson, "International Technology Transfer from the West to the USSR", in *Soviet Economy in a New Perspective – A Compendium of Papers, op. cit.,* p. 810.
4 *Ibid.* The concerns expressed above cover primarily the NATO countries.
5 Recent discussions concerning US government policy toward transfers of technology highlight the issue of "how much" technology and know-how the United States should "share" in order to protect the US technological lead (considered to be key to US international economic and security issues) while at the same time using technology to further foreign policy aims. Consequently, three basic questions are being asked: a) "what kinds of technology can the US permit business to transfer?; b) at what price should technology be transferred? and, c) what will be the cost of transfers to US economic and strategic positions?". "Washington International Business Report", *Technology Transfer: US Government Policy and its Impact on International Business,* Special Report, 77.3, December, 1977.

I. THE EASTERN PERSPECTIVE

1. Importance of Western Technology for the East

a) *Growing Need for Western Technology*

Chapters 2 and 3 illustrated how Eastern needs and demands for Western technology have been growing. Table A-11 indicates that imports of machinery and transport equipment by the CMEA countries increased 15 times (in current prices) between 1961 and 1977.[6] Between 1965 and 1977 these imports increased 10 times.[7] The share which machinery and transport equipment represented in total CMEA imports from the OECD countries also increased: from 29.5 per cent in 1965 to 34.5 per cent in 1976 and 38.4 per cent in 1977.[8]

The industrial co-operation agreements which have been concluded illustrate that transfer from West to East of disembodied technology has also greatly increased since the beginning of the 1970s. By the end of 1976, there were nine joint equity ventures located in the East European countries and 852 substantive non-equity cooperative agreements had been signed with Westerns firms.[9]

b) *Primary Factors Motivating Increased Demands/Needs for Western Technology*

i) *Growth Retardation and the Need to Increase Factor Productivity*

Since the 1960s all of the CMEA countries have experienced a slowing down in GNP growth rates.[10] This trend can be observed from Western as well as from official evaluations (see Table 34). According to some recent Western estimates reproduced by the *Neue Zürcher Zeitung,* economic expansion for all the CMEA countries was 4.2 per cent in 1977 as compared to 5.2 per cent in 1976.[11] The production and investment targets were not fulfilled by several of the CMEA countries in either 1978 or 1979. A reduction in growth rates was clearly given in official statistics. See Table 34.

This phenomenon of decreasing GNP growth rates is most striking in the case of the Soviet Union. For 1951-1960, average annual growth was 5.8 per cent. Average annual growth decreased to 5.1 per cent for 1961-1970 and to 3.7 per cent for 1971-1975 and 1976.[12]

---

6   For Poland the increase was 17 and for Hungary 20 times. See Tables A-21 and A-24 respectively.

7   See Table 9. Polish imports of machinery and transport equipment increased 16 times during this period and the Soviet Union's 12 times.

8   But fluctuations in Western grain exports may skew the statistics, and much of the machinery and transport equipment imported by the East is low technology.

9   Excluding Yugoslavia. See Table 22.

10   See *Selected Trade and Economic Data of the Centrally Planned Economies,* US Department of Commerce, Industry and Trade Administration, Bureau of East-West Trade, Washington, December 1977-January 1978, pp. 2-3.

11   "Vorsichtiegere Planung in der Ostblockwirtschaft", *Neue Zürcher Zeitung,* Fernausgabe Nr. 67, 22 März, 1978, p. 9.

12   *Soviet Economic Problems and Prospects – A Study,* prepared for the Sub-Committee on Priorities and Economy in Government of the Joint Economic Committee, Congress of the United States, 95th Congress, 1st Session, US Government Printing Office, Washington, 8th August, 1977, p. 2; and *Selected Trade and Economic Data of the Centrally Planned Economies, op. cit.,* p. 2 for 1976.

The growth rate also declined in 1977.[13] Decreasing population growths are primarily responsible for the decline.[14] The slow-down in population growth together with the inability to shift more labour from rural to urban areas and from non-employed (domestic) to hired labour accounts for the decline in growth of the non-agricultural labour force from 4.0 per cent in 1959-1965 to 3.0 per cent in 1966-1970, and 2.8 per cent in 1971-1975. Projections for the future are not bright: growth for the total labour force is estimated to be 1.5, 0.9 and 0.5 per cent for the years 1976-1980, 1981-1985 and 1986-1990 respectively. For the same years, growth rates of non-agricultural labour are expected to be 2.5, 1.6 and 1.1 per cent.[15] Industrial labour productivity has also declined. This "was slated to rise by 4.8 per cent in 1978 but actual growth is only 3.5 per cent".[16] Dissatisfaction with economic performance has been signalled at the highest government level. For instance Mr. Brezhnev recently warned ministers in inefficient industries that they would be held accountable for the pervasive "lack of control, mismanagement and gross oversight".[17] According to some analysts top-level dissatisfaction was partially reflected in the shakeup last year of Politburo members.[18]

Shortages of labour, cheap and easily available fuels and other natural resources, caused the decline in growth of over-all productivity. Growth in output per man-hour slowed by nearly one-half between the 1960s and the first half of the 1970s.[19]

The problem of capital-output ratio is also very alarming. This ratio can be calculated in several ways: as output produced per ruble of fixed capital and as the ratio of gross capital investment to the increase of gross national product. The first indicator (increase in gross national product in relation to the increase of fixed capital — five year moving average)

13 Official evaluations are more optimistic. Nevertheless, Soviet and East European sources also indicate a sizeable slow down in growth rates, despite the fact that absolute growth is said to be higher. In the case of Poland, for instance, growth of 6.9 per cent is given for 1976, 2.8 per cent for 1978 and less than 2.0 per cent for 1979 (see Table 34). Soviet official statistics indicate that the average annual growth of national income for 1951-1955 was 11.3 per cent, 9.1 per cent for 1956-1960, 6.5 per cent for 1961-1965, 7.7 per cent for 1966-1970, 5.7 per cent for 1971-1975, 5.3 per cent in 1976, 3.5 per cent in 1977, 4.0 per cent in 1978 and 2.0 for 1979 (in constant prices). Calculated from *Narodnogo Khosjajstvo SSSR za 60 let* (National Economy of the USSR for 60 Years), Moscow, 1977, p. 485. For the 1977 figure see *Ekonomicheskaja Gazeta* (Economic Journal), Moscow, No. 6, 1978, p. 5. For the 1978 and 1979 figures see Table 34.

14 In the case of the Soviet Union, the annual population growth fell from 17.8 per thousand in 1960 to 9.2 in 1970. A small increase (9.9) is expected to occur in 1980. However, it is also estimated that growth rates will fall again in 1990 (5.8) and in the year 2000. Murray Feshbach and Stephen Rapawy, "Soviet Population and Manpower Trends and Policies", in *Soviet Economy in a New Perspective, op. cit.*, p. 122.

15 *Ibid.*, p. 133.

16 "Soviet Economy Loses Momentum", *Business Eastern Europe*, Vol. 7, No. 49, December 8, 1978, p. 387. According to official figures industrial labour productivity grew 3.6 per cent in 1978 and 2.4 per cent in 1979. *Ekonomicheskaya Gazeta*, No. 5, 1979, p. 8 and No. 5, 1980, p. 8. The annual objective was 3.8 per cent and the average annual growth fixed by the tenth five year plan was 5.6 per cent. See Daniel Vernet, "Le rythme de développement de l'industrie s'est ralenti au cours des derniers mois de 1978", *Le Monde*, Paris, 24 janvier 1979, p. 5.

17 Given in "Soviet Economy Loses Momentum", *Business Eastern Europe, art. cit.*

18 "The Business Outlook-Soviet Union", *Business Eastern Europe*, Vol. 8, No. 6, February 9, 1979, p. 45.

19 According to Nikolai Baibakov, Chairman of the USSR State Planning Committee, "one of the main tasks of the Plan for 1979 was speeding up the growth of labour productivity, increasing it by 4.7 per cent in industry, as against 3.6 per cent in 1978, by 4 per cent in construction, as against 2.8 per cent in 1978 and by 2.1 per cent in railway transport, as against 1.4 per cent in 1978". "Higher Growth Planned for this Year, Nikolai Baibakov Reports on State Plan for 1979", *Soviet News*, No. 5956, published by the Press Department of the Soviet Embassy in London, January 23, 1979, p. 11.

Table 34
Economic Performance 1976-1980 — Growth Indicators
Official evaluations
% increase or decrease

	1976 actual[1]	1977 plan[2]	1977 actual[2]	1978 plan[3]	1978 actual[3]	1979 plan[3]	1979 actual[4]	1980 plan[4]
Bulgaria								
National Income	6.2	8-9	6.4	6.8	6.0	7.0	6.5	5.7
Industrial output	7.1	..	10.2	7.7	7.0	7.0	6.6	6.3
Investment	0.7	..	14.5	..	..	..	5.0	..
Czechoslovakia	..	..						
National Income	3.8	..	4.4	5.0	4.0	4.3	2.6-2.8	3.7
Industrial output	5.2	..	1.4	5.0	5.0	4.5	3.7	4.0
Investment	4.0	..	2.6	6.6	6.6	..	1.6	..
German Democratic Republic								
National Income	3.8	..	5.2	..	..	4.3	4.0	4.8
Industrial Output	6.1	..	4.3	5.2	4.0	5.5	4.8	4.7
Investment	8.8	..	4.4	..	..	5.6	..	0.0
Hungary								
National Income	3.0	..	9.4	5.0	4.0-4.5	3.0-4.0	1.0-1.5	3.0-3.5
Industrial Output	5.5	..	5.9	5.5-6.0	5.5-6.0	4.0	2.8	3.5-4.0
Investment	—0.7	..	12.9	2-3	4.0	1.0	0.0	—4.5
Poland								
National Income	6.9	..	5.3	5.4	2.8	2.8	—2.0	1.4-1.8
Industrial Output	9.0	..	7.7	6.8	5.8	4.9	2.8	
Investment	2.1	..	4.2	0.0	—0.2	—10.0	—8.2	—8.1
Rumania								
National Income	10.5		8.5	..	..	8.8	6.2	8.8
Industrial Output	12.4		11.0	..	9.0	10.0	9.1	11.5
Investment	8.1	..	11.8	..	..	..	..	4.9
Soviet Union								
National Income	5.3	4.1	5.0	4.0	4.0	4.3	2.0	4.0
Industrial Output	6.4	5.6	5.5	4.5	4.8	5.7	3.4	4.5
Investment	4.3	4.6	4.1	3.8	5.0	4.9	1.0	2.7

Sources:
1 *Statisticheskij Ezhegodnik Stran Chlenov Soveta Ekonomicheskoj Vzaimopomoshchi, 1978* (Statistical Yearbook of Countries Members of the CMEA, 1978). Moscow, 1978. pp. 27-38.
2 *Bulgaria:* "Vorsichtigere Planung in der Ostblockwirtschaft". *art. cit.,* 2 March 1978, p. 9.
   Soviet Union: Ekonomicheskaya Gazeta, No. 2, Moscow, January 1977, p. 1.
3 *Bulgaria:* "1978 Plan Failure Hits Bulgarian Export Items". *Business Eastern Europe,* Vol. 8, No. 11, 16th March 1979, p. 84.
   Czechoslovakia: "The Business Outlook Czechoslovakia". *Business Eastern Europe.* Vol. 8, No. 11, 16th March 1979, p. 85 (quoted from *Rude Pravo).*
   German Democratic Republic and *Rumania:* Murray Seeger. "Comecon's Growth Below 1978 and 5-Year Targets". *International Herald Tribune,* Paris. 8th February 1979, p. 7.
   Hungary: "The Business Outlook Hungary". *Business Eastern Europe,* Vol. 8, No. 8, 23rd February 1979, p. 61, and *Népszabadság,* December 8, 1978, p. 2.
   Poland: "The Business Outlook Poland". *Business Eastern Europe,* Vol. 8, No. 13, 30th March 1979, p. 101.
   Soviet Union: Plan for 1978: *Ekonomicheskaya Gazeta,* No. 52, 1977, p. 1.
               1978 actual: *Ekonomicheskaya Gazeta,* No. 5, 1979, p. 8.
               Plan for 1979: *Ekonomicheskaya Gazeta,* No. 50, 1978, p. 1.
4 *Bulgaria:* 1979 actual: *Rabotnichesko Delo,* 23 January, 1980, p. 1.
               1980 Plan: *Business Eastern Europe,* Vol. 9, No. 15, 1st February, 1980, p. 44.
   German Democratic Republic: 1979 actual: *Business Eastern Europe,* Vol. 9, No. 6, 8 February, 1980, p. 44.
   Hungary: Business Eastern Europe, Vol. 9, No. 3, 18 January, 1980, p. 18.
   Poland: 1979 actual: *Trybuna Ludu,* 9/10. February, 1980.
               1980 Plan: *Monitor Polski,* No. 30/1979 of 28 December, 1979.
   Rumania: 1980 Plan: *Business Eastern Europe,* Vol. 9, No. 3, p. 21.
   Soviet Union: 1979 actual: *Ekonomicheskaya Gazeta,* No. 5, 1980, p. 8.
               1980 Plan: *Ekonomicheskaya Gazeta,* No. 50, 1979, p. 3.

shows a reduction from roughly 0.50 per cent in 1955 to 0.20 per cent in 1975.[20] A reduced return from new fixed equipment is very apparent. The second indicator as shown in Table 35 (ratio of total gross capital investment to the increase in gross national product) shows the same reduced return from capital in a different way: the need to invest more in order to increase a unit of output. The corresponding targets of the 1971-1975 and 1976-1980 plans are less and less ambitious and do not seem to have been fulfilled.

Table 35

Ratio of Gross Capital Investment to Increased Output[a]

	1971-75 plan	1976-80 plan	1976-79 actual[b]
Bulgaria	4.5	4.7	5.8
Czechoslovakia	5.9	6.9	8.4
German Democratic Republic	5.3	5.8	7.4
Hungary	5.7	6.1	8.1
Poland	3.7	4.8	8.7
Rumania	3.0	3.8	4.2
Soviet Union	5.2	6.2	6.3

a) Increased net material product.
b) Data for 1979 refers to plan targets.

Source: UN Economic Commission for Europe.
Taken from: "EE Investments to Improve Use of Capital and Labor", *Business Eastern Europe*, Vol. 8, No. 14, April 6, 1979, p. 107.

ii)   *Lack of Incentives for Domestic Innovation*

Increasing needs for investment and technology to stimulate economic growth could, of course, be substantially met through domestic efforts and by help from the other CMEA countries. However, the Soviet system of centralised planning does not help efficient innovation at the enterprise level especially in the civilian sector. Innovation appears to require some freedom of choice and creative imagination at the enterprise level. These requirements cannot be met in a system where the enterprise is bound by detailed planning

20   *Soviet Economic Problems and Prospects, op. cit.*, pp. 3-4. Diminishing return on investment is a problem often called attention to by Soviet authorities. It was highlighted as a problem by Kosygin in his speech concerning the 1965 economic reforms. See *Pravda*, 28th September, 1965. See also T. S. Khachaturov, *Sovetskaya ekonomika na sovremennom etape* (Soviet Economy at Present), Moscow, 1975, pp. 318-345. The author concluded that "investments in fixed funds are inefficiently utilised in our country", *Ibid.*, p. 327. This problem seems to be a continuing one. While the 1978 investment plan was overfulfilled (planned increase of 3.8 per cent and an actual increase of 5 per cent), a report of the Central Statistical Directorate "discloses failures in bringing new plants on stream, particularly in the power, chemical fertiliser and plastics sectors. Furthermore, the fragmentation of available funds and the high proportion of unfinished projects, which planners had sought to correct, have further increased. Half the plants that fell behind schedule... were in chemicals, a sector with high priority in the five-year plan and on which many others depend". According to Mikhail Pertsev, Gosplan Deputy Chairman, "the failures in the fuel and metal industries particularly held back national economic growth". "USSR 1978 Plan Results Pinpoint Problem Areas", *Business Eastern Europe, art. cit.* "Investments in 1979 will seek to remedy serious 1978 shortfalls, particularly in the fuel and metal industries. Specifically, the priorities of the current investment strategy include the rolled steel, fertiliser, textile and food industries. Moreover, energy production is seen as particularly urgent — spending on geological prospecting and oil exploration is to be increased by 11 per cent and 20 per cent respectively". "The Business Outlook-Soviet Union", *Business Eastern Europe, art. cit.*

targets imposed from above. Lack of incentives for innovation is a feature common to all centrally planned economies.[21] Imports of Western technology were seen by Eastern planners as a direct remedy for the innovation problem.[22]

iii) *Gains from Trade*

As has been noted in Chapter 1, the traditional Soviet approach to foreign trade, particularly following the introduction of the Stalinist system, was rather autarkic and thus supportive of "self-sufficiency". This approach was imposed on the other CMEA countries after the Second World War.

Judging from recent developments, it seems that Soviet leaders have recognised that the "world economy has become technologically very interdependent and that it therefore is time to terminate the policy of denying themselves access to the world market".[23] The new approach to world trade was essentially formulated in order to gain access to the most modern technologies and equipment which "could add a critical margin of effectiveness to Soviet investment".[24] The advantages anticipated from this approach were already made clear in Gosplan in 1970: "foreign technology could be expected to advance the application of innovative techniques by two to five years... This could be accomplished while economising on domestic R & D expenditures and expanding hard currency earnings through exports of finished products".[25]

Currently a more general approach toward efficiency calculations of foreign trade has become evident: the partial rehabilitation by Soviet leaders of the Ricardian theory of comparative costs. It seems that the Soviet Union is now willing to admit some of the ideas already proposed earlier in other CMEA countries, for instance, that Western technology is needed not only to add to existing capital stock and production capacities, but as an instrument to obtain substantial gains from the international division of labour.[26]

c)   *Role of Western Technology in Soviet Economic Development*

i)   *Western Technology in the Ninth Five Year Plan (1971-1975)*

The primary objectives of the Ninth Five Year Plan were to increase economic growth through modernisation of the civilian economy, to improve the consumer sector of the economy and to raise the efficiency of economic planning and management. Western technology was seen as a crucial element in meeting these goals.

The Plan explicitly stated that "consideration is being given to mutually beneficial co-operation with foreign firms and banks in working out a number of very important

21   See E. Zaleski, J. P. Kozlowski, H. Wienert, R. W. Davies, M. J. Berry and R. Amann, *Science Policy in the USSR*, OECD, Paris, 1969. Part 5 deals with the consequences of the difficulties experienced by the Soviet Union to innovate.

22   See points c) and d) of this section.

23   *Technology Transfer and Scientific Co-operation between the United States and the Soviet Union: A Review, op. cit.*, p. 80. It is believed by some Western analysts that Soviet leaders have decided to "modify or abandon the Stalinist principle of technological and economic independence and turn toward a policy of selective interdependence with the industrially developed nations of the West". See for instance, John P. Hardt and George D. Holliday, "Technology Transfer and Change in the Soviet Economic System", in J. Frederic Fleron, Jr., ed., *Technology and Communist Culture, the Socio-Cultural Impact of Technology under Socialism*, Praeger, New York-London, 1977, p. 185. The issue of dependence/interdependence is discussed in Part II of this Chapter.

24   *Ibid.*

25   Lawrence H. Theriot, "US Governmental and Private Industry Co-operation with the Soviet Union in the Fields of Science and Technology", *art. cit.*, p. 751.

26   S. N. Zakharov, *Raschety effektivnosti vneshne ekonomicheskikh svjazei* (Efficiency Calculations of Foreign Economic Relations), Moscow, 1975. For a corresponding approach in the CMEA countries see Witold Trzeciakowski, "Evolution de la planification et de la gestion du commerce extérieur en Pologne", *Revue de l'Est*, Paris, No. 1, 1977, pp. 5-37 and Agota Dezsenyi-Gueullette, "Les calculs d'efficacité du commerce extérieur en Hongrie avant la réforme économique de 1968", *Revue d'études comparatives Est-Ouest*, No. 2, 1977, pp. 21-95.

economic questions associated with the use of the Soviet Union's natural resources, construction of industrial enterprises, and exploration for new technical solutions".[27]

Some analysts believe that the import of automotive technology, particularly truck technology, is central to this plan. They feel that this strategy was formulated initially in connection with the Eighth Five Year Plan whose centre piece was the agreement with Fiat (1966) to build a passenger car plant. "The impact of the Fiat contract and related agreements on Soviet foreign trade was felt later as Soviet imports of Western machinery increased sharply".[28] In the Ninth Plan the Kama River Truck Plant became the major focus of Western machinery imports and the "selected areas for special attention were widened from automotive technology to include a) natural gas, oil, timber, metal extraction, processing and distribution technology, b) chemical processes ranging from fertiliser to petrochemicals, c) computer assisted systems technology, d) agribusiness technology and e) tourist facility technology".[29] The characteristic of the Ninth Plan "is the overall emphasis on equipment reducing the use of raw materials and increasing labour productivity, that is, equipment embodying, if possible, the latest technological developments in this respect".[30]

ii) *Western Technology in the Tenth Five Year Plan (1976-1980)*

The basic directives of the Tenth Five Year Plan illustrate how Soviet leaders now see the role of Western technology in the fulfilment of their economic plans. Described as the plan of "quality and efficiency", it focuses on the rate of technical change as one of the key problem areas in current Soviet economic development.[31] The provisions of the Plan (for instance, "curtailing the growth of new construction starts in favour of investing in advanced machinery and equipment") substantiate the belief that Soviet leaders do indeed intend to "emphasise concentration and modernisation at the expense of traditional patterns of economic growth".[32]

This development can also be deduced from the composition and mix of items which have predominated in Soviet imports during the last few years — particularly imports from the United States. Special emphasis has been placed once again on equipment for energy extraction and processing, heavy industry (particularly foundries for the manufacture of trucks), computers and computer assisted systems technology, chemical processes (fertilisers, petrochemicals, etc.) and for the expansion of the oil industry and steel production. These are all areas where Western companies can expect to do business.[33]

The Tenth Five Year Plan also calls for "measures aimed at the broader participation of the Soviet Union in the international division of labour and at enhancing the role of foreign

27   Cited by Franklyn D. Holzman, *International Trade Under Communism, op. cit.*, p. 162.
28   John P. Hardt, "The Role of Western Technology in Soviet Economic Plans", in *East-West Technological Co-operation, NATO, Colloquium 1976, op. cit.*, p. 316.
29   *Ibid.*
30   Alexander Woroniak, "Economic Aspects of Soviet-American Détente", paper prepared for the Third Atlantic Economic Conference, Washington, D.C., 12-13 September, 1975, p. 17. See also *Science, Technology, and American Diplomacy – An Extended Study of the Interactions of Science and Technology with United States Foreign Policy,* Vol. I, Committee on International Relations, US House of Representatives, US Government Printing Office, Washington, 1977, pp. 544-547. See Chapter 2 and Annex Tables for the breakdown of machinery imports particularly in relation to industry-branch user.
31   Philip Hanson, *USSR: Foreign Trade Implications of the 1976-1980 Plan,* EIU Special Report No. 36, The Economic Intelligence Unit, Ltd., London, October, 1976, pp. 44-47. See also *Technology Transfer and Scientific Co-operation between the United States and the Soviet Union: A Review, op. cit.,* pp. 77-80.
32   *Soviet Economic Problems and Prospects, op. cit.,* p. 9.
33   See G. F. Ray, "Thoughts on Innovation and the Transfer of Technology", discussion paper, Workshop on East-West European Economic Interactions, Vienna Institute for Comparative Economic Studies, 3-7 April, 1977.

economic ties in the accomplishment of national economic tasks and the acceleration of scientific and technical progress".[34]

d) *Prospects for Future Imports of Technology by the Soviet Union*[35]

   i) *Energy Resources*

Future Soviet needs for Western technology are closely related to the energy problem. Energy exports are very important to the Soviet leadership because they are a major hard currency earning source and, therefore, a means of financing Western imports of high technology, equipment and grain.

The prospective exports for hard currency earnings will have a critical effect on the ability of the Soviet Union to buy the equipment necessary to expand their oil and gas production. As stated by one source, "imports related to energy conservation and production will take precedence, as failure to obtain such equipment and technology would only exacerbate Soviet oil problems and increase Soviet hard currency expenditures for oil over the long run".[36]

There appears to be an insufficiency of drilling equipment — in terms of both quantity and quality — in the Soviet Union, that will continue to be a problem for a long time. "This should ensure an on-going flow from abroad of products and technology well into the 1980s. In refineries and distribution there has been co-operation in the past. In fact, the supply of natural gas to Germany, France and Italy on co-operation contracts has been of great importance to both sides and quite successful. There are additional supplies of gas available that could provide the basis of expansion of the existing programme. This will mean continued imports of pipe and pipe line laying equipment and, perhaps some kind of joint production of pumping equipment".[37] According to the same author, oil and gas production are so vital to the continued growth of the Soviet economy that adjustments will be made elsewhere, even sacrifices to ensure the fulfilment of the Tenth Five Year Plan and future plans.[38]

While there may be no argument about Soviet energy requirements and priorities, particularly as regards oil and gas, growth prospects for energy production are a matter of doubt and debate. The various estimates that have been made of the CMEA countries' oil production, consumption and foreign trade are shown in Table 36.

The first thing this table shows is that the CMEA countries have recently (1975-1977) strengthened their position as *net oil exporters* — and will no doubt continue to do so up to 1979 — but that they owe this situation to the Soviet Union, currently the world's biggest oil producer. At the same time the *net oil deficit* of the CMEA countries, other than the USSR, has risen. It should also be noted that the natural gas surplus in the CMEA countries was

34   Joseph S. Berliner, "Prospects for Technological Progress", in *Soviet Economy in a New Perspective – A Compendium of Papers, op. cit.*, pp. 434-435.
35   In view of the sluggish domestic economic performance in recent years, some Western analysts have concluded that Soviet "reliance on Western imports is likely to increase in 1979 and into the 1980s, as Soviet industry will be hard pressed to meet or even approach key five-year plan production goals for oil, plastics, fertilisers and various steel products". See "The Business Outlook – Soviet Union", *Business Eastern Europe, art. cit.* It is also noted that "production shortfalls in basic industries... may mean increased purchases of Western equipment and technology". See "Soviet Economy Loses Momentum", *Business Eastern Europe, art. cit.* See also "Soviet Economy Worsening, CIA Says". *International Herald Tribune*, Paris, October 5, 1978, p. 5.
36   *Soviet Economic Problems and Prospects, op. cit.*, p. 23.
37   Robert J. McMenamin, "Western Technology and the Soviets in the 1980s", paper presented at the NATO colloquium *The USSR in the 1980s: Economic Growth and the Role of Foreign Trade*, Directorate of Economic Affairs, NATO, Brussels, 17-19 January, 1978, p. 161
38   *Ibid.*

Table 36

Various Estimates of CMEA Countries' Oil and Natural Gas Production, Consumption and Foreign Trade for the Period 1975-1990

	1975 Actual	1976 Actual	1977 Actual	1978 Actual	1979 Plan	1979 Actual		1980 Forecasts	1985 Forecasts	1990 Forecasts
Crude Oil (million tons)										
Soviet Union										
Production	490.7[1]	520[2]	546[2]	572[4]	593[5]	586[31]	10th 5-Year Plan	620-640[18]	..	..
							Annual Plan	606[32]	..	..
							Econ. Comm. for Europe (1978)	640[19]	719[19]	780[19]
							Econ. Comm. for Europe (1980)	..	+700[33]	−800[33]
							Emily E. Jack *et al.*	590[20]	..	..
							CIA April 1977	550-590-600[21]	450-500[24]	769[25]
							CIA July 1977	550-600[22]	..	..
							CIA August 1977	..	400-500[30]	..
							Institut Petro Studies	15[1]	..	1.144[26]
Imports	7.6[1,6]	7.2[6]	(8.0)	..	..	..		10[23]	10[23]	..
Availabilities	498.3[1]	527.2	(554)	..	..	..		560-650[20]	410-720	..
Consumption	368.0[1,7]	378.7[7]	(390)[8]	..	..	..		470[20] 443[27]	CIA: 470-505[20] 552[27]	..
Available for Export	130.3[1,7]	148.5[7]	(154)[8]	..	..	..		135[20] 207[27]	..	..
of which: CMEA Countries	63.3[1]	68.4[7]	69.0[13]	71.0[13]	74.0[13]	..		75[20] 90[27]	90[27]*	..
Other Socialist Countries	14.4[1]	..	..	..	..	..		15[20]	..	..
All Socialist Countries	77.7	..	(92)	..	..	..		90[20]	..	..
Total Western Countries	52.6[1]	..	62[9]	..	..	..		45.0[20]	..	..
of which: Convertible Currency Countr.	38.6[1]	50.0[10]	..	..	..	..		35.0[20]	..	..
Other (European) CMEA Countries										
Production	17.4[11]	17.5[15]	17.5[15]	..	..	..		20.0[27]	20.0[27]	..
Consumption	87.0[16]	93.7[16]	100.2[16]	..	..	..		115.0[27]	140.0[27]	..
Imports - Total[31]	69.6[12]	76.2[12]	82.7[12]	..	..	..		95.0[27]	120.0[27]	..
of which: From USSR	63.3[1]	68.4[8]	69.0[13]	71.0[13]	74.0[13]	..	{ Initial plan / Period plan	78.0[28] 90.0[27]	90.0[27]	..
From other countries								41.0[8] 13.0[28]		

Consumption[17]								
455.0	472.4	490.2	..	..	..	558-585 Imp. 15 or Exp. 102	610-692 Imports: 272 or Exp. 120[34]	800?? 796[33] Exp. 4[33]
Net Exports (or imports)	53.1	65.1	73.3	..	..			

1 Emily E. Jack. J. Richard Lee. Harold H. Kent. "Outlook for Soviet Energy". in: *Soviet Economy in a New Perspective, art. cit.*, p. 473.

2 *SSSR v cifrakh 1977 godu* (The USSR in Figures in 1977). Moscow. 1978. p. 103.

3 *The International Energy Situation: Outlook to 1985.* CIA. April. 1977 (3.5-4.5 million barrels per day including Yugoslavia). Given in *The Soviet Oil Situation: An Evaluation of CIA Analyses of Soviet Oil Production.* Staff Report of the Senate Select Committee on Intelligence. United States Senate. 95th Congress. 2nd Session. US Government Printing Office. Washington. D.C.. May. 1978. p. 7.

4 *Ekonomicheskaya Gazeta.* No. 5. Moscow. January. 1979. p. 8.

5 N. K. Bajbakov. "O gosudarstvennom plane ekonomicheskogo i social'nogo razvitija SSSR na 1979 god" (Comments on the USSR Economic and Social Development Plan for 1979). *Ekonomicheskaya Gazeta.* No. 50. Moscow. December. 1978. p. 11.

6 Including oil products and liquid fuels *Vneshnyaya Torgovlya SSSR v 1976 godu* (USSR Foreign Trade in 1976). Moscow. 1977. p. 38.

7 Total exports from *Vneshnyaya Torgovlya SSSR v 1976 g., op. cit.*, p. 63. CMEA countries see *Ibid.* Sum of exports to Bulgaria, the German Democratic Republic. Hungary, Poland and Czechoslovakia. No exports to Rumania quoted. (Estimated: availabilities less exports).

8 Estimated consumption allowing for a 4 per cent growth in GNP less 1 per cent for energy savings. This method is used by *Soviet Economic Problems and Prospects, op. cit.*, p. 14. This study foresees energy savings at a rate of 2.5 per cent a year up to 1985 but, in view of the fact that 1977 is the first year to which the estimate applies. savings of only 1 per cent have been allowed for. The amount available for export has been calculated by deducting estimated consumption (390 million tons) from availabilities (554 million tons).

9 *Economic Bulletin for Europe.* Vol. 20. No. 1. Pre-publication. Trade (XXVII). United Nations. Geneva. p. 49.

10 *Soviet Economic Problems and Prospects, op. cit.*, p. 22.

11 John R. Haberstroh. "Eastern Europe: Growing Energy Problems". in *East European Economies Post Helsinki, art. cit.*, p. 382.

12 Based on: *Statisticheskij Ezhegodnik Stran Chlenov Soveta Ekonomicheskoj Vzaimopomoshchi – 1978* (Statistical Yearbook for CMEA Countries for 1978). Moscow. 1978. pp. 333-390. Oil and oil products. New imports. Rumanian exports totalling 5.8 million tons in 1975. 7.5 million tons in 1976 and 6.4 million tons in 1977 have been deducted.

13 Estimates made by John R. Haberstroh. *art. cit.*, p. 386. No provision is made for Soviet exports to Rumania.

14 Sum of Soviet production and that of the other CMEA countries.

15 *Statisticheskij Ezhegodnik Stran Chlenov Soveta Ekonomicheskoj Vzaimopomoshchi, 1978, op. cit.*, p. 76. Sum of country statistics.

16 Estimate: production plus imports.

17 Sum of Soviet production and that of the other CMEA countries.

18 *XXV S'ezd Kommunisticheskoj Partii Sovetskogo Sojuza* (25th Congress. Communist Party of the Soviet Union). Vol. 2. Moscow. 1976. p. 244.

19 See *New Issues Affecting the Energy Economy of the ECE Region in the Medium and Long Term (Preliminary Version)*, ECE. 18 January. 1978.

20 Emily E. Jack. *et al. art. cit.*, p. 473.

21 *Prospects for Soviet Oil Production.* Central Intelligence Agency. Washington. D.C.. April. 1977. pp. 2 and 9.

22 *Prospects for Soviet Oil Production. A Supplemental Analysis.* Central Intelligence Agency. Washington D.C.. April. 1977. p. 4.

23 Estimate: See *New Issues Affecting the Energy Economy of the ECE Region in the Medium and Long Term, op. cit.*

24 See *Ibid.*

25 Emily E. Jack. *et al. art. cit.*, p. 474. The authors quote a forecast for 1990 of 1 100 million tons standard coal equivalent (SCE – 7 000 kilocalories per kilogramme). For 1975. a year for which the authors give oil production in tons (491 million) and in SCE (702 million). the ratio is 1.43. This ratio has been used to convert the crude oil production forecast of 1 100 million tons SCE into crude oil tonnage ($\frac{1\ 100}{1.43} = 769$).

26 Production in 1990 will be twice what it was in 1978. Estimates made by Institut Petro Studies. Malmö. given in "Wird Moskau die Erdölproduktion weiter erhöhen?". *Neue Zürcher Zeitung. art. cit.*

27 See *New Issues Affecting the Energy Economy of the ECE Region in the Medium and Long Term. op. cit.*

28 John R. Haberstroh. *art. cit.*, p. 387.

29 30 million tons to be imported if the Soviet Union supplies the 90 million tons. and 120 million to be imported if the Soviet Union becomes a net oil importer.

30 *Soviet Economic Problems and Prospects. op. cit.*, p. 14.

31 *Ekonomicheskaya Gazeta.* No. 5. Moscow. 1980. p. 8.

32 N. K. Bajbakov. "O gosudarstvennom plane ekonomicheskogo i social'nogo razvitija SSSR na 1980 god" (Comments on the USSR Economic and Social Development Plan for 1980). *Ekonomicheskaya Gazeta.* No. 50. Moscow. 1979. p. 12.

33 *Les réserves et les approvisionnements énergétiques dans la région de la CEE. Situation actuelle et perspectives.* United Nations. New York. 1980. pp. 20 and 21.

34 According to *The International Energy Situation: Outlook to 1985.* CIA. April. 1977 (see footnote 3 of this table). The CMEA Countries will probably have to import from the outside. in 1985. from 175 to 225 million tons of oil.

149

about 12 billion³ in 1976 and is intended to grow to 23 billion³ in 1980.³⁹ This is equivalent to about 10.1 million tons of oil for 1976 and 19.6 million for 1980.⁴⁰ These exports bring in substantial hard currency earnings to the Soviet Union.⁴¹

Opinions as to future prospects for Soviet oil and gas production differ widely. Even for the immediate future (1980), predictions differ. For instance, Soviet official forecasts (the Tenth Five Year Plan) put Soviet oil production at 620-640 million tons (these forecasts are also used by the Economic Commission for Europe). However, the annual plan for 1980 foresees only a production of 606 million tons. (See Table 36). American CIA estimates give a figure of 550 and 600 million tons.⁴² Furthermore, estimates as to possible Soviet oil consumption also differ and these differences lead in turn to differing estimates of likely prospects for foreign oil trade for 1980: ranging from 15 million tons net imports to 102 million tons net exports for the CMEA countries as a whole.

Debate centres on Soviet medium- and long-term oil prospects. For 1985, the first point which is debated is what Soviet production is likely to be. The CIA experts, whose forecasts have been very widely discussed, think that Soviet oil production will reach a ceiling in 1980-1981 and then fall off, to a figure of about 400-500 million tons, in 1985. Other experts, in the United Nations Economic Commission for Europe, the World Energy Conference (figures worked out in 1974 with the help of Soviet delegates)⁴³ and the Institut Petro Studies, Malmö, Sweden, predict a sustained increase (World Energy Conference) or even a spectacular increase (Institut Petro Studies) in Soviet oil production by 1990.

The differences in these estimates affect assessments made of the CMEA countries' oil deficit or surplus in 1985, the gap between the most pessimistic and optimistic forecasts being some 400 million tons (see Table 36). We have no CIA estimates for 1990 but if they were to follow the trend forecasts for the years 1980 and 1985, the differences would climb to very much greater proportions.

During a discussion in April 1977 of the CIA estimates it became clear that the analytical techniques used by CIA analysts greatly conditioned their conclusion that the CMEA countries would become net importers of oil by 1985. As was noted, the CIA conclusion was based on the caveat that there would be no substantial improvements in conservation practices, no major cutbacks in the rates of economic growth and no changes in energy consumption patterns.⁴⁴ Nevertheless a study released after this discussion⁴⁵ repeated

39   *Economic Bulletin for Europe,* Vol. 30, No. 1, pre-publication, p. 48.
40   One billion m³ of gas is equivalent to 16,800 barrels of oil a day or 840,000 tons a year. *USSR: Development of the Gas Industry,* National Foreign Assessment Center, CIA, July 1978, p. 22. Hence the figure of 10.1 million tons of oil.
41   In 1975, oil exports brought the Soviet Union $3.2 billion in convertible currency and gas exports $52.5 million. See John P. Hardt, Ronda A. Bresnick, David Levine, "Soviet Oil and Gas in the Global Perspective", in *Project Interdependence: US and World Energy Outlook Through 1990. A* report printed at the request of John D. Dingell, Chairman, Subcommittee on Energy and Power, Committee on Interstate and Foreign Commerce, United States House of Representatives and Henry M. Jackson, Chairman, Committee on Energy and Natural Resources, Ernest F. Hollings, Vice-Chairman, The National Ocean Policy Study of the Committee on Commerce, Science and Transportation, United States Senate, by the Congressional Research Service, Library of Congress, 95th Congress, 1st Session, Washington, D.C., November 1977, p. 812. Soviet convertible currency earnings from oil exports were $4.5 billion in 1976, $5.7 billion in 1977 as reported in *Petroleum Economist,* September, 1978, p. 369 and $6.4 billion in 1978 as reported in "Wird Moskau die Erdölproduktion weiter erhöhen?", *Neue Zürcher Zeitung,* February, 1979. See also John R. Haberstroh, "Eastern Europe: Growing Energy Problems", *art. cit.*
42   Actual Soviet performance in 1979, for which the target in the annual plan is 593 million tons, will be decisive here.
43   Emily E. Jack, J. Richard Lee, Harold H. Lent, "Outlook for Soviet Energy", in *Soviet Economy in a New Perspective, op. cit.,* p. 474.
44   *The Soviet Oil Situation: An Evaluation of CIA Analyses of Soviet Oil Production, op. cit.,* pp. 7-8.
45   *Soviet Economic Problems and Prospects, op. cit.*

the earlier estimate of Soviet oil production in 1985 of 8 to 10 million barrels per day, but estimated some possible effects of conservation and fuel substitution policies on oil consumption.[46] At the same time it was underlined that the Soviet Union would incur substantial currency losses should it become a net oil importer.[47]

CIA estimates were based on assumptions made aboet US export restrictions. For instance, the estimate made in 1970 for 1975 (8.81-9.21 million barrels per day) proved to be incorrect (production in fact was 9.82 million barrels per day) because they were based on the assumption that Western bans on submersible pumps and drill bits would be retained. "As those export restrictions were lifted, the Soviets purchased large amounts of the needed equipment and their production was greater than the CIA forecast".[48] The forecasts for 1975 were revised in 1971 to give a figure of "9.81-10.01 million barrels per day, if the Soviets gained access to Western technology" — a forecast which was fairly close to actual production.[49]

The main point at issue seems to be the CIA prediction that Soviet oil production will level off in 1980-1981 and then decline.[50] In the United States the CIA's estimates of Soviet oil reserves were publicly criticised.[51] They were also questioned or refuted by non-American experts. For instance, a study by the United Nations Economic Commission for Europe even forecast a continuous increase in Soviet oil production from 640 million tons in 1980 (figure given in the Soviet Tenth Five Year Plan and adopted without objection by the Commission) to 710 million in 1985 and 780 million in 1990.[52] These optimistic estimates are still too pessimistic for the Institut Petro Studies in Sweden. The Institute predicts that Soviet oil production will double by 1990 as compared with 1978 and that exports in 1985 will be three times what they were in 1977.[53] Hard currency earnings from Soviet oil should, it is said, increase from $6.4 billion in 1978 to about $20 million in 1985.[54] These forecasts seem to be based on an expected increase in co-operation with the West as regards oil, an optimistic assessment of growth prospects in oil production in West Siberia (Samotlor) and the fact that the Soviet Union is preparing (reported by the Institut) a large-scale plan for the

46   *Ibid.,* p. 14. This study discusses a number of options open to the Soviet Union and takes into account that aspects such as conservation and substitution are important variables which affect estimates that are made. As has been pointed out, "the Agency did not, in effect, retreat from its earlier projections but now states with greater clarity the range of alternatives and the assumptions on which those alternatives rest". *The Soviet Oil Situation: An Evaluation of CIA Analyses of Soviet Oil Production, op. cit.,* p. 10. See also Emily E. Jack, J. Richard Lee and Harold H. Lent, *art. cit.,* for a detailed discussion of Soviet objectives contained in the Fifteen Year Plan period for 1976-1990 for the development of energy, the problems in attaining the stated goals and commitments and the varieties of policy alternatives open to the Soviet Union.
47   "The difference between selling 1 million b/d (as in 1976) and buying 2.7 million b/d [comprising 1.6 million for re-export to Eastern Europe and 1.1 million b/d for domestic consumption] — the projection for 1985 which assumes no unusual conservation efforts — is $17 billion in 1977 prices, more than the USSR's total 1976 hardcurrency imports. A substantial rise in real oil prices, which is likely to occur at the very time when the USSR is becoming a net oil importer, would further increase the hard currency drain". *Soviet Economic Problems and Prospects, op. cit.,* p. 22.
48   *The Soviet Oil Situation: An Evaluation of CIA Analyses of Soviet Oil Production, op. cit.,* p. 11.
49   *Ibid.* "In 1973, 1974, 1975 actual production (according to CIA collections of Soviet statistics) was 8.58, 9.18 and 9.83 respectively". *Ibid.*
50   Pessimistic evaluation on trends in Soviet oil production are based on: "a) limitations on the capacity of the drilling industry; b) relatively inefficient technology and low labour productivity; c) increasing water cut and projected deliverability declines in some Soviet fields, in particular Samotlor field and, d) a lack of recent new discoveries to replace declining production from c)". See *New Issues Affecting the Energy Economy of the ECE Region in the Medium and Long Term, op. cit.*
51   For instance, Marshall I. Goldman "did not believe that the Soviet Union and Eastern Europe would become a net importer of oil since he did not accept the Agency's projection of declining Soviet production. Second, even if production failed, the Soviet Union would not be able to obtain the hard currency necessary to finance imports on the scale projected by the CIA studies". *The Soviet Oil Situation: An Evaluation of CIA Analyses of Soviet Oil Production, op. cit.,* pp. 5-7.
52   See Table 36.
53   Given in "Wird Moskau die Erdölproduktion weiter erhöhen?", *art. cit.*
54   *Ibid.*

development of the USSR's oil industry.[55] It is for these reasons that it predicts that the slow growth rate in the Soviet oil industry will not last.[56] This conclusion has also been disputed. Taking as the base potential future gas production (there is less argument about forecasts in this field), there are analysts who believe that even if in the future gas exports eventually take over from oil exports (as has been argued), gas exports from the CMEA countries — estimated at about 46 million tons for 1985 — could most likely not, over the next decade, make up for a possible Soviet oil shortage.[57] A less spectacular assessment is that made by the Deutsche Institut für Wirtschaftsforschung, Berlin, which predicts that until 1985 there will be a stable production growth which will not only take care of the oil needs of the Soviet Union and that of the other CMEA countries, but will enable oil shipments to the West — even though the volume of such shipments could possibly decrease.[58]

It is quite evident that the different estimates that have been made of Soviet oil and gas production in the 1980s do not coincide. This is an issue that cannot be resolved in this report. Whatever its ultimate resolution, the answer will have a decisive effect on East-West economic relations in general and on technology transfer in particular. In any event, the stress placed by Soviet authorities on technology imports necessary to ensure increased oil production is indisputable.[59] Soviet success or failure in this field will have a direct consequence for its economic growth.[60]

55 *Ibid.*

56 *Ibid.* According to the *Neue Zürcher Zeitung*, the Petro Studies' Report stresses the fact that the Soviet Union intends to "conquer Western markets". Because of "Soviet interest in furthering exports which earn hard currency, the Soviets would reduce their oil exports to the East European countries but would increase gas exports to them". It is pointed out in this regard that "since the 1960s the growth potential of Soviet energy has not kept pace with the growing energy needs of the CMEA countries. While the volume of exports of energy to the West has increased, the share going to the East has decreased. In addition, the Soviet Union needs the hard currency earnings which result from exports to the West (where, in addition, higher prices are charged) to finance imports of machinery and grain. It is for these reasons that the Soviet Union is interested in slowing down its energy exports to the Eastern countries. It is therefore assumed that the USSR will, as of 1980, not augment its oil deliveries to the CMEA partners". See "Die 32te Tagung des Rates für Gegenseitige Wirtschaftshilfe", *op. cit.,* p. 10. The East European countries have also been subjected to a price increase for Soviet oil — to a level just 12 per cent below OPEC prices. See *Business Eastern Europe,* Vol. 8, No. 8, 23rd February 1979, p. 62.

57 See Emily E. Jack, J. Richard Lee and Harold H. Lent, *art. cit.* The forecast for natural gas production in the Tenth Five Year Plan is 400-435 billion m³ in 1980. However, it is thought that "production in 1980 is unlikely to exceed 390 billion cubic metres". See Emily E. Jack, *et al., art. cit.,* pp. 463-464.

58 Given in "Wird Moskau die Erdölproduktion weiter erhöhen?", *art. cit.*

59 Western technology is needed to tap Soviet energy reserves. For instance, the development of the Azerbaidzhan oil fields "has been hampered by a lack of technology for developing the deeper regions beyond the shoreline. Soviet oilmen hope, with the help of US equipment, to send probes into the deeper water of the southern and southeastern Caspian..." "Soviet Oil Industry Eager to get US Technology", *International Herald Tribune,* Paris, 19th February 1979, p. 4. It is pointed out that access to "US submersible pumps or any high volume lifting equipment for increased petroleum output in the Samotlor field in West Siberia and offshore platforms in the Caspian Sea, is especially important". John P. Hardt, Ronda A. Bresnick, David Levine, "Soviet Oil and Gas in the Global Perspective", *art. cit.,* p. 790. "US seismic equipment is said to be among the most technologically advanced equipment now available to Soviet petroleum geophysicists". *Ibid.,* p. 793. "Energy related technology and equipment imports have ... been rising rapidly throughout the 1970s ... The type of Western technology being imported by the Soviets includes: large diameter pipe for the transmission of oil and gas, oil refining equipment, exploration equipment (especially for offshore oil and gas reserves), drill bits, good quality pipe for drilling and casing, and modern seismic equipment". *Ibid.,* p. 798. See also *USSR: Development of the Gas Industry, op. cit.* The conclusion of this analysis is that "until Western-type equipment is purchased and used more widely by the Soviets, most USSR offshore zones probably will remain relatively unexplored and untapped. The Soviet gas industry indeed will remain dependent on imports of Western equipment into the 1980s". See pages 12 and 27. See also James P. Lister, "Siberia and the Soviet Far East: Development Policies and the Yakutia Gas Project", *Regional Development in the USSR – Colloquium 1979,* NATO, Economics Directorate, Brussels, 25th-27th April, 1979. This analysis discusses in detail the USSR's development plans for Siberia (hydrocarbon fuels, particularly oil and natural gas) and the potential role to be played by foreign companies in fulfilling these plans.

60 Besides relying on Western technology, the Eastern countries will be increasingly forced to develop their own energy reserves. This is certainly the reason why the development of the energy sector — including atomic energy — is stressed in all the individual economic plans. It may also explain the

(Cont'd on next page)

152

ii) *Plant and Equipment*

It is estimated that Soviet imports of plant and equipment designed to further increase export capacity will be given a priority second only to energy related imports. Furthermore Soviet reliance on compensation agreements guaranteeing future export earnings will increase.[61]

It is also anticipated that Soviet needs for construction equipment will be particularly acute. As one Soviet businessman noted, "our construction equipment needs are so great that we could take the full output of International Harvester and of Caterpillar together and we still could not have enough to satisfy our builders. It's a matter of what can be ordered in line with our resources".[62]

iii) *Agriculture*

Fulfilment of Soviet agricultural production plans are dependent on the acquisition of Western technology in the fertiliser and meat production area. Little emphasis is placed on import of machinery.[63]

iv) *Development of New Lands*

The development of new lands for agricultural use is given great emphasis in the Tenth Five Year Plan. The huge projects of the Ministry of Land Reclamation and Irrigation will continue well into the 1980s. One such project — connecting the Ob river by canal with the Aral Sea — will constitute one of the largest earth-moving efforts ever undertaken.[64] This and related efforts will require huge quantities of earth-moving machinery and will greatly influence Soviet imports of Western machinery.[65]

v) *Consumer Goods*

Given the above, one can anticipate that Soviet imports of consumer goods and non-essential foodstuffs will most likely be cut drastically.[66] Soviet imports of Western machinery for these industries will therefore most likely decline.

2. Promotion of Western Technology Transfer

a) *A New Approach to the International Division of Labour*

The notion that trade may procure substantial gains has not only found support among East European scholars, but has become the subject of numerous official or semi-official

(Cont'd note 60)

emphasis given to the Long Term Target Programmes, in particular, the importance given to energy, fuel and raw material development. In this regard it has been noted that likely curtailments of Soviet energy exports to the East European countries "have stimulated efforts within COMECON for joint planning and development of the region's energy resources, especially those located in the USSR". The Long Term Target Programme for energy development "will entail substantial joint investments in the form of low-interest investment credits, manpower, equipment, technical assistance and, in some cases, hard currency by COMECON members for the development of raw-material and energy reserves located primarily in the Soviet Union". John M. Kramer, "Between Scylla and Charybdis: The Politics of Eastern Europe's Energy Problem", *ORBIS*, Vol. 22, No. 4, Foreign Policy Research Institute, Winter 1979, p. 939. See also Chapter 1.

61   *Soviet Economic Problems and Prospects, op. cit.*, p. 25. See also section 2 below.

62   Cited by Robert J. McMenamin, *art. cit.*, p. 161. It is noted that "as in the past, machinery and transport equipment will account for about 40 per cent of Soviet imports from the West". Given in "The Business Outlook — Soviet Union", *Business Eastern Europe, art. cit.*, p. 46. See also Chapter 2.

63   Contrary to the Soviet Union, Poland and Hungary are emphasising agricultural machinery imports. Robert J. McMenamin, *art. cit.*, p. 159.

64   *Ibid.*, p. 160.

65   Apart from the specific problem of new land development, it is thought that the Soviet Union has chosen to further regional development, particularly Siberia and the Far East, by means of Western know-how and technology. James P. Lister, *art. cit.*, p. 4.

66   *Soviet Economic Problems and Prospects, op. cit.*, p. 25.

policy statements. Speaking in 1971, Brezhnev noted that "... we can be sure that the expansion of international trade will have an advantageous effect on the improvement of all our industrial performance".[67] Even more to the point, he stated that "our plans — these are not plans calculated for autarky. We do not have a policy of isolating our country from the outside world. On the contrary, we forecast that it will develop on the basis of extensive co-operation with the outside world, not only with socialist countries but also in a significant manner with those countries having a contrary social system".[68]

The "nurturing" of foreign economic relations was declared as being a "key" consideration in setting domestic economic policy. One consequence of this was the formulation of an economic growth strategy which was export oriented since "foreign trade has become an important branch of the economy".[69] Thus at the 25th Party Congress in 1976 it was argued that the development of foreign economic relations was becoming increasingly important for the solution of the country's basic economic problems:

"The Soviet Union's foreign economic activity acquires wider aspects. It aims to include the country in a more active international division of labour and toward the resolution of such actual problems as developing long-term industrial co-operation between East and West and participation of socialist countries in the solution of world energy and raw materials problems. As is shown by [the] practice during the last years, it exists independently of ideological controversy, an objective interest of socialist and capitalist countries, in the development of stable and long-run mutual co-operation".[70]

This trend was emphasised in a resolution adopted by the Central Committee of the Communist Party on 31st January, 1977 in connection with the 60th anniversary of the October Revolution:

"at the base of a profound modification in the balance of power in the world intervenes a profound change of the whole system of international relations".[71]

These new attitudes and ideas were integrated as policy objectives in the Tenth Five Year Plan as can be seen from a statement made by Prime Minister A. N. Kosygin:

"Given the condition of the easing of international tensions, our economic relations with the developed capitalist countries, which can expand rapidly on the basis of the principles set forth in the Final Act of the Conference on Security and Co-operation in Europe, are taking new qualitative aspects. We shall continue the practice of concluding large-scale agreements on co-operation in the erection of industrial projects in our country and on the participation of Soviet organisations in the construction of industrial enterprises in Western countries. Agreements on a compensatory basis, especially with short reimbursement periods for new enterprises, various forms of industrial co-operation and joint scientific research and design work are promising forms of co-operation. Needless to say, our trade and economic ties will develop faster with countries that show a sincere willingness for co-operation and concern for ensuring normal and equitable conditions for its development. Only in this event is it possible to

67 Given by Peter Wiles, "On the Prevention of Technology Transfer" in *East-West Technological Co-operation, NATO Colloquium 1976, op. cit.,* p. 27.

68 *Pravda,* 2nd May, 1973 as quoted in A. M. Voinov, V. Ja. Iokhin, L. A. Rodina, *Ekonomicheskie otnoshenija mezhdu socialisticheskimi i razvitymi kapitalisticheskimi stranami* (Economic Relations Among Socialist and Developed Capitalist Countries), Moscow, 1975, p. 10.

69 "Probleme der sowjetischen Aussenwirtschaftspolitik", *Neue Zürcher Zeitung,* Fernausgabe Nr. 32, 9 Februar 1978, p. 11. The source quotes from a speech given by Mr. Kosygin given at the 25th Party Congress.

70 B. S. Vaganov, Introduction to the collective work *Vneshne ekonomicheskie svjazi Sovetskogo Sojuza na novom etape* (Foreign Economic Relations of the Soviet Union at a New Stage), Moscow, 1977, pp. 6-7.

71 *Ibid.*

have truly broad and lasting economic relations which will be reflected in our national economic plans".[72]

This new approach towards the international division of labour was applied to the CMEA countries as a "bloc". Thus it has been said that "socialist economic integration is not only conducive to the development of the CMEA nations' productive forces and their mutual co-operation, it also provided broad opportunities for extending trade and economic relations with the developing countries and advanced industrial capitalist States. In line with the fraternal countries' policy of peaceful coexistence and proceeding from the fact that the international socialist division of labour is connected with the world division of labour, the CMEA nations develop economic, scientific and technical co-operation based on equality, mutual benefit and respect the sovereignty with all other countries irrespective of their social and State system. This is the long-term policy of the Comprehensive Programme".[73]

Soviet exports to the Western industrialised countries consist primarily of energy products and raw materials. Some authors believe that East-West trade cannot be expanded nor can the international division of labour be developed, on the basis of this export structure. Both require a major expansion of manufactured exports and the consequent development of industrial capacity specifically oriented to the external market.[74] Soviet leaders are aware of this, as can be seen from a statement made by Kosygin at the 25th Party Congress: "since foreign trade has become an important sector of the national economy, the question also arises of the organisation, in a number of cases of special production facilities oriented towards export and the satisfaction of the specific requirements of foreign markets".[75] This aspect was repeatedly stressed during the Party Congress. For instance not only were production facilities to be oriented towards foreign market requirements in order to "contribute to export potential thereby giving a maximum contribution to economic growth... [but] imports were to be regarded as an instrument for the realisation of the economic strategy as given in the State Plan; particularly for maximally expediting technical progress".[76] As noted by a Western analyst, Soviet industrial leaders were consequently expected to study foreign needs and to create or expand specialised lines of production in order to "increase the present range of the kind of export products that are in demand on the world market, in particular, machinery and equipment".[77] As indicated by Mr. Brezhnev in his report to the 25th Party Congress "... in our exports there must be a substantial increase in the share of the output of the manufacturing industry. To this end it is necessary to enlarge

72 A. N. Kosygin, "Main Directions of the Development of the National Economy of the USSR for the Years 1976-1980", in *XXV S'jezd Kommunisticheskoi Partii Sovetskogo Sojuza* (25th Congress of the Communist Party of the Soviet Union), Vol. 2, Moscow, 1976, p. 30.
73 N. Patolichev, (USSR Minister of Foreign Trade, Chariman of the CMEA Standing Commission for Foreign Trade), "Thirty Years of Co-operation in Foreign Trade", *Foreign Trade*, No. 4, USSR Ministry of Foreign Trade, (English translation), 1979, p. 7.
74 Carl H. McMillan, "Direct Soviet and East European Investment in the Industrialised Western Economies", Working Paper No. 7, East-West Commercial Relations Series, Institute of Soviet and East European Studies, Carleton University, Ottawa, Canada, February 1977, p. 25.
However, past trading patterns and Soviet endowments would seem to indicate that the Soviet Union has a comparative advantage in the fields of energy products and raw materials. Furthermore, several large current projects under construction involve Soviet purchase of energy-related technology to be compensated by future Soviet deliveries of energy resources to the West. Nevertheless, there are analysts who believe that Soviet energy resources and raw materials can only be exploited with the help of Western technology.
75 A. N. Kosygin, "Main Directives of the Development of the National Economy of the USSR for the Years 1976-1980", *op. cit.* This official policy shift greatly helped the East European countries which, as noted in Chapter 1 (see Part IV), had supported closer trade relations with the West as early as the mid-1960s.
76 "Probleme der sowjetischen Aussenwirtschaftspolitik", *art. cit.*
77 *Vneshnyaya Torgovlya* (Foreign Trade), Moscow, May 1966, p. 4 as given by Samuel Pisar, *Co-existence and Commerce, op. cit.*, p. 33.

the output of goods in demand in foreign markets and make them more competitive".[78]

This shift pertains also to the other CMEA countries. Their trade with the Soviet Union must be complemented by trade with the West and their exports and production structure reoriented. Reliance on Western trade has varied widely from country to country. Czechoslovakian imports from the CMEA countries have grown more rapidly than those from the West. For Rumania, more than half of her imports come from the West. For the others, the fraction varies between one-third and one-half.[79] Nevertheless all the CMEA countries are "trying to increase their hard currency earnings by exporting complementary goods (raw materials and foodstuffs) as well as substitutional (manufactured) goods".[80]

b)   *Technology Imports as a Factor Motivating the New Approach Toward Trade*

The shift in Soviet attitudes towards Western technology is, partially, ideologically based. It is a development that has had profound repercussions in the post-Khruschev era. Thus the designation of science, in the 1961 Party Programme, as a "direct productive force" caused its shift out of the Marxian "superstructure into the base".[81] It is this shift which enhanced the role that science and technology was to play in economic development.

This new and enhanced role of S & T appears in the various economic reforms which have been attempted and in the succession of economic plans adopted. Thus, the economic reforms instituted in 1965 had as one important aim the encouragement of the growth of technology. At the 23rd Party Congress in 1966, the desirability of importing technology was specifically mentioned.[82] During the 24th Party Congress in 1971, expanded East-West economic relations were officially encouraged. By the time of the 25th Party Congress in 1976, the critical role of Western technology in improving Soviet economic performance was stressed. In the opinion of some, this event signalled a new Soviet attitude to technology transfer; namely, that Western technology has a direct role to play in the fulfilment of Soviet economy plans.[83]

All the recent Five Year Plans reflect this change in the role of Western technology. The Eighth Five Year Plan (1966-1970) provided an important role for Western technology in Soviet economic planning. The Ninth (1971-1975) and Tenth Plan (1976-1980) are believed to reflect the strategy of selective technology transfer.[84] The Tenth Plan in particular is

---

78   D. Petrov, "Export of Soviet-Made Industrial Equipment", *Foreign Trade*, No. 1, USSR Ministry of Foreign Trade, (English translation), 1979, p. 13. The article notes that "in pursuit of this task, especially over the last few years, the foreign trade associations concerned have spared no pains to increase the output of machines and other equipment for export. In the 1976-1980 period the USSR is to increase its industrial exports by 30-35 per cent". *Ibid.*, pp. 13 and 23. A similar effort is being made by the East European countries. As regards Poland, for instance, it is said that "the thrust of Polish foreign trade policy thinking is to narrow the scope of export specialisation from some 17 branches which enjoyed top priority rating in the past to some half a dozen areas which have proven themselves to be internationally competitive... Major emphasis will be placed on developing the export potential of the electrical engineering and service sectors". "Poland to Update FT Code: Co-operation No Panacea", *Business Eastern Europe*, Vol. 8, No. 5, February 2, 1979, p. 34.

79   See the Annex Tables.

80   Max Baumer and Hanns-Dieter Jacobsen, "CMEA and the World Economy: Institutional Concepts", *art. cit.*, p. 1001. The potential effect of this development is treated in Chapter 7. As was noted in Chapter 1 (V, section 2 a) the Soviet Union is actively supporting CMEA integration. This could, according to some observers, lead to a reduction of trade with the West. Whether and/or the extent to which the goal of CMEA integration and the intent to orient Eastern production to Western markets could pose a "policy conflict", is an interesting issue.

81   Cf. Arnold Buchholz in *Ost-Europa*, May 1972 as given in Peter Wiles, "On the Prevention of Technology Transfer", *art. cit.*, pp. 24-25.

82   It is felt that this objective was encouraged by the decision to motorise; Soviet imports from the West rose sharply in the late 1960s.

83   See for instance John P. Hardt, "The Role of Western Technology in Soviet Economic Plans", in *East-West Technological Co-operation, NATO Colloquium 1976, op. cit.*, p. 315.

84   *Ibid.* See also *Technology Transfer and Scientific Co-operation Between the United States and the Soviet Union: A Review, op. cit.*, pp. 77-80.

considered to be particularly important in that it reflects "the change in Soviet economic policy from the classical growth strategy based on increases in factor inputs, to the modern growth strategy based on high rates of technological progress".[85] The issue of feasibility aside, the evidence indicates that Soviet authorities believe that importing modern equipment and technology can be an efficient way of increasing their productive capacity.

It is worthwhile noting that the above mentioned shifts have been the focus of a good deal of dissension among Soviet leaders. Issues such as the need for more rapid technological progress in order to meet Soviet growth targets, the evaluation of the Soviet economy's success (and in particular, the pace of technological innovation by Soviet industry) and the transition to more intensive growth, in general, and more rapid technological advance, in particular, have been debated at the highest Party and government levels.[86] Since Soviet leaders have also variously appraised the USSR's economic and technological performance they have come to different conclusions about the importance of drawing on Western technology. It is of course difficult to accurately evaluate the different opinions. One Western author, having carefully assembled and studied a larger number of official statements, comes to the conclusion that the evaluation of the need for Western technology was the subject of serious disagreement among top Soviet leaders. He believes that this disagreement was sharp and overt in the 1960s, whereas in the 1970s it became more muted. He also believes that this disagreement likewise persisted among secondary elite groups which could influence opinions held by the top leadership.[87] Such discussions reflect disagreement as to the alternative paths to be followed in order to accelerate technological progress. Their outcome will have an important effect on the future role of Western technology in Soviet development, and on Western policy towards the East — based partly on assumptions about how the Eastern systems will evolve.[88]

c)   *Significance of the Arrangements Made to Transfer Western Technology*

The new policy toward technology imports has been accompanied by changes in the forms of transfer.[89] The most important change has been the shift in emphasis from non-negotiable to negotiable transfer channels. This is "primarily a shift in favour of embodied transfer, exemplified by machinery imports, and the commercial purchase of licences and know-how".[90] The reasons for this shift have been made quite clear by Kosygin:

85   Joseph S. Berliner, "Prospects for Technological Progress", *art. cit.*, p. 446.

86   A sampling is given by Bruce Parrott, "Technological Progress and Soviet Politics", in John R. Thomas and Ursula M. Kruse-Vaucienne, editors, *Soviet Science and Technology, Domestic and Foreign Perspectives,* National Science Foundation, George Washington University Press, Washington, D.C., 1977.

87   See Bruce Parrott, *Soviet Technological Progress and Western Technology Transfer to the USSR: An Analysis of Soviet Attitudes,* paper prepared for the Office of External Research, Bureau of Intelligence and Research, US Department of State, Washington, D.C., July, 1978, p. iv.

88   In this regard, one analysis conjectures that while a number of prominent Soviet leaders, among them Mr. Brezhnev, have "concentrated their attention on the new economic problems [and] see these problems as historically novel and unsusceptible to the institutional and foreign-policy solutions which succeeded in the past... Mr. Brezhnev has been unable to build a solid leadership coalition in favour of this view. The gravity of the USSR's technological shortcomings and economic inefficiency has not yet been accepted by a dominant segment of the top leadership... The secondary elites are also divided about the seriousness of the technological problem and the advisability of a "Western solution"... Soviet technological problems are very real. But given the divisions within the USSR in assessing their significance, there are limits on the value of Western technology as a tool of American policy toward that country. The value of technology transfer in such dealings is by no means negligible. But if pressed too far, the attempt to use it as a lever on other issues is likely to shift the political balance within the Soviet elite in favour of those who think the USSR's technological needs are not critical enough to warrant large foreign policy concessions". *Ibid.,* pp. 53-55. This aspect is taken up in Part II, section 4 of the Chapter.

89   Transfer forms have been described in Chapter 3. The purpose here is to indicate some of their policy implications.

90   Philip Hanson, "International Technology Transfer from the West to the USSR", in *Soviet Economy in A New Perspective – A Compendium of Papers, op. cit.,* p. 794.

"Until recently we tended to under-estimate the importance of trading in patents and licences. Such trade is playing an increasingly prominent role in the world today and is developing faster than commerce in industrial goods. Our scientific and technical personnel are able to create — and this can be proven in practice — up-to-date machines and equipment. We can and must, therefore, assume our due place in the world's licence market. In some cases, we too could profit by purchasing licences, rather than resolving the problems concerned ourselves. Purchases of patent rights will enable us to save hundreds of millions of rubles on scientific research during the coming five years".[91]

The new transfer forms not only transmit knowledge through the delivery of entire factories but include provisions for continuous Western technical assistance. This represents a departure from the "once and for all catching up" approach toward technology imports characteristic of the Stalinist era when technology was acquired through machinery imports and purchase of entire factories. The number of foreign specialists and engineers required to instal equipment was kept to a minimum. Only a few dozen specialists were sent by the Ford Motor Company to work at the Gorky automobile plant (in the 1930s), whereas some 2,500 Western personnel were sent to help construct the Zhiguli plant at Togliatta in 1966.[92] A similar number of Soviet technicians went to Italy for training and technical work. Furthermore, Fiat's contract provided not only "for it to sell machinery and equipment, but for Fiat to act as a consultant for other Soviet purchases in the West".[93] It is thought therefore, that the Fiat contract "represents a significant political concession and an important new development in Soviet economic relations with the West".[94]

The Fiat type arrangement represents only an example of the new approach toward the delivery of turnkey plants.[95] Other arrangements, such as co-production and specialisation, joint tendering or joint construction and joint ventures (where permitted) represent a combination of embodied and disembodied technology transfer.[96]

All of the above-mentioned arrangements are very flexible acquisition mechanisms. They facilitate the procurement of whole production systems, provide the possibility of obtaining new, non-standardised technology generally not available through the market, or provide the possibility that the full technological capability will be assimilated through continuing co-operation with the Western firm from which the technology is obtained.[97]

Based on available information, it seems that industrial co-operation agreements emphasise technology intensive industries. A survey of major US companies substantiates the belief that the Soviets are interested primarily in having access to the most modern technology and equipment. The survey indicates that the Soviets have mainly sought out US companies possessing sophisticated technology and that almost fifty per cent of these companies' negotiations were in sensitive high-technology product lines.[98]

Furthermore, the new modes of industrial co-operation are said to help the Soviets import high-technology machinery and equipment while minimizing hard currency outlays.

91 As given by Samual Pisar, *Co-Existence and Commerce, op. cit.*, p. 38.
92 Marshall I. Goldman, "Autarchy or Integration – the USSR and the World Economy", in *Soviet Economy in a New Perspective – A Compendium of Papers, op. cit.*, p. 94.
93 John P. Hardt and George D. Holliday, "Technology Transfer and Change in the Soviet Economic System", *art. cit.*, p. 204.
94 *Ibid.* See also Chapter 5.
95 The Fiat arrangement, while certainly a departure from previous practice, has not been followed in the construction of the Kama river truck plant.
96 See Chapters 3 and 6.
97 Carl H. McMillan, "East-West Industrial Co-operation", *art. cit.*, p. 1193. See also Chapter 3.
98 Robert W. Clawson and William F. Kolarik, Jr., "Trade, Technology and Soviet R & D: US Corporate Executives' View", paper presented at the 1976 AAASS Annual Conference Panel, "The State of Soviet Research and Development", Kent State University, Kent, Ohio, October 8, 1976, p. 13.

For instance, "most of the co-production agreements involve technical and financial participation of Western firms in the exploitation of natural resources or in the construction of plants. The Western partners generally supply equipment and technical services on credit and are repaid by deliveries of raw materials or commodities produced in the co-production agreements".[99]

A number of Western analysts believe that the CMEA countries see the involvement of private enterprise in their economies as one means of guaranteeing more "efficient application of modern technology than could be achieved through their own R & D or by state imports of Western capital goods".[100] This is one explanation of the willingness of CMEA countries to incur a high degree of indebtedness. The provision of needed technology helps overcome the built-in restraints of centrally planned systems with their unfavourable effects on growth.[101]

The Eastern countries have certain preferences for different types of co-operative agreements and their interest in industrial co-operation differs.[102] Bulgaria for instance, had only signed 56 non-equity co-operation agreements by 1976.[103] Czechoslovakia places a great deal of importance on industrial co-operation and views it as a "potential means to improve the structure of the country's industrial production".[104] As a matter of policy, the German Democratic Republic has limited its involvement with the West. However, interest in obtaining industrial raw materials and intermediate products as well as Western labour-saving technology has recently led to a growing interest by the German Democratic Republic in co-operation agreements.[105] Hungary has been most active in concluding industrial co-operation agreements. A large share of these agreements have been concentrated on forms of production specialisation. The primary aim of Hungarian enterprises seems to be to find import substitutes.[106] Poland follows Hungary in terms of the number of industrial co-operation agreements concluded. However, it is generally believed that the Polish agreements are of greater average monetary value. Poland's interest in industrial co-operation is "like that of the other socialist countries, linked to a perceived need to improve the composition of trade with the West".[107] Rumania's interest in industrial co-operation agreements is high. "Acquisition of plant and equipment with at least partial payment in resulting product, plays a major role in Rumanian agreements".[108]

Several of the East European countries have allowed some form of foreign investment in joint venture, participation in management and in profit sharing. Hungary, for instance, "allows Western investment in the joint company participation in management and sharing in company profits. Bulgaria permits profit sharing from joint co-operation activities and Poland has passed legislation which opens the door to foreign investment in certain domestic

99   John P. Hardt and George D. Holliday, *US-Soviet Commercial Relations: The Interplay of Economics, Technology Transfer, and Diplomacy*, prepared for the Sub-Committee on National Security Policy and Scientific Developments of the Committee on Foreign Affairs, US House of Representatives, US Government Printing Office, Washington, D.C., 10th June, 1973, pp. 38-39.
100   Peter Knirsch, *art. cit.*, p. 13.
101   *Ibid.*, p. 20.
102   See Chapter 3.
103   See Table 21. However, there is an "increasing demand to link Western business with counterpurchases, product buy-back or co-operation agreements. Countertrade is requested for deals over $1 million". "Berlin Trade Unit Issues Guide to Countertrade", *Business Eastern Europe*, Vol. 8, No. 5, February 2, 1979, p. 37. See also Chapters 6 and 7.
104   Carl H. McMillan, "East-West Industrial Co-operation", *art. cit.*, p. 1202.
105   *Ibid.*, p. 1205. "Countertrade demands are now frequently as high as 40 per cent (compared to 20 per cent formerly)". "Berlin Trade Unit Issues Guide to Countertrade", *Business Eastern Europe, art. cit.*, p. 37. See also Chapters 6 and 7.
106   Carl H. McMillan, "East-West Industrial Co-operation", *art. cit.*, pp. 1206-1208.
107   *Ibid.*, pp. 1209 and 1210. See also Zbigniew M. Fallenbuchl, "The Polish Economy in the 1970s", in *East European Economies Post-Helsinki — A Compendium of Papers, op. cit.* "Present countertrade levels of 25-30 per cent are expected to rise to 50 per cent by 1980". See "Berlin Trade Unit Issues Guide to Countertrade", *Business Eastern Europe, art. cit.*, p. 37. See also Chapters 6 and 7.
108   Carl H. McMillan, "East-West Industrial Co-operation", *art. cit.*, pp. 1213 and 1214.

industrial enterprises".[109] Such practices, at variance with Marxist-Leninist doctrine "seem to represent a compromise between ideological principles and practical objectives".[110] It is felt that "Eastern investment in fixed capital assets in the industrialised Western economies constitutes important, additional evidence of the commitment of the socialist countries to a long-term, pragmatic expansion of relations with the developed capitalist countries".[111] This pragmatic approach towards investment is thought to "parallel policy measures (increased import of Western technology, scientific and industrial co-operation) taken to improve the marketability of Eastern manufactured exports".[112]

d) Détente

Détente is part of the attempt, initiated largely by the West, to transform "controlled competition". as exercised during the Cold War, into a process of more "relaxed co-operation". It is believed that scientific and technological co-operation can be an important tool with which to achieve this transition.

Détente is an off-shoot from the policy of "peaceful co-existence" inaugurated by Khruschev. Peaceful co-existence itself is much less of an innovation in recent Soviet foreign policy than is often believed. It was an "essential ingredient of Lenin's political strategy both before and after 1917 that when operating from a position of weakness one had to exploit contradictions in the enemy camp, and this entailed a readiness to make compacts with any government or political grouping, whatever its ideology".[113] For example:

"... in its Leninist understanding peaceful coexistence signifies neither the preservation of the social or political status quo nor the moderation of the ideological struggle. In fact, it has facilitated and facilitates the development of the class struggle against imperialism inside individual countries as well as on a world scale".[114]

Translated into the present, some Western analysts believe that in its relations with Europe the "Soviet State defines the concept of peaceful co-existence as a function of its own power to coerce, to intervene and to defend".[115]

---

109 *East European Economies Post-Helsinki – A Compendium of Papers, op. cit.,* p. XIX. "Poland's below-par foreign trade performance has prompted the country's leadership to revise the foreign trade laws... According to Ministry of Foreign Trade officials, the aim is to eliminate the numerous barriers standing in the way of Polish foreign trade development... The new code will provide a uniform legal framework for industrial co-operation agreements, joint-stock companies with Western equity in Poland, Polish JVs [joint ventures] abroad and Western company offices in Poland... According to official Polish sources, the most far-reaching changes in the proposed legislation will deal with rationalising and streamlining provisions for industrial co-operation agreements". "Poland to Update FT Code: Co-operation No Panacea", *Business Eastern Europe,* Vol. 8, No. 5, February 2, 1979, p. 33. See also Chapter 3.
110 Carl H. McMillan, "Direct Soviet and East European Investment in the Industrialised Western Economies", *art. cit.,* pp. 5-6.
111 *Ibid.,* p. 6. The same author notes that "283 cases of direct investment by the USSR and the six East European members of CMEA in 16 Western industrial countries, have been documented. Of the total number of individual Eastern investments, 84% are in jointly owned companies. The remaining 16% is made up of wholly owned companies concentrated principally in banking, insurance and shipping". See *Ibid.,* p. 11.
112 *Ibid.,* p. 23.
113 Richard Pipes, "Détente: Moscow's View", in Richard Pipes, editor, *Soviet Strategy in Europe,* Strategic Studies Centre, Stanford Research Institute, Crane, Russak and Company Inc., New York, 1976, p. 13.
114 See Central Committee of the CPSU, "Theses for the Centenary of Lenin's Birth", *Pravda,* 23rd December, 1969 as given in Mose L. Harvey, Leon Goure and Vladimir Prokofieff, *Science and Technology as an Instrument of Soviet Policy,* Centre for Advanced International Studies, University of Miami, 1972, p. 3.
115 Lothar Ruehl, "Soviet Policy and the Domestic Politics of Western Europe", in Richard Pipes, ed., *op. cit.,* p. 81.

The link between scientific and technological co-operation and co-existence was made by Mr. Brezhnev at the 25th Party Congress. He noted that "economic and scientific-technical ties with the capitalist states, strengthen and broaden the material basis of the policy of peaceful co-existence".[116] At the same Congress, Mr. Brezhnev also emphasised that the Soviet Union could not refrain from aiding the resolution of such important economically-related matters of international concern as the availability of natural and energy resources or the mastery of outer-space related problems.

With the transition from peaceful co-existence to détente, statements similar to the one above form a recurrent motif in Soviet pronouncements. For instance, in 1971, in connection with the preparations for the European Security Conference, Mr. Kosygin stated that: "Certainly the scale of our economic relations with the Western States could be quite different if we could succeed in making constructive steps towards those current problems which at present disturb the international situation. It is well known that the Soviet Union attaches a great importance to the convocation of the Conference on European Security. For us it is perfectly clear that this conference will strengthen the confidence of Europeans and can open the way to broad economic and scientific and technical co-operation".[117] This theme continues to be supported since the Helsinki Conference took place. It is said, for instance, that "the Helsinki accords open up new prospects for progress both in bilateral and multilateral trade and economic relations. International détente and the CMEA countries' rapidly growing productive forces have brought about significant changes in their economic ties with the advanced capitalist countries... New forms of co-operation in the field of industrial, scientific and technological exchange have developed. All this has lent their economic relations a large-scale, stable and long-term character".[118]

While the advantages of foreign trade have been stressed in terms of furthering détente, Soviet policy towards détente must be placed into perspective. First it must be underscored that Western and Eastern interpretations of détente differ. The United States, in particular, has attempted to use increased trade in order to extract political concessions.[119] The Soviet Union is interested in détente because it has eased the way for closer trade relationships and hence wider access to Western technology and know-how. That the Soviet interpretation of détente does not imply wide-reaching political concessions has been shown, for example, by the reaction to the Jackson-Vanik amendment. As noted by one Western writer the "only word that is agreed on between the United States and the Soviet Union in the definition of détente is the term peaceful co-existence".[120] The détente relationship is therefore based upon Western acceptance of the Soviet foreign policy of peaceful co-existence.[121] The Soviets seem to view détente as "a dynamic opportunity to pursue such policies as support for national liberation and to become stronger utilising the very incentives — trade and scientific and technological exchanges — intended by the United States to be the inhibiting influence".[122]

Détente is supported by the Soviet Union because a less hostile environment is necessary for closer relations. It is believed that closer relations were sought for essentially two reasons. First, "isolation could not be a long-term policy in the emerging era of global communications, cultural and scientific exchanges and raw materials interdependence".[123]

116   L. I. Brezhnev, *Ekonomicheskaya gazeta*, No. 9, February, 1976, p. 11.
117   A. N. Kosygin, "Main Directives of the Development of the National Economy of the USSR for the Years 1976-1980", *op. cit.*, p. 63.
118   N. Patolichev, "Thirty Years of Co-operation in Foreign Trade", *art. cit.*, p. 8.
119   This aspect is discussed in Part II of this Chapter.
120   *Détente*, Issues Series No. 1, American Bar Association, ABA Press, 1977, p. 12.
121   *Ibid.*, p. 35.
122   *Ibid.*, p. 12.
123   Lothar Ruehl, *art. cit.*, pp. 81-82.

Second, the "need of the US for raw materials and markets could be exploited so as to induce the United States to help with the modernisation of the Soviet economy".[124]

In the East, détente is officially presented as a matter of mutual convenience. Thus it is argued in official circles that the "capitalist ruling classes have accepted the bankruptcy of the cold war and begun to understand that their political and economic fortunes can be preserved only in co-operation with the forces of socialism".[125] The Western assumption that through détente and hence increased co-operation, political concessions can be extracted has been explicitly rejected. For instance: "... certain circles have not yet abandoned their ambitions to use expanded economic relations as a vehicle by which to interfere with the internal affairs of individual socialist states or the socialist community at large. Attempts still are being made to abuse such relations as means of political and economic pressure or politico-ideological subversion. Some of those attempts are made under the pretext of wishing to facilitate East-West trade through some transformation of the socialist system by ways of "radical reforms" or "decentralisation of planning". Such activities, however, are in plain contradiction to the principles of peaceful co-existence and have nothing in common with the letter and spirit of the Final Act of Helsinki...".[126] While underlining that the Eastern countries will spare no effort to honour and fulfil the final Helsinki Act, it is stressed that "economic, scientific and technological co-operation can only be developed in full consciousness of the difference in the economic and social systems... Co-operation between States, having different social systems, in economic matters cannot be linked to any attempt to interfere in the affairs of the others".[127]

Furthermore, Soviet leaders exclude the related assumption that expanded economic relations will influence or lead to fundamental changes within the socialist system: "In the West they have not given up, for example, attempts to prove that international détente is impossible supposedly without the 'convergence' of the two systems. It is impossible not to see the ill intention and lies of this assertion, which is being portrayed as a 'sign of the times' and obviously is aimed at the naive, the politically unsophisticated people. No, communists will never give up their ideas and principles: they have always struggled and in the future will energetically struggle for their world-wide triumph".[128]

It is clear that Soviet viewpoints can give rise to different interpretations. It is equally clear, however, that Soviet policy has served to intimately link science and technology with détente. This linkage has had profound repercussions on Western policy. Their implication is taken up in Part II of this Chapter.

---

124 Richard Pipes, *art. cit.*, p. 22. There are analysts who believe that the current support of the détente policy rests on a somewhat fragile basis. In this regard it is said "that Soviet leaders criticise détente because the Helsinki agreement has nourished subversive ideas in the East, the bargaining price (placed by the Americans) in the strategic arms negotiations is too high and because the US Congress is attempting to exert pressure on the internal affairs of the Soviet Union. Both the Kremlin old guard as well as younger and more progressive groups are asking whether we wish to play the game with them or not". Samuel Pisar, "Comment sauver la détente? La croisade de M. Carter", *Le Monde*, Paris, 1er october 1977, p. 6.

125 Abraham S. Becker, "Discussion", in Paul Marer, editor, *US Financing of East-West Trade, The Political Economy of Government Credits and the National Interest, op. cit.*, pp. 324-325.

126 Max Schmidt (Director of the Institute of International Politics and Economics of the German Democratic Republic), "East-West Economic Relations against the Background of New Trends and Developments in World Economy and in the International Distribution of Activities", paper for IEA Round Table Conference, Dresden, German Democratic Republic, 29th June to 3rd July, 1976, p. 20.

127 Max Schmidt, "Der Entspannungsprozess und Probleme der ökonomischen Zusammenarbeit von Staaten unterschiedlicher Gesellschaftsordnung", *IPW Berichte*, Heft 2, Institut für Internationale Politik und Wirtschaft der DDR, Februar, 1979, p. 7.

128 *Kommunist*, No. 9, June 1972, p. 79 as given in Mose L. Harvey, Leon Goure, Vladimir Prokofieff, *Science and Technology as an Instrument of Soviet Policy, op. cit.*, pp. 3-4.

II. THE WESTERN PERSPECTIVE

In analysing and assessing Western attitudes to closer trading relations with the East it is convenient to distinguish between two levels of interest: the inter-governmental, involving differing national perspectives, and the national, where differences among government, business and industrial interests exist. Here government reconciliation is frequently sought between political, economic, trade and industrial policy considerations.

Western policies toward the East are strongly conditioned by whatever assumptions are made about the advantages or disadvantages of technology transfer to the East. These assumptions are affected by national security, political and economic considerations. Western policies are also influenced by a variety of assumptions about the future evolution of the Eastern systems. In the following discussion the various viewpoints regarding these are highlighted, without seeking to evaluate them.[1]

1. Security Concerns: Military-Strategic Aspects

The expansion of East-West relations, manifested in the form of transfer of technology and know-how, has led some in the West to examine concerns about security. Simply put, the acquisition of such technology is thought to be detrimental to the Western capability to maintain the existing military balance. The following are some of the questions which are often asked:

  i) To what extent do expanded East-West commercial relations — trade, technology transfer, and investment — provide strategic assistance to the Soviet infra-structure by contributing indirectly to a military build-up?

  ii) What is the likelihood that technology acquired from the West will be applied directly to Soviet military capabilities?

  iii) When technological capabilities are acquired from the West for civilian purposes, is the spill-over to the defence sector likely to contribute significantly to the Soviet military effort?

To determine what technologies or production processes might significantly contribute to the military capability of a foreign power is a complex question. Only limited expertise is available for assessing the potential military uses of technology transferred from West to East.[2] In a recent US Committee report for instance, concern was expressed about "the adequacy of existing US government procedures and institutions to assess the nature and extent of Soviet acquisition of high technology, its absorption into the Soviet military sector and its long-term potential impact on Soviet military capabilities".[3] One judgement stated that there "is government-wide inability to accurately assess the impact and implications of

---

1 A further nuance must be kept in mind here. Namely, the bulk of citations are from secondary sources and have not been evaluated from an "official" governmental viewpoint. The very obvious fact that opinions expressed by private analysts may very well not coincide with those of government officials or that private opinion does not necessarily coincide with government policy, must be kept in mind.

2 Such assessments require estimates of end-use, the assimilative and innovative capacity of the foreign power and so forth.

3 *Transfer of Technology to the Soviet Union and Eastern Europe, Selected Papers,* compiled by the Permanent Sub-committee on Investigations, Committee on Governmental Affairs, United States Senate, 95th Congress, 1st Session, US Government Printing Office, Washington, D.C., September 1977.

the aggregate accumulation by the Soviets of technology and know-how of American and allied origin".[4]

The need of the West to maintain effective regulation of the export of strategic technology in order to protect and maintain military balance has never been at issue. The primary objective of the restrictions placed on exports to the East by the Western countries is precisely to hamper the build-up of military power. Particular emphasis is placed on relevant advanced technology and related plant, equipment and materials. However, what constitutes a "strategic technology" or "strategic potential" is widely debated and has led to disagreement at the national level, as well as to differences in approach among the Western allies. As will be seen, there are also differences of opinion among policy-makers regarding what constitutes "military balance". These disagreements and differences have led to questions about the utility and effectiveness of measures adopted to control technology exports to the East.

a)   *Problems Related to Defining Strategic and Strategic Potential*

To differentiate between strategic and non-strategic goods is difficult. It involves evaluating technology in terms of its potential impact. The concept of "strategic impact" is itself obviously very broad, and includes a dynamic element that "evolves with new developments in science and technology no less than it evolves with changing concepts of warfare and survival".[5] The difficulty of evaluating "strategic impact" was underscored in a recent US study which notes that "even if technical agreement is reached on what technologies are critical, it will remain difficult to quantify how much the effectiveness of weapons and other military systems of possible adversaries would be enhanced if such technologies were transferred".[6]

One can certainly postulate that certain items such as implements of war and many atomic energy products can be easily identified as having military significance. However, when one moves beyond items which are obviously useful in military operations to those which are "supporting goods", the issue immediately becomes complex and opaque. As has been pointed out, the decision whether a commodity is strategic or not strategic "runs afoul of such extreme cases as Khruschev's reference to buttons on soldiers' trousers".[7]

---

4   *Ibid.* For a general discussion of the question of national interest and the transfer of technology see Maurice J. Mountain, "Technology Exports and National Security", in *Issues in East-West Commercial Relations – A Compendium of Papers, op. cit.*, pp. 22-23.

5   *A Background Study on East-West Trade, op. cit.*, p. 12.

6   *International Transfer of Technology*, Report of the President to the Congress together with Assessment of the Report by the Congressional Research Service, Library of Congress, prepared for the Subcommittee on International Security and Scientific Affairs of the Committee on International Relations, US House of Representatives, 95th Congress, 2nd Session, December 1978, US Government Printing Office, Washington, D.C., 1979, p. 21. The report responds to the International Security Assistance Act of 1977 (enacted by Congress in August 1977 and signed into law by the President), Public Law 95-92, requiring the President under Section 24 to conduct "a comprehensive study of the policies and practices of the US Government with respect to the national security and military implications of international transfers of technology in order to determine whether such policies and practices should be changed". *Ibid.*, p. III. The report points out that "from the national security standpoint, current interest in identifying critical technologies and keystone production equipment suggests the possibility that more effective controls might result from a sharper focus on a limited number of highly significant technologies". *Ibid,* p. 21. However, the above-quoted conclusion seems to indicate that in fact the critical issue of evaluating "strategic impact" would still remain.

7   *Ibid.* A Western version of the same point is that in "a sense everything is strategic — including bubble gum". This aspect is related to what is referred to as the "dual-use" problem. As is pointed out, "there are almost no militarily significant technologies which do not also have important peaceful uses. Indeed, in the highly industrialised modern world, while arms and ammunition can still be identified, the distinction between implements of war and peaceful goods as well as the technologies for their manufacture has become so blurred that whether an item is a sword or a plowshare depends today not so much on how it is made but on how and by whom it is used". Maurice J. Mountain, "Technology Exports and National Security", *art. cit.*, p. 30.

Looked at in terms of impact, how is one to define what technology is "strategic"? Is a "strategic" technology one that is old, one that is new or one which can be adapted? Is a sophisticated computer "more strategic" than the capability to build and run a computer system? The problem is compounded if one remembers that technology is not just embodied in products but also in processes, and that technology transfer involves not only hardware but software.

A number of Western commentators think that it is not meaningful to make export control decisions on items based on strategic/non-strategic distinctions. Such analysts, among them Antony Sutton, believe that *most* of the items which are exported have some strategic military potential. Sutton argues that the Soviet Union can release domestic resources by importing Western technology, and by substitution at the margin can then devote such resources to achieving its political objectives.[8] According to him "this substitution is of major importance to military objectives because while domestic resources are being devoted to military development the broader industrial base is being updated and fortified from abroad. The industrial base of any country is the prime determinant of its military strength and ultimately the determinant of success in military operations".[9] Some believe that "all goods are strategic, since all are in the long run substitutes".[10]

Other analysts contest the relevancy of the concept of "militarily strategic". One view, for instance, is that: "In present circumstances, the risks borne by the United States in allowing its enterprises to sell their technologically advanced goods and services to the Soviet Union seem to us to be much more obviously economic and political than military".[11]

It is evidently difficult to reach an agreement on how to determine "strategic potential" and on the relevancy of the concept. The frontier between the military and non-military benefits of a technology is a moving one, and one which is particularly difficult to determine in the case of a centrally planned economy where substitutions between production factors are decided in the course of drawing up the plans.[12]

This difficulty is implicitly recognised by those who acknowledge that the "lack of in-depth information concerning Soviet technological needs leaves many grey areas in making decisions concerning the permissibility of certain exchanges... The assumption that we could draw a convincing line between economic exchanges that did not have strategic relevance and economic exchanges that did have strategic relevance seems to have broken down completely".[13]

b)   *Problems Related to Military Balance*

Differing views as to what constitutes military balance, and how to maintain it, are by and large focused on two closely related issues. The first relates to the technological "gap" and the second to Soviet military requirements. Both are linked to the fear that through technology imports the Soviets will surpass the West militarily, or gain the potential to effectively challenge the West's military strategy of deterrence.

8   Antony C. Sutton, *Western Technology and Soviet Economic Development* 1945-1965, *op. cit.*, pp. 398 and 418.
9   *Ibid.*
10   See for instance Peter Wiles, "On the Prevention of Technology Transfer", in *East-West Technological Co-operation, NATO, Colloquium 1976, op. cit.*, p. 37. See also Samuel Pisar, *Co-existence and Communism, op. cit.*, Chapter 4.
11   Given in Raymond Vernon and Marshall I. Goldman, "US Policies in the Sale of Technology to the USSR", prepared for the Department of Commerce, mimeograph, 30th October, 1974, p. 54.
12   Nevertheless, the strategic potential concept has been found to be practical by many Western countries.
13   *Détente, op. cit.*, p. 34.

i) *Issues Related to the Technology Gap*[14]

Most Western analysts agree that the Soviet Union lags behind the West (particularly the United States) in developing many technologies which have a direct military applicability. For instance: "... the Soviet Union is far behind the United States in technology. In space technology, in semi-conductor devices, in precision machinery, in integrated circuitry and other electronics, computers in high technology which plays such an important part in modern warfare, we are far ahead of the Soviet Union. Thus, in most co-operative ventures, we are going to give more than we get".[15]

Those who maintain this view believe that the present strong position of the West (particularly that of the United States) is due in large measure to the lead in design, development, production and operation of technologically advanced weapons and in the technology of its military industries. Maintenance of lead time is a crucial consideration here.[16] Those who argue from the existence of the "gap" believe that the West will maintain its lead because of the relative slowness of the Soviet system to innovate. Thus it is stated that: "... to those who are concerned about the possibility of transfers of technology providing the USSR and other socialist countries a significant means of narrowing the East-West technology gap, or even of leap-frogging ahead of the West, is the fact that the impact of imported technology is limited in the USSR and other socialist countries by the difficulties inherent in a centrally planned economic system. By reason of their severe incentive, organisational and infrastructure deficiencies, these countries have only a moderate capacity to absorb technology imported from abroad. Certainly, the net effect of acquiring Western technology is beneficial to the Soviet Union and the other socialist countries, or they would not spend valuable hard currency for this purpose. But just as certainly, the Western market economy system provides the incentive, organisation and infrastructure to develop and implement still more advanced technology...".[17] Some believe that since sensitive transfers comprise only a small fraction of trade as a whole, little danger is posed to the military balance. "The most sensitive technology militarily is not the most important commercially. The most sensitive areas are transfers of design and manufacturing know-how and transfers of whole production facilities. Transferring these capabilities to the Soviet Union and other nations is likely to enhance competition more than profits. Thus, there exists a happy harmony between ordinary business motives and arms control objectives".[18]

An argument opposed to the above type of reasoning maintains that in the absence of enough detailed and comparative sector studies of actual lead times, it is problematic, at best, to base policy on the assumption that the relative gain to the East of acquiring Western technologies is insufficient to upset the military balance.

Some believe that "the transfer of military significant technology has been of major proportions, and that the Soviet Union has narrowed the gap in its relative weapons

---

14   Given the context of this discussion the treatment here is necessarily narrow. See Chapter 5 for a detailed summary and discussion of the issues involved.

15   *Technology Transfer and Scientific Co-operation Between the United States and the Soviet Union: A Review, op. cit.,* p. 31.

16   "Lead time" is defined in this context as the difference in time between the availability of a certain level of technology in the West and the East. It is a function of the level and rate of diffusion of design and manufacturing know-how, as opposed to either basic scientific knowledge or the specific design of a particular item.

17   Rogers Morton, *op. cit.,* p. 47. Authors such as Berliner relate the weakness of the Soviet innovation system specifically to the enterprise sector. He concludes that the main obstacles to innovation are often located in the research-production cycle. See Joseph S. Berliner, *The Innovation Decision in Soviet Industry,* the MIT Press, 1976. The need to stay ahead in basic scientific knowledge in order to maintain lead time is not disregarded as it is in the above-given definition on which opinions such as those illustrated by the Morton study seem to be based.

18   *Transfer of Technology to the Soviet Union and Eastern Europe, Selected Papers, op. cit.,* p. 5.

capability with the United States to our detriment".[19] Others state that the relatively uncontrolled transfer of technology, particularly of "design and manufacturing technology related to military, nuclear and commercial technology-intensive products, has minimized the ability of the US and its allies to maintain the technological superiority over the USSR we tradidionally had".[20] It is noted that the "Soviets and their allies have been able to acquire technology that bears importantly on the military balance between East and West... Our current conditions can best be described as acute hemorrhaging".[21] These writers believe that because "effective technology transfer depends on the active participation of the donor organisations, key technologies are being transferred through the sale of turnkey factories and manufacturing know-how, through licenses, through the training of technical person-nel" and so forth.[22] Consequently they recommend that technology of military significance be protected until "R & D can make further significant advances, which, one hopes, will continually offset the Soviet Union's assimilation of prior technology".[23] Proponents of this viewpoint also believe that the Soviet interest "in closing the technology gap is being served by trade and scientific and technological exchanges".[24]

The above-mentioned concerns are mirrored, at least in the United States, in recent legislation or in recommendations made to re-examine the issue of military balance.[25] It is, of course, true that strategic control is the business of government, while trade is primarily conducted by private corporations. It is this consideration which gives rise to the need to

---

19   *Ibid.*, p. 3.
20   Betsy Ancker Johnson and David B. Chang, *US Technology Policy – A Draft Study*, Office of the Assistant Secretary for Science and Technology, US Department of Commerce, National Technical Information Service, March 1977, p. 83.
21   Senator Henry M. Jackson as given by Henry S. Bradsher, "Do Soviets Turn US Technology to Military Use?", *The Washington Star*, September 22nd, 1977, pp. A-1 and A-8.
22   *Transfer of Technology to the Soviet Union and Eastern Europe*, Hearings before the Permanent Subcommittee on Investigations of the Committee on Governmental Affairs, United States Senate, 95th Congress, 1st Session, Part 2, US Government Printing Office, Washington, D.C., 25th May, 1977, p. 13.
23   *Ibid.*, p. 13. In the interest of preserving lead time, it has been suggested that "the test for technology export should be whether or not it would significantly advance the military potential of the receiving country, not if the item is obsolete by US standards". August 1977 memorandum from the US Defense Secretary Harold Brown, "Interim DOD Policy Statement on Export Controls" given in *Special Report, Technology Transfer: US Government Policy and Its Impact on International Business, op. cit.*, p. 4.
It is also suggested that "the true measure of effectiveness of controls over technology is how long the catch-up process takes", Maurice J. Mountain, "Technology Exports and National Security", *art. cit.*, p. 31.
24   *Détente, op. cit.*, p. 35.
25   See for instance the International Security Assistance Act of 1977 (Public Law 95-92) as discussed in *International Transfer of Technology, op. cit.* The nine questions in Section 24 of the Act are addressed, for the purpose of identifying the issues implied by them, in *International Transfer of Technology: An Agenda of National Security Issues*, prepared for the Sub-committee on International Security and Scientific Affairs of the Committee on International Relations, US House of Represen-tatives, 95th Congress, 2nd Session, Congressional Research Service, Library of Congress, US Govern-ment Printing Office, Washington, D.C., February 13, 1978. Based on the assessment made of the Executive response to Section 24 of the International Security Assistance Act of 1977, it would seem that the nine questions which were addressed have not been satisfactorily clarified. The assessment notes that while "the report addresses most of the issues raised by the law and by the [above-mentioned] CRS agenda of issues, the lack of data and analytical depth indicate that this report does not reflect a comprehensive study", *International Transfer of Technology, op. cit.*, p. 37. Relative to the military and national security issues, it is stated that "Congress focused special attention on the need to examine technology transfer from a broad context of national security. The Presidential report arguably does not respond to this guidance. Rather it stresses Government control over the transfer of military technology and gives little attention to the transfer of those civil technologies which could affect future balances of industrial and military strength. Likewise, the full impact on US security of modifications in the economic, military, and social systems of countries receiving US technology and in US economic competitiveness, industrial strength, employment rates, and the research environment are not fully considered". *Ibid.*

define "national interest" in some coherent and logical manner.[26] The definition requires consistency with respect to the transfer of technology since a balanced view of national interest must include considerations of military strength and political and economic gains. This will remain a continuing problem for policy makers.

ii)   *Issues Related to Soviet Military Requirements*

We have seen that there is a good deal of divergence of opinion about the relationship between Western technology transfers and the strengthening of Soviet military capabilities.

Those Western analysts who do not believe that Soviet military potential is significantly increased by having access to Western technology and know-how generally base their argument on the belief that the Soviet Union would never rely on imported technologies to secure and maintain its military strength. Since the Soviets can expect little help from the West in those areas of critical concern to national security (computers, electro-optical devices, aircraft propulsion and so forth), they can be expected to develop their own advanced technology in such areas regardless of availability in the West.

It is also argued that given the stage of military development attained by the Soviet Union, there is little to be obtained from the West which would significantly alter the military balance. The military balance is not sensitive, it is believed, to narrowly defined technological inputs, even to such strategic technologies as semi-conductors or computers, because Soviet military capabilities are not believed to rest on them.[27] Some commentators even go so far as to say that it is in the interest of the West to make technologies, such as computers, available to the East so as to reduce the incentives for autonomous development. They say that "without doubt, US export controls have forced the acceleration of the development of the Communists' computer industry, thus causing economic and military capability to be created at earlier dates and outside of our control".[28]

The argument is also put forward that the military balance is little affected by a tightly restrictive technology policy since the Soviet Union has always pursued its military goals in spite of the difficulties or costs involved in acquiring essential technologies. It is maintained that military balance rests on a host of non-technical factors and, by focusing on narrow technological issues, the problem is skewed out of perspective. Thus, it is said that technology export is only a concern when an important Western advantage rests heavily on the technology in question, but even in these cases it is unclear that the Soviets lack the essential military capabilities that rest on these technologies. Consequently, it is maintained that there is no Western technology, military or otherwise, that the Soviet Union is ultimately incapable of duplicating internally. The only militarily relevant technological advantage that the West has over the Soviet Union is that the "state of the art" of Western technology may at any given time be more advanced than that of the Soviet Union. In fact, it is thought that "except for the United States, the industrial West possesses little in the way of military goods or know-how that the Soviet Union has not already achieved and surpassed".[29]

It is clear that the reasoning outlined above is based on a number of assumptions. One concerns the issue of intent: the possible uses to which Western technologies may be put. The

26   The term "national security" can be defined in many ways. For a discussion of this term in connection with the issue of government control of technology transfer (at least from the US perspective) see Maurice J. Mountain, "Technology Exports and National Security", *art. cit.*

27   Relative to computers it has been concluded that Soviet military capabilities would only be modestly enhanced by relaxing export controls on large computer systems. See Charles Wolf, Jr., *US Technology Exchange with the Soviet Union: A Summary Report*, R-1510/1-ARPA, report prepared for Defense Advanced Research Projects Agency, Rand, Santa Monica, California, August 1974.

28   William C. Norris, "High Technology Trade with the Communists", *Datamation*, January 1978, p. 102.

29   Samuel Pisar, *Coexistence and Commerce, op. cit.*, p. 68.

argument seems to assume that Western technologies transferred for civil purposes are not diverted to military applications, as the Soviet Union does not rely on such technologies for military purposes. A related assumption is that given the characteristics of the Soviet system, its innovative capacities are sufficiently limited for the West to maintain its "lead".

These assumptions have been challenged by those who either disagree with the assessment made as to the usefulness of Western technology in enhancing Soviet military capabilities, or believe that the Soviet Union does need Western technology in order to maintain military strength. Using the same arguments outlined above, it is maintained that because any country can, over time, acquire a given military capability, any delay to the growth of production capabilities in the Soviet Union will increase the West's military security. Delay is here considered to be of the essence.[30]

These analysts believe that posing the problem in terms of a country's ability to duplicate any technology internally is irrelevant. Rather, the significant problem is whether a country can divert enough resources to develop the required technology itself; and what the impact of such diversion will be on other aspects of its research and development. The principal concern in technology transfer is then the extent to which such transfers are resource releasing or resource demanding. Thus the question of technology control must, it is thought, rest on assumptions about the resources required to duplicate such technology and the opportunity costs in terms of alternative uses for these resources.[31]

c)   *Effect on Export Controls*

Disagreements about what constitutes "strategic potential" or maintaining military balance have bedeviled the task of defining a workable concept of "strategic goods" to serve as a common basis for unilateral and multinational export controls. The following give some illustrative examples.

   i)   *Difficulties Encountered on the National Level: the United States as an Example*

One result of the failure to agree on a definition of what constitutes a strategic good or strategic potential has been the persistence of "gray areas" where uncertainties seem to be resolved on other than objective grounds. It has been alleged that "in many cases, the scientific engineering, and economic factors are usually so evenly balanced that the decision (whether or not to export an item) is made mainly on political grounds".[32]

A second result is that the potential utility of the embargo instrument has been questioned. Some critics feel that it is the lack of "clearly defined objectives, in addition to a control list that is excessively concerned with splitting hairs over performance specifications and end-use statements for each product reviewed, which leads to ineffective export controls on East-West commercial trade".[33] Other critics feel that control measures are not germane to the actual problem at hand. It has been recommended, for instance, that design and manufacturing know-how should be the principal subjects of strategic technology control and "that deterrents meant to discourage diversion of products to military applications are not a meaningful control mechanism when applied to design and manufacturing know-how".[34] A similar line of reasoning is offered by those who argue that even if a successful

---

30   The corollary to this argument is that denial of technology forces systemic change in order to promote indigenous innovation. It is asserted that in the long-term this will transform organisational structures and attitudes toward the West. This aspect is treated in Section 4 below.

31   See also Chapters 5 and 6.

32   *A Background Study on East-West Trade, op. cit.,* p. 13. One example cited here is the US decision to sell chemical equipment to Poland and Rumania in order to "demonstrate US satisfaction with growing economic nationalism and independence from Moscow"-

33   *Transfer of Technology to the Soviet Union and Eastern Europe, Selected Papers, op. cit.,* p. 14.

34   *An Analysis of Export Control of US Technology – A DOD Perspective, op. cit.,* p. XV.

embargo on the sale of technology to the Soviet Union were possible, this would not necessarily lead to a reduction in Soviet military effectiveness. The technologies needed most by the Soviet Union to improve factor productivity are mature, standard and available from numerous other sources. Hence, it is felt that preoccupation with the military aspects of the transfer of technology obscures more fundamental non-military problems: "the essential choice is not whether non-strategic technology is to be sold to the Soviet Union and other socialist countries, but rather, whether American companies should participate in the benefits from these sales that are going to occur in any event".[35]

Those analysts who disagree, and who believe that export control measures are useful (though recognising their present short-comings), base one argument on the fact that controls do serve to preserve lead-time. It is suggested, therefore, that present controls be adjusted and "export be denied if a technology represents a revolutionary advance to the receiving nation, but could be approved if it represents only an evolutionary advance".[36] Another argument mentioned is that significant technology transfer occurs when know-how rather than hardware is sold.

The complexity of these issues led Congress to enact the "Export Administration Amendments of 1977" — Public Law 95-52.[37] This law seeks to assess and reconcile the criticisms which have been directed at the Export Administration Act. Among these criticisms are an insufficiency in "protecting US technological lead and allowing leakage of critical technologies enhancing Soviet military capabilities" and controls that are "too excessive and go well beyond what is needed to protect national security".[38]

Under Section 118 of the Law a study of technology transfer policy is required.[39] Among the issues to be investigated are whether and to what extent items and know-how transferred "from the United States or from any country with which the United States participates in multilateral controls would make a significant contribution to the military potential of any country threatening or potentially threatening the national security of the United States".[40]

---

35   Rogers Morton, *op. cit.*, pp. 46-47. See also Section 3.
36   *An Analysis of Export Control of US Technology – A DOD Perspective, op. cit.*, p. 14. See also pp. 4-8 of the publication for a discussion of active versus passive transfer mechanisms. Other suggestions go even further. For instance, in the previously mentioned August 1977 memorandum from the US Secretary of Defense which lists critical technologies and certain "keystone products", the recommendation is made that export of these items to other (non-Eastern) nations should be controlled to stop leakage once the item has left US shores. In those instances where Western allied countries re-export certain types of critical items to the East, the prohibition of the receipt of further strategic know-how from the United States would follow. See *Special Report, Technology Transfer: US Government Policy and Its Impact on International Business, op. cit.*, p. 4 and Richard Burt, "US Seeks to Guard Technological Edge", *The New York Times*, December 11, 1977. See also *Transfer of Technology to the Soviet Union and Eastern Europe, Selected Papers, op. cit.*, p. 16.
It is pointed out that the stated purpose of the Defense Department proposal is to "enable the US Government to regulate technology exports to non-Communist countries in terms of the recipient's intent and ability to prevent either the compromise or the unauthorised re-export of that technology. Defense is recommending that this policy be applied... without regard to whether the exporter is a government department or agency, a commercial enterprise, an academic or non-profit institution, an individual entrepreneur, or in the case of re-export requests, a foreign government or an international organisation; and without regard to the transfer mechanism involved, e.g. turnkey factories, licenses, joint ventures, training, consulting, engineering documents and technical data. Although these recommendations have been promulgated by the Secretary of Defense as internal guidance to the Department of Defense with regard to its role in support of US export controls, it is not yet clear whether they will be adopted as government policy", Maurice J. Mountain, "Technology Exports and National Security", *art. cit.*, p. 32.
37   *Public Law 95-52*, 95th Congress, 91 Stat. 235, "Export Administration Amendments of 1977", H.R. 5840, June 22, 1977.
38   *Special Report Technology Transfer: US Government Policy and Its Impact on International Business, op. cit.*, p. 3.
39   In this connection see *International Transfer of Technology, op. cit.*
40   See *Public Law 95-52, June 22, 1977, op. cit.*

170

However, the law does not define what specific products or technologies should be controlled. That responsibility has been delegated to the Secretary of Commerce.[41]

In summary, it may be stated that the absence of an agreed definition of what is and what is not "strategic" is one of the principal factors which lead critics to cast doubt upon the effectiveness and utility of export control measures. Another source of criticism is no doubt the related problem of finding a basis for reconciling conflict in those instances where national interest as defined by government conflicts with corporate interests.[42]

ii) *Difficulties Encountered on the Multi-national Level: the Example of the Multi-lateral Lists*

The growing opportunities for profitable East-West trade and the easing of international tensions led to the desire to dismantle trade barriers between East and West. There has consequently been a general wish to reduce to a minimum the number of items controlled for strategic reasons, and this has been expressed by periodically reviewing the Lists. The Western member countries however, have different views as to how their national interest may best be served and these differences have emerged from time to time during the course of review and agreement on the Lists.

The United States and the West European countries have had, for instance, different viewpoints as to what constitutes "strategic". Until the passing of the US Export Administration Act of 1969, the United States maintained that "strategic" included both items which have military implication and those which have economic importance. The West European countries had excluded from control items intended to "slow down Communist growth". There is now an agreed list which is shorter and more flexible in export restrictions.[43] Differences in viewpoint are also reflected in the national lists which are maintained. These lists drawn up by the individual Western countries by and large reflect national considerations although they are drawn up in accordance with the partners. An effort has been made by the United States Government to analyse "the uniformity of interpretation and enforcement by the participating countries of the export controls agreed to" by the co-ordinating mechanism.[44] Acknowledging that differences in interpretation of the export controls agreed to by the Committee do exist, the "Special Report on Multilateral

---

41   *Special Report Technology Transfer: US Government Policy and Its Impact on International Business, op. cit.,* p. 3. It seems that in future the Administration wishes to involve itself in the process that leads to "approval or denial of technological export licenses to Communist countries. The new White House role is spelled out in Presidential Review Memorandum 31 ... The White House Office of Science and Technology Policy and the National Security Council will act as 'observers' in all technology-related export license requests by Communist nations". Thomas O'Toole, "White House to 'Observe' Technology Export Deals", *International Herald Tribune,* Paris, October 28, 1978.

42   The probability of such conflicts occurring had been anticipated, though guidelines as to their resolution were lacking. For example, the 1971 amendment to the Export-Import Act requires that the Export-Import Bank's facilities only be extended to Communist nations on a country-by-country basis following a Presidential determination of national interest. The legislative requirement was based on the general belief that trade with Communist countries requires special attention to the divergencies between national and private commercial interests. See John P. Hardt and George D. Holliday, "East-West Financing by Ex-Imbank and National Interest Criteria", *art. cit.,* pp. 294-300.

43   "About 1,000 'exception cases', valued at about $200 million, are now reviewed [by the co-ordinating mechanism] annually (the figure for 1977 was $214 million). The US submits almost half of these cases, and many of the others include US — origin components. Approximately 2 to 4 per cent of these requests are disapproved, another 3 to 5 per cent are withdrawn, and many more are revised to take into account changes recommended during the Committee review process". *Special Report on Multilateral Export Controls,* pursuant to Section 117 of Public Law 95-52, the Export Administration Amendments of 1977, The White House, July 10, 1978, mimeograph, p. 3.

"In considering an exceptions request, the Committee examines the particular circumstances of the case in the context of the strategic criteria for export controls. Domestic political and economic factors in the exporting country of international political factors in the importing country may be considered if they are relevant to the security of member countries. However, the principal determining factor is a judgement of minimal risk of diversion to significant military use". *Ibid.,* pp. 2-3.

44   See Section 117 of *Public Law 95-52.*

Export Controls" notes that "where participating countries have acted contrary to US interpretations of Committee agreements, the US has sought to restore an agreed interpretation. In some instances, our view has prevailed. In others, uniformity of interpretation has eventually been achieved through an agreed relaxation of the embargo".[45]

The disparity of viewpoint between the United States and its West European partners in the co-ordinating mechanism relative to export controls is also exemplified by the fact that many West European officials believe that "the Americans are overly concerned with export controls of items that might be used for military purposes".[46] A French official, "though he acknowledged that military equipment should not be sold", felt that despite Lists and embargoes, "the Soviets can probably get what they need from one source or another, and the West would be better served by concentrating on the economic rather than political dynamics of trade".[47] To substantiate this point examples can be cited "of sales from Western firms to third parties, who in turn, delivered items to the Soviet Union or to East European countries".[48] In contrast to the United States where there seems to be a great deal of concern that trade could reinforce Soviet military potential, stand some West European officials who do not believe that government "needs to justify trade. Trade justifies itself on economic terms".[49]

2. Political Considerations Influencing Technology Transfer: Linking S & T Co-operation with the Achievement of Foreign Policy Goals

Unlike those of the East, where economic considerations have always seemed to dominate the policy of seeking closer economic relations with the West, Western objectives have been and remain mixed. For Western Europe, gaining access to Eastern markets has always been the primary reason for establishing closer trading links with the East. Political considerations have, by and large, been secondary.[50] For the United States, the picture is somewhat mixed. Economic interests and the promise of commercial gains have, of course, been the key elements motivating business and industry to seek closer commercial relations

45 *Special Report on Multilateral Export Contrpls, op. cit.,* p. 3.

46 *Western Perceptions of Soviet Economic Trends.* A Staff Study, Sub-committee on Priorities and Economy in Government, Joint Economic Committee, Congress of the United States, US Government Printing Office, Washington, D.C., March 6, 1978, p. 11. The previously cited *Special Report on Multilateral Export Controls* notes, in this regard, that members of the co-ordinating mechanism "consider that US policy on controls is overly restrictive. They have not committed themselves by treaty or formal international agreement to the embargo, and there are no internationally agreed sanctions for deviations from it". See p. 3 of the *Special Report.*

47 *Western Perceptions of Soviet Economic Trends, op. cit.,* p. 11. The complexity of the problem is illustrated by the Cyril Bath Company, a US manufacturer of metal-forming machines. "Bath's problem in exporting to the Soviet Union arises from the fact that its machines can be used not only in the automotive sector — deemed acceptable — but also in the manufacture of aircraft components. On these grounds, Bath was denied a US export license, despite the fact that an order is now being filled by a French firm for nine comparable machines. In response to Bath's complaints that its French rival, Ateliers et Chantiers de Bretagne (ACB), has an unfair advantage, the US then asked.... [its partners] to approve Bath's metal-forming machines for export to the Soviet aircraft industry. Britain and Germany refused, however, on the grounds that the French government would not grant such approval for its own supplier's machines. The French contend that ACB's machines are for the automotive sector and therefore no permission is required". *Business Eastern Europe,* Vol. 7, No. 48, December 1, 1978, p. 381.

48 *Western Perceptions of Soviet Economic Trends, op. cit.,* p. 11.

49 *Ibid.,* p. 12. In general, for those OECD countries which are not members of any military alliance it is the commercial aspects of East-West technology transfer which are of greatest concern. These are of course affected by "multi-lateral rules" in those instances where items are imported from a country which is a member of the multi-lateral co-ordinating mechanism.

50 As indicated in Chapter 1, notable exceptions to this general "rule" were the French policy (current in the mid-1960s) of "Europe to the Urals" and Germany's policy of "Ost Politik" initiated by the Brandt administration. The latter continues to be an important factor in German policy towards the East. On the whole, Western Europe and Japan have viewed trade relations as elements supporting their political objectives.

with the East. For the US government however, political considerations, in addition to economic ones, have been a very important — if not primary — reason for seeking closer commercial relations with the East and, in particular, with the Soviet Union. Increased trade has frequently been viewed as a "bargaining chip" in the attempt to achieve foreign policy goals such as a moderation of the arms race or an improvement in détente. Since the United States is one example of a country in which the attempt has been made to link commercial relations and technology transfer with the achievement of foreign policy aims, the following discussion concentrates upon the evolution and purpose of US policy in this area.

a)   *Evolution of the "Linkage" Policy*

In the United States the attempt to link trade to the achievement of foreign policy goals is not new. Elements of such an approach were already evident in the 1930s.[51] Attempts to use economic policy as a lever to achieve specific foreign policy goals have been a recurrent feature of US diplomacy.[52]

A new aspect of this long-standing practice appeared in the late 1960s. In the US Export Administration Act of 1969, the policy emphasis was shifted from trade control to trade promotion.[53] In August 1971, the Nixon Administration inaugurated its "New Economic Policy" to attack foreign as well as domestic economic problems. The resulting programme not only emphasised stronger US interests in East-West trade, but also "closely linked the issue of expanding East-West trade ties with the broader range of security and political issues that were to make up the agenda of the May 1972 Summit meeting between President Nixon and Party Secretary Brezhnev".[54] A third factor was a government report published in December 1971 which explicitly called for improved trade relations with the East. The report called attention to the potential advantages of seeking commercial agreements with the East: "commercial agreements will give the political and military agreements already reached a firm basis in economic self-interest...".[55] The role which US technology and

51   See Chapter 1, Section II, 3.
52   Examples which have been cited are the proposal to include the Soviet Union in the Marshall Plan in order to facilitate political settlements after World War II and the economic concessions given to Rumania in recognition of its relatively independent foreign policy. See *Science, Technology and American Diplomacy,* Vol. I, *op. cit.,* p. 539. In later years, the policy of "building bridges", articulated by the Johnson Administration, was an expression of the hope that commercial exchanges would lead to better political relations. A recent example of the attempt to apply economic leverage to achieve foreign policy goals is the Jackson-Vanik amendment to the 1974 Trade Act. Even the previously mentioned "Stevenson Bill" does not "grant MFN status and credits without qualification or condition. It replaces an explicit linkage to emigration policy with a procedural formulation that implies that the continued availability of credits and MFN will be subject to periodic review and an evaluation of... Russian conduct across the entire range of US interests, including emigration policies". "US MFN and Exim Credits Proposed for the USSR", *Business Eastern Europe, art. cit.*
53   See Chapter I, Section V. This policy has been supported by the Carter Administration. For instance it has been stated that "trade between the United States and the Soviet Union has multiplied four-fold between 1971 and 1974. But it has since grown at an unsatisfactory pace. The President has instructed... to make clear to our Soviet counterparts and to Prime Minister Kosygin and President Brezhnev that the United States wants our trading relationship to expand... We will be guided by our commitment to increase US-USSR trade...". W. Michael Blumenthal, *op. cit.*
54   John P. Hardt, George D. Holliday and Young C. Kim, *Western Investment in Communist Economies, A Selected Survey on Economic Interdependence,* prepared for the Sub-committee on Multinational Corporations of the Committee on Foreign Relations, United States Senate, 93rd Congress, 2nd Session, US Government Printing Office, Washington, D.C., August 5, 1974, p. 6. This linkage was made explicit. For instance, speaking after the Moscow Summit in 1972 Dr. Kissinger noted that "we hoped that the Soviet Union would acquire a stake in a wide spectrum of negotiations and that it would become convinced that its interests would be best served if the entire process unfolded. We have sought, in short, to create a vested interest in mutual restraint". Kissinger Briefing to Congressional Leaders", *Congressional Record,* June 19, 1972, p. S-9600, as given by John P. Hardt, "Military-Economic Implications of Soviet Regional Policy", in *Regional Development in the USSR,* NATO Colloquium 1979, *op. cit.,* p. 249.
55   Peter G. Peterson, *US-Soviet Commercial Relationships in a New Era,* Department of Commerce, Washington, D.C., August 1972, p. 3.

know-how were thought to play in furthering "mutual interest" was likewise made explicit: "... commercial advantages ... would accrue to both countries from trade in goods in which they presently excel — raw materials in the case of the USSR, agricultural products and advanced civilian technology in the case of the US".[56]

By the time of the Summit meeting of May 1972, co-operation in science and technology had become an important input into the policy officially called "linkage". Thus the acceptance by the signatories of the "Basic Principles of Relations Between the United States and the Union of Soviet Socialist Republics" gave rise, among others, to the Scientific and Technical Co-operative Agreement and its ten companion agreements.[57] The aim of these agreements was to provide for "increased exchange and co-operation ... designed to benefit both nations in certain areas of scientific research and détente".[58] In the United States "the most important factor influencing the move toward a Science and Technology Agreement with the Soviet Union was political".[59] This factor "overshadowed any thought of scientific or technical benefit".[60] The purpose was made explicit. "... the agreements on scientific and technical co-operation were concluded at the highest political level as part of the process of normalising relations between our two countries ...".[61] The quantum jump in the number of technological exchanges between the United States and the Soviet Union, which resulted from the S & T Co-operative Agreements, was thus not the result of specific requests by scientists and engineers. Rather, it resulted from the basic foreign policy approach of the United States in its movement towards détente.

Two related motives are said to have led the United States to seek closer relations with the Soviet Union. Both are subordinate to the overall goal of attaining greater world stability, itself dependent on improved US-Soviet relations. One was the "hope that more extensive interactions with the West would open Soviet society". The second was "the desire to have greater access to the Soviet economic and political leadership".[62] Both were to further the policy of détente and both implied linking security, economic and political factors. The aims and significance of the policy of "linkage", particularly as it pertains to détente, are discussed below.

b) *Purpose and Assumptions Underlying Linkage Policy*

Co-operation in trade, in general, and specifically in science and technology, was believed to support the policy of promoting world stability and of aiding the peaceable

---

56 *Ibid.,* p. 6.
57 See Chapter 1, Section V, 3 (d).
58 *Background Materials on US-USSR Co-operative Agreements in Science and Technology, op. cit.,* p. 3. The S & T Co-operative Agreement provided for government to government co-operation in certain areas deemed beneficial to both nations. Co-operation was intended for areas where both nations could significantly enhance the other's research effort. *Ibid.*
59 *Review of the US/USSR Agreement on Co-operation in the Fields of Science and Technology,* Board on International Scientific Exchange, Commission on International Relations, National Research Council, National Academy of Sciences, Washington, D.C., May 1977, p. 30.
60 *Ibid.*
61 Testimony of Myron Kratzer, Acting Assistant Secretary for Oceans and International Environmental and Scientific Affairs, Department of State, as given in *Review of US-USSR Co-operative Agreements on Science and Technology, Special Oversight Report No. 6, op. cit.,* p. 15. Though not explicitly tied to the policy of linkage, the interconnection between scientific co-operation and foreign policy considerations is also evident in assessments made as to the influence which the successful conclusion of the second Strategic Arms Limitation Treaty (Salt 2) would have on scientific exchanges. For instance, Dr. Frank Press, director of the Office of Science and Technology and science adviser to President Carter has stated that "if Salt 2 succeeds, other forms of co-operation — such as scientific exchanges — could succeed. But if the agreement failed, it was difficult to see much chance for other kinds of co-operation to succeed". "Science Exchanges at Stake in Salt Talks, Says Adviser", *Nature,* Vol. 278, March 15, 1979, p. 198.
62 *Review of the US/USSR Agreement on Co-operation in the Fields of Science and Technology, op. cit.,* p. 30.

settlement of political disputes.[63] The overall purposes of linkage policy were therefore to encourage détente, to bring about the relaxation of restrictive Soviet domestic policies and promote internal economic reforms, and to encourage polycentrism in the Eastern bloc.[64]

It was hoped that "incentives could be created for the Soviets to restrain their expansionism in the interest of obtaining imports (products such as grain as well as technology) needed from the West for internal development".[65] Trade, particularly in technology, was thought to fit this strategy of developing incentives for restraint particularly well because of the "growing Soviet dependence on Western technology to close the developmental gap".[66] Concerning the ability of science and technology to reduce tensions, it was thought that one "advantage of extensive US-USSR interchange is that it helps to reduce uncertainties and creates conditions conducive to rational consideration of foreign economic and political policy alternatives. It fosters a recognition of mutual dependence and movements towards mutually assured simultaneous survival".[67] This was thought to be possible because as ties between East and West "develop and expand, the number of people in leadership roles in the collectivist countries who have personal, vested interest in stable, positive relations with the US will increase".[68]

In summary, it was assumed that while "expanded economic relations which facilitate massive technology transfer from the United States to the USSR may create new, potentially dangerous dimensions in US diplomacy,... there is at least a possibility that the process of integrating the centrally planned Soviet economy into the market economy of the United States and the non-Communist world might unleash irreversible forces of constructive change which could, in turn, contribute to international interdependence and stability".[69] It was presumably for this reason that the United States "approached the question of economic relations [with the Soviet Union] with deliberation and circumspection and as an act of policy not primarily commercial opportunity".[70]

c)   *The Link with Détente*

Improvement of political relations is of course the basic objective of détente. Loosely defined, détente is seen as providing the possibility for a joint slowing down of military

---

63   "Linkage policy" has been most clearly articulated and followed by the United States. However, a similar association was generally made by the signatories of the Helsinki Final Act. As already pointed out in Chapter 1 (Section V), the global need for economic interdependence was recognised and agreed to. The underlying rationale was the perception of trade relations as a factor in improving political ties. See Report of the Congress of the United States on *Implementation of the Final Act of the Conference on Security and Co-operation in Europe: Findings and Recommendations Two Years After Helsinki, op. cit.,* p. 66.

64   See *Science, Technology and Diplomacy in the Age of Interdependence, op. cit.,* p. 82 and M. Christopher Kwiecinski, "Should Eximbank Finance East-West Trade: Summary of Issues and National Debate", *art. cit.,* in Paul Marer, editor, *US Financing of East-West Trade, the Political Economy of Government Credits and the National Interest, op. cit.,* p. 245. See also Section 4.

65   *Détente, op. cit.,* p. 9.

66   *Ibid.,* p. 22.

67   Gary Fromm, Director of the Washington Office of the National Bureau of Economic Research, as given in *Review of US-USSR Co-operative Agreements on Science and Technology, Special Oversight Report No. 6, op. cit.,* p. 16. While stressing the fragile nature of the détente relationship, some analysts believe that the "only chance of engaging the least dogmatic elements of the Soviet regime in a constructive co-operative dialogue rests on scientific, technological and commercial exchanges". Samuel Pisar, "Comment sauver la détente", *Le Monde,* Paris, 1er octobre 1977 (I. La croisade de M. Carter), p. 6. See also *Le Monde,* Paris, 2-3 octobre 1977 (II. L'équation de la coexistence), p. 4.

68   William J. Casey, "Current Status and Outlook" (Technology Exchange with the USSR), *Research Management,* Vol. 17, July 1974, p. 7, as given in Arthur H. Hausman, "East-West Technological Co-operation Likely Transfer Patterns (1976-1980)", *art. cit.,* p. 341.

69   *Science, Technology and Diplomacy in the Age of Interdependence, op. cit.,* p. 87.

70   Henry Kissinger (then Secretary of State), as given in Susanne S. Lotarski, *art. cit.,* p. 1038.

production. By placing two systems with vastly different social, political and economic values in more intimate proximity with each other, it was believed that "a framework in which each side will be in a position to interpret the capabilities and intentions of the other with somewhat greater accuracy, thereby reducing the risk that mankind will destroy itself" could be provided.[71] Put more pragmatically, détente has also been viewed as "containment plus linkage or adding the carrot to the stick".[72]

Through linkage, the "détente objective" of the United States was to try to link the Soviet Union to trade and to technology as a means of increasing the political cost of moves which might disrupt the détente relationship.[73] It is presumably for these reasons that one of the "major aims of the policy of East-West détente pursued by the Nixon-Ford Administrations has been a rapid expansion of commercial relations between the United States, the Soviet Union and the countries of Eastern Europe".[74] The major role to be played by science and technology in support of détente has been emphasised. It is said for example, "that the export of US advanced technology to the Soviet Union has been a keystone of the political-economic policy of détente".[75] While détente is not based solely on science and technology, scientific and technological exchanges have been intimately linked to the possibility of its achievement. Thus those supporting the conclusion and expansion of the scientific agreements offer the following argument:

"Scientific exchange beneficial as it it might be for science, is not an end in itself, because science is not an end in itself. The goals are international understanding, détente and peace. Thus, the broad question before this committee is not whether to pursue détente, but how to pursue it, and the broad question before the scientific community is how to manage its affairs so as to help the world toward peace".[76]

Trade and science and technology as elements of leverage to achieve political goals are themselves a subject of controversy. Three aspects have generally been singled out for scrutiny.[77] The first concerns trade per se and the trend of associating scientific undertakings in any direct way with political and diplomatic priorities. There are those who question the possibility that more extensive trade with the USSR will improve overall relations, thereby leading to a reduction of international tension. Trade itself, it is pointed out, can generate competition, friction and conflict. This attitude is perhaps best summed up by the following statement: "I doubt that there are close connections between trade and ideas, trade and peace, or indeed trade and anything".[78] Others point out that "if the improvements in relations are superficial or transitory the United States may be incurring risks to its national security by transferring valuable technology to potential adversaries without commensurate

71   Raymond Vernon and Marshall I. Goldmann, "US Policies in the Sale of Technology to the USSR", *art. cit.*, pp. 6 and 25.
72   *Détente, op. cit.*, p. 9. The policy of linkage goes well beyond the policy of "containment" current in the early 1950s.
73   Tying the Soviet Union into greater dependence on the West, by providing them with larger markets for their primary-product exports in the West and new opportunities to import machinery from the West could also, it was thought, bring about shifts in bloc trade patterns because East Europe's commercial bargaining power with the Soviet Union might be weakened. See Paul Marer, "Soviet Economic Policy in Eastern Europe", in *Reorientation and Commercial Relations of the Economies of Eastern Europe – A Compendium of Papers,* Joint Economic Committee, Congress of the United States, 93rd Congress, 2nd Session, US Government Printing Office, Washington, D.C., August 16, 1974, p. 161.
74   *Science, Technology and Diplomacy in the Age of Interdependence, op. cit.*, p. 88.
75   Arthur H. Hausman, "East-West Technological Co-operation – Likely Transfer Patterns (1976-1980)", *art. cit.*, p. 335.
76   *Technology Transfer and Scientific Co-operation Between the United States and the Soviet Union: A Review, op. cit.*, p. 31.
77   The wisdom of détente is, of course, itself an issue and subject of debate. It is a subject which falls outside the scope of this study.
78   Peter Wiles, *Communist International Economics,* Praeger, New York, 1969, Chapter 18 as given in Stanislaw Wasowski, ed., *East-West Trade and the Technology Gap, A Political and Economic Appraisal,* Praeger Publishers, New York, 1970, p. 200.

benefits".[79] Finally, a number of analysts point to the limits of using trade to achieve political goals. It is said for instance, that "where trade in technology is concerned, the needs of the communist countries who might prefer to be self reliant do not appear to us so fundamental that these countries would be prepared to trade political or security interests which they consider essential in order to obtain technology... Exports of manufactured goods to the Soviet Union last year amounted to 3/100s of 1 per cent of the Soviet GNP — hardly a vital element in their economy. Once again, it is hardly a significant factor in that trade. It does not give us much of a lever...".[80]

The second aspect concerns technology transfer (where the arguments are similar to those given above). For one thing, it is maintained that the Soviet Union and the East European countries have simply "used détente as a means to pursue access to the latest technology from the United States and other industrialised nations".[81] The fundamental Soviet motive for détente is seen here as using Western technology, particularly computers, for a "continuing build-up of its military forces".[82] Furthermore, it is felt that "State monopoly of foreign trade and ownership of all technology facilitate Soviet political and technological exploitation of trade and exchange relationships with the assurance that there are adequate controls built into the system to protect the Soviet Union from disadvantageous exchanges".[83]

The third aspect relates to the large capital flows (credits) which have become available as a result, in part, of the policy of détente.[84] Some analysts believe that "a large and lasting creditor position vis-à-vis the Soviets may become a considerable problem... By its very existence the debt will create an interest group, which the Soviets at some point may be tempted to manipulate to their own advantage... Secondly, the same circumstances may well limit our freedom of action in international relations... It must be remembered that in the era of superpowers it is the debtor country that has the upper hand".[85] Others doubt whether the United States can buy détente with credit, and question the policy of détente to the degree that it explicitly links the commercial and political aspects of East-West relations. It is pointed out here that a "genuine and lasting easing of tensions requires resolution of the difficult issues which divide the United States and the Soviet Union — in the Middle East... with respect to nuclear weapons and human rights,... These problems... will not be resolved overnight and most certainly will not vanish at the first sign of American cash".[86]

In sum, the link that has been made between closer scientific and technological

79   *International Transfer of Technology: An Agenda of National Security Issues, op. cit.,* p. 9. A recent Russian emigrant said to have had access to Politburo documents, maintains that Soviet "détente doctrine contained both a peaceful goal (East-West trade and co-operation) and a military threat (Western aid was meant to allow the Soviets to sustain their high level of arms spending)". Boris Rabbot, "The Debate over Détente", *The Washington Post,* July 10th, 1977, p. B5.

80   *Détente, op. cit.,* p. 25. A similar argument is that "given the relatively modest gains accruing to the East, particularly the Soviet Union, from imported technology and its strong commitment to political objectives, it seems unlikely that significant concessions on non-economic issues could be gained by using technology as a bargaining chip". *Economic Relations between East and West: Prospects and Problems,* A Tripartite Report from Fifteen Experts from the European Community, Japan and North America, The Brookings Institution, Washington, D.C., 1978, p. 19.

81   *Transfer of Technology to the Soviet Union and Eastern Europe, Hearing, op. cit.,* p. 4.

82   *Ibid.,* p. 16.

83   *Détente, op. cit.,* p. 34.

84   As noted in Chapter 1, détente has served to legitimise East-West trade and this trade is conducted largely on the basis of Western credit. See also Chapter 6.

85   Robert W. Campbell and Paul Marer, *East-West Trade and Technology Transfer, An Agenda of Research Needs, op. cit.,* pp. 76 and 77. This notion has been supported by one of the principal architects of the détente policy. While stating that "future leverage is still there if it can be organised", Mr. Kissinger acknowledged that the "Russians now have more leverage as a debtor than the West as a creditor". "Kissinger Defends the Selling of Détente", *International Herald Tribune,* Paris, December 6, 1978, p. 4.

86   Adlai E. Stevenson III, "Views on Eximbank Credits to the USSR", in Paul Marer, *US Financing of East-West Trade, The Political Economy of Government Credits and the National Interest, op. cit.,* p. 250.

relations, trade and détente, raises as many problems as such linkage was hoped to solve. However, one development is evident. While the lessening of political tensions which resulted from abatement of the Cold War was a natural precursor for closer trade relations, it has, in turn, led some to conclude that the linkage between science and technology and détente will lead to closer international relations in general. Technology transfer has thus not only become linked to foreign policy considerations but is very much influenced by them. There are also some who feel that "foreign policy considerations dominate the entire structure of technology exchanges with Communist countries. Technical problems — the degree of reciprocity, impact of transfer, monitoring and coordinating transfers in compliance with export control, private technology exchange protocols, inadvertent or indirect transfer and marketing implications — are largely ignored... Commerce Department proposals to require submission of protocols (private sector agreements) to the government for review within 15 days of signature have been vigorously opposed by exporters and no reporting requirement probably exists".[87]

In conclusion, it must be noted that the West Europeans have always tended to stress the economic rather than the political opportunities in trade with the USSR. The "idea that the West can use trade as a leverage against the Soviet Union in order to influence Soviet policy is discounted as unrealistic and counterproductive".[88] In this regard some "argue that the West is too pluralistic and trade oriented to effectively use leverage anyway, and that the large number of countries and business firms make it impractical to control trade to the extent implied in a leverage policy".[89] The consensus seems to be that "Western trade with the Soviet Union will continue to grow regardless of steps the US might take to restrict or liberalise its trade policies, provided a major recession does not occur in the West and Moscow does not run out of foreign exchange or credits".[90]

Perhaps one reason why Western policy to date has been marked by inconsistency and disagreement (for instance, the need to build new economic and trade relations versus the need to minimise such relations in the interest of military security) is that it attempts simultaneously to address different, often conflicting, dimensions of a most complex problem. These dimensions must be reconciled if any policy is to be effective.

3. Economic and Commercial Considerations Influencing Technology Transfer Policy

The economic and commercial considerations which have led the Western countries to seek closer trading relationships with the East are varied. Given the relatively small share which, on average, trade with the East represents in the total trade of the OECD countries, such trade has not been seen primarily as a means for improving the general economic situation.[91]

The search for closer trade relations with the East can, however, be directly interpreted from commercial considerations. From the point of view of those enterprises actually engaged in trade, co-operation offers the promise of profit. From the point of view of national policy, there can be a variety of reasons for seeking it. Some countries, such as Canada, seem to seek co-operation in order to "strengthen differentiation in the foreign trade structure".[92] Japan has sought it in order to "secure access to raw materials and fuel products".[93]

87　*Summary Statement of Report to the Congress, The Government's Role in East-West Trade, Problems and Issues, op. cit.,* pp. 36-37.
88　*Western Perceptions of Soviet Economic Trends, op. cit.,* p. 13.
89　*Ibid.,* p. 12.
90　*Ibid.*
91　For those countries, for instance Austria or Finland, where trade with the East does represent a large share of overall trade, co-operation in certain industrial branches could, of course, have an effect on their general economic situation. See F. Levcik and J. Stankovsky, *Industrielle Kooperation zwischen Ost und West, op. cit.,* p. 62. See also Chapter 7 of this volume.
92　F. Levcik and J. Stankovsky, *op. cit.,* pp. 60-61.
93　*Ibid.,* p. 60. The prospect of securing access to raw materials and fuel products is rapidly becoming a primary motive for all of the Western countries. *Ibid.*

As has been noted in Chapter 1, Japan and the countries of Western Europe much earlier pressed for more liberalised trade policies toward the East than did the United States. As the opportunity for individual gains by means of sales of technology to the East European countries increased, as a result of long-term economic agreements providing for scientific and technological co-operation, increased pressure was placed on the US government by the business community to liberalise its trade policy with the East. West European penetration of Eastern markets, based on domestic and non-domestically acquired technology (primarily US) was seen by the US business community as constituting an increasing threat to their competitive position.[94]

In the gradual transition from a policy of export control to a policy of export promotion, US policy-makers were faced with a number of questions. Among them, the following might be mentioned:

a) In engaging in trade and technology exchange with the Soviet Union, is the United States helping to establish a Soviet potential that will affect the US position in the world market adversely?;

b) Will the expansion of economic relations with the USSR provide an opportunity to obtain new markets for US goods in Eastern Europe? and,

c) Can the USSR be regarded as a reliable source of supply for industrial raw materials and as a consistent customer for US industry?[95]

While the questions listed above may differ in nuance, pertinence or relative weight attached to them, they do appear to be fairly representative of Western concerns as a whole. This is surely equally true with respect to the various pressures placed on the US governement to adopt a more liberal commercial policy toward the East. Given the fact that the US transition to a more liberal trade policy with the East is of recent origin and that the effects of technology transfer to the East on the US economy are in the process of being re-examined, the following discussion focuses on the United States.

The purpose of this discussion is to list the types of considerations which have led the US government to seek closer commercial relations with the East. The viewpoints opposing this policy are also given. The assumptions which have been made relative to the advantages of closer commercial relations with the East have led to the present, relatively liberal, trade policy. These assumptions can be assumed to be illustrative of the reasons why the individual OECD countries are actively engaged in trade with the East.[96]

a) *The Lure of the Market*

   i) *Fear of Competition*

   — *Eastern Markets*

Those officials in government circles who supported increased trade, argued that it would help to reduce balance of payments deficits and provide a stimulus to selected sectors of the economy. This view, articulated most explicitly during the Nixon Administration,

94 As West European economic strength increased, US firms became increasingly preoccupied by the fear of being driven "out of business". Competition came not only from West European industries and firms, but also from US-owned subsidiaries in Western Europe.

95 See Herbert S. Levine, *et al., Transfer of US Technology to the Soviet Union: Impact on US Commercial Interests, op. cit.,* p. 24. See also Alan L. Otten, "Selling Know-How", *Wall Street Journal,* 15 September, 1977, p. 24.

96 See Chapter 7 for the discussion of the effect of closer trading relationships and technology transfer on the general economic situation of the OECD countries. In this Chapter an attempt is also made to evaluate the arguments and assumptions which have resulted in closer commercial relations with the East.

found support in academic and business circles.[97] Given a more favourable political climate, and government-provided means for at least partially sanctioning and facilitating entry into Eastern markets, business and industrial circles exerted growing pressure on the US government to relax those export control measures viewed as depriving American firms of numerous export opportunities.[98] Fear of competition provided a common theme for those who supported a more liberal policy.[99]

A common feeling on the part of corporate executives, particularly those whose firms deal in conventional or "stable" technology, was "that the Soviets would inevitably buy most of what they want. If US vendors do not sell, then foreign competition will benefit".[100] Thus, many industrialists maintained "that the United States cannot by itself control trade with the USSR. The countries of Western Europe and Japan are frequently willing to sell items to the USSR which we might be more reluctant to make available. The United States is a unique supplier in only a few areas, and can be expected to lose some of this edge as other Western nations improve their technology. In most areas for which export approval is even a remote possibility, technology comparable to that of the United States is already available to the USSR from other Western sources".[101]

The theme "if we don't others will" also found support in government circles. It was argued, for instance, that "the increased availability of high technology products elsewhere rendered some of our original curbs on exports to the Soviet Union increasingly anachronistic. The real loser from these particular restraints would have increasingly been the US producer and worker... There comes a point at which we must face the fact that business is business, and, if it is going to go on in any event, we might as well have a piece of the action".[102]

97   See for instance, Marshall I. Goldman, "US Policies on Technology Sales to the USSR", in *East-West Technological Co-operation, NATO Colloquium 1976, op. cit.,* p. 114 and William C. Norris, "High Technology Trade with the Communists", *art. cit.,* p. 102. The notion that increased trade with the East, and in particular with the Soviet Union, is beneficial to the US economy has also been endorsed by the Carter Administration. It is said, for instance that "we believe that there are important markets which can be developed in the United States for Soviet products. We know that the United States can benefit significantly through greater imports of Soviet products and technology. An example of what has already been achieved is the successful marketing in the United States of surgical suturing instruments based on Soviet technology...", W. Michael Blumenthal, *op. cit.*

98   The US-Soviet Agreement for Co-operation in the Fields of Science and Technology established a basic format and organisational structure for subsequent inter-governmental science and technology agreements. Article 4 of the basic Agreement sanctioned the search for appropriate means of establishing and developing "direct contacts and co-operation between agencies, organisations and firms of both countries and the conclusion of implementing agreements for particular co-operative activities engaged in under this Agreement". See *Background Materials on US-USSR Co-operative Agreements in Science and Technology, Report, op. cit.,* p. 16. As was noted in Chapters 1 and 3, a relatively large number of agreements have been concluded on the basis of this Article. While the major portion of US trade with the East continues to be negotiated on the basis of individually negotiated export-import transactions, longer term industrial co-operation arrangements are becoming more popular. It is thought that US firms are beginning to favour these because such arrangements are believed to provide a more effective means of assessing Soviet markets. In many instances they appear to be the best, if not the only, entry vehicle into Eastern markets. See Lawrence H. Theriot, *art. cit.,* p. 750; and Rogers Morton, *op. cit.,* p. 14.

99   In the United States concern was being expressed that the Soviet Union might conceivably achieve its economic aims through expanded economic relations that involved solely trade and technology exchange with Western Europe and Japan. See Herbert S. Levine, *et al., Transfer of US Technology to the Soviet Union: Impact on US Commercial Interests, op. cit.,* p. 24. This concern was evidently related to the fact that many of the US firms which had established subsidiaries in Western Europe did so, for instance, in high technology areas — chemicals, electronics and sophisticated machinery. The presence of American subsidiaries in Europe gave the CMEA countries a new way of obtaining US technology. Through contacts with US-owned subsidiaries, this opportunity was eagerly exploited. For instance, "between 1959 and 1962, of 70 applications for permits to export to Eastern Europe received by US authorities from foreign subsidiaries of US corporations, 57 were approved". Joseph Wilczynski, *The Economics and Politics of East-West Trade, op. cit.,* p. 292.

100   Robert W. Clawson and William F. Kolarik, Jr., *art. cit.,* p. 32.

101   As given in *Détente, op. cit., p. 24.*

102   Peter G. Peterson, *op. cit.,* p. 13.

It can be assumed that those who supported the above viewpoints undoubtedly found support in Soviet tactics employed to exploit readily available non-US marketing opportunities: either by simply doing business elsewhere or threatening to do so. For instance: "At any rate, the USSR's foreign economic ties are broad enough so that it can find partners who are genuinely interested in setting up stable, mutually advantageous trade and economic co-operation. For instance, it is no accident that a few days ago the Paris newspaper *Les Echos* expressed the opinion that in the present atmosphere possibly it is a convenient time for Western Europe to play its cards and step up co-operation with the Soviet Union".[103]

— Western Markets

The possibility that through trade the Eastern countries, especially the Soviet Union, might themselves become competitors, was discounted.[104] It was maintained for instance, that since the Soviet system is slow to innovate — unlike those systems which are geared to the profit motive to guide and stimulate their efforts to foster innovation — "the Soviet Union is less likely to emerge as a competitor than most of the countries we sell to".[105] An interview of US corporate executives indicated that the majority believe that Soviet products based on US know-how will not be competitive in industrialised nations. This pertains in particular to "stable" products (products which do not become obsolete rapidly due to technical development) and to high-technology products. The reasons cited were the "inability of the Soviet Union to absorb and reproduce transplanted technology, vast internal needs which preclude any large scale decision to penetrate industrial markets — or any export markets — and inability of the Soviets to deal with "dynamic" product lines".[106] Another argument which is made relates to the volume of commodities exported by the US and the Industrialised West (IW) to the Eastern countries. In this regard it is said that in view of the fact that the share of "IW exports destined for communist countries is small... this suggests that communist countries have not been and are not likely to become such a dominant force in the marketplace that they could exert significant pressure on Western suppliers of advanced technology, even assuming that communist countries would (or could) act collectively... *In short, by world trading standards, high technology products do not*

---

103   A. Bovin, *Izvestia,* January 19, 1975, p. 2.

104   See Chapters 5 and 6. The advantages thought to accrue from expanded trade with the East have been linked to the notion that "to the extent that imported technology does increase exports of goods and services from the East, the Eastern countries will be better able to pay for additional imports. The effect will be to increase trade in both directions, and any loss of markets through Eastern competition is apt to be offset by the expenditure of increased export earnings in the West". According to this analysis concerns about competition "seem to be largely unwarranted". *Economic Relations Between East and West: Prospects and Problems, op. cit.*

105   Marshall I. Goldman, "US Policies on Technology Sales to the USSR", *art. cit.*, p. 115. As has been noted, one of the basic pre-occupations of Western opponents of expanded East-West trade is the concern that most of the gains from trade go to the East. It is said for instance, that "foreign trade of the nonmarket economies is conducted by state monopsonies and monopolies, which can supposedly exercise strong market power against competing sellers and buyers in the West", *Economic Relations Between East and West: Prospects and Problems, op. cit.*, p. 13. In opposition to this, it is maintained — an argument not unlike that related to innovative capacity in that it focuses on weaknesses in the Eastern systems — that "available empirical evidence certainly does not support the belief that the Eastern state trading bodies are successful monopsonists. [Thus] when everything is taken into account, the commonsense conclusion about East-West trade is that both sides gain — otherwise trade would not continue — but that in normal circumstances neither side can reap disproportionate benefits. This conclusion also applies to trade in technology, a field of exchange sometimes considered exceptionally beneficial to the East". *Ibid.,* pp. 13 and 14. See also Chapter 7 of this volume.

106   Robert W. Clawson and William F. Kolarik, Jr., *art. cit.,* pp. 22 and 23. See also Chapter 7 where the issue of impact is discussed in detail.

dominate in exports to communist countries, are not large in volume, and are not experiencing any marked shift in relative importance. [107]

The possibility of adverse competition occurring as a result of expanded East-West trade in general, and from technology transfer in particular, provides a primary focal point for those who oppose trade expansion. Critics of a more liberal commercial policy with the Soviet Union assert that the Soviet Union not only may become competitive with the West in the future, but is already competing.[108]

It is felt, for instance, "that in those areas where technology is not "fast-moving" but where the Soviet Union can copy long-existing mass-production operations, the Soviets may be able to break into Western markets simply because they are able to cut prices".[109]

Other critics fear that the Soviet Union will become competitive also in high technology areas. Referring to the fact that the export of US advanced technology has become a keystone of the policy of détente, it is said that "under competitive pressure and in the face of an opportunity for a short-term gain... it is possible for a company to give away overnight something which it has spent years of difficult effort to obtain. This is especially true in many high-technology areas which affect our national security as well as our position in world trade".[110] This is believed to be particularly true for high-technology firms which have "in-house capability to engineer and license know-how for sophisticated turnkey plants in the USSR which would manufacture stable products that their companies sell in the industrialised world — for example, automotives and chemicals".[111]

Fears of competition seem to be particularly pronounced among firms dealing in products such as chemicals and plastics.[112] With regard to "competition" from manufactured capital goods, consumer durables, or industrial raw materials there appears to be a widespread belief among (certain) executives that the USSR will be able to take existing transplanted technology and divert it for manufacturing.[113] Ability to "reproduce trans-

---

107   Hedija Kravalis, Allen J. Lenz, Helen Raffel and John Young, "Quantification of Western Exports of High Technology Products to Communist Countries", in *Issues in East-West Commercial Relations – A Compendium of Papers, op. cit.*, p. 38, As given in this article, the industrialised Western countries include: USA, Canada, Japan, Belgium-Luxembourg, France, Germany, Italy, Netherlands, Austria, Norway, Sweden, Switzerland, United Kingdom and Denmark (listed in the order given in the source). The communist countries are: Bulgaria, Cuba, Czechoslovakia, the German Democratic Republic, Hungary, Poland, the People's Republic of China, Rumania, USSR, Yugoslavia (given in the order listed in the source).

108   See Chapter 7 of this volume.

109   Marshall I. Goldman, "US Policies on Technology Sales to the USSR", *art. cit.*, p. 105.

110   Arthur H. Hausman, "East-West Technological Co-operation, Likely Transfer Patterns (1976-1980)", *art. cit.*, p. 335. The author quotes Malcolm R. Currie, then Deputy Director for Research and Engineering, US Department of Defense. For instance, with respect to the decision taken (September, 1978) to approve the sale of sophisticated drilling technology to the Soviet Union it is said that there is "no doubt that the controversial export of a complete plant for making rock drilling bits will give the Soviet Union an enhanced capability for producing oil". The major policy question is the "profound economic and political impact of the plant on the Soviet future rather than a red herring issue of diversion to military use". Don Oberdorfer, "Sale of Drill Plant to Russia is Assailed", *International Herald Tribune*, Paris, October 5, 1978, p. 3. The article quotes William Perry, Undersecretary of Defense for Research and Engineering. The same article quotes Fred Bucy, President of Texas Instruments Inc., to the effect that "the drilling technology can assist the Soviet Union to develop its energy resources independent of further US and Western support and may give the Russians the capability in the late 1990's to compete aggressively with the United States in drilling operations in the major oil-producing areas of the world". See also Thomas O'Toole, "White House to 'Observe' Technology Export Deals", *art. cit.*

111   Robert W. Clawson and William F. Kolarik, Jr., *art. cit.*, p. 22.

112   *Ibid.* It is said, for instance, that "compensation and buy-back deals under which COMECON countries pay in chemicals for plants bought from the West, are now producing an uncomfortably large harvest in the depressed, glutted Western European chemicals market". Some European chemical firms are seeking ways to "protect themselves from cheap imports of COMECON plastics, synthetic rubber and fertilisers at prices up to 40 per cent below those of Western European producers". "A European Selling Price System for Chemicals?", *The Economist*, June 24, 1978, p. 89. See also Chapter 7 of this volume.

113   Robert W. Clawson and William F. Kolarik, Jr., *art. cit.*

---

planted technology is also feared as is Soviet capability to develop product service networks in the industrial nations".[114]

Apart from the concern that the Soviet Union will compete on Western markets, some industrial leaders fear that the Soviet Union will compete with the West on Third World markets, both in "slow-moving" as well as in high technology goods.[115]

The adoption of Public Law 92-52 illustrates that the US government also shares some of these concerns, though these are not only confined to fears of competition from the Eastern countries. Section 119 of this Law requires that a study be conducted "of the domestic economic impact of exports from the United States of industrial technology whose export requires a licence under the Export Administration Act of 1969. Such a study shall include an evaluation of current export patterns on the industrial competitive position of the United States in advanced industrial technology fields and an evaluation of the present and future effect of the exports on domestic employment".[116]

The analysis which was prepared pursuant to Section 119 of Public Law 92-52 concludes that "although not enough evidence was found to estimate the impact of technology transfers on US trade, production and employment, several important conclusions were reached:

"There is evidence that the United States is losing its competitive position (as measured by market shares) in international markets for several products that have been characterised as technologically intensive.

There is some evidence that high technology content is important to the competitiveness of products in international markets, but the evidence is not overwhelming.

There is no conclusive evidence that US exports of technology are hurting or helping the competitive position or the overall economic position of the United States. However, certain domestic employees and firms, and their communities may experience dislocation costs when technology is transferred abroad.

It is very difficult to estimate impacts of US technology transfers on the US economy".[117]

ii) *Considerations Related to Balance of Payments Problems*

Increasing difficulties encountered by US firms in domestic as well as international markets were perceived by some to have resulted in sizeable balance of payments problems.[118] Increased trade with the East was seen as one means of improving the balance of

---

114   *Ibid.*
115   See *Ibid.,* p. 23 and Marshall I. Goldman, "US Policies on Technology Sales to the USSR", *art. cit.,* p. 105. With respect to plastics, some go so far as to say that "the build-up of plastics production capacity in Eastern Europe and in the Far East — plus the inevitability that the Middle Eastern states will eventually move "downstream" into plastics — is now calling into question the viability of producing plastics at all in Western Europe". John Murphy, "Plastic Makers Feel the Heat", *New Scientist,* March 8, 1979, p. 773. See also Chapter 7 of this volume.
116   See Public Law 95-52, June 22nd, 1977. In countries such as Sweden, firms have entered market sharing arrangements in order to avoid competition, particularly in Western and Southern markets.
117   *Technology Transfer, A Review of Economic Issues,* US International Trade Commission, US Department of Commerce, US Department of Labor. A Study Pursuant to Section 119 of the Export Administration Amendments of 1977 (Public Law 95-52), Washington, D.C., June 1978, Executive Summary, p. i. See pages 10-11 of the publication for a definition of "technology intensity". See also *International Transfer of Technology, op. cit.,* which analyses the benefits and risks of technology transfer. The report itself "suggests that the risks of technology transfer to the domestic economy and to the US competitive position in international markets are either nonexistent or quite limited". The assessment of the report however maintains that in effect these risks are not dealt with. The response of the report is that the risk is not great because such technologies are subject to licensing. No effort is made to provide evidence to the reader that the licensing process is effective". See p. 43 of report. See also Chapter 7 of this volume.
118   See Marshall I. Goldman, *Détente and Dollars, Doing Business with the Soviets,* Basic Books Inc., New York, 1975, pp. 73-74.

payments. It was maintained, therefore that since "high technology products, plus agricultural products, account for the bulk of US exports today, it is particularly important at this time, with trade deficits mounting, to exert every reasonable effort to increase exports".[119]

A number of related beliefs was subsumed in this contention. One was that the dollar value of US exports would be increased. "Dollar income, made possible by the export of technology to the East, not only helps the balance of payments, but supports the exchange rate of the dollar".[120]

Sales to the Soviet Union were also thought to help the economy by reducing unit costs. It was said that "at a time when government investment, subsidies and tax incentives are being used to ensure that US prices are competitive in the world market, an expansion of foreign markets is a factor that may facilitate reductions in cost and presumably prices. The opening of the Soviet market to US business may provide the basis for a larger, more economical scale of domestic output".[121] The semi-conductor industry was believed to provide an unprecedented opportunity in this regard. In view of the fact that Eastern Europe represented a large consumer market not yet penetrated by the US semi-conductor industry, it was thought that if "US-owned companies could capture a significant share of the 1980 market, our world competitive position would be greatly enhanced because semi-conductor prices are highly reliant on volume. What would happen if the United States elects not to pursue the Eastern European market? Quite simply, our competitors in Japan and Western Europe will".[122]

Using the arguments outlined above, opponents to expanded trade relations with the East demonstrate a certain scepticism with respect to the commercial gains and balance of payments advantages that are said to accrue from greater technological exports to the East. They suggest that economic growth and other economic benefits associated with increased trade would be affected only modestly by Soviet or East-West trade.[123] In fact, the assumption that gains will be obtained at all is said to be a product of "our eagerness to correct rapidly our deficit in trade balance and by our sometimes naive acceptance of the availability of vast new markets... The market may be significantly less than advertised, the difficulties of doing business extensive and the ability to pay questionable".[124]

b)   *The Promise of Greater Employment Possibilities*

The prospect of greater employment generated by increased trade with the East, provided another argument for those supporting a more liberal commercial policy towards the CMEA countries.[125] In fact, some saw a direct correlation between increased trade and creation of employment. In the computer industry, for example, it was estimated that "sales of large computers and peripherals... could in 10 years build up to an annual level

---

119   William C. Norris, *art. cit.,* p. 102.
120   Rogers Morton, *op. cit.,* p. 44.
121   *Science, Technology and American Diplomacy,* Volume 1, *op. cit.,* p. 572.
122   *Proceedings of the East-West Technological Trade Symposium,* Sponsored by the US Department of Commerce, Washington, D.C., November 19, 1975, p. 28.
123   John P. Hardt and George D. Holliday, *US-Soviet Commercial Relations: The Interplay of Economics, Technology Transfer, and Diplomacy, op. cit.,* p. 42.
124   Malcolm R. Currie, as given in Arthur H. Hausman, "East-West Technological Co-operation, Likely Transfer Patterns (1976-1980)", *art. cit.,* p. 336. See also *Technology Transfer: A Review of the Economic Issues, op. cit.,* and *International Transfer of Technology, op. cit.*
125   The report *Technology Transfer: A Review of the Economic Issues, op. cit.,* pp. 21-22, does not come to a conclusive judgment on this issue. The report *International Transfer of Technology, op. cit.,* concludes that "it is not clear that the transfer of US technology overseas has, historically, resulted in a net loss of jobs in the United States... Unless US firms begin to license, sell, or otherwise transfer recently developed techniques — well in advance of competitive development of similar techniques elsewhere — US transfer of technology will continue to have only limited overall impact on loss of US jobs". See p. 26. The assessment of the report points out that "there is some evidence not cited in the report that US firms have begun to export advanced technologies". See p. 48. See also Chapter 7 of this volume.

representing 150,000 jobs".[126] Concerning transport equipment, it was noted that the "export of Caterpillar earth-moving equipment to the Soviet Union and Eastern Europe... created thousands of jobs for Americans".[127] It was thought that growth in US exports of goods and services to the East generated additional employment even in such periods (for instance 1973-1974) when employment rates were high.

Support for the above was also found in government circles. For instance: "does (increased) trade... give us anything? Sure, it gives us jobs. We exported 1.8 billion to the Soviet Union last year. That is the rough equivalent to 65,000 jobs in the United States".[128] Some believed in fact that the reduction in employment by certain industries was partially due to unnecessary restrictions on US exports of technology-intensive products.[129]

Those who question that employment would benefit from expanded trade point out that any such gains would be modest.[130] Organised labour has been the most vociferous critic of the employment gain argument. Thus it is maintained that a liberalised trade relationship allowing more imports into the United States would bring massive layoffs and production cutbacks. Citing Rumania as one example, it was said that "Rumania imports technology from US firms under contract to export from Rumania, thus jobs from US exports and jobs at home will be curtailed".[131] A similar argument is made with respect to Eximbank credit policy. It is claimed that "Eximbank helps US exporters to concentrate on selling abroad production capacity rather than products... Exported production capacity helps foreigners compete successfully with US products thereby contributing further to the erosion of jobs in the US".[132]

c) *Access to Products Needed for Industrial Growth*

Access to Eastern products deemed to be necessary for industrial growth provided an important motive for those who believed that increased trade would be beneficial. In the past, most Soviet exports to the United States have consisted of raw materials and primary products. Thus access to those basic industrial raw materials, in short supply in the United States and vital to several sectors of the US economy, could be cut off, it was argued, if US trade policy towards the Soviet Union was not more forthcoming.[133]

Benefits were also perceived in the potential acquisition of commercially useful technology from the Soviet Union. Soviet technical capabilities were thought to be particularly "attractive in areas where R & D investment in the West has been minimal because of insufficient commercial interest in the output of that research".[134] It was thought that by "facilitating imports of Soviet machinery and industrial products, the United States might reap an unexpected benefit from expanded trade ties with the Soviet Union, namely, the acquisition of new Soviet technology in a few industrial sectors. In certain high priority industries, the Soviet Union has devoted considerable resources to research and development. Some Soviet industries have made important technological innovations which prove very valuable to US firms. The steel and aluminium industries and certain mining industries

126   William C. Norris, *art. cit.*, pp. 101-102.
127   *Ibid.*, p. 102.
128   Arthur T. Downey, then Deputy Assistant Secretary of Commerce for East-West Trade, given in *Détente, op. cit.*, p. 22. See also Rogers Morton, *op. cit.*, pp. 19 and 49.
129   Betsy Ancker-Johnson and David B. Chang, *op. cit.*, p. 83.
130   John P. Hardt and George D. Holliday, *US-Soviet Commercial Relations: The Interplay of Economics, Technology Transfer, and Diplomacy, op. cit.*, p. 42.
131   AFL-CIO Statement on S. Con. Res. 35, June 6, 1975.
132   M. Christopher Kwiecinski, *art. cit.*, p. 237.
133   See John P. Hardt and George D. Holliday, *US-Soviet Commercial Relations: The Interplay of Economics, Technology Transfer and Diplomacy, op. cit.*, p. 54; Rogers Morton, *op. cit.*, p. 15; and *Détente, op. cit.*, p. 22.
134   Lawrence H. Theriot, *art. cit.*, p. 751.

are examples of US sectors which could benefit from such an exchange of technology".[135]

The above is contested by those who believe that the United States will derive little benefit from trade with the East. First it is maintained that the Soviet Union's technological level is too small in order to gain any advantage.[136] Second it is pointed out that the co-operative ventures entered into with the Soviet Union were concluded out of political motives with little or no consideration given to the potential scientific or technical benefits.[137] It is felt, therefore, "that on the whole, but with some exceptions for specific fields, we have been and are currently teaching the Soviets more in the course of our exchanges than we are learning from them".[138]

In sum, it may be said that Western commercial policies with respect to the East are marked by disagreements about the major issues involved. The policy to expand trade with, and technology transfer to, the East — despite misgivings expressed by many in both the private and government community — is based on the assumption that benefits will be derived from such expansion. Little documented information of an evaluative character concerning the long-range impact of East-West trade is available.[139] Nevertheless, some Western governments have met with considerable resistance from interested parties and pressure groups when they have attempted to restrict economic relations with the East.

On the whole, restrictions on East-West commercial relations have been greatly reduced. Yet how to strike a balance between the desire to promote technology exports considered necessary for growth, and the need to prevent the accelerated diffusion of technological skills and know-how to economic competitors which might threaten the national interest, still remains a dilemma.

4. **Assumptions about the Effects of Technology Transfer on East-West Economic Relations and on the Evolution of Eastern Systems**

Some rather fundamental assumptions have been made about the relationship between trade and technology transfer as it affects both the future evolution of East-West economic relations and the internal evolution of the Eastern systems. Two central assumptions, closely related one to the other, are currently the subject of keen debate. Argument turns about the propositions:

a) That closer economic relations will/will not lead the East to increased interdependence/interrelatedness with the West and ultimately to a reordering of Soviet economic priorities, and,

b) That closer economic relations with the West will/will not bring about a change in the basic features of Eastern societies, and the Eastern system as a whole.

In what follows, an attempt is made to set out the arguments on both sides.

135   John P. Hardt and George D. Holliday, *US-Soviet Commercial Relations: The Interplay of Economics, Technology Transfer and Diplomacy, op. cit.,* p. 55. See also Rogers Morton, *op. cit.,* p. 48.
136   See Section 1 of this Part and Chapter 5.
137   See Section 2 of this Part.
138   Carl Kaysen, *Review of US-USSR Exchanges and Relations, Executive Summary,* National Academy of Sciences, Washington, D.C., p. 11 (n.d.). See also *Review of the US/USSR Agreement on Co-operation in the Fields of Science and Technology, op. cit.* The report *International Transfer of Technology, op. cit.,* pp. 17-18, is inconclusive on this issue. It is assumed that scientific exchanges are useful because neither... "agencies nor individual scientists would wish to participate if useful results were not forthcoming". This evaluation is questioned in the assessment on the grounds that "the evaluation of US-USSR co-operative programs includes no discussion of complaints from some US participants that the US side is benefiting less than the Soviet side". *Ibid.,* p. 44. See also Loren Graham, "How Valuable Are Scientific Exchanges with the Soviet Union?", *Science,* Vol. 202, No. 4366, October 27, 1978. While citing evidence which on balance supports the belief that scientific exchanges are worthwhile, "the majority of scientists who were polled... considered the United States to be ahead of the Soviet Union in most scientific fields". See p. 384 of cited article.
139   See Chapter 7. One outcome of this at least in the United States, is passage of the previously mentioned Public Law 95-52 and Public Law 95-92 designed to correct this situation.

a) Interdependence/Interrelatedness

One important, if not totally cohesive school of thought believes that as technology continues to be transferred, Eastern relations with the industrialised West will evolve in the direction of greater "interdependence" or "interrelatedness".[140] The basic assumption is that as the Soviet Union "progresses along the path of industrialisation it will become increasingly under the influence of the general pattern of industrial development. This general pattern for industrialised nations appears to be one of increasing international involvement as economic development proceeds (the interwar period can be viewed as a recession-ridden anomaly in regard to the role of international trade)".[141] It is believed that the expansion of economic relations with the West represents a (tolerable and desirable) degree of "interrelatedness" for the Soviet Union since, at this moment, closer economic relations with the West "do not involve capitalist countries in the basic core of the Soviet economy (which is now sufficiently developed to guarantee the independence of the Soviet Union), but operate at the margin to improve the performance of the Soviet economy".[142] Some authors establish a direct link between interdependence and technology transfer: "technology and industrialisation have created an unprecedented degree of economic interdependence among nations, [because] no nation is self-sufficient either in the raw materials and foods it requires to feed its industries and its people, or in the advanced technologies that are vital to the most efficient production of goods and services".[143]

Support of the notion of interdependence or interrelatedness has, in turn, led to the assumption that the Soviet Union will become even more interested in co-operation and thereby less willing to risk a serious rupture with the West. Thus it is thought that "the long-term interlocking of economies... while perhaps somewhat risky, nevertheless adds an element of stability to détente and renders economic relations more of a bargaining chip in other areas than it would otherwise be".[144]

Proponents of the "interdependence/interrelatedness" thesis invoke a number of developments in its support. One is Eastern acceptance of certain aspects of the theory of comparative costs. Another is Eastern support of the principle of intra-industrial specialisation and of the notion of the international division of labour.[145] Other general

---

140 Some Western authors distinguish between "interdependence", "interrelatedness" and "technological interrelatedness". Often, however, the terms are used interchangeably, and this usage is adopted here. However, it should be pointed out that for some authors the distinction is fundamental. These analysts reject the term "interdependence" as it implies that the Soviet Union has given up its traditional objective of remaining economically independent from the West: which they feel is not the case. Soviet policy toward foreign economic relations has also often been described as a policy of autarky. This is thought to be as inappropriate as the term "interdependence" because it implies a policy of reducing involvement with other nations to zero. "A policy of non-dependence does not require zero involvement but only that a country be not dependent on any other country for any goods crucial to its existence". These authors therefore prefer the term "interrelatedness" because it better reflects the range of international involvement (on the part of the Soviet Union) which lies between the lowest on the scale (autarky) and the highest (interdependency). See Herbert S. Levine, *et al., Transfer of US Technology to the Soviet Union: Impact on US Commercial Interests, op. cit.,* p. 42. Reference is frequently made to the fact that traditional Soviet policy toward foreign trade was one of "autarky". In the case of a direct quotation from a source the word will be maintained. However, the above-mentioned nuance should be kept in mind.

141 Herbert S. Levine, *et al., Transfer of US Technology to the Soviet Union: Impact on US Commercial Interests, op. cit.,* p. 43.

142 *Ibid.,* p. 47.

143 Rogers Morton, *op. cit.,* p. 25.

144 Franklyn D. Holzman, *International Trade Under Communism, op. cit.,* p. 164.

145 See Part I of this Chapter. Given the economic problems besetting the CMEA countries, increased reliance on the international division of labour is thought to be a long-term development. It is also believed that the reliance on Western credit to finance East-West trade makes it "imperative that these countries succeed in participating more effectively in international division of labour to have the capacity to repay Western credits". Robert W. Campbell and Paul Marer, editors, *East-West Trade and Technology Transfer, op. cit.,* p. 51.

---

developments which are invoked in this context are East European and Soviet participation in certain international economic organisations, the conclusion of many trade, long-term economic, scientific and technical co-operation agreements and the development of new forms of technology transfer.[146] Attention is also called to Soviet pronouncements which are believed to reflect "official renouncement of the former policy of economic isolation in the new environment of détente", and to Eastern reliance on Western technology to promote economic development.[147] Both, it is suggested, reflect the apparent willingness of the Soviet leadership to "reassess the ideological underpinnings of their traditional policy of autarky".[148] Finally, it is thought that while Soviet acceptance of global interdependence must be measured in terms of resistance to change and "how the Soviet leadership will deal with some fundamental structural problems in the Soviet economy (the Soviet Union is on the threshold of a wholesale generational turnover at the upper levels of its power structure — whether they will tend to move toward nationalism and orthodoxy, or toward Western-style modernization cannot be predicted), we should not underestimate the capability of the Soviet system to manage its problems on a day-to-day basis without any clear-cut solutions to these choices, it may have some relevance for our policy choices that the development of economic relations with the advanced industrial societies of the West is bound to have some influence on the directions that will emerge".[149]

Certain kinds of institutional changes are cited as evidence that the Soviet Union has accepted long-term Western involvement in its economic development. One example often quoted is the Togliatti plant, because the project involved Western assistance on an industry-wide basis and in its operation is said to be much less integrated than is usually the case in Soviet industrial undertakings.[150] It is believed that the earlier policy "of producing a Soviet plant in the indigenous administrative setting has been challenged and modified. There appears to be increasing acceptance of the idea that improved performance requires... broad Western involvement in the entire cycle of technology transfer...".[151] The "case studies of automotive technology transfer in the 1930s and in the 1966-75 period indicate some movement in overall policy from independence to technological interdependence...".[152] The case studies are also said to suggest that "traditional methods are perceived by Soviet leaders as inadequate and in need of change. Some of the broad outlines of that change are emerging and the absorption of Western technology appears to be influencing the direction of the change".[153] Thus, "long-term commercial transfers of technology to Soviet end users by private American firms may be among the most effective US vehicles for influencing Soviet economic performance".[154]

Finally, some believe that the Soviet Union seems to be reordering its economic priorities in order to facilitate access to Western technologies and know-how. This conclusion is based on a US study of the problems and functional inter-actions of science, technology and American diplomacy. This "report strongly suggests that the thawing of the Soviets toward trade relations with the United States is attributable to an apparent reordering of Soviet priorities, in favour of technological change and an improvement in the availability

---

146   See Chapters 1 and 3.
147   *Western Investment in Communist Economies, op. cit.*, p. 3. See also Part I of this Chapter.
148   *US-Soviet Commercial Relations: The Interplay of Economics, Technology Transfer, and Diplomacy, op. cit.*, p. 44.
149   Quoted with adaptation from Marshall Shulman as given in John P. Hardt, "Military-Economic Implications of Soviet Regional Policy", *art. cit.*, p. 248.
150   John P. Hardt and George D. Holliday, "Technology Transfer and Change in the Soviet Economic System", *art. cit.*, p. 205.
151   *Ibid.*, p. 212.
152   *Ibid.*
153   *Ibid.*, p. 213.
154   *International Transfer of Technology: An Agenda of National Security Issues, op. cit.*, p. 18.

of desirable consumer goods to the Soviet workers and peasants".[155] It is thought, furthermore, that judging from the "mellowing of the other socialist States whose ties with the West have been greater than the Soviet Union's, there is reason to believe that pressures can be created through increased contacts to divert resources to the consumer sector. Growing need to satisfy consumer demands... has played an important part in the need the Soviets feel to cut their expenditures for... weapons".[156]

These views however, are strongly contested by those who do not share them. It is argued instead that the Soviet Union and the East European countries have "aggressively pursued access to Western technology... and have purchased sophisticated products which they are unable to produce themselves... [with the intention] of gaining access to Western know-how to make the products themselves under conditions that enable them to control ownership and use of the technologies. Their ultimate goal is self-sufficiency".[157] Consequently, it is suggested that "the current emphasis on buying Western technology is no more than a medium-term tactical expedient that will prove as reversible as the Concessions policy of the 1920s. The Soviet Union, or the CMEA group as a whole, are and will remain closer to economic and technical autonomy than dependence; the commercial or political price that can be charged for Western technology is strictly limited, and the vision of a single, closely interdependent international economy of advanced industrial States is illusory".[158] The East European countries, will, it is thought, "remain heavily involved in CMEA, both because of Soviet pressure and because their ability in the short- or medium-term to restructure their economies toward more trade with the West is limited. Also, after their sobering experience with fluctuations in prices and quantities on the world market since 1973, they appreciate more keenly the advantages provided by the stability of CMEA trade agreements".[159]

Furthermore, it is held that Soviet leaders "feel strongly about self-sufficiency, and if there is any hint that they are becoming dependent upon Western sources or that they are being required to make concessions for Western technology, they will choose to cut off imports even if it means postponing a new programme".[160]

Much the same developments as are cited to support the belief in interdependence/interrelatedness, are also used to dispute it. Some authors note that Soviet support of the principle of intra-industrial specialisation explicitly excludes economic integration with the West.[161] Indeed, it is pointed out that Soviet and East European authorities have repeatedly stressed that the principle implies that "trade between countries can expand without necessitating substantial changes in the internal structure of production of the trading partners".[162] Eastern support of the notion of the international division of labour is equally qualified: "... the two social systems are different by virtue of their very nature, for which

155   *Science, Technology and Diplomacy in the Age of Interdependence, op. cit.,* p. 86.
156   *Détente, op. cit.,* p. 11.
157   *Transfer of Technology to the Soviet Union and Eastern Europe, Hearings, op. cit.,* p. 16.
158   Given in Philip Hanson, "External Influences on the Soviet Economy since the Mid-1950s: The Import of Western Technology", *art. cit.*
159   Morris Bornstein, "Economic Reform in Eastern Europe", in *East European Economies Post-Helsinki — A Compendium of Papers, op. cit.,* p. 132. The CMEA may be considering a change in voting procedure. (Reported in the Yugoslav national news agency and transmitted in the *International Herald Tribune,* Paris, June 17-18, 1978, p. 9. According to the article in the *Herald Tribune,* such a change could greatly increase the influence of the Soviet Union over other Member countries' economic policies. If the change is adopted, this would (it is maintained in the article) certainly make it easier for the Soviet Union to control dissent within CMEA, and to influence the future evolution of CMEA trade and the internal allocation of resources to conform to Soviet views of intra-bloc specialisation and division of labour. See also Chapter 1, Part V.
160   *Western Perceptions of Soviet Economic Trends, op. cit.,* p. 12.
161   See Chapter 1, Part V, Section 2 (b).
162   Given in Herbert S. Levine, *et al., Transfer of US Technology to the Soviet Union: Impact on US Commercial Interests, op. cit.,* p. 48. This is the corollary to the previously mentioned nuance that expanded economic relations with the West do not mean Western involvement in the "core" of the Soviet economy.

reason a common socio-economic basis cannot even be created by most advanced sharing of activities".[163]

It is said that Western help has always been sought by the East on a sporadic basis. This suggests to some analysts that East-West relations should be "interpreted as a result of Soviet self-interest in economic improvement more than a result of détente politics" or anything else.[164] Past precedent is also said to indicate that East-West trade will remain limited, thereby precluding anything but the most limited kind of interdependence. It is pointed out that while Western credit is serving in the short term to finance trade expansion, imports from the West by the USSR can, in the long term, only be increased if there is a corresponding increase to the West of Soviet exports. Inability to export has already been disruptive to Soviet-Western commercial activities. During the 1930s for instance, the Soviet Union was unable to pay for imports of machinery and Western technical assistance. As a consequence, Soviet purchases in the West dropped sharply. For the next 25 years (excluding the War years) the Soviet economy functioned on the basis of self sufficiency.[165] Thus it can be argued that a return to this former policy is in the realm of the possible. Prevention of such a return requires that Soviet exports be sharply increased; a problematic prospect and one likely to raise problems in some Western countries.[166]

The tenuous nature of East-West relations — one that has always been directly affected by political developments — also leads some to view the possibility of achieving stable, long-term interdependence with a good deal of scepticism.[167] In this regard it is pointed out that current Soviet support for closer economic relations with the West has been forged over a great deal of Soviet internal opposition and dissent among policy makers and in particular concerning the best means of achieving technological progress.[168] It is pointed out that the current policy towards the West is strongly associated with Mr. Brezhnev. A review of "East-West economic interactions and of the role of foreign trade as an instrument of Soviet national purpose, [shows] that economic as well as political relations with the West will depend on the Soviet leadership... and is tied to the question of leadership succession".[169] Political divisions between East and West also affect Western perceptions of the desirability and/or possibility of achieving closer relations with the East. "Should relations return to those at the height of the Cold War era, it is unlikely that Soviet exports would be well

163   Max Schmidt, "East-West Economic Relations Against the Background of New Trends and Developments in World Economy and in International Distribution of Activities", *art. cit.,* p. 21.

164   John P. Hardt, "Soviet Commercial Relations and Political Change", in Robert Bauer, editor, *The Interaction of Economics and Foreign Policy,* University of Virginia Press, Charlottesville, Virginia, 1975, p. 81.

165   See Chapter 1, Part II.

166   See Chapter 7.

167   US-Soviet relations appear to be particularly prone to such perturbations. The decision by President Carter to place all US exports of oil technology to the Soviet Union under government review and the cancellation of a high level US team's visit to discuss technology exchanges were decisions taken in response to the trials and conviction of two Soviet dissidents, A. Shcharansky and A. Ginsburg. Reported by Oswald Johnston, "US Seeks Ways to Press Russia on Dissent Trials", *International Herald Tribune,* Paris, July 12, 1978, p. 1 and by Richard Burt, "US to Use Oil Expertise as Rights Prod on Russia", *International Herald Tribune,* Paris, July 20, 1978, p. 1. Since that decision was taken there had been a "warming trend" in US-Soviet Science Co-operation. By November 1978 plans were already in process to reschedule the previously cancelled trip (which took place in February 1979) and licenses for oil and gas exploration and production equipment were approved. See "Warming Trend in US-Soviet Science Co-operation", *Science,* Vol. 202, November 17, 1978; W. Michael Blumenthal, *op. cit.;* and Kevin Klose, "US Carrot, Stick Start Soviet Trade Talks", *International Herald Tribune,* Paris, December 5, 1978, p. 2. The recent suspension of all exports of sophisticated technology and machinery to the Soviet Union as well as the imposition of controls limiting the export of industrial know-how to the Soviet Union — decisions taken in response to the Soviet Union's invasion of Afghanistan — would indicate another shift towards "cooler relationships". See Michael Getler, "US Tightens Control on Export of Strategic Technology to Russia", *International Herald Tribune,* Paris, March 19, 1980, p. 1. See also Chapter 1.

168   See Part I of this Chapter.

169   Herbert S. Levine, *et al., Transfer of US Technology to the Soviet Union: Impact on US Commercial Interests, op. cit.,* p. 125.

received in the West, that Soviet needs would be fulfilled by Western firms, or that the Soviet Union or the Western nations would opt for a greater degree of interrelatedness".[170]

Finally, it is disputed that there has been any reordering of Soviet economic priorities. It is pointed out that a "firm commitment to new priorities runs counter to the traditional policy of the Party and is also uncharacteristic of Party Secretary Brezhnev's past record".[171] In the period preceding the announcement of the Ninth Five Year Plan, "Brezhnev voiced his displeasure over the performance of the economy, but committed himself firmly to neither a reform of planning and management, nor a new set of priorities".[172] An important body of opinion believes there is no evidence that during this period the "high military priority of the 1960s was being scaled down".[173] The approach chosen by the Soviet Union to improve civilian R&D is said to "show the reluctance — or the inability — of the leadership to make any significant changes in the priority system".[174] This approach was announced by Brezhnev during the 24th Party Congress: "in view of the high scientific and technical level of the defense industry, the transmission of its experience, inventions, and discoveries to all spheres of the economy assumes paramount importance".[175] In view of the fact that in the past there has never been a significant change made in Soviet priorities with respect to the overall allocation of resources, despite the influx of Western technology, it appears unlikely that such re-allocation would be made in the future.

b)   *Assumptions Made about the Future Evolution of the Eastern Systems*

A number of Western analysts believe that closer economic ties with the West will bring about fundamental changes in the Eastern systems. The types of scientific and technological agreements which have been concluded are believed to play an important role in this process because they involve Western technicians and businessmen in key aspects of economic life and generally provide access to the Eastern economies. "Through specific private commercial and governmental channels the traditional general system of secrecy is being breached, however selectively and modestly".[176] It is supposed that present institutional arrangements will need to be adjusted in order to profit from the inflow of Western technology and thus could lead to more fundamental internal changes. Such changes could implicitly challenge the role of the Party.

The essence of the argument is that "imported machinery, consumer goods, and managerial techniques reflect ideas which in turn reflect the views and culture of societies in which these imports originate. Thus, a country cannot borrow technology... and use Western commodities... while at the same time keeping out the political and cultural influences that accompany them".[177] This belief seems to have been an important consideration on the US side in the co-operative agreements concluded with the Soviet Union. A long-range objective of the United States is said to be to "impact Soviet priorities as a result of the increased contacts and exposure to the West".[178] "When you look down the categories of agreements — space, health, energy, environment, transportation, housing — these contacts that we developed with the various parts of the Soviet bureaucracy and society

---

170   *Ibid.*, p. 124.
171   *US-Soviet Commercial Relations: The Interplay of Economics, Technology Transfer and Diplomacy, op. cit.*, pp. 24-25.
172   *Ibid.*, p. 25.
173   *Ibid.*
174   Gur Ofer, *The Opportunity Cost of Non-monetary Advantages of the Soviet Military R&D Effort*, a report prepared for the Director of Defense Research and Engineering, R-1741-DDRE, RAND, Santa Monica, California, August 1975, p. 45.
175   *Pravda*, March 31, 1971, p. 5, as given in *Ibid.*
176   John P. Hardt and George D. Holliday, "Technology Transfer and Change in the Soviet Economic System", *art. cit.*, p. 216.
177   See Robert W. Campbell and Paul Marer, editors, *East-West Trade and Technology Transfer, op. cit.*, p. 9.
178   *Détente, op. cit.*, p. 10.

191

are invaluable... in terms of their exposure to some influences which in the past they have been kept away from... We are, we think, making satisfactory progress towards our political objectives of improved access to the Soviet system. Co-operative activities are making possible an expansion and intensification of our access to an increasing number of Soviet individuals, organisations and geographic areas".[179]

Soviet Party leaders it is agreed, do recognise that there is a risk in allowing closer contact with the West, and in making institutional adjustments to facilitate the initiation of projects proposed under the co-operative agreements. Nevertheless, it is believed that the continuing reorganisation of Soviet industry, instituted in part to accommodate Western technological systems into the Soviet economic system, "may facilitate the removal of Western-assisted projects from control by traditional ministerial authorities and Party officials... Appropriate changes in the role of the Party to foster efficiency in economic management, modernisation, and effective absorption of Western technology is not a settled issue".[180]

Some authors believe there is evidence available to show some "movement from the traditional Soviet model of technology transfer to a modified systems approach and increasing acceptance of the idea that improved performance requires... new kinds of production facilities that more fully adapt Western managerial and technical methods to Soviet conditions".[181]

In addition to the Togliatti project, the Kama River Truck Complex is cited by some as an example of selective reform. For instance "the newly formed association organised for the Kama truck plant operates outside as well as within the traditional management and planning bureaucracy: perhaps coincident with expanding Western ties, some economic sectors appear to be exploring new administrative forms or variants of old. What seems to be involved is a removal of these Western connected enterprises from the traditional bureaucracy and relaxation of old ministerial ties, and control by the local Party organisation. The regional complexes and production associations seem to fit this new trend. The regional complexes, such as the Tyumen petroleum complex, West Siberian gas development, and the Baikal-Amur railroad development, appear to require considerable Western economic involvement and seem to be moving away from traditional lines of control".[182] In short, what is suggested is that "technology absorption, diffusion, and domestic innovation — the whole Western cycle of technological interchange — compel the Soviet Union toward the systems approach, including Western involvement in management".[183]

Others go even further and propose that, due to increased technology transfer, the Eastern systems will be "obliged to modify themselves in the direction of the enterprise system".[184] Thus it is postulated that "as technology transfer expands and the impact of Western technologies becomes more effective, the Soviet consumer will enjoy more consumer goods and particularly consumer goods of Western origin. When that happens, Soviet citizens will come to recognise the superiority of Western systems and will encourage their government to adopt political and economic institutions on the Western pattern".[185]

179   *Ibid.,* pp. 10-11.
180   *Technology Transfer and Scientific Co-operation Between the United States and the Soviet Union: A Review, op. cit.,* pp. 86-87.
181   John P. Hardt and George D. Holliday, "Technology Transfer and Change in the Soviet Economic System", *art. cit.,* pp. 205 and 212.
182   John P. Hardt, "The Role of Western Technology in Soviet Economic Plans", *art. cit.,* p. 323.
183   John P. Hardt and George D. Holliday, "Technology Transfer and Change in the Soviet Economic System", *art. cit.,* p. 190. On page 216 of the same article the authors refer to a "modified systems approach".
184   Given in and adapted from Raymond Vernon and Marshall I. Goldman, "US Policies in the Sale of Technology to the USSR", *art. cit.,* p. 10. This is the much discussed "convergence theory".
185   *Ibid.*

Concerning the Party, some believe its role could change, though not necessarily be challenged.[186] It is suggested that the "emerging pluralism within the Soviet elite has produced conflict between contending groups, most evident in decisions on resource allocation".[187] This conflict "cuts across institutional lines and ranges some professionals and Party generalists on one side against a similar grouping on the other side".[188] While "in the short-run, these changes and debates may well be contained within the present system and insulated from those parts of the system not intended for change, the long-run effect is open to more speculation... Whether progressive forms of Western contacts and techno-logical transfer can be contained within traditional Party political and governmental bureaucratic frameworks remains to be seen. The ripple effect of modest institutional change may have more profound substantive change, especially in the long-run".[189]

In summary, it is proposed that if one considers "change more narrowly, within the parameters of reviving economic performance as a primary goal of Soviet leadership, the steps necessary to attain the desired results through flexible and pragmatic change become increasingly likely over time. This is not to suggest that the Soviet leadership must change the Soviet economic system, and revise resource allocation priorities, but rather that the logic may have increasing force and that the leaders' own self-interest dictates this sort of institutional and political change".[190]

The foregoing views, however, meet with strong opposition from those who do not believe that increased contact with the West will lead to far-reaching Eastern economic and institutional reforms. These individuals base their arguments on the belief that Western economic help is not incompatible with a conservative Eastern domestic policy nor do they believe that the adoption of Western technology can bring about a change in "values of the Soviet leadership or in the most important characteristics of the Soviet political system".[191] Rather, it is thought that technology "is likely to be integrated or absorbed into the existing bureaucratic value systems and behavioural patterns rather than precipitate fundamental changes in Soviet politics".[192]

It is argued that the Soviet Union has always resisted too close contact with the West and that every opening to the West has created the need for greater security. This "explains why in the Soviet Union, and before that in Russia, the times of commercial interaction on the largest scale with the West have also been the periods of the tightest forms of repression".[193]

To judge by past experience, it seems that the Soviet government has the ability to turn off Western influence whenever it chooses. Visiting engineers, technicians and businessmen tend to be isolated in the Soviet Union, preventing the Soviet population from meaningful contact with them. It is said that "while the Soviets have made it plain that they want to tune into American science as part of efforts to modernise their industrial systems, they have never loosened up to the point where scientific collaboration between the two countries is free and easy... The main difficulty with this scheme to liberalise Soviet society from the outside is that so far there is little evidence that it works. Soviet dealings with the outside world — both for scientific and other purposes — have increased in recent years, but mainly

186   John P. Hardt, "The Role of Western Technology in Soviet Economic Plans", art. cit., p. 323.
187   Ibid., p. 325.
188   Ibid.
189   Ibid.
190   John P. Hardt, "Soviet Commercial Relations and Political Change", art. cit., p. 81.
191   Erik P. Hoffman, "Technology, Values, and Political Power in the Soviet Union: Do Computers Matter?", in Frederic J. Fleron, Jr., ed., Technology and Communist Culture: The Socio-Cultural Impact of Technology under Socialism, op. cit., p. 401. The author uses modern information technology as the basis for his thesis.
192   Ibid., p. 402.
193   Robert W. Campbell and Paul Marer, ed., East-West Trade and Technology Transfer, An Agenda of Research Needs, op. cit., p. 9.

in response to Soviet economic needs".[194] Some US businessmen believe that the co-operation agreements which have been concluded between private US companies and the State Committee for Science and Technology are used by the Soviets as a device to reach their technology import goals while limiting the impact of foreign contacts on the domestic system. For instance, of the companies sampled to estimate the progress that had been made since implementation of Basket II of the Helsinki Agreement, the majority noted that there had been no improvement in gaining access to end-users.[195]

Despite the revival of discussion since 1973 of economic reforms, little else appears to have happened, and far-reaching reforms do not even seem to be contemplated.[196] Some believe that the re-organisations undertaken in the late sixties, designed to stimulate production, seem to have increased the authority of central organisations. It is noted for instance, that "there seems to be some evidence that research institutes currently are subject to tighter controls by the agencies to which they are accountable".[197] No fundamental reform of the economic system appears to be currently under active discussion. "At the 25th Party Congress, Brezhnev stressed the importance of rewarding enterprises and workers for net results rather than gross output, and experiments to test such measures are continuing. Although further modifications of success criteria are likely, the benefits will be inconsequential as long as incentives remain tied to fulfilling plans for whatever target or targets".[198]

It is thought that since "decentralisation of economic planning and management can only be brought about by a revision, perhaps diminution, in the Party's control of society, it can be credibly argued that General Secretary Brezhnev will prefer to minimise the effects of internal and external policy changes required to modernise the Soviet economy and revive its growth rate".[199] In the countries of Eastern Europe widespread reforms have been opposed: on ideological as well as pragmatic grounds. The Party apparatus and the bureaucracies opposed them because they threatened entrenched power positions. It was also feared that reforms might lead to inflation and unemployment. They were opposed on ideological grounds because reforms threatened the paramount role of the Party. And finally, the reforms were opposed because of the possibility that they could easily spill over into wider political demands.[200]

194 "US-Soviet Ties: The Implications of Severence", *Science and Government Report*, Vol. VIII, No. 11, June 15, 1978, pp. 2-3.
195 Report to the Congress of the United States on *Implementation of the Final Act of the Conference on Security and Co-operation in Europe: Findings and Recommendations Two Years After Helsinki, op. cit.*, p. 73. Furthermore, it has been noted that growth in trade and co-operative work in the sciences (called for in Basket II), including joint research activities (which has moved ahead) remains largely unrelated to implementation of the final Act. At the Belgrade meeting (1977-1978) which examined the implementation of the Helsinki Final Act, Basket II was very prominent in the discussions. However, it is thought that the enhancement of the role of Basket II is mainly related to recognition by the CMEA countries that their economies cannot develop without Western technology. Such Western requests as improved personal contacts were left unmet. See Bettina Hass-Hürni, "Economic Issues at Belgrade", *Journal of World Trade Law*, July/August, 1978, pp. 289-301.
196 Reforms have, of course, been attempted (and are being attempted in all the CMEA countries). However, the aims of these reforms continue to be thwarted by what is believed to be a cardinal principle: the unchallenged hegemony of the Communist parties in all areas of social, political and economic life. Thad P. Alton, "Comparative Structure and Growth of Economic Activity in Eastern Europe", in *East European Economies Post-Helsinki – A Compendium of Papers, op. cit.*, p. 253.
197 Eugène Zaleski, "Planning and Financing of Research and Development in the USSR", in John R. Thomas and Ursula M. Kruse-Vaucienne, ed., *Soviet Science and Technology, Domestic and Foreign Perspectives, op. cit.*, pp. 276-304.
198 Central Intelligence Agency, *Organisation and Management in the Soviet Economy: The Ceaseless Search for Panaceas*, National Foreign Assessment Centre, ER 77-10769, December 1977, p. 20. The somewhat limited experiment with the new incentive system and the lukewarm reaction to it on the part of enterprise managers are discussed by Bruce Parrott, *art. cit.*
199 John P. Hardt, "Soviet Commercial Relations and Political Change", *art. cit.*, p. 52.
200 See Morris Bornstein, "Economic Reform in Eastern Europe", *art. cit.*; and R. V. Burks, "The Political Hazards of Economic Reforms", in *Reorientation and Commercial Relations of the Economies of Eastern Europe, op. cit.*

Scepticism is also expressed concerning the notion that effective use of Western technology requires a drastic restructuring of domestic economic institutions, though some adjustments may be required. On the contrary, some believe that "the very promise of additional resources and higher growth rates made possible through closer economic and technological relations with the West can be employed to placate the more sceptical elements in the regime".[201] A number of analysts postulate that technology transfer is viewed by the East as a substitute for reforms. The reasons for turning economically toward the West have been clearly indicated, it is believed, by Mr. Brezhnev. His statements are thought to mean that turning to the West is seen as an "alternative to the politically more dangerous economic reforms and as a promise or bait of a better standard of living for the elite. At the same time, it has been made clear by Brezhnev, implicitly and explicitly (e.g. promotions to the Politburo) that he intends to keep the political dangers in check".[202] Judging from the emphasis placed on Western technology it seems that Soviet leaders have decided that internal economic problems can best be dealt with by aid from the West.[203] Thus it is thought that Soviet leaders have deliberately chosen technology transfer from the West as the major strategy in solving their economic problems.

An analysis of the Polish economy indicated that "the transfer of technology from the West on a relatively large scale, in the form of the most advanced machines and licences, contributed to the achievement of the high growth rates registered in the first half of the 1970s. It was accepted, together with the import of foreign capital, as a substitute for systemic reforms".[204]

With respect to the role of the Party, it is thought that while "economic necessity may pressure the leadership for change, established institutional pressures will resist any aggregative improvement that may adversely influence either Party power or political performance".[205] Any improvement in economic efficiency which would endanger the central features of the Party control system would, it is thought, be avoided.[206] Nor is it likely "that technocrats will replace the *apparatchiki* because of the effect of necessities imposed by science and technology on the Soviet system... It is not so much the case of suborning the politicians by scientists but of continuation of the process of the adaptation of science to the Soviet system and of ongoing domestication of scientists and technologists by Party authorities".[207]

Attention is also called to the 24th Party Congress in 1971. During that Congress, the "Party Statutes were revised to give Party organisations in fundamental research institutions greater control over research administration".[208] This, as has been pointed out, has not been fundamentally changed. Finally, "the large regional complexes (previously mentioned) tend to upgrade the role of the central Party and governmental organs... Probably more direct involvement of important ministries and Party leaders is characteristic of these projects. In fact, this is borne out by discussions with corporate leaders of major American firms accredited to the Soviet Union".[209]

201 Robert W. Campbell and Paul Marer, editors, *East-West Technology Transfer, An Agenda of Research Needs, op. cit.*, p. 71.
202 *Ibid.*, pp. 73-74.
203 See, for instance, Boris Rabbot, *art. cit.*
204 Zbigniew M. Fallenbuchl, "The Polish Economy in the 1970s", *art. cit.*, p. 856.
205 J. Hardt, D. Gallik, and V. Treml, "Institutional Stagnation and Changing Economic Strategy in the Soviet Union", in *New Directions in the Soviet Economy*, US Congress, Joint Economic Committee, 89th Congress, 2nd Session, 1966, pp. 19-62, as given in John P. Hardt, "Soviet Commercial Relations and Political Change", *art. cit.*, p. 73.
206 John P. Hardt, "Soviet Commercial Relations and Political Change", *art. cit.*, p. 76.
207 Leopold Labedz, "Soviet Science", in B. Barber, *Sociology of Science*, Columbia University Press, 1962, as given in Leopold Labedz, "Science and the Soviet System", in John R. Thomas and Ursula M. Kruse-Vaucienne, ed., *op. cit.*, p. 151.
208 John R. Thomas and Ursula M. Kruse-Vaucienne, editors, *op. cit.*, p. xxiii.
209 John P. Hardt, "The Role of Western Technology in Soviet Economic Plans", *art. cit.*, pp. 323-324.

As for the possibility that adopting Western technology will lead to some convergence towards Western systems, it is said that the historical record speaks against its likelihood. First, it is noted that the Soviet government has never responded to "grass root pressure. Both before and after the revolution, control has tended to rest in the hands of an absolute leader. Pluralism does not exist and when change comes, it is almost always imposed from above and it tends to be discontinuous and not evolutionary".[210]

Second, Soviet and West European leaders have explicitly and most vehemently rejected convergence on ideological grounds.[211] On the whole, Western analysts seem to agree with the notion that "increasing commercial relations, emphasising technological exchange, might provide a vehicle for change in the international environment and might lead to increased communications between the Communist and non-Communist systems. Both economic systems might well change, but not necessarily toward convergence".[212]

There is, clearly, no way of reconciling the conflicting views which have been described above. By way of conclusion, however, a number of unknowns should be mentioned which affect both sides of the argument. One, already alluded to, is the effect of current political differences between the two systems. These differences, coupled with the fundamental ideological schism between the two, are likely to set the limits to East-West "interrelatedness". The nature and degree of East-West co-operation depends a good deal on future Soviet economic needs, and on whatever steps are taken to improve economic performance. Here Soviet military policy will play a decisive part. "Significant changes in priority can only be brought about by shifts away from new military resource claimants and to civilian ones".[213] Party opposition to such shifts can be expected to be great, even though Soviet policy has swung towards closer ties with the West since the 1970s. However, the "leadership has coupled its overtures to the West with a policy of stepped-up struggle against social and political ideas of Western origin, and writers dealing with the subject of propaganda continue to emphasise the importance of combatting imperialist "ideological sabotage" especially within the scientific-technical intelligentsia".[214] An important stratum of Soviet society, it must be remembered, influential in the policy-making process, remains deeply sceptical of the value of Western scientific and technical assistance. Given divisions within the elite over the advisability of Western technological ties, it is very difficult to predict how Soviet policy will evolve in the future. According to one author, future trends may "depend in considerable measure on the degree of compatibility which emerges between established Soviet economic institutions and the strategy of extensive foreign technological borrowing".[215]

Most Western analysts believe that a large scale infusion of technology into the Soviet system will not yield the expected benefits without some fundamental changes in that system. If this is so, two results are said to be possible. One, "that the policy of expanded Western borrowing will be defeated by a political alignment which emphasises the policy's comparatively small benefits and which is convinced that any far-reaching economic reform is still unjustified".[216] The second "is that the need to make good use of Western technology will become one argument supporting the case of a less traditional political alignment powerful enough to force through a major restructuring of the country's economic institutions".[217] A third possibility of course, is that the Eastern countries will attempt to maximize Western technology imports while not changing their economic systems.

210 Raymond Vernon and Marshall I. Goldman, art. cit., p. 10.
211 See Part I of this Chapter.
212 John P. Hardt, "Soviet Commercial Relations and Political Change", art. cit., p. 75.
213 Ibid., p. 52. Parentheticall, it is possible that a shift in favour of military resource claimants can occur.
214 See Bruce Parrott, "Technological Progress and Soviet Politics", art. cit., p. 321.
215 Ibid., p. 322.
216 Ibid., p. 322.
217 Ibid., pp. 322-323.

Chapter 5

THE INFLUENCE OF TECHNOLOGY TRANSFERS ON EASTERN ECONOMIES

1. CENTRAL PLANNING, INNOVATION AND TECHNICAL PROGRESS IN THE USSR

a) **Central Planning and Innovation**

Scientific and technical progress has been one of the fundamental goals of the Soviet Communist Party ever since it came to power. The accession of most Central European countries to CMEA after the Second World War has made this problem more acute and given it a new dimension.

The insistence on scientific and technical progress resulted in the introduction of a system of R & D planning and management. The various means used have been described at length in an OECD study on science policy in the USSR.[1] It should be stressed, however, that the administrative procedures and agencies for introducing scientific and technical progress in the USSR have recently been strengthened.[2]

In the USSR, more prominence has been given to the plans for developing science and technology within the framework of the national plans. The science and technology plans now appear in the first chapter of the national plan, before production goals. At the same time, 10 to 15 year forward plans for science and technology development have been formulated together with a plan for the next five years (1976-1980). In the latter, forecasts relating to a number of sectors were incorporated into "co-ordination plans" under the authority of a "principal" ministry, and provision was made for the necessary financial and material resources. These "co-ordination" plans have recently been replaced by "scientific and technical programmes" (about 200 in number) designed essentially to make R & D achievement-oriented and to co-ordinate R & D plans with plans for investment and supply of equipment and technology. More detailed supervision of priority projects also seems to be envisaged.

---

1   E. Zaleski, J.P. Kozlowski, H. Wienert, R.W. Davies, M.J. Berry, R. Amann, *Science Policy in the USSR, op. cit.*
2   Louven E. Nolting, *The 1968 Reform of Scientific Research, Development and Innovation in the USSR,* US Department of Commerce, September 1976. See also Eugène Zaleski, "Planning and Financing of Research and Development in the USSR", in *Soviet Science and Technology, Domestic and Foreign Perspectives,* ed. John R. Thomas and Ursula M. Kruse-Vaucienne, *op. cit,* pp. 276-304.

These planning changes were accompanied by management reforms. Steps were taken to set up complex research organisations, combining several research bodies attached to a ministry, in order to perfect the technology desired by a particular ministerial section. Another move was to organise science/production unions by merging the various research bodies and production units, in an attempt to overcome the traditional administrative barriers. Finally, the Soviet government has made most of the research organisations subordinate to the production units by establishing "production unions" and "industrial unions". This has served to strengthen the supervisory powers of the central agencies in regard to research and innovation.

Other measures taken in the field of R & D concern the definition of R & D efficiency (it is currently proposed to introduce a set of criteria that take account of technical, technical/economic and statistical indicators, etc.) and the strengthening of material incentives to innovation.

Is this continuous pressure from the authorities of the Eastern countries to promote scientific and technical progress enough to ensure effective innovation? A number of Western experts do not think so. In 1969, R.W. Davies, M. J. Berry and R. Amann stressed the obstacles to innovation in those countries: shortage of development facilities, reluctance of plants to innovate, administrative compartmentalisation and lack of sectoral co-ordination.[3] The risks which a Soviet firm takes in innovating and the weakness of material incentives were clearly stressed.

The same criticisms of innovation in the USSR are being made today. As Joseph Berliner points out, the decision to innovate increases the risk element in a firm's external dealings — with suppliers, customers, price inspectors, etc. The strategy of the Soviet entrepreneur is therefore to minimise the rate and scale of changes in the production process.[4] The recent economic reforms have removed some of the biggest obstacles to innovation; for instance, the criterion of gross output as a gauge of a firm's success.[5] However, other barriers integral to the administrative planning system remain, e.g. the calculation of investment efficiency, which is still entirely arbitrary and which testifies not to the efficiency of the innovation but to the tenaciousness of the pressure brought to bear on the authorities in favour of the efficiency criteria proposed. The administrative structure is still not conducive to innovation, nor is the system of prices and incentives. Joseph Berliner estimates that the amount of the premiums for innovation is such that it barely offsets the loss of the principal premiums.[6] More generally, he feels that what is lacking, and which is so important in any economy is independent innovative activity on personal initiative.[7] David Granick puts forward a similar, though more qualified, view.[8] Although he blames the Soviet system of incentives for these innovation deficiencies, he thinks that planned economies could have mandatory incentive systems better suited to innovation, as is already the case in the German Democratic Republic.[9]

Whilst recognising the part played by the environment in which the Soviet enterprise operates, other authors also stress the technological obstacles to innovation. In John P. Young's opinion, the Soviet industry's erratic performance is due to the wide differences in equipment and internal enterprise structures, especially as between basic and ancillary

3   R. Amann, M.J. Berry and R.W. Davies: "Science and Industry in the USSR", in *Science Policy in the USSR, op. cit.,* pp. 425-434.
4   Joseph S. Berliner, *The Innovation Decision in Soviet Industry, op. cit.* p. 524.
5   *Ibid.*
6   Joseph S. Berliner, "Prospects for Technological Progress" in *Soviet Economy in a New Perspective, op cit.,* p. 445.
7   *Ibid.*
8   David Granick, "Soviet Research and Development Implementation in Products: A Comparison with the German Democratic Republic", in *International Economics – Comparisons and Interdependences,* Festschrift für Franz Nemschak, edited by Friedrich Levcik, Vienna, New York, 1978, pp. 37-56.
9   *Ibid.*

activities. Under Stalin, the policy of acquiring the latest technology abroad and the heterogeneous origins of Soviet equipment (imports, leasing, dismantling of German factories, domestic production) made these differences even greater. More recently, local investment measures and selective buying of plant abroad have led to wide variations in production, cost, profitability and labour productivity levels. These "objective" reasons for the differences make it difficult to distinguish between negative performances due to poor management and that caused by technological factors. In any case, Soviet authorities are in no way inclined to shut down inefficient factories which are helping to meet unsatisfied demand, neither are they in a position to apply the standards achieved by the most efficient factories to all the others: each enterprise (and its performance) is thus a separate case.[10]

The Soviet authorities' attitude towards innovations does not make them any easier to introduce. The overriding tendency to set objectives requiring incremental improvement hampers the introduction of major innovations, and the refusal to lower the targets to allow for foreseeable innovation difficulties leads to increasingly strong resistance the greater the extent of the innovation. There comes a point when the innovation is so all-embracing that it means "rebuilding" the factory. The enterprise management then ceases to be responsible but the involvement of the many administrative departments and construction enterprises often causes considerable delay.[11]

Other factors have also hindered innovation in the USSR. For instance, the country's relative isolation from the outside world. This is an important factor because currently technical progress is increasingly dependent on advances made in other countries.[12]

Thus, in the Soviet Union there would seem to be two forces pulling against each other: the government which, as Joseph Berliner points out, is adopting in the Tenth Five Year Plan, a new growth strategy based on scientific and technological progress and technology imports, and the inertial force of the administrative planning system itself which discourages innovation.[13] Western technology imports are therefore part of the new government policy, the success of which is of crucial significance.

b) Problem of the General Scientific Level in the USSR

Very few studies are available on this subject and all the conclusions quoted here are tentative. Nevertheless, those concerning the comparison between the Soviet Union and the United States[14] are of interest.

According to Thane Gustafson, the Soviet Union spends as much on science as the United States but the results achieved are not as good. Whether in terms of Nobel prizes, frequency of citations by fellow specialists or merely the number of papers published, American scientists are ahead of their Soviet colleagues in most fields.

10   John P. Young, *Impact of Soviet Ministry Management Practices on the Assimilation of Imported Process Technology (with examples from the motor vehicle sector),* Paper presented at the Joint Annual Meeting of the South-Western and Rocky Mountain Associations of Slavic Studies, Houston, Texas, April, 13, 1978, pp. 4-13.

11   *Ibid.,* pp. 14-15.

12   Joseph S. Berliner, "Some International Aspects of Soviet Technological Progress", *The South Atlantic Quarterly,* Summer, 1973, p. 340. Berliner quotes Edward Denison, in *The Sources of Economic Growth in United States and the Alternative before Us.* Denison feels that more than half the know-how responsible for American growth comes from other countries.

13   In this connection, see also Philip Hanson, *USSR: Foreign Trade Implications of the 1976-1980 Plan, op cit.*

14   Thane Gustafson, *Why does the Soviet Union Lag Behind the United States in Basic Science?,* Kennedy School of Government, Center for Science and International Affairs, Harvard University, September 1978. The following pages [Section (b)] are entirely based on and adapted from the above-quoted author.

Gustafson warns his readers not to be too hasty in concluding that this is due to political difficulties; Lysenko and Sakharov being the examples that spring immediately to mind. In the USSR, science enjoys very high prestige and Soviet scientists may be able to count on steadier support for their work than their American counterparts. With the major exception of controls on foreign travel, the difficulties facing Soviet science are not therefore due to political or ideological interference.[15] It is a matter of management and organisation: the solutions found in the United States are more effective and more favourable for scientific progress. So long as the environment remains unchanged, the gap between the United States and the Soviet Union is unlikely to be bridged.[16]

Gustafson describes five different patterns in support of his conclusions:

i) Soviet science is strongest in fields which depend the least on material support (instrumentation, sophisticated materials and equipment). The most outstanding example is mathematics. Soviet scientists are very strong on the "blackboard side" and in other fields such as condensed-matter physics, theoretical astrophysics, theoretical seismology, mathematical psychology, elementary particle theory and plasma physics. On the experimental side, Soviet science is less strong. The advent of sophisticated and rapidly-evolving instrumentation has brought about a Soviet lag in fields such as organic chemistry, molecular biology and other biological sciences, and in several branches of oceanography and atmospheric science.

ii) Soviet science is often slow to accept conceptual changes, especially if they depend on observational data from other fields. For example, radio astronomy has only recently achieved a status equal to optical astronomy.

iii) In some fields, Soviet scientists do leading work by maintaining a steady effort in traditional specialities such as electrochemistry, biology, geology and oceanography.

iv) In areas where Soviet scientists have made crucial breakthroughs, they have not been able to maintain their lead. Examples are the work of Kapitsa and Landau in low temperature physics, and in the field of acoustic electronics.

v) Soviet science holds leading positions in priority fields of which nuclear physics and laser research are two outstanding examples. Another is biological research. Soviet scientists also lead the world in fields requiring large facilities such as plasma physics. However, high priority does not always ensure Soviet scientists a leading position because the West catches up when it builds its own facilities. Soviet physicists were the first to develop fusion processes (Tokamak magnetic-bottle devices) but since 1970 they have lost their lead. The largest accelerator was built at Serpukhov, but Fermilab, United States, has now moved ahead.

As Gustafson points out, these are generalisations and within each country, every discipline has varying success. Nevertheless, three major features of Soviet science need to be stressed:[17]

— the large, block-funded Soviet institute produces inflexibility and conservatism and is influenced by considerations that are frequently not concerned with science. In the United States, the decentralised, project-funded system is more flexible although it can also be stressful and unstable;

— the Soviet system of education separates undergraduate education from research and creates relations of dependence that lessen young researchers' initiative;

— both the Soviet and the American systems face a crisis in instrumentation due to revolutionary developments in the last decade.

15 *Ibid.* p. 1.
16 *Ibid.,* p. 3.
17 Thane Gustafson, *op. cit.,* pp. 6-7.

c) Attempts at an Overall Assessment of Soviet Technological Progress

Considerable difficulties arise in an overall assessment of technical progress. The results of any research on the growth and productivity of such factors as capital, work and "technological progress" are inevitably approximate (see Table A-40). In spite of the number of studies available and the huge amount of work they represent, the "technological progress" identified in terms of production functions is more in the nature of a residual (in the sense that technological progress does not emerge until the contributions of other growth factors such as capital and labour have been estimated) and in many instances corresponds more with the relative level of economic development of the various countries.[18]

In the United States and the United Kingdom, however, two interesting attempts have been made to assess Soviet technological progress in terms of a number of indicators of technological performance. The first, by Michael Boretsky, covers the period 1940-1962 and compares Soviet and American levels of performance.[19] The second, produced by the Birmingham Centre under R.W. Davies, compares Soviet performance with that of the United States, Japan, the United Kingdom and Germany.

Table 37

Indicators of Technological Change: Comparison of Growth Rates of Soviet and United States Industries from 1940 to 1962

Indicator Rate of growth of:	1940-55	1955-62	1940-62
Consumption of electric power per production worker	=	=	=
Maximum capacity of steam turbines for electricity production	+	—	=
Length of HVAC transmission lines (over 400 kV)	n.a.	+	+
Proportion of aluminium and magnesium in total basic metal consumption	—	—	—
Percentage of steel output by electric arc or O_2 process	—	—	—
Percentage of metalforming machine tools in total stock of metalworking machine tools	—	—	—
Output of NC machine tools	nil	—	—
Output of synthetic resins and plastics	—	=	—
Output of chemical fibres	+	+	+
Number of telephones in the economy	—	+	—
Total +	1	2	1
—	4	3	4
=	2	2	2

a) + more rapid Soviet progress ; — more rapid United States progress; = approximately equal progress.

Source: M. Boretsky, art. cit., pp. 133-256, as adapted by R. W. Davies, "The Technological Level of Soviet Industry: an Overview", in The Technological Level of Soviet Industry, op. cit., p. 49.

Table 37 shows the results of Michael Boretsky's research adjusted by the Birmingham Centre.[20] The general conclusion which can be drawn is that the rate of technological change was faster in the United States than in the USSR throughout the period 1940-1962.

18   Ronald Amann "Some Approaches to the Comparative Assessment of Soviet Technology: its Level and Rate of Development", in R. Amann, J.M. Cooper and R.W. Davies, ed, The Technological Level of Soviet Industry, Yale University Press, New Haven and London, 1977, pp. 12-15.
19   M. Boretsky, "Comparative Progress in Technology, Productivity and Economic Efficiency: USSR vs USA", in New Directions in Soviet Economy, op. cit., pp. 133-256.
20   Boretsky's method has been criticised in the United States, in particular by Joseph Berliner (ASTE Bulletin, Autumn, 1971, pp. 18-24). The Birmingham Centre has taken these criticisms into account when adjusting the results and has retained only those indicators which in its opinion are unambiguously acceptable.

Table 38

Indicators of Technological Change: Comparison of USSR Industry Growth Rate with those of the United States, Japan, the United Kingdom and Germany, 1955-1973

Rate of growth of:	1955-1960				1960-1965				1965-1970				1970-1973				1955-1973			
	USA	JPN	UK	Ger.	USA	JPN	UK	Ger.	USA	JPN	UK	Ger.	USA	JPN	UK	Ger.	USA	JPN	UK	Ger.
Electricity consumed per person employed in industry and construction	+	-	-	+	+	+	+	+	=	-	-	-	-[a]	-[a]	-	-	+[a]	-[a]	-	+[a]
AC transmission lines of 300 kV and above as % of total AC lines	+		+	+	+	+	+	=									-[b]	-[b]	-[b]	-[b]
Output of nuclear power stations as % of total electricity output									+	-	+	+	+	+[a]	+	+	-[ac]	-[ac]	+[c]	-[c]
O_2 steel as % of total steel output					+	-	-	-	+	-	+	+		+[a]	+	+	-[ac]	-[d]	+[d]	-[d]
Continuously cast steel as % of total steel output					-	-	-	-	+	+	+	+	+	-	-	-	[f]	-[e]	+[e]	-[e]
Metal-forming machine tools as % of total stock in machine-building and metal-working					+[g]		+[g]		+[h]		-[h]		+[h]		-[h]		+[j]	+[j]	+[j]	
Metalcutting machine tools 10 years old or less as % of total stock in machine building and metalworking																	+[jk]	+[jk]	+[jk]	+[jk]
NC machines as % of total m.c. machine tool output	+	-	=		-		+		+	+	+	=	+	+	+	=	+	+	+	+
Output of plastics and synthetic resins per capita	+	-	+	-	+	+	+	+	+	+	+	+	+	+	+		+	+	+	+
Output of chemical fibres per capita	=	+	+	+	=	+	+	+	=	+	+	+		+	=	=	+	+	+	+
Output of synthetic fibres per capita		-	-	-	+	+	+	+		-	+		=	+	=	=	+	+	=	+
Output of synthetic rubber per capita	+	-	-	-	+	+	+	-	+	-	-	-	+[a]	-[a]	-[b]	=	+[b]	-[b]	-[b]	-[b]
Number of telephones per capita	=	-	=	-	=	-	+	=	+	-	-	+	=[a]	-[a]	=[a]	=[a]	+[a]	-[a]	+[a]	+[a]
Number of groups in which USSR more rapid	3	0	1	2	4	2	4	1	3	1	3	2	2	3	3	2	4	1	3	2
Number of groups in which other country more rapid	0	3	1	2	1	4	1	2	1	4	2	3	2	2	2	2	2	5	1	4
Number of groups in which rate approximately equal	1	0	2	0	1	0	1	2	2	1	1	1	2	1	1	2	0	0	2	0
Total number of groups	4	3	4	4	6	6	6	5	6	6	6	6	5	6	6	6	6	6	6	6

+ USSR more rapid: − other country more rapid: = approximately equal development ± 1 percentage point.

General Notes: Two indicators in Table 37 (Maximum capacity of steam turbines for electricity production; proportion of aluminium and magnesium in total consumption of basic metals) have been omitted owing to the lack of adequate information. The following indicators have been added: output of nuclear power stations as a proportion of total electricity output; metal-cutting machine tools 10 years old or less as proportion of total stock of machine tools in machine building and metalworking; numerically controlled machine tools as a proportion of total metal-cutting machine tool output; output of synthetic rubber. Figures for increase in output have been calculated in terms of increases in total output included in Boretsky and in Table 37.

Notes:
a) To 1972.
b) 1960-70.
c) From 1965: these years have been selected because in 1960 the proportion in the USSR was very low and no nuclear power was produced in Japan or Germany.
d) From 1960.
e) From 1965.
f) 1965-70 (other years not available).
g) 1962-66 in USSR: 1963-68 in USA: 1961-66 in UK.
h) 1966-72 in USSR: 1968-73 in USA: 1966-71 in UK.
j) 1962-72 in USSR: 1963-73 in USA: 1961-71 in UK.
k) Soviet figures are for total stock.
m) To 1970.

American progress was more rapid during 1940-1955 because the United States had not suffered any war time destruction of industrial capacity while the USSR had been devastated. Although Soviet reconstruction of civilian industry was extremely rapid in 1946-1950 and in 1951-1955, it was to a considerable extent based on pre-war technology.[21] The Soviet performance improved slightly from 1955 to 1962.

The Birmingham Centre study, the results of which are set out in Table 38, extends Boretsky's research to 1973 (the 1955-1962 period was studied by both Boretsky and the Birmingham Centre) and covers three other countries: Japan, the United Kingdom and Germany.

On the whole, the rate of technological progress was slightly faster in the USSR than in the United States and the United Kingdom, but slower than in Germany and much slower than in Japan. The slightly slower Soviet rate in the 1965-1979 period improved from 1970 to 1973.

An interesting point is that this improvement coincided with a reduction in Soviet industrial production growth rates. Although these remained higher than those of the leading industrial countries, apart from Japan, the gap narrowed. As pointed out by R.W. Davies, however, this "convergence" of growth rates was accompanied by appreciable differences in the diffusion of new products and manufacturing processes.[22] This is already apparent from the statistics for Soviet industrial production: during the past 20 years, growth in the traditional industries in the USSR has continued to be steep whereas elsewhere it has slowed considerably. The same applies to per capita production of crude steel, cotton and other industrial products.[23]

d) **Soviet Technological Progress in Some Advanced Technology Industries**

Western research in this field is still sketchy, the only information available being the pathfinding study by the Birmingham Centre. Table 39 shows the Centre's results for a number of products.

The study is based on the Nabseth and Ray method applied to Western countries[24] and compares the rates of diffusion of new technology in various industries.

Table 39 shows that traditional methods continued to be used in Soviet steelworks, even after the introduction of oxygen smelting and continuous casting, whilst in the Western industrialised countries, the new processes tended to replace the old. It took sixteen years for the oxygen process to account for 20 per cent of steel production in the USSR compared to five to twelve years for the countries listed in Table 39 and two years for other Western countries.[25] Each stage in the introduction of the oxygen smelting process was slower in the USSR than in Western countries. The initial two-year lag behind the United States increased to six years in the subsequent stages and although the USSR was initially one year ahead of Japan in oxygen steelmaking it subsequently fell ten years behind.[26]

21 G. Warren Nutter, "The Structure and Growth of Soviet Industry: A Comparison with the United States", in *Comparisons of the United States and Soviet Economies,* Joint Economic Committee, 86th Congress, 1st Session, Part I, US Government Printing Office, Washington, D.C., 1959, p. 105.
22 R.W. Davies, "The Technological Level of Soviet Industry: an Overview", *art. cit.,* p. 52.
23 *Ibid.*
24 L. Nabseth and G.F. Ray (ed.), *The Diffusion of New Industrial Processes,* Cambridge, 1974.
25 R.W. Davies, *art. cit.,* pp. 52-53 and 60.
26 See Table 39.

Table 39

Rate of Diffusion of New Technology in a Few USSR Industries Compared to that of the United States, Japan, the United Kingdom and Germany

A. Oxygen steelmaking

	First industrial installation year	Year output of O_2 steel as proportion of total output reached:		Number of years between:		
		5%	20%	First indl. instn. and 5%	5% and 20%	First indl. instn. and 20%
USSR	1956	1966	1972	10	6	16
USA	1954 —2	1962 —4	1966 — 6	8	4	12
Japan	1957 +1	1960 —6	1962 —10	3	2	5
UK	1960 +4	1963 —3	1965 — 7	3	2	5
Germany	1955 —1	1962 —4	1966 — 6	7	4	11

B. Continuous casting of steel

	First industrial installation year	Year output of continuously cast steel as proportion of total output reached:		Number of years between:		
		1.5%	5%	First indl. instn. and 1.5%	1.5% and 5%	First indl. instn. and 5%
USSR	1955	1966	1972	11	6	17
USA	1962 + 7	1967 + 1	1969 —3	5	2	7
Japan	1960 + 5	(1960) =	1970 —2	6	4	10
UK	1958 + 3	1966 =	1974 + 2	8	8+	16
Germany	1954 —1	1965 —1	1968 —4	11	3	14

C. Synthetic fibres

	Year first produced commercially (nylon)	Year output of synthetic fibres as a proportion of total chemical fibres output reached:			Number of years between:		
		10%	20%	33%	First prodn. and 10%	10% and 33%	First prodn. and 33%
USSR	1948	1962	1966	1973	14	11	25
USA	1938 —10	1951 —11	1954 —12	1959 —14	13	8	21
Japan	1942 — 6	1958 — 4	1960 — 6	1963 —10	16	5	21
UK	1941 — 7	1957 — 5	1960 — 6	1964 — 9	16	7	23
Germany	1941 — 7	1958 — 4	1961 — 5	1964 — 9	17	6	23

D. Polyolefins

	Year first produced commercially	Year output of polyolefins as a proportion of total plastics output reached: 15%	Number of years between first production and 15%
USSR	(1953)[a]	1970	17
USA	1941 —12	(1956) —14	15
Japan	1954 + 1	1963 — 7	9
UK	1937 —16	(1955) —15	18
Germany	1944 — 9	1965 — 5	21

a) Estimate — polyethylene first produced in 'early 1950s'.

(Cont'd on next page)

Table 39 (cont'd)

E. HVAC transmission lines

	Year first 300 kV line	Year first 500 kV line	Year first 750 kV line	Year lines over 300 kV as a proportion of total lines (over 100 kV) reached:	
				5%	10%
USSR	1956	1959	1967	1960	1970
USA	1954 —2	1965 + 6	1969 + 2	1966 + 6	1970 =
UK	1963 + 7	— +	— +	1966 + 6	1970 =
Germany	(1955) —1	— +	— +	1971 + 11	— +

F. Nuclear power

	Year first commercial power station	Year output of nuclear power as a proportion of total electricity output reached:			Number of years between:		
		0.5%	1.0%	2.0%	First stn. and 0.5%	0.5 and 2.0%	First stn. and 2.0%
USSR	1954	1971	1973	(1975)	17	4	21
USA	1957 + 3	1967 — 4	1970 — 3	1971 — 4	10	4	14
Japan		1970 — 1	1970 — 3	1971 — 4		1	
UK	1956 + 2	1959 —12	1959 —14	1962 —13	3	3	6
Germany	1961 + 7	1967 — 4	1969 — 4	1970 — 5	6	3	9

G. NC machine tools

	Year first prototype	Year output of NC mc. ts. reached:		Number of years between:		
		50 units a year	1% of total mc. t. output	Prototype and 50 units	50 units and 1%	Prototype and 1%
USSR	1958	1965	1971	7	6	13
USA	1952 —6	1967 —8	1965 —6	5	8	13
Japan	1958 =	1966 + 1	1973 + 2	8	7	15
UK	1956 —2	1963 —2	1968 —3	7	5	12
Germany	1958 =	1964 —1	— + +	6	9 +	15 +

H. Production of artificial and synthetic fibres in 'comparable' periods: USA and USSR (in thousand tonnes)

	USSR			USA		
	Artificial fibres	Synthetic fibres		Artificial fibres	Synthetic fibres
1965	330	77	1951	587	77
1966	362	96	1952	515	96
1967	395	116	1953	543	112
1968	424	130	1954	492	129
1969	441	142	1955	572	172
1970	456	167	1956	521	182
1971	473	203	1957	517	234
1972	507	239			
1973	543	287	1960	466	307

Source: R. W. Davies, art. cit., pp. 55-57.

Continuous casting accounted for 5 per cent of total output in the USSR two to four years later than in the United States, Japan and Germany, although the USSR had introduced the process seven years before the United States and five before Japan. A slight advance was maintained over the United Kingdom (see Table 39).

A similar pattern may be observed in the relationships between the production of artificial fibres and the new synthetic fibres. Although the total production of chemical fibres expanded more rapidly in the USSR than in the main Western countries (see Table 38), the introduction of synthetic fibres was distinctly slower. Soviet industry was six to ten years behind in 1948 and even further behind in 1973 (see Table 39). Table 39 also shows that growth of synthetic fibre production in the USSR was accompanied by an increase in "traditional" artificial fibre output, while the latter tended to fall off in the United States during the "comparable" years.

In electric power transmission, the USSR is also lagging behind in the use of the most modern processes. During the 1960s, it was well ahead in the development of HVAC (high voltage alternating current) lines (as measured in terms of voltage), moving even further ahead during the first stage of diffusion. By the second stage, however, the United States had caught up with the USSR.[27]

Soviet slowness in the introduction of advanced technology is particularly apparent in nuclear energy (see Table 39). Although the USSR was the first country to produce nuclear energy commercially, it took twenty-one years to raise the share of nuclear power to two per cent of total electricity output compared with six years for the United Kingdom, nine for Germany and fourteen for the United States. By 1975, the USSR was much behind leading Western countries.

Soviet performance in NC (numerical control) machine tools production seems to be better. The USSR was six years behind the United States when it produced its first prototype in 1958, and by the time NC machine tools had reached one per cent of the total machine tool output in 1971, the lag was still only six years; rapidly increasing production has now put the USSR ahead of Japan and Germany (see Table 39).

It should be noted that there are still no data for several research-intensive and high technology products. Nor are there any production data available for control instruments and for the armaments industry comparable to those available for the Western industrialised countries.[28]

The slow Soviet diffusion of new technology is not only the result of inferior technological capability as compared to the leading Western industrialised countries. Largely it is due to the fact that the USSR maintains existing facilities in use far longer than the West. Equipment in steelworks and textile plants has been kept in service in the USSR when it would have been withdrawn elsewhere. Even intercontinental ballistic missiles and other weapons of earlier generations, withdrawn from the United States' arsenal, are still retained in service in the USSR.[29]

This retention of out-of-date equipment may be economical if there is an excess of manpower; although it leads to obsolescence.[30] In the case of the Soviet Union, the reason for keeping such equipment is probably due to routine management. As R.W. Davies points out, in the process control industry, for example, "it would be difficult to find an adequate

27 See Table 39 and *Technological Level of Soviet Industry, op cit.,* pp. 60 and 210.
28 R.W. Davies, *art. cit.,* p. 60.
29 *Ibid.,* p. 58.
30 In the USSR, total fixed capital rose by 58.9 per cent over the period 1956-1960, 50.2 per cent between 1961 and 1965 and 42.7 per cent between 1966 and 1970. But its useful life decreased only from 31 years for the period 1956-1960, to 25 years for 1961-1965 and to 22 years for 1960-1970. The rate at which industrial plant was taken out of service declined only from 3 per cent in 1967 to 1.8 per cent in 1970. Gerhard Fink and Jiri Slama, "Le problème du renouvellement du capital fixe dans quelques pays socialistes", *Revue d'études comparatives Est-Ouest,* No. 4, Editions du CNRS, Paris, December 1976, pp. 119 and 121.

economic justification for the continued production of five major and six minor systems, some of which are antiquated even by the standards of the most advanced Soviet production, which has itself been basically unchanged for ten years".[31]

2. IMPORTANCE OF THE CONTRIBUTION OF WESTERN TECHNOLOGIES

a) The Central Planning System and its Capacity for Assimilating Foreign Technology

By comparison with the market economies, the Soviet-type system has advantages and disadvantages with regard to the assimilation of technology[32]: The advantages include:

— small incidence of the time factor;
— a high ratio of investment to national income;
— rapid growth of capital (diminished somewhat over the past years);
— reduced duplication;
— absence of internal commercial secrecy;
— utilisation of several forms of transfer to balance external payments.

The disadvantages include:

— absence of foreign direct investment;
— restriction of personal foreign contacts;
— initiative confined to a limited number of persons ("oligarchic astigmatism");
— small role played by foreign trade;
— difficulty in obtaining hard currency;
— difficulties regarding internal stimuli, prices and dissemination of information;
— inhibition of transactions because of the intermediary role played by the foreign trade enterprises.

Some of these disadvantages can be avoided by importing turnkey plants or industrial complexes such as VAZ (Volga car works) or Kama (Kama river truck plant). These complexes have highly integrated and mechanised production lines designed for optimum performance in specific operations for a specific product and operating in tandem with companion equipment. Since these complexes imported from the West are indivisible units, it would be very complicated to copy equipment and adapt it to suit other plant. In many cases the whole works would have to be rebuilt — a decision calling for approval by higher authorities. For John P. Young, therefore, it is more difficult in the Soviet Union to disseminate imported Western technology requiring considerable expert help, than to diffuse domestic technology, and the bigger the transfer, the greater this difficulty becomes.[33]

These features of the Soviet system apply by and large to the other socialist countries of Eastern Europe. However, it is difficult to assess the real importance of Western technology. To quote Philip Hanson: "the only honest conclusion that can be drawn about the contribution made in the past by Western technology transfers to the USSR through negotiable channels is that the importance of that contribution is unknown, probably not dominant but possibly appreciable".[34]

31 R.W. Davies, *art. cit.*, p. 58.
32 Philip Hanson, *The Soviet System as a Recipient of Foreign Technology*, Birmingham, 1977, p. 30.
33 John P. Young, *Impact of Soviet Ministry Management Practices on the Assimilation of Imported Process Technology, op. cit.*, pp. 17-19.
34 Philip Hanson, "External Influences on the Soviet Economy Since the Mid-1950's", *The Impact of Western Technology, op. cit.*, p. 12.

This would therefore seem to be a field which has been little explored and merits particular attention. The approach to it should be broadened by considering such theoretical problems as the effectiveness of foreign trade in technology, and business and economic policy concerns. This research should be supplemented by studies of particular sectors. This report will confine itself to reviewing the different aspects of assimilation of Western technologies, considering in turn their entry into service, direct and indirect effects and diffusion.

b) Attempts at an Overall Assessment of the Contribution of Western Technology

Any attempt to assess the contribution of Western technology to Eastern economies comes up against several difficulties.

It is practically impossible to ascertain the effect of technologies which are not incorporated in the products. Even the effects of technologies acquired through machinery and equipment purchases are very difficult to identify since it is practically impossible to distinguish between the effects of national and of imported technology. However, results in practical terms depend not only on whether technologies are old or new but also on the total economic situation as a whole.

As a first approximation, the relative importance of Western technology can be measured as the ratio between total machinery imports and investment in plant. All that can be done here is to present some data for the USSR based on the Birmingham Centre studies. A special study would be required for other CMEA countries to be included in this research.

In 1970, machinery imports from the West represented approximately 4 per cent of all Soviet investment in capital equipment. Imports of machinery into the USSR including imports from other CMEA countries amounted to 12 per cent of all Soviet investment in capital equipment.[35] This is approximately the same figure for machinery imports into the United States. However, machinery imports play a more important role for the USSR [36].

As suggested by R. W. Davies, USSR industries may be divided into three groups according to their dependence on imported technology.

The first group includes strong indigenous technology in which imports of machinery are of minor importance. It covers rockets, weapons, the nuclear industry and certain civilian industries such as electric power. The amount of imported machinery in these industries is very low. For example, in 1970, power stations received only 0.1 per cent of their machinery from Western countries. In 1971, such imports accounted for 6.1% of USSR investment.[37]

The second group is an intermediate category and includes well established indigenous technology. The latter has been responsible for pioneering important innovations. Imported technology however has played a major role in certain fields. In this group R. W. Davies includes the steel and machine tool industries. [38]

In *the third group,* Soviet dependence on machinery is very high. This category chiefly consists of the computer and chemical industries. [39]

35   R. W. Davies, *art. cit.,* p. 63.
36   *Ibid.*
37   *Ibid.,* p. 64. R. W. Davies, however, acknowledges that in this group some of the Western machinery imported was of crucial importance. In 1957, for instance, the purchase of air-blast circuit breakers from the French firm Delle Alsthom was a decisive factor in the successful introduction of 500 kV HVAC transmission lines.
38   *Ibid.,* p. 64.
39   *Ibid.,* p. 65.

c) Importance by Sector of Activity

It would be useful to extend the classification to other branches of industry and economic sectors and to have a finer breakdown. This has been attempted by Philip Hanson for the period 1955-1971 and for 1975.

Table 40

Soviet Hard Currency Imports of Machinery and Equipement[a] in 1975

STN number[b]		1 From hard currency sources ($ mn)[c]	2 As percentage of same category of imports from all sources	3 As percentage of all hard currency machinery imports	4 Branch investment share (%) 1974[d]
1	All machinery and equipment	4 577	36.5	100.0	—
100-105	For machine building and metalworking industries	868	66.6	19.0	9.5
10514	(of which, identified as for motor industry)	345	86.5	7.5	..
110	For the electric power industry	5	4.7	0.1	4.2
111	For electrical engineering	48	16.5	1.0	..
120-121	For mining and ore treatment	88	65.3	1.9	2.1[e]
123	For metallurgical industries	219	42.5	4.8	3.5[e]
127	For oil refining	60	43.6	1.3	..
128	For oil and gas drilling, prospecting, etc.	148	69.6	2.8	6.0
140	For food-processing	103	32.2	2.2	3.5
144-146	For the textile, clothing, footwear, etc. industries	210	44.0	4.6	1.8
150	For the chemical industry	502	56.8	11.0	4.1
151-152	For the timber, woodworking, pulp and paper industries	98	44.1	2.1	2.0
162-181	For agriculture	27	5.8	0.6	25.6
192	For shipping	212	14.1	4.6	..

a) Identified imports from all hard currency trade partners of goods in the Soviet trade classification "machinery, equipment and transport equipment". Non hard currency sources for such imports are the rest of CMEA, Yugoslavia and Finland. The total figure in column 1 is based on Soviet data and is not precisely comparable with the series in Table 6 (of the source quoted below), which is based on OECD data.
b) Numbers used in the Soviet trade nomenclature, 1971 edition.
c) Converted to US dollars from roubles at 0.7219r = $ 1, the average of monthly Gosbank exchange quotations for 1975.
d) Percentage shares of all "productive" investment. This excludes investment in housing and construction work for educational, scientific and cultural institutions. The figures relate to structure, etc., as well as to machinery.
e) Only partial data available: for coal mining only, in the case of mining; for iron and steel only in the case of metallurgy.
Columns 1-3 derived from *Vneshnyaya Torgovlya SSSR v 1975g;* Column 4 from *Narkhoz 1974.*

Source: Philip Hanson, *USSR: Foreign Trade Implications of the 1976-80 Plan, op. cit.*, p. 64.

In order to estimate the differences between sectors of industrial activity, Philip Hanson tries to compare, by branch, the percentage of machinery imported from the West with the percentage of domestic investment for the years 1955-56, 1960-61, 1965-66 and 1970-71.[40] Where the import share is higher than the domestic investment share, Hanson concludes that the industry concerned is dependent on Western technology. Accordingly, he finds that in the Soviet Union the "most dependent" industries in those years were chemicals, computers, shipbuilding, motor vehicles, wood, paper and cellulose and light industry. The chemicals industry seems to have been more dependent than the others. The agricultural and food industries showed less dependence probably because they are less research dependent and also because imports from East Europe and the Soviet national effort make a contribution.

40 Philip Hanson, "External Influences on the Soviet Economy Since the Mid-1950s; The Import of Western Technology", *art. cit.,* pp. 20-26.

Hanson concludes his findings by observing that heavy (and increasing) use of Western machinery may result in a rapid improvement of an industry's performance, but that the general level of machine imports in the USSR has not been sufficiently high (except in a few cases) to make a noteworthy contribution vis-à-vis other factors influencing labour productivity.[41] Hanson's study for the most recent year — 1975 — confirms these results (see Table 40).

The comparative shares of imports from the West in total USSR imports for various categories of machinery show that several categories have a higher than average (36.5%) share — see Table 40, column 2: equipment for the motor industry, 86.5 per cent, for drilling and prospecting, 69.6 per cent, for the metallurgy industry, 66.6 per cent, and for ore treatment, 65.3 per cent, etc.

In another approach, Hanson compares the various branches' shares in total machinery imports from the West (column 3) with their shares in total USSR productive investment (column 4). The figures show the major role of Western imports in machine building and metal working (including motor/manufacturing), the chemical and textile industries and mining and ore treatment.

At the same time, Hanson attempts to identify the branches where machinery imports from the West are likely to rise during the Tenth Five Year Plan (1976-1980). The main ones are the chemical, oil and gas, and coal industries, gold extraction and metal working. In Hanson's opinion, however, consumer goods industries are unlikely to receive large quantities of imported machinery from the West during that period.[42]

The usefulness of the approach by sector of activity is undeniable, but few in-depth studies of this kind have been published to date. A case in point is the excellent study by Richard W. Judy on computers and transfers of computer technology from West to East.[43] Even in this very thorough investigation, the author does not attempt to calculate the direct or indirect effects of the transfers made either by way of information flows or by way of imports. However, he concludes that all the significant inventions in the computer field do derive from the West.[44]

Richard Judy's research has been expanded by Martin Cave. In a 1977 study,[45] Cave concluded that "it is certain at the present moment, as it was in 1968, that all the significant inventions have been produced in the West, although this does not apply to programming, where the Soviets have taken a theoretical lead, even if that lead does not have a great practical significance".[46]

d) **Calculating the Effects and Significance of Technology Imports from the West**

The internal effects produced by technology transfers form part of a wider subject, namely that of calculating the effectiveness of foreign trade. The difficulties involved as a result of the domestic pricing system, multiple exchange rates and the subsidies given to the different transactions with foreign enterprises are such that no correct method of calculation has yet been found. Hungarian authors of a paper on this subject have come to the conclusion that the main requisite for increased effectiveness of foreign trade is reform of the domestic

41 Philip Hanson, "International Technology Transfer from the West to the USSR", *art. cit.,* p. 801.
42 Philip Hanson, *USSR: Foreign Trade Implications of the 1976-1980 Plan, op. cit.,* pp. 65-69.
43 Richard W. Judy, "The Case of Computer Technology" in *East-West Trade and the Technology Gap,* edited by Stanislaw Wasowski, New York, Washington, London, 1970, pp. 43-72.
44 *Ibid.,* p. 64.
45 Martin Cave, "Soviet Computer Technology", in R. Amann, J. M. Cooper and R. W. Davies, *The Technological Level of Soviet industry, op. cit.*
46 *Ibid.*

pricing system.[47] In present circumstances, all calculations of the effectiveness of Western technology imports can be no more than approximate.

The most comprehensive study of the subject (not including, however, the past decade) which also discusses Soviet experience in great detail, has been made by Antony C. Sutton.[48] The influence of Western technolgy is studied systematically by industrial sectors from 1917 to 1965. This study is as detailed as it possibly could be, given the length of the period covered. Sutton concludes that there has been virtually no internal mechanism for stimulating innovation and that almost all the technological progress in the civilian sector and much of the progress in the military sector has been "borrowed" from the West.

Antony Sutton's investigations, although based on an impressive body of documentation, have the disadvantage of not providing any measure of transfer effects. Recently, a study has been carried out by Herbert S. Levine and Donald W. Green[49] on the Soviet Union and Stanislaw Gomulka has published a paper on Poland.[50]

Green and Levine assess the total contribution (direct and indirect) made by imports of Western plant to Soviet production in recent years. They do so on the basis of the Cobb-Douglas function including labour, the capital stock produced in the Soviet Union and imported from CMEA countries, and the capital stock imported from the West (computed by converting foreign trade prices into external prices). Thus they obtain different coefficients of output elasticity (marginal productivity) for the "domestic" capital stock and the "Western" capital stock. They initially came to the conclusion that the marginal productivity of the capital goods imported from the West is fifteen times higher than that of the home-produced capital stock.[51] They also estimated, that without imports of Western plant and machinery resulting from détente, Soviet industrial growth would have been 19 per cent less than it was during the period from 1968 to 1973.[52] This estimate was later revised to about 5 per cent.[53]

Green and Levine's calculations are based on a number of estimates which the authors themselves admit might incorporate a certain margin of error: Western capital is calculated with the aid of import statistics converted into 1955 roubles and the depreciation rates are somewhat vague. But more serious objections have been raised by Philip Hanson.[54]

Hanson considers first of all that Green and Levine over estimate the proportion of Western plant in Soviet capital stock. Instead of 5 per cent, he suggests only 2 per cent. He also questions the general applicability of their conclusions. Imports of Western plant necessarily occur in the sectors where productivity is highest. Any increase in the proportion

47 On Soviet calculations of foreign trade effectiveness, see Lawrence J. Brainard, "Soviet Foreign Trade Planning", in *Soviet Economy in a New Perspective, op. cit.,* pp. 695-708. See also S. N. Zakharov, *Raschety effektivnosti vneshne ekonomicheskikh svjazej* (Calculation of the Effectiveness of External Economic Relations), Moscow, 1975. An overall approach to the problem of calculating the effectiveness of the Socialist countries' foreign trade is presented in the article by Mme Agota Dezsenyj-Gueullette "Les calculs d'efficacité du commerce extérieur en Hongrie avant la réforme économique de 1968", *Revue d'études comparatives Est-Ouest,* No. 2, Editions du CNRS, Paris, 1977, pp. 21-95.

48 Antony C. Sutton, *Western Technology and Soviet Economic Development,* 3 vols. (from 1917 to 1965), Stanford, 1968, 1971 and 1973.

49 H. S. Levine and D. W. Green, "Implications of Technology Transfers for the USSR", in *East-West Technological Co-operation, NATO Colloquium 1976, op. cit.,* pp. 43-78.

50 Stanislaw Gomulka, "Investment Imports, Technical Change and Economic Growth, Poland, 1971-1980", paper presented at annual conference of the National Association for Soviet and East European Studies, Cambridge, England, April 1977.

51 These figures were subsequently revised to give a factor of only 8 to 10. Donald W. Green and Herbert S. Levine, "Macroeconomic Evidence of the Value of Machinery Imports to the Soviet Union", in *Soviet Science and Technology. Domestic and Foreign Perspectives,* edited by John R. Thomas and Ursula M. Kruse-Vaucienne, *op. cit.,* pp. 394-423.

52 *Ibid.*

53 *Ibid.,* p. 394.

54 Philip Hanson, "International Technology Transfer from the West to the USSR", *art. cit.,* pp. 799-800.

of this plant would cause the marginal productivity of imported Western plant to fall. Hanson also points out that the results obtained by Green and Levine represent the combined contribution of domestic and imported technology which distorts estimates of the indirect gains. Allowance should also be made for the domestic resources which might be released in the absence of imports (costly in terms of products exported). Finally, Hanson asks if the Soviet system of management, cumbersome and bureaucratic as it is, can really produce such different performances in the assimilation of domestic technology on the one hand and Western technology on the other. In any case he remains sceptical on this point.

Stanislaw Gomulka, for his part, tries to measure the aggregate effect of rising imports of machinery (from both East and West) on Polish industrial growth during the years from 1970 to 1975. He concludes that these imports will make it possible to increase the annual rate of industrial growth from 5.7 per cent to 7.7-8.0 per cent approximately (that is, from 35-40 per cent for 5 years).[55]

These calculations are of course subject to wide margins of error: not all machinery imports are included, no account is taken of licence imports and the conversion of dollar import prices into domestic prices is not wholly reliable. Finally, the aggregate estimate conceals appreciable differences as between importing sectors.

An attempt to measure the effects of Western technology transfers on the Soviet mineral fertilizer industry was recently made by Philip Hanson.[56] In his paper Hanson provides an excellent review of the industry and the policy in regard to imports of Western turnkey plants. These imports have been particularly high for the production of fertilizers (ammonia, nitric acid, sulphuric acid, phosphoric acid) and complex fertilizers[57] and Hanson gives some very interesting details. He then examines the influence of these imports on agricultural production and calculates that the net incremental farm output of some 4 billion roubles in the period 1970-1975 was obtained by means of 2 billion roubles' worth of imported Western plant installed in the USSR between 1960 and 1975.

These figures are given as approximations and therefore it would not be right to criticise them as being inaccurate. Hanson himself adds that the Soviet mineral fertilizer industry has certainly received more than an above average amount of plant and know-how from the West[58], and that this particular case cannot be taken to apply generally. It seems however, that many of the criticisms he makes of Green and Levine's calculations could be made of his own: difficulty of converting import prices into domestic prices, the sector's sensitivity to imports, the contribution of domestic R & D and the dilatoriness of the Soviet system. Moreover, the "chance elements" in Soviet collectivised farming are particularly difficult to gauge.

55   Stanislaw Gomulka, art. cit., p. 11.
56   Philip Hanson, "The Impact of Western Technology: A Case Study of the Soviet Mineral Fertilizer Industry", paper presented at the Conference on Integration in Eastern Europe and East-West Trade, Bloomington, Indiana, 28-31 October, 1976.
57   Mineral fertilizers comprise: 1) nitrogen, 2) phosphate (P_2O_5) and 3) potash (K_2O) fertilizers. It is advantageous to supply them to agriculture in proportions having a high nutritive content or combining these principal elements in a single fertilizer. The latter, more technically advanced type is called "complex". Philip Hanson, "The Impact of Western Technology: A Case Study of the Soviet Mineral Fertilizer Industry", art. cit., p. 6. In 1975 the proportion of complex fertilizers produced by plants imported from the West was 40 per cent, whereas it was very small for domestic plant, Ibid., pp. 22 and 36.
58   Ibid., p. 27.

e) Diffusion of Technologies Imported from the West [59]

There is reason to wonder whether the problem of "diffusion" of Western technology in the East European countries has been correctly stated. The term "diffusion" applies to the know-how built into a design, a licence or a product. Once this know-how has been imported it becomes part of the nation's assets and its diffusion obeys the same rules as the diffusion of domestic discoveries or technologies applied in advanced production plants.

Western studies on the subject of "diffusion" stress this issue. Philip Hanson [60], moreover, mentions several shortcomings in the diffusion of both domestic and foreign technology in the USSR: institutional barriers, information difficulties, paucity of equipment for processing information [61], lack of competition and motivation. He points out, too, that pressure from the authorities and even from the Politburo is very influential in the diffusion of imported technology. He also considers that the diffusion problem will become more and more difficult as technology imports increase. The resources available to support "pressure from above" will become increasingly scarce. [62]

The diffusion of imported technologies is also becoming more difficult because of the increasing complexity of new products. The technical progress built into a new truck or lathe is fairly easy to detect from the product itself. But the transition from mechanical engineering to electronics and petrochemicals poses more complicated problems. Yet even if the Eastern countries acquire a licence, it is not certain that they will know the most advanced techniques, Western firms being reluctant to sell the very latest technologies. [63]

A certain reluctance to introduce new Western technologies has also been reported in the USSR. [64] In fact, the number of specific studies on diffusion of Western technology in the East European countries is very small indeed. It is proposed here to compare the three best-known of them, dealing with computers, chemical fertilizers and motor vehicles. [65]

In the computer industry transfers of Western technology have largely come under a strategic embargo, although this was recently relaxed. The conclusions of a recent study cited earlier are very clear, viz., the technology which has been acquired, often in spite of the embargo, has not been able to be widely circulated in the USSR. [66]

The study of the mineral fertilizer industry by Philip Hanson reaches some fairly guarded conclusions. Soviet officials do not seem to stress the contribution made by domestic technology to the production of ammonia, urea and complex fertilizers (around 1970). But

---

59   As it is difficult, in practice, to distinguish between national and foreign technology diffusion, many of the conclusions in Chapter 5, Section 1 (c) (especially those based on Table 39) also apply to the diffusion of technologies imported from the West.

60   Philip Hanson, "The Diffusion of Imported Technology in the USSR", in *East-West Technological Co-operation, op. cit.,* pp. 143-164.

61   Hanson cites the findings of a survey among 300 institutes, design offices and enterprises in the different industries, which show that in 85 per cent of them designs and blueprints were copied by hand, with tracing paper and India ink. In some cases as many as 120 copiers were employed just to carry out these tasks. Hanson also says that between 1966 and 1974 Rank Xerox sold only 4 000 photocopying and duplicating machines to the USSR. *Ibid.,* p. 146.

62   Philip Hanson, "International Technology Transfer from the West to the USSR", *art. cit.,* pp. 807-809.

63   Joseph S. Berliner, *The Innovation Decision in Soviet Industry, op. cit.,* p. 517.

64   Concealment of foreign discoveries may enable Soviet researchers to gain considerable advantages. As Mark Perakh points out in "Utilization of Western Technological Advances in Soviet Industry", in *East-West Technological Co-operation, op. cit.,* p. 180, "when a scientist in the USSR finds a description in the literature of a Western invention in his field, he as a rule will verify the results in his own laboratory. If he is able to obtain the results described in the patent, he will include the relevant subject in his application for funds for the following year. The original patent (or other source of information) is generally not mentioned".

65   Comments regarding the diffusion of imported Western technology in the other industries are quoted later on in the section on the technological gap. They are taken from the major study recently published by the Birmingham Centre, *The Technological Level of Soviet Industry, op. cit.*

66   Ron Scheiderman, "High Technology Flow", *Electronica,* 8th January, 1976.

the percentage of Soviet output represented by complex fertilizers rose steeply between 1970 and 1975. This implies substantial diffusion.[67] Then again, the Soviet authorities seem to be sounding out the possibilities of acquiring other plants to manufacture complex mineral fertilizers.[68]

More generally, Western equipment appears to have made a substantial contribution to growth in Soviet production of chemicals and has reduced material and labour costs. Soviet technology however has made little improvement and there has been no progress in the production of equipment for the chemical industry.[69]

The diffusion of advanced technologies imported for the construction of the Volga automobile plant (Vaz) (or Fiat) at Togliatti is certain. These technologies were applied, for example, in the Cheboksary tractor plant, the Magnitogorsk steelworks and elsewhere.[70] But the problem of the quality of the different components supplied by sub-contractors soon came up. A round table organised by *Izvestia* in 1974 revealed that of 170 main *Zhiguli* components quality-tested by Gosstandart, 50 (or 29.4 per cent) failed to come up to standard.

The authors of Soviet papers on the subject seem to be concerned about this problem and consider that the new technologies applied at Togliatti have not been diffused or brought up to date.[71] The main options before the plant at present are the following: large-scale resumption of purchases of foreign machines and licences or construction of a big domestic facility capable of supplying Togliatti with machines of comparable quality. Given that already 70 per cent of the plant's equipment is imported, the problem is clearly an acute one.[72]

The problem of the Togliatti plant is even more significant if it is looked at in its historical context. In May 1929 the Ford Corporation signed a contract with the Soviet government to co-operate in the building of an automobile plant at Niznyj Novgorod (called Gorkij since 1932) with a capacity of 100,000 vehicles. The 1929 contract was followed by supplementary agreements with Ford and other Western companies which provided for technical assistance but no management assistance. The Western firms showed the Soviet specialists how to build the plant and operate the machinery and left the plant's management to the Soviets, once building was completed. The agreements also provided that Ford would make available to Gorkij all the recent developments in production technology introduced in the Ford plants throughout the duration of the contract (scheduled for 9 years). But the Soviets did not act on the proposals made by Ford in fulfilment of its pledge.[73]

When the Soviets signed the contract with Fiat in 1966, they seemed to have learned their lesson. The provisions written into the contract included supply of licences, technical and organisational studies and Western assistance over a long period. The Fiat engineers

67   Part of this output might come from plants imported from the CMEA countries.
68   Philip Hanson, "The Diffusion of Imported Technology in the USSR", *art. cit.*, pp. 147-164.
69   *Soviet Chemical Equipment Purchases from the West: Impact on Production and Foreign Trade*, National Foreign Assessment Centre, Washington, October 1978, page 1. See also Chapter 4 of this volume.
– Soviet plants with Western equipment accounted for:
– 40 per cent of complex mineral fertilizer (NPK) production in 1975;
– 60 per cent of polyethyl production in 1975;
– 75-85 per cent of polyester fibre production in 1975;
– 72 per cent of new ammonia capacity in 1971-1975 and
– 85 per cent of new ammonia plant capacity for 1976-1980. *Ibid.*
70   *Ibid.*, p. 147.
71   E. Golland, *Ekonomika i organizacija promyshlennogo proizvodstva*, (Economics and Organisation of Industrial Production), Moscow, 1976, No. 1, cited by Philip Hanson, "The Diffusion of Imported Technology in the USSR", *art. cit.*, p. 808.
72   *Ibid.*
73   John P. Hardt and George D. Holliday, "Technology Transfer and Change in the Soviet Economic System", December 1975, version of the paper presented at the NATO Colloquium, 17-19 March, 1976 in Brussels, mimeographed, pp. 36-61.

were obliged to introduce major changes to adapt the Lada to Soviet requirements. Instead of the few dozen specialists sent to Gorkij during the thirties, 2 500 persons went to Togliatti, 1 500 of whom were from Fiat. Over the same period more than 2 500 Soviet technicians went to Italy to learn methods there.

Transposition of Western technology to the Soviet Union is thus seen to be particularly difficult, and even prolonged assistance from Western technicians is not always sufficient to ensure technical progress equivalent to that in the West. The case just described serves as a reminder that technology is not a product, but the application of science to the production of goods and services. It is a specific know-how necessary in order to define, design and produce a product that meets a need.[74] Diffusion as such within a country has only a limited significance if imported technology is not continually readjusted to fit advanced techniques. Borrowed technology therefore cannot be a substitute for internal innovation.

3. TECHNOLOGICAL GAPS BETWEEN EAST AND WEST

a) The Problem of Measurement

The first step in assessing the technological gap is to choose between global macro-economic measurements[75] and those for the various technological progress indicators.

There is an abundance of *partial* indicators of technological level, but they are not always easy to identify and often difficult to interpret. Here, a distinction could be made between indirect and direct indicators and, among the latter, between those relating to advances in technological knowledge and those relating to the production, diffusion and marketing of the more advanced products.

Examples of *indirect* indicators are education budgets, number of graduates and number of persons employed in research and development. R & D expenditure is another indirect indicator.

Direct indicators, which reflect advances in knowledge, include patent statistics and numbers of scientific papers published, quotations and Nobel prizes received. However, the problem of how to interpret these figures has yet to be solved.

Other *direct,* although partial, indicators relate to the quality of products or manufacturing processes. One such indicator results from the attempt to ascertain "empirically" the technological level of an industry. This is done by calculating the relative proportions of both imported and exported "advanced" and "backward" technology incorporated in products and processes.

The rate of technological development is then the rapidity with which a firm or a nation rises to a higher technological level. Several criteria are used for this purpose and they include:[76]

— the dates of the first commercial application of main technologies;
— the rate of diffusion of main technologies;
— the transfer of technology (including import of key processes);
— the pattern of foreign trade by product category;
— the comparative parameters of key items of equipment.

74   J. Fred Bucy, "On Strategic Technology Transfer to the Soviet Union", *Current News,* special edition, 11 August, 1977, p. 3.
75   Re Global Assessments of Technological Progress, see Chapter 5, Section 1 (b) above.
76   Ronald Amann, "Some Approaches to the Comparative Assessment of Soviet Technology: Its Level and Rate of Development", in *The Technological Level of Soviet Industry, op. cit.,* pp. 24-26.

All these criteria are used in case studies. But there have not been many of these case studies and they do not lend themselves readily to composite conclusions. For that reason, the tendency is to put emphasis on the operational characteristics of the end products or on their quality. In the case of machines, for instance, this means comparing indices like reliability, power/weight ratio or operating speed. But this can only be done for key products that decisively influence the general level of an industry.[77]

Table 41

Comparative Statements on Soviet Technological Lags as of 1970

Technology	OECD Report[a]	Western industrial delegation[b]	Sutton[c]
Coal mining - underground operations	—	"Ten years behind"[d]	"Ten-year lag"
Atomic energy	"Equal or in the lead"	"Competent", "Lack of experimental equipment"[e]	"10 to 15-year lag as of 1970"
Blast furnaces	"Equal or in the lead" (1959)	"No lag"[f]	"No lag"
Steel rolling	"Equal or in the lead" (1959)	"20 to 30-year lag"	"30-year lag"
Ore beneficiation	"USSR lagging" (1960)	"Patterned after early American models"[f]	"20-year lag"
Oil well drilling	"USSR equal or in the lead" (1959)	—	"Depth limitations"
Pipeline compressors	—	"Far behind"[g]	"20-year lag"
Large-diameter pipe	—	"Far behind"[g]	"20-year lag"
Chemical engineering (all phases)	"USSR lagging" (1959)	—	"Minimum 30-year lag"

a) E. Zaleski, *et al.*, *Science Policy in the USSR, op. cit.*, pp. 496 to 499.
b) See Antony C. Sutton, *Western Technology and Soviet Economic Development, 1945 to 1965, op. cit.*, pp. 372 and 373.
c) *Ibid.*, pp. 369-370.
d) Private letter from Vasiliiy Strishkov, former Russian coal mining engineer, now with US Bureau of Mines, Washington, D.C.
e) *Atomic Energy in the Soviet Union*, Trip Report of the US Atomic Energy Delegation, May 1963 (Oak Ridge, Tenn., AEC Division of Technical Information Extension, n.d.).
f) *Steel in the Soviet Union*, Report of the American Steel and Iron Ore Delegation's Visit to the Soviet Union, May and June 1958 (New York, American Iron and Steel Institute, 1959).
g) "USSR Natural Gas Industry", Report of the 1961 US Delegation to the Soviet Natural Gas Industry, 1961 (n.p. American Gas Association, n.d.).

Source: Antony C. Sutton, *Western Technology and Soviet Economic Development, 1945 to 1965, op. cit.*, p. 379.

As case studies of this kind are not sufficiently numerous nor widely circulated, assessments of technological gaps are mainly based on the characteristics of certain key products. But clearly this approach is inadequate.

The determination of a "lead" or a "lag" is often derived from such information as is available and not from the information that is desired. This is apparent from one of the comparisons, made by the Birmingham Centre in 1969, of output and productivity, and of indications relating to size, number of machines in service, construction lead times and

77   *Ibid.*

certain technical specifications.[78] A whole body of research would be necessary to justify the use of one indicator over another.[79]

The use of very varied criteria is bound to be reflected in the results obtained. An interesting statement of comparative findings was presented by Antony C. Sutton and this has been reproduced in Table 41. In the case of steel rolling, for instance, the Soviet calculations quoted by the Birmingham Centre indicate equality with the West or even a slight Soviet lead whereas the Western calculations show a Soviet lag of 20 to 30 years! The most noteworthy of the very few case studies that exist are those by Robert R. Campbell on the energy sector[80] and by Robert Perry on the aircraft industry.[81] A major step forward was the work recently published by the University of Birmingham's Centre for Russian and East European Studies,[82] the results of which are quoted in detail in the following paragraphs.

The chief interest of this research is that it investigates specific measurements in R & D and diffusion and provides quantitative data on the Soviet lead or lag that can be used in subsequent work.[83]

b) Patent Statistics as an Indicator of Technological Level

Statistics of registered patents constitute one of the most significant and practical means of measuring the level of the technological knowledge of various countries.[84] Table 42 shows the results of some interesting research by Jiri Slama and Heinrich Vogel in this field for the year 1974. It would appear that the East European countries' contribution to world production of know-how (represented by the application of patents) is particularly small (see Table 42, column 18). While six Western industrialised countries (the United States, Germany, Japan, the United Kingdom, France and Switzerland) accounted for 79.3 per cent of total world patent applications in 1974, CMEA countries contributed only 3.3 per cent.

The ratio of a country's patent applications world-wide to the number of applications by that country for patents from the rest of the world (see Table 42, column 21) is also highly significant. The ratio is very high for the United States (2.6), Germany (2.0) and Switzerland (1.8), but less favourable for most CMEA countries. The latter often apply for patents from Western countries, reflecting limited East European know-how.

According to Jiri Slama and Heinrich Vogel, research at micro-economic level concerns the correlation between the number of patents applied for or granted and the technological level of production.[85] In support of this argument a sample of fifty-seven American pharmaceutical firms is cited. Additional research on this subject seems necessary, however, and it is not certain that the number of patents is a good gauge of value or of application. However this remains an interesting, albeit partial, approach.

78   R. Amann, M. J. Berry, R. W. Davies, "Science and Industry in the USSR", in *Science Policy in the USSR, op. cit.,* pp. 496-499.
79   This has since been done by Ronald Amann. See his article entitled "Some Approaches to the Comparative Assessment of Soviet Technology: Its Level and Rate of Development", in *The Technological Level of Soviet Industry, op. cit.,* pp. 1-34.
80   Robert Campbell, "Technological Levels in the Soviet Energy Sector", in *East-West Technological Co-operation, op. cit.,* pp. 241-263.
81   Robert Perry, *Comparisons of Soviet and United States Technology,* Rand Corporation, R-827-PR, June 1973.
82   *The Technological Level of Soviet Industry, op. cit.*
83   *Ibid.,* p. 30.
84   Jiri Slama and Heinrich Vogel, "Technology Advances in Comecon Countries", in *East-West Technological Co-operation, op. cit.,* pp. 197-210.
85   *Ibid.*

Reporting country	USA	Ger.	J	UK	F	CH	I	NL	S	C
Country of origin	1	2	3	4	5	6	7	8	9	10
1. USA		11 036	11 904	12 488	9 264	2 568		4 364	3 384	15 56(
2. Germany	8 897		5 439	7 283	7 646	4 385		3 856	2 894	2 213
3. Japan	5 163	5 122		4 412	2 940	778		1 222	629	1 96(
4. United Kingdom	5 109	3 183	2 302		2 781	665		1 231	1 047	1 70(
5. France	3 157	2 828	1 594	2 576		985		1 234	716	1 077
6. Switzerland	2 057	2 814	1 591	1 892	2 001			1 061	846	747
7. Italy	1 192	1 100	605	928	1 050	451		412	252	41(
8. Netherlands	985	1 266	1 009	812	1 215	371			542	484
9. Sweden	1 357	1 193	794	1 034	855	326		353		561
10. Canada	2 191	308	297	629	224	59		86	133	
11. USSR	728	475	400	382	411	103		55	195	152
12. GDR	—	651	40	199	246	114		40	75	
13. Hungary	108	161	81	126	120	69		71	67	46
14. CSSR	173	272	98	174	140	83		27	41	32
15. Poland	61	76	27	56	51	22		16	28	12
16. Balance	—63 243	—33 868	—5 622	6 825	8 185	—9 591		4 706	1 999	21 032
17. Applic. from world (total)	38 445	33 011	27 810	35 705	30 927	12 033		14 966	11 933	26 14(
18. 17 in %	11.1	9.5	8.0	10.3	8.9	3.5		4.3	3.4	7.(
19. Applic. from world, per mil. of population	181.4	532.1	253.6	638.4	588.2	1 871.1		1 105.2	1 462.9	1 163.(

a) Last column added by authors.

Source: Jiri Slama and Heinrich Vogel, "Technology Advances in Comecon Countries", in *East-West Technological Co-operation,* NA Colloquium 1976, *op. cit.,* p. 226.

c) **Results of Comparisons in Specific Branches of Industry**

High voltage electric power transmission

In electric power transmission, voltage is an approximate indicator of technological progress. The higher the voltage the smaller the power losses along the transmission line. Three main types of electric power transmission are considered: high voltage alternating current, high voltage direct current and ultra-high voltage alternating current (over 1 000 kV).

The main stages in the development of high voltage alternating current lines in the Soviet Union are set out in Table 39 (above) and Tables 43 and 44. The USSR has become a leader in the development of high voltage alternating current. In 1956 a 400 kV line was built, and in 1959 the USSR was the first country to build a 500 kV line. In fact this was an extension of the 400 kV system already installed in Sweden, but the Soviet Union made considerable use of its own technological know-how.[86]

86 W. G. Allinson, "High Voltage Power Transmission", in *The Technological Level of Soviet Industry, op. cit.,* pp. 222-223.

rigin in the World and in Eastern Europe (1974)

USSR	GDR	H	CS	PL	Saldo	Applic. in world (total)	17 in %	World Applic. per million of popul.	Popul. (millions)	Ratio Applic. in the world to the applic. from the world[a]
11	12	13	14	15	16	17	18	19	20	21
216	429	335	365	454	63 243	101 688	29.3	479.8	211 920	264.5
199	921	510	816	753	33 868	66 879	19.3	1 078.0	62 041	202.6
354	104	79	121	68	5 622	33 432	9.6	304.8	109 670	120.2
294	115	136	204	180	—6 825	28 880	8.3	516.3	55 933	80.9
431	173	104	134	128	—8 185	22 742	6.6	432.5	52 577	73.5
486	284	176	243	261	9 591	21 624	6.2	3 362.5	6 431	179.7
174	106	90	127			10 332	3.0	186.6	55 367	
128	66	36	—51	51	—4 786	10 180	2.9	751.8	13 541	68.0
194	81	32	73	100	—1 999	9 934	2.9	1 217.8	8 157	83.2
40	14	3	7	8	—21 032	5 112	1.5	227.4	22 479	19.5
	260	100	196	73	—1 560	4 120	1.2	16.5	249 749	72.5
494		222	379	78	3	3 077	0.9	181.1	16 980	100.1
161	118		127	87	—88	1 945	0.6	185.7	10 473	95.7
86	146	41		25	—1 560	1 548	0.4	105.4	14 690	49.8
51	45	27	49		—1 904	654	0.2	19.4	33 691	25.6
560	—3	88	1 560	1 904	+64 388					
860	3 074	2 033	3 108	2 558		347 050				
1.6	0.9	0.6	0.9	0.7			100.0			
22.7	181.0	194.1	211.6	75.9				89.9		

Table 43 shows that in the construction of HVAC lines of 700 kV and over the USSR ranks second after Canada but is ahead of the United States. Here again, it seems that the USSR relies on its own technology.[87]

According to Table 44, in 1970 the USSR had the longest network of 500 kV HVAC lines, but was behind the United States in 735-765 kV lines. The USSR growth targets for its 750 kV lines are 2 000 km in 1975-76 and 4 500 km for 1980-1982. The corresponding targets for 500 kV are 17 500 km and 24 000 km respectively.[88]

In the 1960s, the United States and Canada were the only countries to build lines for voltages over 700 kV. Most European countries envisaged this as a future possibility. However, in the general trend toward higher voltage, the USSR began to lose its lead, operating a very high proportion of lines at 160 kV and below. In 1970 the share of lines of 300 kV and above (including 500 and 750 kV) in the Soviet Union was only 10.1 per cent compared with 26 per cent in Sweden, 13 per cent in France, 32 per cent in the United Kingdom and 11 per cent in the United States and in Germany.[89]

87 *Ibid.*
88 *Ibid.*, p. 212.
89 *Ibid.*, p. 215 and Table 44.

219

The main events in the introduction of high voltage direct current lines are set out in Table 45. In 1951, the USSR became the second country to introduce high voltage direct current with a 200 kV line. A second 800 kV (± 400 kV) line was introduced in 1962 (which put the USSR in the lead). The United States did not follow suit until 1970. Plans for a 1 500 kV line were launched a few years ago and the equipment is now at the production stage. Another line — for 2 200 kV — is in preparation.[90] Nevertheless, the USSR is not the world leader. Canada was scheduled to complete a 900 kV line in 1976 and Mozambique a 1 076 kV line in 1975.[91] The latter line is of particular significance as it will use thyristor convertors for which the Soviet technology lag is approximately six years.[92]

Table 43

**Main Events in the Development of HVAC (700 kV and above)
Transmission Lines**

	1962	1963	1964	1965	1966	1967	1968	1969
USA			Prototype 750 kV line for AEP power system^c					AEP 765 kV line brought into operation^g
USSR		Projecting Experimental Konakov — Moscow line. (90 km 750 kV)^a		Began construction of Konakov —Moscow line^d		Konakov —Moscow line commissioned^f		
Canada		Decision to uprate 500 kV project to 700 kV^b		735 kV line scheduled for commercial service^e				

Sources:
a) B. P. Lebedev, *Elektrichestvo,* 1963, No. 12, p. 36.
b) *Ibid.,* p. 34.
c) *Electrical Review,* 16 August 1963, p. 287.
d) V. V. Burgsdorf, S. S. Rokotyan and A. N. Sherentsis, *Elektrichestvo,* 1965, No. 1, p. 7.
e) *Electrical Review,* 1 January 1965, p. 2.
f) S. S. Rokotyan, *Elektricheskie Stantsii,* 1967, No. 12, p. 5.
g) *Electrical and Electronics Abstracts,* January-June 1972, No. 15097, *op. cit.,* p. 798.
Quoted from: W. G. Allinson, "High Voltage Power Transmission", in *The Technological Level of Soviet Industry, op. cit.,* p. 210.

In the case of ultra high alternating current (over 1 000 kV), the USSR has put into operation one line of 1 150 kV. Lines 1 000 kV and above are planned in the United States and Canada. European countries have decided to go through a transitional stage at 750-800 kV. It is not currently possible to establish the comparative technological level between East and West in this field.[93]

Applied technology in fuel mining and processing

There is a technology lag in the Soviet coal industry. The major objective here is the development of automated units to mine coal and to convey it to the transport facilities. The

90 R. W. Davies, "The Technological Level of Soviet Industry: an Overview", *art. cit.,* p. 42.
91 *Ibid.,* p. 42.
92 W. G. Allinson, "High Voltage Power Transmission", *art. cit.,* p. 220.
93 *Ibid.,* pp. 221-222.

Table 44
Lengths of HVAC Transmission Lines, 1959-1970 (in km)

		100-160 kV	%	160-299 kV	%	300-400 kV	%	500 kV	%	735-765 kV	%	Total	%
USA[c]	1959	169 697	84	28 270	14	4 970[d]	2	0	0	0	0	202 937	100
	1965	210 649	76	58 398	21	8 656	3	476	0.2	0	0	278 179	100
	1970	249 397	67	83 081	22	25 741	7	11 433	3	950	1	370 602	100
USSR[a]	1959[e]	83 300	80	15 400	15	3 100	3	2 400	2	0	0	104 200	100
	1965	140 000	74	35 100	18	7 110	4	8 170	4	0	0	190 380	100
	1970	190 500	71	50 800	19	13 950	5	13 140	5	160	0.1	268 550	100
UK[b]	1959	12 800	82	2 750	18	0	0	0	0	0	0	15 550	100
	1965	16 500	70	6 300	27	880	3	0	0	0	0	23 680	100
	1970	17 500	56	3 750	12	9 800	32	0	0	0	0	31 050	100
Germany[c]	1959	20 634	74	7 376	26	0	0	0	0	0	0	28 010	100
	1965	28 849	70	10 650	27	1 296	3	0	0	0	0	40 795	100
	1970	34 383	65	12 680	24	6 232	11	0	0	0	0	53 295	100
Sweden[c]	1959	7 589	51	4 286	28	3 140	21	0	0	0	0	15 015	100
	1965	9 364	49	4 878	25	4 975	26	0	0	0	0	19 217	100
	1970	n.a.		n.a.		n.a.		n.a.		n.a.		n.a.	100
France[c]	1959	9 114	40	11 694	52	1 370	8	0	0	0	0	22 695	100
	1965	9 295	32	16 221	57	3 218	11	0	0	0	0	28 734	100
	1970	9 338	28	19 662	59	4 398	13	0	0	0	0	33 398	100

Sources:

a) All USSR figures are taken from (Eds.) A. S. Pavlenko and A. M. Nekrasov, *Energetika SSSR v 1971-1975 godakh*, 1972, p. 186.

b) UK figures for 1959 are for the end of the 1958-59 financial year and are taken from *Electricity Council Report and Accounts 1960-61*, pp. 172-3. The figures have been converted from miles to kilometres. UK figures for 1965 and 1970 are for the financial years ending in those years, and are taken from *Electricity Council Report and Accounts 1970-71*, pp. 146-7. Again, the figures have been converted to kilometres.

c) 1959 figures were taken from *Situation and Prospects of Europe's Electric Power Supply Industry in 1960/61*, UN Economic Commission for Europe. p. 74, Table 31. Since 1965 and 1970 data were not directly available, they were compiled as follows:

Length of lines in existence in 1963 were taken from the *Electric Power Situation in Europe in 1964/65 and its Future Prospects*, New York, 1966. UN Economic Commission for Europe. pp. 42-3. To these figures were added lengths of lines brought into service in 1964 at the various voltages. These figures were taken from the reference just quoted. Then lines brought into service during 1965 were added. The 1965 additions were obtained from *The 20th Survey of Electric Power Equipment*, OECD. 1967. pp. 32-4.

A similar procedure was followed to obtain the position in 1970. Yearly additions taken from the following:

The 21st Survey of Electric Power Equipment, OECD. 1968, pp. 32-4 (for 1966).
The 22nd Survey of Electric Power Equipment, OECD. 1969. pp. 31-3 (for 1967-8 additions).
The 24th Survey of Electric Power Equipment, OECD. 1971, pp. 25-7 (for 1969-70 additions).

The figures are subject to the following reservations:

1. All USA additions from 1967 onwards inclusive are OECD Secretariat estimate.
2. 1970 additions for Germany are OECD Secretariat estimates.
3. 1964, 100-160 kV additions for Germany were not available. Therefore, the 1963 situation was used also for 1964 position at this voltage.

4. The most important distortion in the figures derives from the fact that they do not take account of the lines withdrawn from service or uprated. Since these are likely to be at lower voltages, the proportion of low voltage lines is probably overstated for these countries. Thus from the point of view of seeing which countries have the highest proportions of the highest voltages, the USA, France, Sweden and Germany are cast in a more favourable light than had the correct figures been available.

d) The figure is for lines at 287-345 kV.
e) 1959 USSR data are for 1960.
Quoted from: W. G. Allinson. "High Voltage Power Transmission", *art. cit.*, p. 225.

Table 45
Details of First and Some Later HVDC
Transmissions Brought into Operation

	Country	Date of Full Intro.	Voltage	Capacity	Distance	Places	Notes
1	France	1910	125 kV	20 mW	150 km	Montiers-Lyons	8 kV dynamos in series
2	Germany	Before 2nd WW	200 kV	15 mW	115 km	Elbe-Berlin	Mercury-Arc Convertors
3	Sweden	?	90 kV	6.3 mW	50 km	Trollhatten-Mallerad	MAC[a] Experimental
4	USSR	1951	200 kV	30 mW	112 km	Kashira-Moscow	MAC[a] Experimental cable
5	Sweden	1954	± 100 kV	20 mW	90 km	Sweden-Gotland	MAC[a] Submarine
6	UK-France	1961	± 100 kV	160 mW	64 km	Cross-Channel	MAC[a] Submarine
7	USSR	1962	± 400 kV	750 mW	475 km	Volgograd-Donbass	MAC[a] Overhead line
8	New Zealand	1965	± 250 kV	600 mW	40 km		
9	Italy	1965	200 kV	200 mW	104 km	Italy-Sardinia	
10	Japan	1965	± 125 kV	300 mW			
11	Sweden-Denmark	1965	250 kV	250 mW	88 km		
12	Canada	1968	260 kV	312 mW		Vancouver	
13	USA	1970	± 400 kV	1 440 mW		Celilo Sylmar	
14	Canada	(1975-76)	± 450 kV	700 mW		Nelson River	MAC[a]
15	Mozambique	(1975)	± 533 kV	1 500 mW		Caborra Bassa-Apollo	Thyristor Convertors

a)   MAC = Mercury Arc Convertor.

Sources:
1   *Electric Power Situation in Europe in 1955,* UN Economic Commission for Europe. Geneva. 1957. p. 46.
2   *Ibid.,* p. 47.
3   *Ibid.,* p. 47.
4   V. P. Pimenov and M. R. Sonin. *Elektrichestvo,* 1955. No. 7. pp. 93-9. Quoted in *Electrical Review:* 7 October 1955. p. 705.
5   *Situation and Prospects of Europe's Electric Power Supply Industry, 1962/63,* UN Economic Commission for Europe. New York. 1964. pp. 71-2, Table 28.
6   *Ibid.*
7   S. S. Rokotyan, *Elektricheskie Stantsii,* 1967. No. 12. p. 5.
8   *Situation and Prospects of Europe's Electric Power Supply Industry 1962/63,* UN Economic Commission for Europe. New York. 1964. p. 71-2. Table 28.
9   *Ibid.*
10 *Ibid.*
11 *Ibid.*
12 F. H. Last and R. M. Middleton. in *CIGRE 24th Session 1972.* Vol. 1. paper 14-07. p. 2.
13 *Ibid.*
14 E. Jeffs. *Energy International.* October 1974. p. 17.
15 *Ibid.*
Quoted from: W. G. Allinson. "High Voltage Power Transmission". *art. cit.,* p. 217.

Soviets are unable to design, manufacture and introduce such very sophisticated machinery. Foreign observers have noticed, in particular, how slow the Soviet Union has been to adopt such modern machines as narrow web cutters and movable hydraulic roof supports. All of these were introduced long after they were in operation in Western Europe.[94] In addition, a much larger proportion of the Soviet workforce is employed in manual tasks making the lag in those areas 10 to 15 years.

In oil and gas, the Soviet lag (the USSR is the world's biggest oil producer) was made apparent when deeper-lying reserves located in the more inaccessible regions of Siberia had to

94   Robert Campbell, "Technological Levels in the Soviet Energy Sector", *art. cit.,* pp. 245-246.

Table 46
Soviet Views of Technological Level
of Soviet Machine Tools

Year	Soviet Union backward	Soviet Union ahead or equal
	Type of machine tool. etc.	
1948	Gear shaving, diamond boring, honing machines, gear grinders with abrasive worm, centreless grinders with large working wheels[1]	
1958	Jig boring, grinding, gear grinding machines.[2] Automated lines with PC.[3] Machine tools for car industry.[4] Multispindle automatic honing machines with automatic cycle; automatic loading and removal,[5] gear forming machines,[6] gear hobbing; internal grinders, external broachers[7]	Automation of control[10]
1959		Gear cutting machine, Gleason type[11]
1963	Precision machine tools, grinders, automatic lines using batch produced machines[8]	
1966		1K62 centre lathe[12]
1968		Electrochemical and electrophysical machining,[13] 1K62, 1A616 centre lathes[14]
1970	Ultrasonic and spark erosion machines; programme control with computer[9, 17]	Heavy machine tools[15]
1971		Measuring machines[16]

Sources:
1   Quoted in *The Engineer's Digest*, August 1952, p. 254.
2   P. M. Pen'kov. *Razvitie tipazha i struktura vypuska metallorezhushchikh stankov i komplektuyushchikh prinadlezhnostei k nim*. v 1958 i 1959-65 gody, 1958, p. 21.
3   O. V. Spasskaya. *Osnovnye zadachi tekhonologii stankostroeniya i smezhnoi promyshlennosti na 1959-65 gody*, 1958. pp. 42-3.
4   *Ostraslevoe soveshchanie po stankostroeniya, Minsk, iyul' 1958: Vystupleniya uchastnikov soveshchaniya (sokrashchennaya stenogramma)*, 1958. p. 47.
5   *Ibid.*, p. 113.
6   *Ibid.*, p. 120.
7   M. E. Mardanyan. *Osnovnye napravleniya razvitiya konstrukstii metallorezhushchikh stankov*, 1958, p. 16.
8   *Planovoe Khozyaistvo*, 1963, No. 12. p. 69.
9   *Materialy vsesoyuznoi nauchnoi konferentsii po ekonomicheskim problemam nauchno-tekhnicheskogo progressa, Vypusk pervyi*, 1970, p. 114.
10   P. M. Pen'kov, *op. cit.*, pp. 25-6.
11   P. M. Pen'kov. *Tekhnicheskii progress i sozdanie material'no-proizvodstvennoi bazy kommunizma*, 1959. p. 79.
12   D. S. L'vov. *Osnovy ekonomischeskogo proektivaniya mashin*, 1966, p. 18.
13   *Opyt organizatsii predpriyatii Ministerstva Stankostroitel'noi i Instrumental'noi Promyshlennosti v novykh usloviyakh planirovaniya i izuchenie perspektivnykh voprosov razvitiya otrasli*, 1968, p. 14.
14   Yu. D. Matevosov. *Ekonomicheskie problemy razvitiya stankostroitel'noi promyshlennosti v Zakavkaz'e*, Erevan. 1968. p. 154.
15   *Vestnik Mashinovedeniya*, 1970, No. 3. p. 12; *Stanki i Instrument*, 1970, No. 4, p. 25.
16   I. I. Knyazitskii and G. V. Ust-Shomushskii, *Tochnye, nadezhnye, ekonomichnye*, Odessa. 1971, pp. 21, 39.
17   The quality of these machines are questioned by some in the West. It seems there are few technicians qualified to service them.
    Quoted from: M. J. Berry and M. R. Hill, "Technological Level and Quality of Machine Tools and Passenger Cars", in *The Technological Level of Soviet Industry, op. cit.*, p. 531.

be tapped. The Soviet Union lacked both the experience and the equipment required and the Soviet machine engineering industry proved incapable of meeting the new demand. Drilling remains the worst bottleneck. The Soviets have not been able to introduce electric drilling, improved rotary technology and turbo-drill designs or bits suited to turbo-drill technology. According to Robert Campbell: «... Experimental work with diamond bits showed great promise, but widespread use of diamond bits appears to have been blocked by some obstacle, and virtually nothing is now said about diamond bits. The alternative is bits armoured in a special hard steel and this solution seems to be having some effect on drilling productivity.»[95]

95   *Ibid.*, pp. 253-254.

The only information available which could be used for a general survey of Soviet technology in machine tools comes from Soviet sources and is shown in Table 46.

The USSR was lagging in this area in 1948-1963 but seems to have improved its technological level since that time and is now a leader in several fields.

The only available study on this area in the West is that by M. J. Berry and J. Cooper, Birmingham Centre.[96] It contains some very useful information on numerically controlled machine tools, the most advanced sector of this industry. The main research results are presented in Table 47 (see also Table 39).

Table 47

Leads and Lags: The Innovation Process and Major Developments in USSR Industry of Machine Tools

Indicator	Year achd. by USSR	USSR in relation to:			
		USA	UK	Japan	Germany
a) Stages of the innovation process					
— Start of research	1949	—2	—1	+4	+6
— First prototype	1958	—6	—2	=	=
— Start of industrial production	1965	—8	—2	+1	—1
— Diffusion: NC output as a proportion of total output:					
To 0.1%	1968	(—9)	(—5)	=	—4
To 0.25%	1969	—9	(—5)	—1	—3
To 0.5%	1970	—8	—4	=	+1
To 1.0%	1971	—6	—3	+2	+ +
To 1.5%	1973	+ +	+ +	+ +	+ +
— Diffusion: growth of NC stock:					
To 500 units	1969	(—10)	—5	—2	—2
To 2 000 units	1970	—8	—2	—1	=
To 5 000 units	1972	—8	(+2)	+1	(+2)
To 15 000 units	1974	—5	+ +	+ +	+ +
b) Major developments					
— First machining centre	1971	—12	(—10)	—5	—10
— First third generation control system	1973	—7	(—5)	(—5)	(—5)
— First use of computer for control	1973	—6	(—4)	—5	(—4)

Note: Number of years: USSR in advance, +; behind, –; estimate, ( ); Soviet lead, not yet achieved in other country, ++.
Source: M. J. Berry and Julian Cooper, "Machine Tools", *art. cit.,* p. 195.

According to M. J. Berry and J. Cooper, in the initial stages of the production of machine tools, Soviet performance compared well with that of Germany and Japan but lagged behind that of the United Kingdom and well behind that of the United States. The lag persisted during the 1960s owing to the difficulties involved in producing suitable control systems based at first on semi-conductors and later on integrated circuits. In 1968, in order to prevent the gap from widening any further, the USSR government took steps to spur the industry on and since that time the gap has narrowed considerably. The authors conclude that the USSR now produces good computers and that the manufacture of automatic production systems is likely to improve through co-operation with the German Democratic Republic.[97]

96 M. J. Berry and Julian Cooper, "Machine Tools", in *The Technological Level of Soviet Industry, op. cit.,* pp. 121-198.
97 M. J. Berry and J. Cooper, "Machine Tools", *art. cit.,* p. 198. However, in the Chapter in which he recapitulates the contents of the book, R. W. Davies does not share this optimism and stresses instead the problems of the Soviet electronic industry; see *The Technological Level of Soviet Industry, op. cit.,* p. 41.

Table 48

Stock of Computers Held in the Soviet Union as Compared with that in France, Germany, United Kingdom and United States

	1960	1962	1965	1967	1970	1975 (plan)
France		285[1] (6.1)		2 208[1] (44.3)	(90)[2]	
Germany		548[1] (10.0)		2 963[1] (49.5)	(109)[2]	
UK		312[1] (5.8)		2 252[1] (40.9)	(91)[2]	
USA	5 000[4] (27.7)[4]	7 305[1] (39.0)		39 516[1] (198.5)	70 000 (344)[2]	
USSR	[120][4] (0.56)		[2 000][3] (8.7)		[6 000][3] (24.6)[2]	[22 000][3] (84.6)
Ukraine	10[5] (0.24)	40[5] (0.94)	200[5] (4.4)		600[5] (12.6)	

Note: Figures in brackets give stocks per million population.

Sources:
1 *Gaps in Technology: Electronic Computers*, OECD, Paris, 1969, p. 126.
2 J. Slama *et al.*, *Ost Europa*, 1974, No. 2, p. 122.
3 Estimates, or estimated plans.
4 A. C. Sutton, *Western Technology and Soviet Economic Development, 1945 to 1965*, *op. cit.*, p. 319. (for end of 1950s).
5 Mikulich *et al.*, in *Mekhanizatsiya i Avtomatizatsiya Upravleniya*, 1974, No. 4, p. 7.
Quoted from: Martin Cave, "Computer Technology", *art. cit.*, p. 393.

The excellent research published by Richard Judy in 1968 has now been supplemented by Martin Cave's paper in the Birmingham Centre's publication.[98]

A first indicator of the level reached by the Soviet computer industry is the stock of computers currently held in the USSR compared to that in the leading Western countries. The corresponding data are given in Table 48.

Table 49

Lag of Soviet Computer Output behind the United States

Prototypes or First Commercial Production of Computers

1. Lag of best Soviet computer behind best US computer (in years)[a]

1955	4
1960	9
1965	14
1967	4
1970	7
1973	10

2. Date of first production of comparable Soviet and American computers[b]

American computer	Similar Soviet model	Date of appearance in USSR	Lag (in years)
IBM 650	Ural 1	1955	1
IBM 702	Ural 4	1962	7
IBM 1620	Nairi 1	1964	4
IBM 7094	BESM-6	1966	4
IBM 360 series	ES series	1972-3	6-8

a) Criterion: operations per second. BESM-6 has been included.
b) Comparison of dates of first American commercial installation and first Soviet industrial production.

Source: Martin Cave, "Computer Technology", *art. cit.,* p. 401; R. W. Davies, "The Technological Level of Soviet Industry: an Overview", *art. cit.,* p. 44.

The Soviet stock of computers does not seem very high compared with that of the leading Western countries (especially when related to population figures) and is considerably lower than that of the United States. Martin Cave quotes even lower estimates than those in Table 48.[99] Can this relatively low stock of computers in the USSR be attributed to the country's lower technological level?

Table 49 attempts to summarise a few basic data on the Soviet — US lag which seems to have sharply increased in the 1950s. Martin Cave thinks that both computers and computer equipment (peripherals, characteristics of disc units, characteristics of line printers) are concerned and that this lag has not narrowed during the past five or six years.[100] In software however the situation has improved considerably. The author concludes that the Soviet lag in the computer field should not be over-emphasized. Prior to 1968 Soviet computer

98 Richard Judy, "The Case of Computer Technology", in *East-West Trade and the Technology Gap,* edited by Stanislaw Wasowski, *op. cit.,* pp. 43-72. Martin Cave, "Computer Technology", *art. cit.,* pp. 377-406.

99 According to *Soviet Cybernetics Review,* July 1971, p. 2, the Soviet stock of computers was 5 000 to 6 000 in 1970 and the figure under the 1975 plan was only 18 000. Quoted by Martin Cave, "Computer Technology", *art. cit.,* pp. 391-392.

100 *Ibid.,* pp. 401-402.

technology operated under conditions of rather low priority but since then, it has received considerable government support.[101]

During the 1970s, the Soviet Union and its CMEA partners designed, developed and put into production a series of first generation computers known as the Unified System (*Edinenaja sistema - ES* or *Ryad*). The series is upward-compatible: programs that run on one of the models will run without change on any larger model. By 1980, the Unified System (ES) will probably be second only to the IBM 360-370 in terms of the number of installed mainframes.[102]

In 1975, the Minsk Plant introduced the ES 1022, the first of the interim Ryad Group. It is 2 to 3 times as productive as the ES 1020.[103] The Kazan Plant has recently begun production of another model, the ES 1033.[104] The CMEA countries are currently developing a new, more advanced, group of models, the Ryad-2 (see Table 50). These computers have much the same relationship to the earlier Ryads as the IBM S/370 has to the S/360. New features include much larger primary memory, semi-conductor primary memory, virtual-storage capabilities, block-multiplexor channels, relocatable control storage, improved peripherals and expanded system timing and protection facilities. By early 1977 most of the new models were well into the design stage. The appearance of prototypes and the initiation of serial production will probably occur over the next five years.[105]

During the first two or three years of Ryad production, output was an eighth or a tenth of that for IBM S 360 over a similar period. This difference is probably a reasonable measure of overall relative computer development capabilities.[106] During corresponding periods[107] almost 35 000 IBM S 360 units were produced as compared to approximately 5 000 Ryad-1 computers. Ryad output was only about a third of the planned target and it has not come up to either the quantitative or the qualitative standards of the IBM S 360.

The Unified System provides the CMEA countries with unprecedented quantities of reasonably good hardware although major problems remain. The CMEA semi-conductor industry is backward and dependent on what it can get, either legally or illegally, from the West. The importance of customer convenience and peripherals is just beginning to be appreciated. Supporting technology for large-scale data processing applications, such as ground and satellite communication equipment, is of low standard. Institutional problems still cripple the effective distribution of hardware. Finally, after 10 years in the making, there is nothing about ES hardware that might be described as really innovative by current Western standards.[108]

As already stated, Soviet software has always been of poor quality. Intensive efforts were made to obtain and adapt IBM 360 software for the Ryad (ES) models. By now, almost everything offered by IBM to S 360 installations has been acquired and much of it has been made suitable for Ryad.[109] Since IBM has discontinued systems DOS 360 or OS 360, the Eastern countries are now on their own as far as maintenance and enhancement of the two systems is concerned. Apparently, this has not produced particularly impressive results.[110] In early 1977 OS/ES MFT had already gone through several releases and was on at least the fifth. This probably reflects massive accumulations of errors rather than improvements.[111]

101   *Ibid.,* pp. 402-403.
102   N. C. Davis, S. E. Goodman, "The Soviet Bloc's Unified System of Computers", *Computing Surveys,* June 1978, p. 93.
103   *Ibid.,* p. 105 (Davis and Goodman's estimate).
104   *Ibid.,* p. 106.
105   *Ibid.,* pp. 109-110.
106   N. C. Davis, S. E. Goodman, *op. cit.,* p. 110.
107   Between April 1964 and June-September 1970 for IBM S 360 and S 370, respectively, and between January 1972 and mid-1978 for Ryad-1 and Ryad 2. *Ibid.,* p. 110.
108   N. C. Davis, S. E. Goodman, *op. cit.,* p. 111.
109   *Ibid.,* p. 113.
110   *Ibid.,* p. 114.
111   *Ibid.,* p. 114.

The initial applications of software available on ES systems were standard programs readily obtainable from the West: linear programming, numerical routines, critical-path algorithm, etc. Ryad is following in IBM's footsteps.[112] The development of simple, unambitious software systems seems to be coming along reasonably well in the USSR. The systems include monitoring, test automation and data recording, rather than direct process control, industrial data-processing systems, etc. In summary, the CMEA countries are more attentive to software needs. The authors conclude that it will be difficult, but not impossible, for them to overcome an assortment of complex systemic problems that affect the development of software.[113]

The Soviet turbine engine industry

In the case of turbine engines, up to 1955 Soviet turbine engines were generally more advanced than those of other countries (though less resistant). Between 1955 and 1971 American developments were faster and the gap was bridged by 1971. In aviation, the Soviets continue to lag behind and the American lead was maintained through the 1960s.[114] Certain Soviet fighter aircraft certainly seem as good as the best produced elsewhere and the same could probably be said of Soviet helicopters.[115] But the designs of current passenger and cargo planes are old and the aircraft themselves are not of superior quality.

Table 50

A. Selected Characteristics of Interim Ryad Computer Systems

Model	ES-1022	ES-1032	ES-1033	ES-1012
Responsible Country	Bulgaria USSR	Poland	USSR	Hungary
Processor				
Operating speed (k opns/sec)	80	200	200	—
Selected performance times (μ sec)				
Short operations	9	2.5-4.0	1.4-2.7	2.6
Floating point add/sub.	30	4.5	4.5	n.a
Fixed point multiply	80	9.0	8.5	8.5
Floating point divide	100	14.0	17.7	n.a
Instruction set	......IBM S/360 Instruction Set...........			Special 109 Instructions
Principle of processor control	......Rigid Microprogram...................			
Primary memory				
Capacity (kbytes)	128-512	128-1024	256-512	8-64
Cycle time (μsec)	2.0	1.2	1.2	1.0
Length of accessed word (bytes)	4	4	4	2
Channels				
Selector channels				
Number	2	3	3	—
Transmission rate (kbyte/sec)	500	1100	800	—
Multiplexor channel				
Transmission rate in multiplex mode (kbyte/sec)	40	110	70	40

(Cont'd on next page)

112   *Ibid.*, p. 114.
113   *Ibid.*, p. 117.
114   Robert Perry, *Comparisons of Soviet and US Technology, op. cit.*, p. 39.
115   The Soviet Union has taken the lead, however, in the design and production of heavy transport helicopters.

Table 50 *(cont'd)*

B. Selected Characteristics of the Ryad-2 Computer Systems

Model	ES-1025	ES-1035	ES-1045	ES-1055 (without buffer)	ES-1055 (with buffer)	ES-1065 [b]
Responsible Country	Czechoslovakia	Bulgaria USSR	Poland	GDR	GDR	USSR
Processor						
Operating speed (k opns/sec)	30-40	100-140	400-500	450	750	4 000-5 000
Selected performance times (μsec)						
Short operations	5-18	2.6-4.5	0.6-2.2	0.6-3.9	0.3-2.2	0.12
Floating point add/sub	50-55	9.7	1.9-2.3	1.6-3.6	1.3-1.6	0.24
Fixed point multiply	95-220	23	2.8-3.4	3.4-5.2	3.1	0.6
Floating point divide	225-235	32	8.4-11	4.1-6.0	3.9	1.2
Instruction set	IBM S/360 Instruction Set					
	Some additional IBM S/370-like commands					
Principle of processor control	Microprogram					
Working memory[a]						
Primary memory capacity (kbytes)	128-256	256-512	256-3072	256-4096	256-4096	1-16 Mbyte
Virtual memory			up to 16 Mbytes			
Buffer memory					8 kbytes available	

a) Peripheral configurations that we have seen in print closely resemble those given for Ryad 1. The Czechoslovakia and the German Democratic Republic models do not include papertape equipment.
b) The ES-1060 has gravitated to the Ryad-2 group.

Source: N. C. Davis, S. E. Goodman. "The Soviet Bloc's Unified System of Computers". *Computing Survey*, June 1978, pp. 97 and 106.

The process control industry in the USSR is given less priority than the computer industry, and the technological gap is even wider. A comparison of British and Soviet performance clearly illustrates the situation (Table 51).

Table 51

Comparison of Controller Production in the USSR and the United Kingdom

First commercial production of electronic controllers

	USSR	UK	Soviet lag (in years)
First electronic controller	1951	1949	2
First fully transistorised controller	none by 1974	1959	15

The rate of development of selected Soviet and British control systems

	USSR		UK
	USEPPA	Electr. GSPI	Bailetronic
Idea formulated	About 1957	1958	1956
Development started	1958	About 1961	1957
Development finished	1964	Not finished	1961
Production started	1964	1967	1961
Development of improved version started	1964	About 1971/2 with GSPI never completed	1961

Source: Z. A. Siemiaszko. "Industrial Process Control". in *The Technological Level of Soviet Industry. op. cit.*, pp. 365 and 372.

The Soviet lags are probably due to the low level of automation of common processes, to the near zero-rate innovation and to the antiquated equipment still being produced. The equipment now being developed for the power industry represents only a marginal improvement with respect to the equipment already being produced. There do not seem to be any substantial Soviet counter-balancing leads.[116]

Nevertheless, one original Soviet innovation — the USEPPA miniature modular pneumatic system — was produced in the USSR in 1964, six years ahead of the West. It was developed as a substitute for electronic systems because of the slow Soviet progress in electronics. In 1972 however, USEPPA was superseded by a German pneumatic controller. No new Soviet system, pneumatic or electronic, has been fully developed since 1964, whereas in the United Kingdom new systems are produced every two years on average. In any case, the initial lead in pneumatic controllers has now been lost and the lag in electronic controllers has increased as can be seen in a comparison with United Kingdom performance[117] (see Table 51).

116   Z. A. Siemiaszko, *art. cit.,* p. 371.
117   R. W. Davies, "The Technological Level of Soviet Industry: an Overview", *art. cit.,* p. 44.

Table 52

Comparative Technical Data Relating to the Moskvich 412 and Passenger Cars Produced by Western European Firms

	Moskvich 412	Vauxhall Victor 1600	Opel Rekord	Fiat 125
Engine capacity (cc)	1 478	1 599	1 432	1 608
Maximum power (hp)	75	72	58	90
Maximum speed (km/hr)	145	150	133	160
Time to accelerate from rest to 100 km/hr (sec)	18.8	n.a.	20.5	14.8
Wheel base (mm)	2 400	2 590	2 670	2 506
Minimum road clearance (mm)	178	150	130	120
Number of lubrication points	14	4	none	none
Fuel consumption (litres/100 km)	9.1	10.0	9.9	11.2
Horse power per litre	50.9	45	38.8	56
Specific weight (ratio of the total weight of the car to the maximum power) (kg/hp)	17.9	19.4	23.9	14.9
Service time to first major overhaul (km)	125	n.a.	n.a.	100

Source: B. V. Vlasov, op. cit., p. 45.

Frequency of Detected Faults in Nine Models of Moskvich 412

Location of defect \ Type of defect	Potential defect requiring remedial action	Defect to be put right at owner's convenience	Extreme defect requiring urgent attention	Extreme defect requiring immediate attention
Engine and cooling systems	2	8	0	2
Fuel and exhaust	1	1	2	0
Transmission	2	2	2	2
Braking system	0	1	3	15
Steering mechanism	0	15	3	11
Suspension and wheels	3	11	3	4
Body	14	4	1	0
Miscellaneous	6	3	0	1
Total	28	45	14	35

i.e. 35 extreme defects were observed in the 9 cars tested. 2 in the engine and cooling system. 0 in the fuel and exhaust systems. etc.

Source: Tests conducted by Consumers' Association. 1973 (unpublished).
Quoted from: M. J. Berry and M. R. Hill. "Technological Level and Quality: Machine Tools and Passenger Cars". art. cit., pp. 553 and 555.

The technological level of Soviet passenger cars

The Soviet passenger car industry has a rather narrow product range. Under the 1971-1975 Five-Year Plan the production target was 1.3 million passenger cars,[118] including 150 000 Zaporozhets (45 hp),[119] 420 000 Moskvich (55 or 80 hp),[120] 60 000 Zhiguli (80 hp), 75 000 Volga and Chajka (80 to 240 hp)[121] and an unspecified number of ZIL-114 (330 hp).[122]

118 M. J. Berry and M. R. Hill, "Technological Level and Quality: Machine Tools and Passenger Cars", art. cit., p. 550. It should be noted that in the USSR 1 201 000 passenger cars were actually produced in 1975, 1 239 000 in 1976, 1 280 000 in 1977 and 1 314 000 in 1979. Narodnoe Khozjajstvo SSSR za 60 let (The USSR Economy — 60 years), Moscow, 1977, p. 227; and Economicheskaya Gazeta, N° 6, February 1978, p. 5 and Ekonomicheskaya Gazeta, No. 5, February, 1980, p. 8.
119 The source quoted (M. J. Berry and M. R. Hill), art. cit., p. 552 indicates ranges for car power data, e.g. 45/40 hp, 55/50 hp, etc. We have quoted the higher value.
120 55/50 hp for Moskvich 408 and 80/75 hp for Moskvich 412.
121 80/75 hp for GAZ-21 (Volga), 110/98 for GAZ-24 (Volga) and 240/200 for GAZ-14 (Chaika), M. J. Berry and M. R. Hill, art. cit., p. 552.
122 Ibid., pp. 551-553.

231

The specification of the *Zhiguli* model corresponds to that of the Fiat 124, its Western counterpart. The Togliatti plant, Volgograd produced and designed the car. *Moskvich* models are designed in the USSR but technical assistance is provided by Renault and Peugeot under an agreement signed in October 1966.[123]

The study by M. J. Berry and M. R. Hill is based on a comparison between the 1971 and 1973 performance of the *Moskvich 412*. Similar comparisons would be useful in the case of the *Zhiguli* (Lada), which now has a strong foothold on the Western markets.

Table 52 shows comparative technical data for the *Moskvich 412* as published by Soviet authors in 1971. In their opinion, the latest Soviet models measure up to the best Western models and Soviet design takes account of export market requirements.[124] These views are not fully confirmed by the United Kingdom Consumers' Association, which in October 1973 tested eleven *Moskvich 412* cars with mileages of 0 to 30 000 on the clock. The results of this study are set out in Table 52. The Consumers' Association concluded that in several cases there had been lack of attention to the safety features. The main faults were in steering and braking, usually the result of leaks in hydraulic systems or poor fitting of cylinders. The Soviet agents in London nevertheless defended the safety record of the 20 000 *Moskvich* models sold in the United Kingdom in 1973. In 1970, the *Moskvich* was awarded a certificate of safety in Paris.[125]

The technological level of the Soviet chemical industry.

Heavy organic chemicals are one of the key sectors of modern chemical industry. They provide the essential feedstocks for the manufacture of synthetic materials and other organic end-products.[126] Several industrial processes, especially those based on petroleum (e.g. catalytic cracking) did not become widely diffused until after the Second World War, when production was stimulated by the rapid growth of plastics and synthetic fibres.[127]

The USSR has published very few statistics on this subject. This may suggest that this is a relatively backward area, particularly with respect to heavy organic chemicals and petrochemicals.

By 1948, 40 per cent of organic chemicals in the United States were derived from petroleum and natural gas. By the early 1960s feedstocks for organic chemicals had also changed radically in other Western countries such as the United Kingdom, France and Japan. Table 53 shows that this shift was much slower in the USSR. In 1970, the latter still depended far more than the West on coal and fermentation as sources of organic chemicals.

In synthetic materials, the most modern branch of the chemical industry, the USSR trails well behind the leading Western industrialised countries. According to Table 54, the main new plastic materials were very slow to be introduced in the USSR — not merely by comparison with the United States — and the lag does not seem to have been made up in the post-war years.

In 1972, for instance, the per capita output of synthetic fibres in the USSR was 1.0 kg. compared to 11.6 kg. in the United States, 10.8 kg. in Germany and 10.4 kg. in Japan. In the same year, the share of chemical fibre in total synthetic fibre output was only 32.0 per cent in the USSR compared to 79-80 per cent in the United States and in Germany, 69 per cent in Japan and 60 per cent in the United Kingdom (see Table 55).

123 Antony C. Sutton, *Western Technology and Soviet Economic Development, 1945 to 1965*, *op. cit.*, pp. 197-198.
124 M. J. Berry and M. R. Hill, *art. cit.*, p. 551.
125 *Ibid.*, pp. 554-555.
126 Ronald Amann, "The Chemical Industry: Its Level of Modernity and Technological Sophistication", in *The Technological Level of Soviet Industry, op. cit.*, p. 249.
127 *Ibid.*, p. 249.

The USSR is, however, the world's leading producer of phenol formaldehyde resins, which are well-established and require less advanced technology. In the manufacture of polystyrene, polyolefins and copolymers (see Table 56), the USSR lags behind the advanced industrial countries.

In conclusion, although the USSR now has the second largest chemical industry in the world, most of its progress is in traditional sectors (mineral fertilizers, man-made fibres, etc.). A general indicator of the Soviet chemical industry's performance might be the ratio between the tonnages of synthetic materials and inorganic chemicals produced. Table 57 shows the comparative performance of the USSR and the leading Western industrialised countries. In spite of the considerable growth in the tonnage of synthetic materials produced in the USSR its relative importance compared to the "West", especially for Japan and Germany, is not great.

Table 53

Petrochemicals and Raw Materials Used for Organic Synthesis in the USSR and in Several Western Industrialised Countries

Proportion of Petrochemicals Used in the Manufacture of Synthetic Materials (%)

	1958	1965	1970 (plan)
USSR			
Aliphatic	25	56	75-80
Aromatic	12	20	60
Inorganic	16	50	60

	1958	1963
USA		
Aliphatic	94	95
Aromatic	70	86
Ignorganic	46	61

Source: N. P. Fedorenko, *op. cit.*, 1967, p. 28.

Structure of the Raw Material Base for Organic Synthesis in 1970 (%)

	USSR[b]	Germany[a]	Japan[a]	UK[a]	France[a]
Oil and natural gas	67.0	89.6	94.5	91.9	89.1
Coal	26.0	8.0	5.5	8.1	7.6
Other sources	7.0	n.a.	—	—	3.3

Sources:
a) *The Chemical Industry, 1971-72,* OECD, Paris, 1973, p. 129.
b) G. F. Borisovich, Yu. T. Livshits and L. V. Koronnaya, *Khimicheskaya Promyshlennost',* 1974, No. 3, p. 6.
Quoted from: Ronald Amann, "The Chemical Industry: Its Level of Modernity and Technological Sophistication", *art. cit.,* p. 251.

The technological level of Soviet light industry (and consumer durables)

We have been unable to find any detailed work on the comparative technological level of light industry in the USSR and the Western industrialised countries. The only material available consists of a number of views expressed by Western observers or Soviet authors. These have been summarised in Table 58. Most of the information presented is relatively dated and the lead or lag criteria are fairly general and arbitrary at times.

Table 54

First Commercial Production of Various Plastic Materials in Order of Introduction in the USSR

Major plastics	USSR	USA	Germany (after 1945)	UK	France	Italy	Japan
Phenol-formaldehyde (bakelite; Soviet designation 'Karbolit')	1915 (on semi-handicraft scale: the original 'Karbolit'[1] factory was enlarged in 1925)	1909*	1910	1910	1916	1922	1923
Galalith	1925[3]	1919	1899*	1912	1900	1921	1927
Celluloid	1926[2]	1870*	1878	1877	1875	1924	1908
Cellophane	1936[9]	1924	1925	1930	1917*	1946	1929
Urea-formaldehyde	Soviet designation—'amino-plasts'. Production began during third five year plan (1938-41). But real development occured during early post-war period 1945-50[10]	1929	1929*	1928*	1930	1936	1935
Polyvinylchloride (PVC)	1940[4]	1933	1931*	1940	1940	1951	1939
Polyvinyl butyral	1947[8]	1937*	1948		1942	1944	1944
Silicon polymers	1947[6]	1941*	1950	1952	1954	1955?	1951
Polymethylmethacrylate (perspex)	Late 1940s[19] (wide diffusion of volume and range by 1959)[13]	1936	1930*	1933	1938	1937	1938
High pressure polyethylene	Early 1950s[14]	1941	1944	1937*	1954	1952	1954
Polytetrafluorethylene (PTFE: Teflon)	1949[12]	1943*	1958	1945	1958	1955	1963
Polystyrene	Mid-Late 1950s[7]	1933	1930*	1950	1951	1942	1957
Epoxy resins (Switzerland—the innovating country)	Late 1950s[5]	1947	1955	1955	n.d.	1958?	n.d.
Linear polyethylene	1959	1956	1955	1959	1956	1954*	1958
Polycarbonate	1970 (Produced, then, on a semi-industrial scale)[16]	1957*	1957*	n.d.	n.d.	n.d.	1959
Polyformaldehyde	1970 (Produced on a semi-industrial scale) (polypromyshlennyi masshtab)[17]		n.d.				

Sources:

1 M. I. Garbor. "Promyshlennost plasticheskikh mass i sinteticheskikh smol". in Khimicheskaya promyshlennost SSSR. 1959. p. 80.
2 Ibid. pp. 80-91.
3 Ibid. p. 81.
4 Ibid. p. 82.
5 Ibid. p. 91.
6 Ibid. p. 93.
7 Ibid. p. 96. (The source disclosed that polystyrene was being produced in the USSR at the time of writing. 1959. implied that it was a recent development. but gave no exact indication of the first date of commercial production.)
8 Ibid. p. 101.
9 G. E. Birger and A. A. Kontin. "Promyshlennost khimicheskikh volokon". in Khimicheskaya promyshlennost SSSR. 1959. pp. 119-20.
10 P. M. Luk'yanov and A. S. Solov'eva. Istoriya khimicheskoi promyshlennosti SSSR. 1966. p. 180.
11 G. E. Birger et al. op. cit., p. 124.
12 M. I. Garbor. art. cit. p. 103.
13 Ibid. p. 102.
14 P. M. Luk'yanov. Kratkaya istoriya khimicheskoi promyshlennosti SSSR. 1959. pp. 442-3.
15 P. M. Luk'yanov and A. S. Solov'eva. op. cit. p. 192.
16 M. I. Garbor. op. cit., p. 97.
17 Stroitel'naya sel'skokhozyaistvennaya i meditsinskaya tekhnika. khimicheskaya tekhnologiya. 1971. p. 181.
18 Encyclopedia Americana. 1970. Vol. 22. p. 221 ff.
19 P. M. Luk'yanov. op. cit. p. 420.

General notes:
The items in the table are in the chronological order in which they were introduced commercially in the USSR.
Not all the items in Hufbauer's tables are included because. for many. there is no information in Soviet sources: this may or may not mean that they were not produced.
Western dates can be found in G. C. Hufbauer. Synthetic Materials and the Theory of International Trade. London. 1966. p. 131.
n. d. = not produced in 1966 but no data thereafter.
n. a. = not available.
* = innovating country.
Quoted from: Ronald Amann. "The Chemical Industry: Its Level of Modernity and Technological Sophistication". art. cit., pp. 278-279. (Notes 11. 15 and 18 do not appear in the table reproduced from Amann's article.)

In spite of the obvious USSR lag in some fields, undoubted improvements have also been recorded in Soviet production of consumer goods such as television sets, radio receivers, washing machines, refrigerators, cameras, watches and even textile goods.[128] In this area however, it is more a question of a reduction in the Soviet Union's lag. Yet these improvements are very real, which implies that "pressure from above" is far from ineffectual.

Table 55
Output of Chemical Fibres (1000 tonnes)
Per Capita Figures in Kg

	1950	1953	1957	1960	1963	1967	1970	1972
USSR								
Total output:	24.2[1]	62.3[3]	148.7[3]	211[1]	308[1]	511[1]	623[2]	747[9]
Artificial	22.9[1]	57.9[3]	137[3]	196[1]	266[1]	395[1]	456[2]	508[9]
Synthetic	1.3[1]	4.4[3]	11.7[3]	15[1]	43[1]	116[1]	167[2]	239[9]
Synthetic per capita	0.01	0.02	0.06	0.07	0.19	0.49	0.69	0.96
Synthetic as % of total	5.4	7.1	7.9	7.1	14.0	22.7	26.8	32.0
USA								
Total output:	626	655	751	773	1 189	1 697	2 251	3 061
Artificial	571[8]	543[5]	517[6]	466[6]	612[5]	630[5]	623[5]	632[9]
Synthetic	55.5[8]	112[5]	234[6]	307[6]	577[5]	1 067[5]	1 028[5]	2 429[9]
Synthetic per capita	0.39	0.70	1.36	1.70	3.05	5.36	7.93	11.57
Synthetic as % of total	8.9	17.1	31.2	39.7	48.5	62.9	72.3	79.3
Germany								
Total output:	162	175	258	281	373	495	723	801
Artificial	161[8]	169[5]	239[6]	229[6]	265[5]	243[5]	226[5]	160[9]
Synthetic	0.5[8]	6.5[5]	19[8]	52[6]	108[5]	252[5]	497[5]	641[9]
Synthetic per capita	0.02	0.13	0.37	0.98	1.95	4.38	8.35	10.75
Synthetic as % of total	0.5	3.6	7.4	18.5	28.9	50.9	68.7	80.0
Japan								
Total output:	115.5	242.5	481	551	701	1 101	1 550	1 629
Artificial	115[8]	236[5]	439[6]	433[6]	462[5]	523[5]	522[5]	512[9]
Synthetic	0.5[8]	6.5[5]	42[6]	118[6]	239[5]	578[5]	1 028[5]	1 117[9]
Synthetic per capita	0.01	0.07	0.46	1.26	2.49	5.78	9.93	10.35
Synthetic as % of total	0.4	2.7	8.7	21.4	34.1	52.5	66.3	68.6
UK								
Total output:	172.4	189.8	225	269	326	433	599	627
Artificial	168[8]	181[5]	193[6]	208[6]	221[5]	239[5]	262[5]	253[9]
Synthetic	4.4[8]	8.8[5]	32[6]	61[6]	105[5]	194[5]	337[5]	374[9]
Synthetic per capita	0.09	0.17	0.62	1.16	1.96	3.53	6.05	6.72
Synthetic as % of total	2.5	4.6	14.2	22.7	32.2	44.8	56.3	59.6

Sources:
1 Narodnoe khozyaistvo SSSR v 1969 godu, 1970, p. 214.
2 Ibid., 1922-1972, 1972, p. 172.
3 Promyshlennost SSSR, 1964, pp. 148-9.
4 Ibid., p. 112.
5 UN Statistical Yearbook, 1971, pp. 240-1 and 243-4.
6 Ibid., 1966, pp. 263-4 and 266-7.
7 Ibid., 1948, pp. 202-4.
8 Ibid., 1959, pp. 207-209.
9 Ibid., 1973, pp. 246-50.
Quoted from: Ronald Amann, "The Chemical Industry: Its Level of Modernity and Technological Sophistication", art. cit., p. 309.
(Notes 4 and 7 do not appear in the table reproduced from Amann's article.)

128   Martin C. Spechler, "The Pattern of Technological Achievement in the Soviet Enterprise", in *The Association for Comparative Studies Bulletin,* Summer 1975, pp. 67-69.

The technological lag cannot be measured in the same way as the difference in height of two persons. Even if, on the whole, one country may be in the lead, it is still possible for the laggard to "beat" its opponent in many fields!

Furthermore, the technological lag cannot be considered on a static basis. It is constantly changing as both knowledge and the state-of-the-art advance. That no satisfactory assessment is possible and that the sources we have presented are far from exhaustive or even adequate is not, therefore, surprising. Examples of fields where knowledge is scarce are technology in light industry and in the food industry, the agricultural machinery sector and performance of strategic weapons.

The best that can be done therefore is to present the reasonable opinions of certain experts although it is impossible to confirm their accuracy.

Table 56

Output of Key Plastic Materials

	USSR	USA	Germany	Japan	UK	France
Phenol-formaldehyde resins (1 000 tonnes)						
1960	112[1]	215.7[2]	77.3[2]	n.a.	45.3[2]	23.4[2]
1965	177[1]	418[3]	100[3]	76[3]	52.2[3]	31[3]
1970	265[1]	500[4]	164[4]	219[4]	46.1[4]	54.4[4]
1972	285[1]	n.a.	173.5[5]	251[5]	41.8[5]	66.1[5]
Polyolefins (1 000 tonnes)						
1960	1.2[1]	542.5[2]	80.8[2]	n.a.	125.7[2]	37.1[2]
1965	57[1]	1 552[3]	300.9[3]	454[3]	256.2[3]	116.2[3]
1970	267[1]	3 160[4]	900[4]	1 885[4]	496.8[4]	429.2[4]
1972	307[1]	n.a.	1 176[5]	2 098[5]	553.3[5]	646.3[5]
PVC and copolymers (1 000 tonnes)						
1960	24.8[1]	401.5[2]	172.7[2]	n.a	107.8[2]	110.5[2]
1965	80.8[1]	n.a.	375.3[3]	483[3]	196.8[3]	213.4[3]
1970	160[1]	1 415[4]	777[4]	1 161[4]	314.5[4]	412.1[4]
1972	207[1]	n.a.	931[5]	1 080[5]	344.3[5]	538.7[5]
Polystyrene and copolymers (1 000 tonnes)						
1960	7.5[1]	407.8[2]	n.a.	n.a.	49.6[2]	43[2]
1965	28.9[1]	n.a.	n.a.	125[3]	92.1[3]	83.5[3]
1970	82.2[1]	1 320[4]	n.a.	668[4]	162[4]	191.9[4]
1972	98.5[1]	n.a	n.a.	796[5]	184[5]	266.6[5]

Sources:
1 *Statisticheskii ezhegodnik stran-chlenov SEV,* 1973, pp. 111-12.
2 Statistical supplement for 1953-64 data. *The Chemical Industry 1965-66,* OECD, Paris, 1967, pp. 58-62. Figures for phenol-formaldehyde resins including cresylic plastics.
3 *The Chemical Industry 1965-66.* OECD, Paris, 1967, p. 190.
4 *Ibid., 1970-71,* OECD, Paris, 1972, p. 153.
5 *Ibid., 1972-73,* OECD, Paris, 1974, p. 153.
Quoted from: Ronald Amann, "The Chemical Industry: Its Level of Modernity and Technological Sophistication". *art. cit.,* p. 308.

Table 57

Weight of Synthetic Materials as a Percentage of Weight of Inorganic Chemicals in the USSR and the Main Western Industrialised Countries

	USSR	USA	Japan	Germany	France
1960	4.48	15.78	17.20	23.70	13.89
1961	5.19	15.87	17.61	27.57	14.20
1962	5.74	18.78	18.33	32.17	16.34
1963	6.47	19.99	23.21	34.71	18.08
1964	7.29	20.65	28.28	39.44	19.25
1965	7.73	21.09	31.02	43.92	19.87
1966	8.65	23.25	36.04	49.46	21.88
1967	9.53	23.78	45.61	58.00	23.63
1968	10.57	29.41	66.23	65.60	25.95
1969	11.29	32.58	75.33	75.62	30.37
1970	11.93	32.11	86.88	83.62	33.21

Source: Ronald Amann, "The Chemical Industry: Its Level of Modernity and Technological Sophistication", *art. cit.,* p. 259.

Table 58

Technical Comparisons between Soviet Light Industry Goods and Consumer Durables and those made in the Western Industrialised Countries

	Date	Country of Source	Basis for Comparison	Quality Characteristic Compared
USSR at same level or in the lead				
Wrist watches (some)	1967	USSR United Kingdom	. .	. .
Photographic equipment	1967?	USSR	. .	. .
Radio receivers	1967?	USSR	. .	. .
Washing machines (some)?!	1967?	USSR	. .	. .
Refrigerators	1967?	USSR	. .	. .
Knitting machine (Ukrainian)	1960	USSR		Productivity per man or per unit area
Spinning machinery	1959	USSR	USA	Capital productivity
USSR behind				
Washing machines	1970	USA	USA	Too small
Television sets	1968-69	USSR United Kingdom	UK	. .
Refrigerators	1969	USSR	. .	Use of electricity
Motorcycles	1969	USSR	. .	Technical characteristics
Razor blades	1968	United Kingdom	UK	Assortment poorer
Men's suits and jackets; women's garments	1968	United Kingdom	UK	. .

Source: Martin C. Spechler, "The Pattern of Technological Achievement in the Soviet Enterprise", *art. cit.,* pp. 73-77.

According to a US study,[129] the USSR is considerably behind the United States in technology of such fields as space, semi-conductor devices, precision machinery, integrated circuitry and other kinds of electronics.

129   *Technology Transfer and Scientific Co-operation Between the United States and the Soviet Union: A Review, op. cit.,* p. 31.

The Director of the Birmingham Centre, producer of the best study on the subject yet, takes an even more definite line: "In most of the technologies we have studied there is no evidence of a substantial diminution of the technological gap between the USSR and the West in the past 15-20 years, either at the prototype/commercial application stages or in the diffusion of advanced technology".[130]

130   R. W. Davies, "The Technological Level of Soviet Industry: an Overview", *art. cit.*, p. 66.

Chapter 6

EFFECT OF ECONOMIC FACTORS
ON EAST-WEST TECHNOLOGY TRANSFER

In Chapters 2 and 3 we tried to assess technology transfer through two channels: products (Chapter 2) and know-how (Chapter 3). It is by no means easy to interpret the data, and any attempt to isolate political, economic and technological factors must be somewhat arbitrary. All we can do, therefore, is to outline the effects of a number of economic factors and possibly explain the more important variations:

— the economic situation;
— trends in prices and terms of trade;
— changes in the debt situation;
— greater use of countertrading and compensation agreements to overcome the problem of payment.

1. TECHNOLOGY TRANSFER AND THE ECONOMIC SITUATION

Technology transfer, like all trade, is particularly sensitive to changes in the economic climate. Unfortunately this aspect has been almost wholly disregarded up to now. One of the few studies dealing with this subject has been carried out by J. Stankovsky.[1]

According to this author, when conditions are buoyant, it is easy for East European countries to find markets for their products — inclusive of manufactures. These countries also find the liberalisation of trade more attractive. Conversely, in periods of recession the East European countries are often badly hit because demand for their exports — consisting mainly of raw materials and intermediates — contracts. Both prices and quantities are affected. East European countries are often "marginal" suppliers that are excluded when times are bad, a situation reinforced by the not always remarkable quality of their products. At the same time, the industrialised countries of the West redouble their efforts to sell to East European countries whose reduced foreign currency earnings force them to retrench or postpone purchases until more credit can be obtained. The economic situation also has a direct effect on the negotiation of industrial cooperation agreements.

During an upswing in the business cycle, Western firms are interested in arrangements enabling them to increase deliveries to the East in the short term (sub-contracting, and

---

1 J. Stankovsky, "Bestimmungsgründe im Handel zwischen Ost und West" (Determinants in East-West Trade), *Forschungsberichte des Wiener Institutes für Internationale Wirtschaftsvergleiche,* No. 7, 1972.

sometimes even joint production and specialisation as well). Long-term co-operation is found less attractive, partly due to credit restrictions and higher rates of interest.

During downswings in the economy, Western firms are generally less keen on co-operation agreements that imply increase of productive capacity which would create potential competitors for them. They are more interested in long-term agreements calling for deliveries (joint ventures, supply of complete plants, licensing and tri-partite co-operation) in the more distant future.[2]

This kind of agreement, like those on scientific and technical co-operation and certain grants of licences, has no direct effect on production capacity and is not sensitive to changes in the economic situation.[3]

In recent years the oil and energy crisis caused a recession in the West. In the OECD countries growth in production practically came to a standstill in 1974 and levelled out at about 3 per cent per year in 1975 and 1976. Towards the end of 1977, for all the Western industrialised countries put together, the United Nations Economic Commission for Europe forecast an increase in Gross Domestic Product of 2.5 per cent in 1977 and 3.5 per cent in 1978.[4] A 3 to 3.5 per cent increase was expected in 1977 for Germany, although in May of that year the government had been banking on 5 per cent. In France, the 1977 forecasts were cut from 4.8 per cent to 3 per cent and in the same year in the United Kingdom, growth was expected to be low or even negative for the third year running. In the United States, the situation improved in 1976 following the 1974-1975 recession, but forecasts of growth in GDP were only 5 per cent for 1977 and 4.5 per cent for 1978.[5]

After a temporary and partial improvement in the situation during 1977 and 1978, the oil crisis is again threatening economic growth in the West. The initial forecast of a 3.5 per cent growth in real income for the OECD countries would seem, now, to need revising downwards to around 1.5 per cent (June 1979).[6]

According to official estimates, in Eastern Europe and in the Soviet Union production increased at a greater rate (5 to 7 per cent on average between 1974-1977). However, slower growth rates have recently been observed, especially in the USSR. From 1971 to 1975, the official growth rate for Soviet national income averaged 5.7 per cent. But in 1976 it was only 5.2 per cent, in 1977 3.5 per cent, in 1978 4 per cent and in 1979 2 per cent. The 1980 plan forecasts a growth rate in Soviet national income of 4.0 per cent.[7]

These trends in the economic situation in Eastern Europe necessarily affected trade relations with the West, trends which are summarised in Table 59.

It would seem that 1977 was clearly a turning point in East-West trade. In that year, Western exports to the Soviet Union fell by 9.2 per cent in volume, although total exports to the East were reduced by only 6.8 per cent. Some recovery in Western exports is apparent in 1978. The first figures available for 1979 seem to suggest that the lag of 1977 has not yet been made up.[8]

2 It has been called to the attention of the authors that this does not generally apply to Sweden.

3 J. Stankovsky, op. cit.; F. Levcik and J. Stankovsky, op. cit., p. 170.

4 Economic Bulletin for Europe, Vol. 29, op. cit., p. 1.

5 Ibid.

6 OECD Economic Outlook, No. 25, Paris, July 1979, p. 3.

7 Narodnoe Khozjajstvo SSSR za 60 let (National Economy of the USSR in the Past 60 Years), Moscow, 1977, p. 485; Ekonomicheskaya Gazeta, No. 6, 1978, p. 5; Economicheskaya Gazeta, No. 52, 1977, p. 3; Ekonomicheskaya Gazeta, No. 50, 1978, p. 3; Ekonomicheskaya Gazeta, No. 5, 1979, p. 8; No. 50, 1979 p. 3 and No. 5, 1980 p. 8.

8 According to Economic Survey of Europe in 1979, Part I: "The European Economy in 1979", Chapter 3, Table 6.2 (Pre-publication, 31st March, 1980). The volume of Western Exports to CMEA countries increased in January-September 1979 by 1.8 per cent compared with the same period of the previous year. The volume of Western imports increased by 11.7 per cent. It should be noted that Polish imports from the industrialised countries fell by 1.9 per cent in 1978 compared with 1977 (current prices). Based on Maly Rocznik Statystyczny, 1979 (Small Statistical Yearbook, 1979), Warsaw, 1979, p. 173.

The decline in exports to the West from the CMEA countries was fairly sharp (12 per cent in volume) in 1974 and in 1977 they were still fluctuating around their 1973 level. In 1978 the increase in volume was only 3.3 per cent[9]. The CMEA countries are still finding it difficult to sell to the West.

These fairly weak trends in East-West trade since 1976 do not have the same significance for the industrialised countries of the West as compared with the CMEA countries. Whereas the latter provide only 3 per cent of the West's total imports and take only 4 per cent of its total exports[10], 30 per cent of total CMEA imports and 26 per cent of these countries' total exports come from or go to the West (1977 figures).[11]

Table 59

The Growth of East-West Trade — 1974-1979

Percentage change over the same period of the preceding year

	Exports of industrialised market economies						Imports of industrialised market economies					
	1974	1975	1976	1977	1978	Jan-May 1979	1974	1975	1976	1977	1978	Jan-May 1979
Eastern Europe and the Soviet Union												
Value ($ US)	40	33	4.5	0.7	18.0	14.0	42	6	16.0	10.7	14.0	18.0
Prices	30	15	—9.1	8.1	9.8	14.0	62	10	—0.3	7.2	10.4	13.1
Volume	8	15	14.9	—6.8	7.5	0.0	—12	—3	16.4	3.2	3.3	4.4
of which:												
Soviet Union												
Value ($ US)	34	66	9.0	—0.3	16.0	10.0	59	6	25.4	14.0	14.0	20.0
Prices	20		—9.2	9.8	9.9	14.0	68	11	3.7	9.7	10.2	13.1
Volume	12	41	20.1	—9.2	5.6	—3.5	—5	—4	20.9	3.9	3.5	6.1

Sources: Statistics of Foreign Trade, Series A, OECD, Paris; Direction of Trade and International Financial Statistics, IMF, Washington, D.C.; National Statistics.
As quoted in: Economic Bulletin for Europe, United Nations, Vol. 30, No. 1, pre-publication, Table 3.1; and in *Economic Bulletin for Europe,* United Nations, Vol. 31, No. 1, pre-publication, Table 3.1.
Note: For the derivation of these price and volume indices, see *Economic Bulletin for Europe,* United Nations, ECE, Vol. 28, pp. 105-108.

In the case of machinery imports, however, East-West trade appears to have been less dependent on the world economic situation. Although world trade in machinery (SITC Group 7) rose by a factor of 2.2 between 1972 and 1976 (the same rate as for the developed industrialised countries), machinery exports from developed countries to centrally planned countries grew by a factor of 3.2 during the same period.[12] A 3.3 per cent reduction was, however, recorded in 1975 and 1976.[13]

2. TRENDS IN PRICES AND TERMS OF TRADE

Terms of trade are difficult to calculate because, except for Poland and Hungary, East European countries do not publish indices for unit values of external trade. On the other hand, Western countries do not make any distinctions between the various destinations and origins in their unit values. The figures in Table 60 are taken from United Nations estimates.

9   *Economic Survey of Europe in 1978,* Part I, *op. cit.*
10   Not including trade between Germany and the German Democratic Republic.
11   *Economic Bulletin for Europe,* Vol. 30, No. 1, Pre-publication, Tables 3.2 and 3.3.
12   *Monthly Bulletin of Statistics,* United Nations, New York, June 1978, p. xlvi.
13   *Ibid.*

243

They show a considerable improvement in the terms of trade since 1974 for the CMEA countries as a whole in their trade with the industrialised West. The advantage obtained by the USSR in this respect has been particularly striking, presumably because of the increase in oil prices. There is as yet no explanation for the deterioration in the situation for the USSR in 1976 and 1977.

Table 60
Unit Values in East-West Trade
Fisher Index, US $ Unit Values — 1965 = 100

	Western Exports to 7 East European Countries		Western Imports from 7 East European Countries		Terms of trade[a] with:	
	All 7 Countries	Soviet Union	All 7 Countries	Soviet Union	All 7 Countries	Soviet Union
1966	99.0	99.0	98.4	101.2	100.6	97.8
1967	105.0	94.4	98.9	102.1	106.1	92.5
1968	96.9	90.3	97.4	102.7	99.5	87.9
1969	102.7	91.6	100.0	101.9	102.7	89.9
1970	113.8	98.6	111.4	113.5	102.1	86.9
1971	118.1	95.5	118.8	127.4	99.4	75.0
1972	132.1	105.8	124.0	127.5	106.5	83.0
1973	169.7	134.4	160.9	162.8	105.5	82.6
1974	219.3	167.8	285.7	379.9	76.6	44.2
1975	236.0	190.6	278.3	389.2	84.8	49.0
1976	226.2	398.3	291.4	415.5	77.6	95.9
1977	248.5	454.7	324.4	460.7	76.6	98.7

a) Arrived at by calculation: unit value of Western exports divided by unit value of Western imports.
Source: Economic Bulletin for Europe. United Nations. Vol. 30. No. 1. pre-publication. Table B.4.

Poland's terms of trade with the Western countries have also improved thanks to its coal exports (1970 = 100 : 1974 = 116.7 and 1975 = 122.6).[14] For Hungary, on the other hand, there has been a deterioration in its terms of trade with the dollar and other hard currency countries (1970 = 100, 1974 = 83.6, 1976 = 83.0 and 1977 = 79.8)[15] mainly accounted for by trade in machinery, raw materials and semi-finished products, and food commodities. For these, the 1977 indices (1970 = 100) are: 79.1 for machinery, 79.8 for raw materials and semi-finished products, and 65.9 for food commodities.[16] Part of the answer to this trend in the terms of trade is to be found in the United Nations statistics reproduced in Table 61.

There is clearly a big difference in the trend in the terms of trade for all exports and imports as compared with only machinery and transport equipment. Whereas the terms of trade in general deteriorated considerably between 1965 and 1977 for the West (mainly between 1971 and 1974 as a result of the higher energy and raw materials prices) the opposite is true for machinery and transport equipment. The terms of trade for these products improved in favour of the West by about 25 per cent between 1965/1968 and 1971/1974, and by 51 per cent between 1968/1969 and 1976/1977. In other words, the West has recovered on machinery some of the losses due to higher raw material and fuel prices but the overall trend is a net loss in the terms of trade.

14 Article by Rafal Kosztirko in *Handel Zagraniczny,* No. 21, 1976, quoted by Anita Tiraspolsky, "Les termes de l'échange des pays de l'Est de 1970 à 1977", *Le courrier des pays de l'Est,* La Documentation française, Centre d'études prospectives et d'informations internationales, Paris, May 1978, p. 26.
15 *Külkereskedelmi statisztickai évkönyv,* 1971 to 1976, quoted by Anita Tiraspolsky, *art. cit.,* p. 28.
16 *Ibid.,* p. 29.

Table 61

The Change in Prices and Terms of Trade in East-West Trade
(Western Exports to and Imports from the East)[a]

	1965-1968 to 1968-1971 Total	1965-1968 to 1968-1971 Soviet Union	1965-1968 to 1971-1974 Total	1965-1968 to 1971-1974 Soviet Union	1965 to 1974 Total	1965 to 1974 Soviet Union	1968-1969 to 1976-1977 Total	1968-1969 to 1976-1977 Soviet Union
CHANGE IN PRICES								
Exports								
Total								
Paasche index	107.6	107.1	155.2	155.0	201.5	208.0	231	213
Fisher index[b]	108.0	108.9	159.9	160.0	213.4	216.5	—	—
of which:								
Machinery and Transport Equipment:								
Paasche index	117.8	119.0	178.7	181.9	226.2	233.3	276	273
Fisher index[b]	117.9	118.8	179.1	181.2	226.3	231.4	—	—
Imports								
Total								
Paasche index	108.5	111.6	165.4	178.5	243.6	275.8	305	358
Fisher index[b]	110.0	112.0	173.0	184.2	263.3	299.9	—	—
of which:								
Machinery and Transport Equipment:								
Paasche index	109.2	111.5	142.4	146.2	175.2	179.4	183	191
Fisher index[b]	109.3	110.8	143.4	146.6	178.4	180.7	—	—
TERMS OF TRADE								
Total								
Paasche index	99.1	96.0	93.8	86.8	82.7	75.4	76	60
Fisher index[b]	99.0	97.3	92.4	86.8	81.0	72.3	—	—
of which:								
Machinery and Transport Equipment:								
Paasche index	107.9	106.7	125.6	124.4	129.1	130.0	151	143
Fisher index[b]	107.9	107.2	124.9	123.7	126.9	128.1	—	—

a) The term «Western countries» used in these United Nations statistics quoted from the *Economic Bulletin for Europe*, Volume 28, *op. cit.*, pp. 77-129 is synonymous with "industrial market economies". Volume 28, *op. cit.*, p. 76, note 1. For Western Europe it includes: Germany, Italy, Yugoslavia, France, Finland, United Kingdom, Sweden, Austria, Netherlands, Belgium-Luxembourg, Denmark, Spain, Switzerland, Greece, Norway, Turkey, Ireland, Portugal and Iceland. It also includes the United States, Canada and Japan. *Ibid.*, p. 80.

b) Geometric average of Paasche and Laspeyres indices.
Sources: Adapted from *Economic Bulletin for Europe*, Volume 28, *op. cit.*, pp. 107 and 113. *Economic Bulletin for Europe*, Volume 31, No. 1, Pre-publication, Tables B.1 and B.2.

Table 62

Prices and Volume Growth of East-West Trade by Major Groups of Commodities — 1975 to 1978
(Percentage change over the preceding year)

| | Exports of industrialised market economies | | | | | | | | Imports of industrialised market economies | | | | | | | |
| | Prices | | | | Volume | | | | Prices | | | | Volume | | | |
	1975	1976	1977	1978	1975	1976	1977	1978	1975	1976	1977	1978	1975	1976	1977	1978
Total	5	−4.5	9	9	25	11	−7	8	8	3	8	11	−2	13	3	3
Food products	7	−7	−9	reduction	62	42	−22	reduction	1	2	11	14	−4	7	−11	0
Beverages and tobacco	11	−10	−1	8	14	22	−1	15	25	15	2	12	21	1	−4	−3
Crude materials, inedible	−5	4	9	17	9	9	3	−5	−1	7	16	1	−8	4	−4	−3
Mineral fuels	18	−2	−3	7	8	9	21	30	17	2	7	11	6	24	−3	2
Animal and vegetable oils and fats	−23	−15	21	19	32	−14	−5	−21	−12	−25	13	0	11	−35	3	−12
Chemicals	−4	−8	5	11	9	6	11	9	−9	−8	0	5	5	23	40	11
Basic manufactures	−1	−12	10	3	20	12	−13	11	0	−6	5	12	−14	18	11	4
Machinery	15	2	13	20	35	0	−3	−3	15	1	4	20	15	19	13	−2
Miscellaneous manufactured goods	11	3	13	18	14	11	−10	−12	9	3	15	17	2	13	3	1

Sources: 1975 and 1976: *Economic Bulletin for Europe*, Vol. 28, *op. cit.*, pp. 105-108, quoting United Nations Data Bank. 1977 and 1978: *Prices*: calculated from Tables 3.4 and 3.5. *Economic Bulletin for Europe*, Vol. 31. No. 1. pre-publication: *Volume*: *Ibid*. Tables 3.4 and 3.5.

Note: Trade by commodities groups in 1978 excludes the United States.

Additional data to those in Table 61 are set out in Table 62. In 1976, the prices of machinery exports from the Western industrialised countries to Eastern Europe rose by two per cent, in contrast to the overall downswing of 4.5 per cent. At the same time a stagnation and even a reduction in volume could be observed since 1976. An interesting point is that the prices of East European imports from the West increased by 9 per cent in 1977 and 1978. But the prices of East European machinery imports rose more steeply (13 per cent in 1977 and 20 per cent in 1978).

What are the reasons for this unfavourable (for the East European countries) trend in the terms of trade for machinery and transport equipment? Are they due to cyclical considerations or are they purely a matter of product quality? Some measure of this quality is represented, partly at least, by the unit value of the machinery and transport equipment bought and sold by the East European countries. The relevant figures are given in the Annex, Table A-43.

It can be seen that the unit value of machinery and transport equipment sold by the Western countries over the period 1965-1968 was twice that of machinery exported by the East European countries. This gap widened between 1971 and 1974 when the factor reached 2.7. The figures, moreover, vary with category of machine, where the highest factors were 2.6 and 3.6 in 1965-1968 and 1971-1974 respectively; for some products, automobiles for example, the factors are lower.

3. THE FINANCING PROBLEM FOR PURCHASES OF MACHINERY AND EQUIPMENT FROM THE WEST

a) Indebtedness of the East European Countries

One of the most important economic problems affecting East-West technology flows is the question of payment. The East European countries' debt has increased steeply in recent years. This raises questions as to future East-West trade prospects.

The total indebtedness of the East European economies is not easy to estimate. Official calculations are not available in the East or in the West. In the East European countries such information is considered to be a State secret. In the West accurate figures are lacking because of the large number (and different kinds) of loans and the competition among countries exporting to the East.

Table 63 gives estimates (mainly those made by the Chase Manhattan Bank) of the indebtedness of the CMEA countries. These estimates are not unanimously agreed to in the West. The experts of the United Nations Economic Commission for Europe, for example, put the debt of the CMEA countries to the industrialised, market-economy countries at $32-35 billion net at the end of 1976.[17] Paul Marer's estimate is $40.8 billion[18] (both estimates are net figures). Other specialists, like William F. Kolarik, Jr., estimate the CMEA countries' net debt for the end of 1976 at $39.9 billion.[19] Estimates for the end of 1977 range

17 *Economic Bulletin for Europe,* Vol. 29, p. 124. *East-West Markets,* 16th May, 1977, puts the CMEA countries' net indebtedness at $37.0 billion at the end of 1976.

18 Paul Marer, Statement in *US Policy Toward Eastern Europe,* Hearings before the Subcommittee on Europe and the Middle East, Committee on International Relations, US House of Representativees, 95th Congress, 2nd Session, September, 7th and 12th, 1978, Washington, 1979, p. 100. Quoted by Morris Borstein, Research Conference on "East-West Relations in the Eighties", Bellagio, June, 18th-23rd, 1979, p. 64.

19 William F. Kolarik, Jr., "Statistical Abstract of East-West Trade Finance", in *Issues in East-West Commercial Relations, op. cit.,* p. 198.

Table 63

Hard Currency Indebtedness of CMEA Countries to the West — 1970-1979

(Estimates in billions of US dollars, at year-ends)[a]

	1970	1973	1974	1975	1976	1977	1978	1979
				Gross indebtedness				
USSR	n.a.	4.0	5.9	11.4	14.1	..	..	..
Bulgaria	n.a.	1.5	1.7	2.4	3.0	..	..	..
Czechoslovakia	n.a.	0.9	1.1	1.5	2.2	..	..	..
German Democratic Republic	n.a.	2.8	3.6	4.9	5.9	..	..	..
Hungary	n.a.	2.0	2.3	3.2	3.4	..	..	..
Poland	n.a.	2.5	4.9	7.8	11.0	..	..	..
Rumania	n.a.	2.1	2.4	2.8	2.9	..	..	..
Total Seven Countries	n.a.	15.8	21.9	34.0	42.5	..	..	..
CMEA banks	n.a.	1.8	2.1	3.3	4.3	..	..	..
CMEA total	n.a.	17.6	24.0	37.3	46.8	..	..	..
				Net indebtedness [b]				
USSR	1.8	n.a.	2.4	8.4	10.3	16.3	17.2	..[d]
Bulgaria	0.9	n.a.	1.4	2.0	2.5	3.0	3.4	3.4
Czechoslovakia	0.3	n.a.	0.7	1.2	1.8	1.9	2.0	3.3
German Democratic Republic	1.1	n.a.	3.1	4.2	5.1	6.6	8.2	9.2
Hungary	0.7	n.a.	1.8	2.3	2.4	4.3	5.7	7.1
Poland	1.0	n.a.	4.4	7.1	10.2	14.0	16.8	18.7
Rumania	1.1	n.a.	2.2	2.3	2.5	3.2	4.4	5.8
Total Seven Countries	6.9	n.a.	16.0	27.6	34.8	49.3	57.9	..
CMEA banks	0.6	n.a.	1.9	3.1	4.0	3.5	3.0	..[d]
CMEA total	7.5	n.a.	17.9	30.7	38.8	52.7[c]	60.7	65

a) Brainard/Chase Manhattan Bank estimates. The estimates were chosen for their comprehensiveness and their consistency over time.
b) Gross indebtedness minus deposits held in Western banks.
c) Differences are due to rounding.
d) Total USSR and CMEA banks: $ 17.5 billion.

Sources: 1973-1974: Chase Manhattan Bank, as quoted in Business Week, 7th March, 1977, p. 40; 1970, 1975 and 1976: Lawrence J. Brainard "Eastern European Indebtedness", Conference on Monetary and Financial Problems East and West, Budapest, 1977; L. J. Brainard, "East Europe Improves its Trade", Euromoney, May, 1978.
1977: Neue Zürcher Zeitung, 21 June, 1978, p. 11: based on estimates of the Chase Manhattan Bank.
1978: M. Karr, Soviet and East European Debt Perspectives, Chase, International Finance, quoted in Jan Stankovsky, Ost-Westhandel 1979 und Aussichten für 1980, Forschungsbericht, Wiener Institut für Internationale Wirtschaftsvergleiche, Wien, March 1980, No 58, Table 12.
1979: Business Week, 25 February, 1980. Based on Jan Stankovsky, Ost-Westhandel 1979 und Aussichten für 1980, op. cit., Table 12.

from $47.4 billion (Kolarik)[20] to $48.0 billion (Lenz)[21], $49.4 billion (Marer)[22] and to $52.7 billion (Chase Manhattan Bank).[23, 24]

The disaggregation of this debt, shown in Table 63, shows some major differences. The Soviet Union and Poland together are responsible for about two-thirds of the debt, whereas the Czechoslovak share is small. There are, incidentally, two ways of assessing the relative

20 Ibid., p. 198.
21 Allen J. Lenz, "Potential 1980 and 1985 Hard Currency Debt in Perspective", in Issues in East-West Commercial Relations, op. cit., p. 190.
22 Paul Marer, op. cit., p. 100.
23 See Table 63.
24 Apart from these differences in estimates, there are also certain differences in the definition of the debt itself. Some authors think that the CMEA countries' foreign currency reserves, the USSR's gold

(Cont'd on page 250)

Table 64

Hard Currency Indebtedness of CMEA Countries to the West, End 1976, by Type of Obligation

Billions US dollars

Type of Obligation	USSR	Bulgaria	CSSR	GDR	Hungary	Poland	Rumania	Total
IMF and IBRD credits	—	—	—	—	—	—	0.54	0.54
Official and officially guaranteed export credits outstanding[a]	6.59	0.42	0.48	1.10[b]	0.09	5.38	0.50	14.56
Supplier credits[c]	3.70	0.40	0.80	1.50	—	2.25	1.15	9.80
Private Western bank credits[d]	9.63[e]	2.20	0.97	3.99	3.40	6.07	0.82	27.07
Bonds outstanding	—	0.02	—	—	0.10	0.03	—	0.15
Gross indebtedness	19.92[e]	3.04	2.25	6.59	3.59	13.73	3.01	52.12
Deposits with Western banks[c]	4.10	0.39	0.39	0.68	0.99	0.71	0.33	7.58
Net indebtedness	15.82[e]	2.65	1.85	5.91	2.60	13.02	2.68	44.53

a) For the derivation of the figures and for additional official data. see Sources.
b) Including Intra-German trade swing credits.
c) Including outstanding obligations in the à forfait market.
d) Banks in Group of Ten Countries and Switzerland and branches of US banks in the Caribbean area and the Far East. BIS data. adjusted for residual.
e) Excluding CMEA banks (IBEC. IIB). estimated at $ 1.9 billion net (Chase Estimate).

Note: With the exception of data on bank credits and deposits which are from BIS. all other data are from Chase Manhattan Bank. Apparently there is some double counting among bank credits and officially guaranteed export credits (buyer credits) as well as supplier credits (a major proportion of non-recourse paper is taken up by banks). Data are estimates.

Sources: *East-West Markets,* March 7. 1977. pp. 7 ff. and March 21. 1977. pp. 6 ff.: Bank for International Settlements. Quarterly Euro-currency tables.

importance of the debt. Namely, to compare it with national income or to relate it to the value of total exports.

The breakdown of the debt by type of obligation is not available for every year. Estimates for 1976 are shown in Table 64. In that year, 52 per cent of the gross debt consisted of private Western bank credit, 28 per cent was official or officially guaranteed and 19 per cent was in the form of supplier credits. Estimates for the end of 1977 are shown in Table 65. Here the breakdown by the various types of loans and credits can be calculated only in relation to the total debt. It is 47.8 per cent for loans granted by Western Banks, 42.3 per cent for government guaranteed loans and only 8.2 per cent for supplier credit.[25]

Table 65

Estimated Composition of Net Hard-Currency Debt of Eastern Europe, USSR and CMEA Banks, December 31, 1977

Millions US dollars

	Drawings on Official Credits	Supplier Credits[a]	Net Liabilities to Western Banks[b]	Outstanding Bonds and Notes	IMF and IBRD Drawings	Total
Bulgaria	798	100	2 065	0	0	2 963
Czechoslovakia	841	200	884	0	0	1 925
German Democratic Republic	2 455	400	3 729[c]	0	0	6 584
Hungary	460	0	3 630	180	0	4 270
Poland	5 775	1 200	6 890	82	0	13 947
Rumania	1 256	200	1 073	0	670	3 199
Total Eastern Europe	11 585	2 100	18 271	262	670	32 888
USSR	10 730	2 200	3 411	0	0	16 341
CMEA banks	0	0	3 500	0	0	3 500
Total	22 315	4 300	25 182	262	670	52 729

a) Including outstanding à forfait obligations.
b) Banks in Group of Ten Countries, Switzerland, and foreign branches of US banks in the Caribbean and Far East.
c) Excluding net liabilities of the German Democratic Republic to banks in Federal Republic of Germany.

Source: East-West Markets, May 15, 1978, pp. 3 and 10, as quoted by Morris Bornstein, Research Conference on "East-West Relations in the Eighties", Bellagio, June 18-23, 1979, p. 64.

We have no details of the breakdown by Western creditor country. As for bank credits, the United Kingdom's share was particularly high (30 per cent of the total) followed by German banks (about DM.20 billion at the end of 1975)[26] and French banks. The share of the United States' banks amounted to only 13 per cent.[27]

(Cont'd note 24)
stocks and the loans granted by the CMEA countries to the developing countries should all be deducted.

The OECD has estimated the Soviet Union's gold reserves at $5 billion (Financial Market Trends, No. 14, December 1976). According to Soviet Economic Problems and Prospects, op. cit., p. 23, the reserves were 1,870 tons in 1976. According to E. Merx of the Dresdner Bank, the USSR's gold reserves were estimated at $10-12 billion.

The developing countries' debt to the planned economy countries (including a small part of the loans granted by China) was $9.0 billion in 1976 and $11.5 billion in 1977. Servicing this debt cost $0.9 billion in 1976 and $1.1 billion in 1977, Economic Bulletin for Europe, Vol. 30, No. 1. Pre-publication, Table 2.11.

25   For more detail on the debt and on credit mechanisms see Françoise Lemoine, "L'endettement des pays de l'Est en devises convertibles", Le courrier des pays de l'Est, La Documentation française, Centre d'études prospectives et d'informations internationales, Paris, October 1978, pp. 3-29; Florence Pertuiset, "La structure et les mécanismes des crédits occidentaux dans les échanges Est-Ouest", Le courrier des pays de l'Est, La Documentation française, Centre d'études prospectives et d'informations internationales, Paris, January 1979, pp. 3-32.

26   Joachim Pfeffer, "Zunehmende Verschuldung Osteuropas gegenüber dem Western", Neue Zürcher Zeitung, 16th August, 1977.

27   Neue Zürcher Zeitung, 21st June, 1978, p. 11. These figures probably relate to 1977.

b) Technology Transfer as the Main Reason for the Debt

In any analysis of the indebtedness of the CMEA countries it is essential also to study their balance of payments figures. The East European countries rarely publish information about this but it is possible to reconstitute the figures, at least in part, from data given in a study by the United Nations Economic Commission for Europe. The result is shown in Table 66. The total trade balance deficit for all the CMEA countries for the period 1965-1976 was over $30 billion.[28] It increased greatly again in 1977 and 1978.

The first conclusion to be drawn from Table 66 is that the deficit in the CMEA countries' balance of payments is mainly due to the deficit in their trade balances. The trade balance division by product group in Table 67 shows that machinery and transport equipment are largely responsible for this deficit. The deficit under this heading was higher than the CMEA countries' aggregate trade deficit in 1974 and 1977 and roughly equal to it in 1975 and 1976.

The connection between trade balance deficit and technology transfer, moreover, is not simply a formal or accounting phenomenon.[29] It stems from the mechanism by which Western machinery is sold to the CMEA countries. Western firms desiring to do business with countries suffering from a chronic strong currency deficit are forced to extend credit. One has only to examine the content of contracts for loans granted to CMEA countries to see the real link between the granting of credit and technology transfers to the East. Examples shown in Table 68 show this clearly and the link has been referred to by a Western expert in these terms: "If, in spite of everything, the risks of extending credit to the East do not give cause for concern the reasons are firstly the good reputation that the East European countries have for paying their bills and the utilisation of Western credit not for consumer purposes but for the financing of imports of Western capital goods and the related technology transfers.[30]

The need to give credit for sales to East European countries is not disputed in the West.[31] An agreement has recently been reached about the terms on which credit should be granted.[32] Nevertheless, there is some competition among Western countries for contracts with the East and in practice this has resulted in the grant of more favourable credit terms.[33]

---

28   Including the deficit of the German Democratic Republic in its trade balance with Germany and those of Japan and Australia in their trade balance with the seven CMEA countries — not included in the total shown in Table 66.

29   For example, the $31.3 billion increase in the net deficit of the CMEA countries between 1970 and 1976 (see Table 63) is only slightly less than the total figure for those countries' machinery imports. For the period 1971-1976, this amounted to $35.8 billion for imports from OECD countries (see Table A-6 in the Annex).

30   Joachim Pfeffer, *art. cit.*; See also Albert Masnata, "Les échanges commerciaux avec l'Est. Dans quelle mesure le financement disproportionné des échanges Est-Ouest par l'Ouest contribue-t-il à leur expansion?" *NATO Review,* August, 1977, pp. 25 and 27.

31   The same point is made by Albert Masnata, *art. cit.*

32   The first payment in an export credit deal has to be 15 per cent of the value of the contract. The maximum term is 5-10 years (depending on the level of development of the importing country) and the minimum rate of interest is from 7.25 to 8 per cent depending on the term of the loan and the classification of the country concerned. *Economic Bulletin for Europe,* Vol. 29, *op. cit.,* p. 125. For recent policy (since the beginning of 1980) after the invasion of Afghanistan by the Soviet Union, see Chapter 1, note 254.

33   At the Göttingen Conference in June 1977, Mr. Watanuki, Director of the Japanese Trade Centre in Vienna, claimed that, according to the information he had, Japanese firms lost 8 major contracts to a total value of $1.3 billion in 1976 because European competitors did not apply the minimum interest rates. Joachim Pfeffer, *art. cit.* British and American complaints about French intentions as regards applying the minimum interest rates were recorded in the *Sunday Times,* London, 27th November, 1977. Judging by the available data, it would seem that the Western countries have indeed not rigidly adhered to the 5-year limit set by the Berne Union for loans with government support. Loans to the East exceeding the five-year ceiling totalled $1.7 billion in 1973 and $3.2 billion in 1974 and 1975. See also Chapter 1.

Table 66

Balance of Payments of Industrial Market Economies with the Seven[a]

Millions US dollars

	1965	1966	1967	1968	1969	1970	1971	1972	1973	1974	1975	1976	1965-1976	1977	1978
Trade balance (f.o.b./f.o.b.)	242	322	488	549	620	890	581	1 662	2 836	3 209	8 255	5 951	25 605	3 652	4 937
Transport, insurance	11	4	−44	−7	−12	1	16	31	17	123	25	20	185	−70	−80
Travel	−52	−66	−71	−72	−74	−114	−152	−212	−292	−332	−430	−480	−2 347	−470	−520
Investment income	79	82	107	121	153	192	292	287	506	762	1 295	1 600	5 476	2 100	2 800
Transfers	−62	−62	−67	−112	−112	−138	−210	−303	−349	−374	−352	−400	−2 541	−805	−820
Non-trade balance (including transfers)	−24	−42	−75	−70	−45	−59	−54	−197	−118	179	538	740	1 220	..	..
Balance on current account (including transfers)	218	280	413	479	575	831	527	1 465	2 718	3 388	8 793	6 691	26 378	4.407	6 317
Identified capital flows	−272	−11	−334	−147	−183	−623	−196	−1 095	−3 243	−2 409	−5 715				
Multilateral settlements												6 691			
and net errors and omissions	54	−269	−79	−332	−392	−208	−331	−370	525	−979	−3 078			..	..
Memorandum Items:															
Trade balance															
West-East Germany	−14	70	55	−4	157	114	52	170	128	162	236	156	1 282	..	..
Japan-Seven	−43	−50	−291	−297	−190	−101	1	95	−330	101	927	1 499	1 322	..	..
Australia-Seven	96	51	47	58	58	86	95	149	339	283	421	583	1 645	..	..

a) Industrial market economies include South Africa and exclude Australia and New Zealand. Japan is excluded due to absence of data at ECE. Transactions between West and East Germany are excluded throughout the period.

Sources: Economic Bulletin for Europe, ECE. Geneva. Vol. 28, 1976, for the period 1965-74; Vol. 29, 1977, for the period 1975-76; and Vol. 31, No. 1. Pre-publication. Table 3.11.

Table 67

The Trade Balance of Eastern Europe and the Soviet Union with the Industrial Market Economies (f.o.b./f.o.b.)

Millions US dollars

	Total				of which: Soviet Union			
	1974	1975	1976[a]	1977[a]	1974	1975	1976[a]	1977[a]
Food and live animals	—49	—1 279	—2 189	—1 001	—425	—1 589	—2 034	—1 345
Beverage and tobacco	—32	—19	—16	—26	—28	—39	—31	—35
Crude materials, inedible, except fuels	1 824	1 565	1 698	1 814	1 791	1 596	1 665	1 837
Mineral fuels, lubricants and related materials	4 503	5 646	7 116	7 500	3 245	4 063	5 390	5 929
Animal and vegetable oils and fats	170	161	49	53	135	109	49	22
Chemicals	—1 834	—2 021	—1 898	—2 022	—483	—679	—540	—542
Manufactured goods, classified chiefly by material	—3 643	—5 385	—4 590	—3 963	—1 749	—3 090	—2 865	—2 546
Machinery and transport equipment	—5 012	—8 038	—7 950	—8 585	—2 128	—4 258	—4 524	—5 038
Miscellaneous manufactured articles	354	283	322	473	—263	—415	—494	—527
Commodities and transactions not classified according to kind	—60	—60	—60	—7	—31	—31	—31	—5
Total	—3 779	—9 148	—7 518	—5 764	65	—4 332	—3 414	—2 250

a) Inconsistencies with other tables are due to rounding differences in coverage and differences in reporting.

Source: Economic Bulletin for Europe. Vol. 30. No. 1. Pre-publication. Table 3.12.

Table 68

Some Credits Accorded to the CMEA Countries and Guaranteed by the Governments of the Creditor Countries in 1975 and 1976

Millions of US dollars

Year and Western nation	Credit available	Interest rate (%)	Length of repayment (years)	Down payment (%)	Description
Bulgaria					
1975 Austria	122	6.5	5	n.a.	Austrian equipment
Sweden	60	n.a.	5	25	2 hotels
1976 Japan	280	7.5	2-8 ½	15-20	Japanese plants, machinery, and equipment
German Democratic Republic					
1975 France	136	n.a.	8	15	Bogie wagons and bogies
1976 Denmark	38	n.a.	8	n.a.	Steel mill equipment
Federal Republic of Germany	340	n.a.	8	n.a.	3 chemical plants. The credit is being raised by a consortium of Federal Republic of Germany banks
Hungary					
1976 Japan	250	n.a.	n.a.	n.a.	Japanese goods
Poland					
1975 Austria	230	7.5	5	10	Heavy-duty trucks
Austria	250	n.a.	5	20	Steel
Belgium	335	n.a.	7	15	Coal mining equipment; consumer goods
Canada	500	7.75-8	8	10	Kwidzyn pulp and paper mill
France	1 700	7.5	8	15	Police fertilizer plant; capital equipment
Italy	300	n.a.	5	15	Italian plant and equipment
Italy	200	n.a.	5	15	Semifinished goods
Japan	180	6.5	8	20	Machinery and equipment
Federal Republic of Germany	425	2.5	25	..	Financial credit
1976 Japan	450	7.5	8	20	Industrial plant and equipment
United Kingdom	310	7.5	8	15	PVC complex
United States	188	8-9	3	..	CCC credits
Federal Republic of Germany	124	..	10	n.a.	Expansion and modernization of Poland's copper industry. Repayment in copper. The funds are to come from a consortium of Federal Republic of Germany banks; the credit carries a 7 percent plus floating interest rate
Rumania					
1976 Austria	110	6.75-7.75	up to 10	10	Austrian goods, including Steyr trucks and transportation equipment; 25 percent of each loan may be allocated to non-Austrian goods and 15 percent to local costs in Rumania
Japan	200	n.a.	n.a.	n.a.	Supplier credit line
Japan	80	7.5	8 ½	n.a.	Modernization of port of Constanza, 2 ½-yr grace period
Italy	240	7.5	8	15	Italian goods

Source: Joan Parpart Zoeter, "Eastern Europe: The Growing Hard Currency Debt", in East European Economies Post-Helsinki, op. cit., p. 1368.

Western technology is also contributing indirectly to increase the indebtedness of the CMEA countries through the payment of "investment income" (see Table 66). Recent increases in loans by East European countries from the West have made the burden of interest payments very much heavier. In addition, purchases of licences and technical assistance by CMEA countries have increased and outstripped by far the value of similar services supplied to Western countries.[34]

c) The Problem of the Eastern Countries' Credit-worthiness

Future propspects for transfers of Western technology and capital goods will largely depend on the credit-worthiness of the CMEA countries. The Eastern countries' ability to continue to obtain large credits is uncertain. On the one hand there has been a steep increase in the CMEA countries' debt to the industrialised countries of the West. On the other hand, however, there are some experts who consider the debt to be "acceptable" — at least for the present.

The debt situation of the CMEA countries is given in Table 63. Western analyses — or predictions — concerning future possible debt positions of the Eastern countries vary. For the most part, the analyses which are made or suggested are based on postulates made as to the alternatives believed to be available to Eastern policy makers.

One of the best studies — by F. Levcik and J. Stankovsky[35] — postulates three possibilities: continuation of present trends, reversal of the present trend (lower rate of increase in imports and higher rate of increase in exports) and lastly, maintenance of the trade balance deficit at its present level. The projections were made to 1980.

The main point arising from these calculations[36] is that the indebtedness of the East European countries to the West will increase considerably in all the alternatives considered. In the first alternative (continuation of present trends) the debt would increase from $41 billion in 1976 to $96 billion in 1980, or — allowing for debt servicing — $110 billion. If the trade balance deficit were held at its present level this would reduce the projected figure for 1980 to $81 billion (or $90 billion including debt servicing). Even if the most restrictive policy were followed and the growth rate in imports were reduced (at the price of a slower rate of growth in national income) this would do no more than reduce the projected deficit for 1980 to $73 billion ($80 billion including servicing). Even in this "most favourable" case, the debt of the East European countries would increase at a rate of 10 per cent (compared with 28 per cent in 1976). The corresponding figures for the Soviet Union would be 9 per cent and 42 per cent.[37]

Detailed calculations on CMEA countries' prospects have been produced more recently by Allen J. Lenz of the Bureau of East-West Trade of the US Department of Commerce. The results for 1980 and 1985 are shown in Tables 69 and 70. Lenz takes several possible import growth rates ranging from −2.5 per cent to +10 per cent per year. According to his calculations the debt will be somewhere between $62 and 65 billion in 1980 and between $85 and 108 billion in 1985.

---

34   For Czechoslovakia, the deficit in payments for the grant of licences totalled $278 million over the period 1968-1975. For Hungary the figure was $267 million for the period 1968-1976. *Economic Bulletin for Europe, United Nations*, Vol. 29, *op. cit.*, p. 125.

35   Friedrich Levcik and Jan Stankovsky, "Kredite des Westens und Österreichs an Osteuropa und die UdSSR". *Monatsbericht*, Österreichisches Institut für Wirtschaftsforschung, No. 5, 1977, offprint, not paginated.

36   Levcik and Standkovsky's estimates are based on the elasticity of imports in relation to national income (1.27 for East Europe as a whole and 1.35 for the USSR) and on the planned increase in national income. (For exports the factors are 1.60 for East Europe and 1.26 for the USSR.) The figures are based on current prices assuming an average inflation rate of 3 per cent a year between 1976 and 1980. F. Levcik and J. Stankovsky, *art. cit.*

37   *Ibid.*

Table 69

**Merchandise Export Annual Growth Rates Required to Achieve Stability in Level
of Eastern Debt by End 1980, under Various Import Growth Rates**
Millions of dollars

	Annual required growth rate (per cent)	1980 imports	1980 exports	1980 trade balance	Current account balance excluding interest	End 1980 debt	1980 interest
				−2.5 per cent import growth			
USSR	+0.4	13 903	11 133	−2 770	+980	18 079	1 330
Poland	+14.9	5 978	5 688	−290	+300	18 059	1 282
Other Eastern European	+8.9	11 484	11 455	−29	+596	25 765	1 839
Total		31 365	28 276	−3 089	+1 886	61 903	4 451
				0 per cent import growth			
USSR	+3.5	15 000	12 196	−2 804	+946	18 180	1 334
Poland	+17.8	6 450	6 083	−367	+233	18 309	1 294
Other Eastern European	+11.5	12 390	12 296	−94	+531	26 033	1 853
Total		33 840	30 575	−3 265	+1 710	62 522	4 481
				5 per cent import growth			
USSR	+9.6	17 364	14 482	−2 882	+868	18 487	1 350
Poland	+23.2	7 467	7 012	−455	+145	18 680	1 313
Other Eastern European	+16.7	14 343	14 097	−246	+379	26 602	1 882
Total		39 174	35 591	−3 583	+1 392	63 769	4 545
				10 per cent import growth			
USSR	+15.6	19 965	16 993	−2 972	+778	18 816	1 367
Poland	+28.6	8 585	7 994	−591	+9	19 138	1 336
Other Eastern European	+21.8	16 491	16 027	−464	+161	27 284	1 915
Total		45 041	41 014	−4 027	+948	65 238	4 618

**1980 Eastern Debt and Interest Payments Assuming 1978-80 Imports at 1977 Levels, Exports
Grow 10 Percent Annually**
Millions of dollars

	1980 imports	1980 exports	1980 balance of trade	End 1980 debt	1980 interest only cost
USSR	15 000	14 641	−359	13 289	1 164
Poland	6 450	4 991	−1 459	20 414	1 364
Other Eastern European	12 340	11 806	−584	26 992	1 886
Total	33 840	31 438	−2 402	60 695	4 414

Source: Allen J. Lenz. "Potential 1980 and 1985 Hard Currency Debt of the USSR and Eastern Europe under Selected Hypotheses" in *Issues in East-West Commercial Relations. op. cit.*. p. 191.

What is the true significance of this debt? Related to GNP of the CMEA countries totalling about $1,000 billion, (in 1976) the aggregate debt of $40-50 billion amounted to only 4-5 per cent of GNP. For comparison, the debt of the developing countries was about $190 billion, while the corresponding GNP amounted to only about one third of that of the CMEA countries.[38] A more precise calculation on this subject is shown in Table 71. This

38   F. Levcik and J. Stankovsky, *art. cit.*

table shows that there is a difference between the USSR — where the debt in 1976 was equivalent to only 1.8 per cent of GNP — and Bulgaria, Hungary and Poland where the figure was somewhere between 10 and 11 per cent of GNP. These percentages are far higher in the case of the developing countries (except Yugoslavia — included in this group — and Venezuela).

Another approach is to compare the total debt with the value of exports to hard currency countries. Estimates by various authors are shown in Table 72. The debt/export ratio has been increasing since 1970 and was about 0.8-4.7 at the end of 1976. It was low for the USSR, Czechoslovakia and Rumania and particularly high for Bulgaria and Poland.

Table 70

Merchandise Export Annual Growth Rates Required to Achieve Stability in Level Eastern Debt by End 1985, under Various Import Growth Rates

Millions of dollars

	Annual required growth rate (per cent)	1985 imports	1985 exports	1985 trade balance	Current account balance excluding interest	End 1985 debt	1985 interest
			−2.5 per cent import growth				
USSR	−1.3	12 250	9 907	−2 344	+1 407	21 837	1 622
Poland	+6.4	5 267	6 160	+893	+1 493	26 620	1 961
Other Eastern European	+3.4	10 118	11 590	+1 472	+2 097	37 490	2 762
Total		27 635	27 657	+21	+4 997	85 947	6 345
			0 per cent import growth				
USSR	+1.8	15 000	12 687	−2 313	+1 437	22 277	1 654
Poland	+8.7	6 450	7 309	+859	+1 459	28 032	2 053
Other Eastern European	+5.8	12 390	13 925	+1 535	+2 160	38 914	2 868
Total		33 840	33 921	+81	+5 056	89 223	6 578
			5 per cent import growth				
USSR	+7.6	22 162	19 765	−2 397	+1 353	24 546	1 807
Poland	+13.5	9 530	10 328	+798	+1 398	30 969	2 258
Other Eastern European	+10.5	18 306	19 716	+1 410	+2 035	43 072	3 147
Total		49 998	49 809	−189	+4 786	98 587	7 212
			10 per cent import growth				
USSR	+13.3	32 154	29 870	−2 284	+1 466	26 781	1 971
Poland	+18.4	13 826	14 483	+657	+1 257	34 451	2 491
Other Eastern European	+15.4	26 559	27 898	+1 339	+1 964	47 240	3 433
Total		72 539	72 251	−288	+4 687	108 472	7 895

Source: Allen J. Lenz. "Potential 1980 and 1985 Hard Currency Debt of the USSR and Eastern Europe under Selected Hypotheses". in *Issues in East-West Commercial Relations, op. cit.*, p. 192.

Another indicator of the meaning of the debt to the hard currency countries is the ratio of debt servicing to merchandise exports to the West. Table 73 shows that, in this respect, the Soviet Union and Czechoslovakia fall into one category, the remaining CMEA countries into

257

another. In the case of the USSR and Czechoslovakia, the figure is approximately 25 per cent, and is regarded as satisfactory by the banks.[39]

The USSR, incidentally, is generally recognised as being in a favoured position. An attempt to estimate Soviet gold sales is made in Table A-44. The yield is put at about $1.4 billion jn 1976 and $1.7 billion in 1977. To this there should be added sales to the hard currency countries of diamonds — $0.5 billion in 1975 and $0.8 billion in 1976[40] — other precious metals — $0.6 billion in 1976[41] — and armaments — $1.5 billion in 1976.[42] However, if there is a reversal of the USSR's position by 1985 with respect to oil sales, this favourable position could be seriously impaired (see Chapter 4).

Table 71

Comparison of External Debt of CMEA Countries and Selected Western Countries
In millions of US dollars

Country	1976 GNP[a]	Net debt 1976[b]	1976 debt as a per cent of GNP	Percentage held by public authorities
CMEA countries:				
Bulgaria	20.9	2.3	11.0	100
Czechoslovakia	58.0	2.1	3.6	100
German Democratic Republic	66.2	4.9	7.4	100
Hungary	26.8	2.8	10.4	100
Poland	92.2	10.2	11.1	100
Rumania	52.6	3.3	6.3	100
Total EE	316.7	25.6	8.1	100
USSR	921.7	16.2	1.8	100
Cuba	10.6	1.3	12.3	100
Total CMEA	1 249.0	43.1	3.5	100
Other developing countries:				
Argentina	37.5	6.7	17.9	n.a.
Brazil	145.9	25.9	17.8	75
Colombia	20.1	2.6	12.9	n.a.
Mexico	79.0	26.0	32.9	75
South Korea	25.1	7.4	29.5	n.a.
Spain	102.3	10.7	10.5	64
Venezuela	31.0	2.6	8.4	n.a.
Yugoslavia	33.1	5.7	5.7	n.a.

a) Current dollars.
b) Estimated CIA.
n. a.: not available.

Source: Lawrence H. Theriot, "Communist Country Hard Currency Debt in Perspective", in *Issues in East-West Commercial Relations, op. cit.*, p. 183.

The other East European countries are distinctly less well-placed than the USSR and Czechoslovakia. It is clear from Table 73 that the ratio between debt servicing and exports to hard currency countries is well over the 25 per cent limit regarded as acceptable by Western extenders of credit. Although the long-term share (over 5 years) of these countries' debt is

39 Mr. Apponyi, Chase Manhattan Bank, quoted by Joachim Pfeffer, *art. cit.*; E. Merx, Dresdner Bank, quoted in *Handelsblatt* of 4th/5th June, 1977, p. 7. See also John T. Farrell, "Soviet Payments Problems in Trade with the West", in *Soviet Economic Prospects for the Seventies — A Compendium of Papers,* Joint Economic Committee, Congress of the United States, 93rd Congress, 1st Session, Washington, D.C., 1973, p. 692.
40 *Soviet Economic Problems and Prospects, op. cit.,* p. 22.
41 Chase Manhattan Bank estimates quoted by Werner Beitel, *Probleme der Entwicklung des Sowjetischen Westhandels, Kreditfinanzierung, Exportstruktur, Kooperationsabkommen,* Stiftung Wissenschaft und Politik, Ebenhausen, September 1977, p. 17.
42 *Soviet Economic Problems and Prospects, op. cit.,* p. 23.

relatively high (32-42 per cent except in the case of Czechoslovakia and the German Democratic Republic) debt servicing in absolute figures — and interest payments, in particular — reach considerable heights. This is the case especially for Poland. Her debt repayment and servicing should reach about $4.5 to 5 billion in 1980, a sum equivalent to the total export receipts from the industrialised West in 1979 (see Table 73).

Other criteria for assessing the significance of the debt have also been proposed. Charles J. Gmür of the Schweizerische Kreditanstalt suggests calculating the ratio between total debt to hard currency countries and the sum of total currency reserves, and available raw materials and other means of acquiring foreign currency.[43]

Table 72

Various Estimates of the Ratio of Debt to Hard Currency Exports, 1970-1977
End of year

		1970	1974	1975	1976	1977
USSR	(a)	0.8	. .	1.1	1.0	. .
	(b)	. .	0.3	1.1	. .	. .
	(c)	. .	. .	. .	1.4	1.5
Bulgaria	(a)	3.0	. .	3.1	3.3	. .
	(b)	. .	2.1	3.2	3.4	. .
	(c)	. .	. .	. .	4.7	. .
Czechoslovakia	(a)	0.4	. .	0.6	0.8	. .
	(b)	. .	0.4	0.7	1.0	. .
	(c)	. .	. .	. .	1.2	. .
German Democratic Republic	(a)	1.4	. .	1.8	2.0	. .
	(b)	. .	1.4	2.0	2.3	. .
	(c)	. .	. .	. .	1.7	. .
Hungary	(a)	1.1	. .	1.7	1.5	. .
	(b)	. .	1.3	1.8	2.0	. .
	(c)	. .	. .	. .	2.1	. .
Poland	(a)	0.9	. .	1.9	2.5	. .
	(b)	. .	1.4	2.1	2.7	. .
	(c)	. .	. .	. .	3.1	3.5
Rumania	(a)	2.2	. .	1.0	1.0	. .
	(b)	. .	1.0	1.3	1.2	. .
	(c)	. .	. .	. .	1.7	. .

Sources:
(a) Lawrence J. Brainard, "Eastern European Indebtedness", Conference on the Monetary and Financial Problems in East and West, Budapest, 1977.
(b) Chase Manhattan Bank, quoted by Richard Portes, "East Europe's Debt to the West: Interdependence is a Two-Way Street", *Foreign Affairs,* July, 1977, p. 761.
(c) Lawrence H. Theriot, "Communist Country Hard Currency Debt in Perspective", *art. cit.,* p. 184 (for 1976). Allen J. Lenz, "Potential 1980 and 1985 Hard Currency Debt of the USSR and Eastern Europe under Selected Hypotheses", *art. cit.,* p. 190 (for 1977).

How do the Western industrialised countries assess the creditworthiness of the CMEA countries? The overriding feeling would seem to be expressed in the conclusions of the Symposium organised by the Göttingen University Institut für Völkerrecht early in June 1977 on the financing and currency problems of East-West trade in the following words: "The creditworthiness and payment ethics of the USSR and, with certain restrictions, the

43   Quoted by Joachim Pfeffer, *art. cit.*

other CMEA Member countries is unquestionable. In spite of a substantial increase in indebtedness towards OECD countries... there is no anxiety at the banks as regards the absolute debt figure or its increase since 1971".[44]

Table 73
Various Estimates of CMEA Countries' Debt Servicing Ratios, 1970-1978[1, 2]

		1970	1973	1974	1975	1976	1977	1978 fore-casts
USSR	(a)	0.18	0.17	0.15	0.22	0.26	0.28	..
	(b)	..	..	..	0.22	0.26	..	..
	(d)	..	..	..	..	..	0.27	0.28
Bulgaria	(a)	0.35	0.35	0.45	0.66	0.75	0.85	..
	(b)	..	..	..	..	..	..	..
	(c)	0.35	..	0.45	0.66	0.75	..	..
	(d)	..	..	..	..	..	1.00	..
Czechoslovakia	(a)	0.08	0.15	0.17	0.22	0.30	0.34	..
	(b)	..	..	..	0.22	0.30	..	..
	(c)	0.08	..	0.17	0.22	0.30	..	..
	(d)	..	..	..	..	..	..	0.25
German Democratic Republic	(a)	0.20	0.25	0.24	0.27	0.33	0.40	..
	(b)	..	..	..	0.22	0.30	..	..
	(c)	0.20	..	0.24	0.27	0.25	..	..
	(d)	..	..	..	..	..	..	3
Hungary	(a)	0.20	0.20	0.24	0.35	0.39	0.44	..
	(b)	..	..	..	0.31	0.37	..	..
	(c)	0.20	..	0.24	0.35	0.40	..	..
	(d)	..	..	..	..	..	..	3
Poland	(a)	0.20	0.21	0.27	0.43	0.49	0.60	..
	(b)	..	..	..	0.43	0.47	..	..
	(c)	0.20	..	0.27	0.43	0.50	..	..
	(d)	..	..	..	..	..	0.45	0.50 (e)
Rumania	(a)	0.36	0.35	0.29	0.42	0.41	0.42	..
	(b)	..	..	..	0.45	0.46	..	..
	(c)	0.36	..	0.29	0.42	0.42	..	..
	(d)	..	..	..	..	..	..	3

1 Net debt to hard currency exports to the West.
2 Repayments of principal on medium and long-term debt and of interest on all debt (interest on medium and long-term debt only) as a fraction of merchandise exports to the West.
3 Between 35 and 45% for Hungary, Rumania and the German Democratic Republic.

Sources:
(a) William F. Kolarik, Jr., "Statistical Abstract of East-West Trade Finance", in *Issues in East-West Commercial Relations, op. cit.,* p. 201.
(b) Chase Manhattan Bank, cited by Richard Portes, "East Europe's Debt to the West: Interdependence is a Two-Way Street", *art. cit.,* p. 761.
(c) Joan Parpart Zoeter, *art. cit.,* p. 1367.
(d) *Neue Zürcher Zeitung,* of 21st June, 1978, p. 11.
(e) According to *Business Eastern Europe,* Vol. 9, No. 8, February 22, 1980, p. 57, Polish debt repayment and servicing will, by 1980, eat up some $4.5–5 billion of export revenues; this is equivalent of total 1979 sales to the industrialised West.

This optimistic verdict applies in particular to credit for the Soviet Union. As Werner Beitel notes, the present level of indebtedness is not currently regarded as alarming by Western lenders or by the Soviet Union itself.[45] According to Beitel, there will probably be a

44 Conference Notice, International Symposium on the Financing and Currency Problems of East-West Trade – Institut für Völkerrecht, Göttingen University, Department of International Economic Law, in conjunction with the Berliner Institut für Weiterbildung von Führungskraften der Wirtschaft, Göttingen, 1st-3rd June, 1977.
45 Werner Beitel, *op. cit.,* p. 32.

fresh increase in loans to the Soviet Union but the rate of future indebtedness is likely to fall.[46] In another paper written jointly with Jürgen Nötzold, he says that "the potential of the European loan markets and their loan-granting capacity is incomparably greater than it was in the period between the wars".[47] The Soviet Union is able to obtain a large share of its loans on a long-term basis. Interests both in the West and in the USSR weigh in favour of an increase in credit.[48] Other opinions tending the same way have been voiced by E. Merx of the Dresdner Bank.[49]

Table 74

Long-Term (over 5 years) Share of Debt and Cost of Debt Servicing in $ million, 1970-1975

1. Eastern Europe: Estimated Long-Term Debt[a]
In per cent of total net indebtedness

	1970	1973	1974	1975
Bulgaria	15	25	25	33
Czechoslovakia	15	15	13	13
German Democratic Republic	10	19	14	16
Hungary	19	37	37	42
Poland	62	42	36	32
Rumania	20	25	30	40

a) Consists of credits of 5 years and more.

2. Eastern Europe: Estimated Hard Currency Repayments and Interest, 1975[a]
In millions of US dollars

	Repayment of medium- and long-term liabilities	Interest	Total debt service
Bulgaria	140	100	240
Czechoslovakia	260	90	350
German Democratic Republic	500	210	710
Hungary	250	130	380
Poland	900	400	1 300
Rumania	500	200	700

a) Interest represents aggregate interest payments on all hard currency debt.

Source: Joan Parpart Zoeter. art. cit., pp. 1367-1368.

Opinions about the creditworthiness of CMEA countries other than the Soviet Union are far less definite and certain reservations have been reported in the press.[50] A major subject in the discussion on this issue is the "umbrella theory" according to which the Soviet Union would make itself responsible for any failure on the part of its CMEA partners. Opinions on this subject were divided at the Göttingen Symposium.[51]

46 *Ibid.*, p. 31.
47 Werner Beitel and Jürgen Nötzold, *Entwicklungstendenzen und Perspektiven der Wirtschaftsbeziehungen der Bundesrepublik Deutschland mit der Sowjetunion,* Stiftung Wissenschaft und Politik, Ebenhausen, March 1978, pp. 41-42.
48 *Ibid.*, p. 42.
49 «Positives Urteil über Sowjets als Schuldner. Aber Tempo der Neuverschuldung beunruhigt», *Handelsblatt,* 5th-6th June, 1977, p. 7.
50 Joachim Pfeffer, *art. cit., Wirtschaft,* 8th-9th January, 1978.
51 Joachim Pfeffer, *art. cit.* This theory, however, did not operate in favour of North Korea when it suspended payment of its debts.

Difficulties for CMEA countries other than the Soviet Union have already been forecasted by Western experts. The effect of these difficulties could be a reduction in imports, and reduce trade balance deficits in order to prevent further debt increases.[52] "Under the most favourable conditions — a strong recovery in exports and good harvests — most East European countries should be able to import the necessary industrial materials without economic or financial assistance. But they can do this only if they are willing to curb their imports of machinery and equipment. All of the East European countries have, in fact, indicated their intention of allowing little, if any, growth in (or even of cutting) imports of Western capital equipment".[53]

The first effects of these decisions were already apparent in 1977. According to OECD statistics, the trade balance deficit of the CMEA countries was cut from $7,791 million in 1976 to $5,901 million in 1977. The biggest reduction was achieved by the USSR (from $3,670 to $2,221 million) and Poland (from $2,165 million to $1,430 million). Hungary, and Rumania on the other hand, increased their deficits from $484 to $773 million and $171 to $553 million, respectively.

The East European countries make no secret of their desire to reduce the deficit in their trade balance. Hungarian experts envisage a slowdown in imports from the West[54] and Polish reactions appear to be similar. In his projections of East-West trade for 1980, A. Czepurko, a Polish economist, makes this most likely assumption: minimal expansion of CMEA countries' imports from the West and maximal expansion of their exports. This, he says, would restore equilibrium to the trade balance between the CMEA and Western countries.[55] Recently published Soviet figures also appear to envisage greater growth in trade inside the CMEA rather than with the Western countries for the period 1976-1980.[56] It may therefore be assumed that the East European countries are about to reduce the flow of imports from the West to some extent. Such action would have an immediate effect on the volume of technology imported. But the real problem lies in the effects that Western credit will have on production. If the latter leads to an increase in exports subsequent repayment will be facilitated.[57]

On the American side the stress is put on the need to increase imports from the East European countries. It is said that "until this is done, and without a steady growth in Western credit, exports from the West must inevitably decrease. It is unlikely that Western credit will be available indefinitely, but even so the West must, in the end, facilitate payments for its exports by being prepared to import goods and services from the East. This means that, at a given moment, if Western firms are to be paid — or even if their outstanding credits are merely to be kept at the present level — Western trade balance surpluses must be turned into deficits".[58]

More concrete fears about the ability of the CMEA countries to meet their debts have also been expressed. Richard Portes, for example, believes that the West should already be considering its response to "any East European request for debt relief which might follow the negotiation of such arrangements for the LDCs. Western central banks, bank regulatory authorities, and finance ministers should also explore various lines of approach to

52   Joan Parpart Zoeter, *art. cit.*, p. 1361.
53   *Ibid.*, p. 1362.
54   Report presented at the CESES Colloquium organised jointly with the Wroclaw College of Economic Science (Poland), July, 1976.
55   A. Tiraspolsky and B. Despiney, "Les échanges des pays européens du CAEM de 1971 à 1975 – Perspectives 1980", *Le courrier des pays de l'Est*, La Documentation française et Centre d'études prospectives et d'informations internationales, Paris, November 1976, p. 14.
56   *Ibid.*, p. 16.
57   M. Lavigne, "Une nouvelle phase des relations Est-Ouest: Espoirs et difficultés d'une intensification des échanges", *Le Monde diplomatique*, Paris, September 1976. Quoted in *Problèmes économiques*, La Documentation française, Paris, 3rd November, 1976, p. 29.
58   On this subject see: Allen J. Lenz, "Potential 1980 and 1985 Hard Currency Debt of the USSR and Eastern Europe Under Selected Hypotheses", *art. cit.*, p. 186.

rescheduling the debts of an East European country or coping with the consequences of a default". Portes also warns against the pressures being exerted in the West for the introduction of greater protective measures or trade discrimination against the East since these countries can only cope with their debts by selling more goods to the West. Finally, Portes suggests that there should be some general control of Euromarket lending and suggests that some way be found of "setting aggregate limits for East European countries which are related to their economic capabilities and governmental policies".[59] How could all these policies suggested by various authors be implemented? The CMEA countries seem to prefer the solution of pressing for barter deals in their contracts with the West. This is clear from their use of counter-purchase and compensation agreements.

4. COUNTERPURCHASE AND COMPENSATION AGREEMENTS AS A VEHICLE FOR TECHNOLOGY TRANSFER TO THE EAST

a) **Definition of Countertrade**

The term countertrade covers a range of transactions in which the amounts owing to a country for the goods it has sold are offset by the purchase of products from the importing country. These are therefore tied transactions for which the settlement of all or part of the hard currency debt is made in the form of products from the East.[60]

There are three main types of countertrade: barter agreements, counterpurchases and compensation agreements.

Barter agreements are individual, short-term (two years at maximum) transactions in which the Eastern goods to be purchased are specified at the time the contract is signed. There is no flow of money and, as a form of countertrade, the barter agreement is relatively rare. The sequence of events is generally as follows:[61]

— Western firm contracts with Eastern partner for the exchange of commodities;
— Western firm delivers commodities to Eastern partner;
— as payment for Western imports, Eastern partner delivers commodities to Western firm or other designated Western party;
— if Eastern commodities are delivered to a third party firm Western firm receives payment from that designated party.

Counterpurchase transactions are agreements in which a Western seller provides the Eastern buyer with technology, plant or equipment, and agrees to purchase Eastern goods equal to a percentage of the sales contract value. A counterpurchase transaction involves two separate, but inherently linked contracts — one for the sale of Western products, and a second for the purchase of the Eastern products. Credit is an integral part of counterpurchase; the Eastern partner pays for purchased merchandise with Western credit. The value of the Eastern goods offered as counterdeliveries is generally less than 100 per cent of the original sale contract value. The goods are normally "non-resultant products", i.e., they are not derived from or related to the Western export of technology, plant or equipment. The sequence of events is as follows:[62]

59 Richard Portes, "East Europe's Debt to the West: Interdependence is a Two-Way Street", *Foreign Affairs*, July 1977, pp. 781-782.

60 Jenelle Matheson, Paul McCarthy, Steven Flanders, "Countertrade Practices in Eastern Europe", in *East European Economies Post-Helsinki — A Compendium of Papers, op. cit.*, p. 1278.

61 *Ibid.*, pp. 1279-1280.

62 *Ibid.*, pp. 1279-1282.

- Western firm contracts for the sale of plant and equipment to Eastern partner;
- Eastern partner negotiates with Western bank for credit;
- Western bank extends credit to Eastern partner;
- Western firm delivers commodities to Eastern partner;
- Western bank makes payment to the Western firm (full or part depending on the terms of the agreement);
- Western firm contracts with Eastern partner for the purchase of Eastern commodities;
- Western firm pays Eastern partner for commodity;
- Eastern firm repays Western credit;
- If the Western firm cannot use the Eastern commodities it may negotiate with a trading house to sell them;
- Eastern firm delivers commodities to either the Western partner or to another designated Western party;
- Western firm receives payment from firm designated to handle the Eastern commodities.

Compensation agreements involve two separate but inherently linked contracts providing for the sale by a Western firm of technology, plant or equipment and the reciprocal purchase by the Western firm of Eastern goods. The values involved in compensation transactions are usually much higher than in barter or counterpurchase agreements and the compensation arrangement covers a far longer period (10-20 years). Usually, the Western partner purchases products derived from the technology, plant or equipment that was supplied. The sequence of events is as follows: [63]

- Western firm contracts to sell plant and equipment to an Eastern partner;
- Western firm contracts to purchase some of the plant output (resultant products) once production has begun;
- Eastern partner negotiates with Western bank for credits with which to purchase Western plant and equipment;
- Western bank extends purchase credits to Eastern partner;
- Western firm delivers plant and equipment to Eastern partner;
- Western bank pays Western firm for deliveries;
- When production has begun, Eastern partner delivers part of the output to the Western firm;
- Western firm pays Eastern firm for deliveries of product;
- Eastern partner repays Western bank credit.

It is clear from the above that countertrade is mainly a question of arrangement for paying for goods and services supplied by the West. In practically all cases this relates to supplies covered by the industrial co-operation agreements described in Chapter 3. In fact, for counterpurchase and compensation agreements, it is a question of ways of repaying credit obtained in connection with the supply of technology, plant and equipment. Counterpurchase and compensation agreements are not additional instruments to existing East-West industrial co-operation agreements but form *one of the possible solutions* in connection with these agreements.

In actual fact, with the increase in the East European countries' hard currency debt, Eastern partners are pressing for an ever-increasing proportion of settlements to be in the form of counterpurchase or compensation arrangements in industrial co-operation contracts. For the CMEA countries, "co-operation" is beginning to have the same meaning as "compensation" in which, of course, the Western partner has to find the credit which is then repayable by counterpurchase or in the form of products made from imported plant or

63 *Ibid.*, pp. 1281-1284.

Table 75

Identified Value of Compensated Agreements Concluded Between 1969-1979

Value of Western exports in millions of dollars

	1969	1970	1971	1972	1973	1974	1975	1976	1977	Total 1969-1977	Total 1978-1979[a]	Total 1969-1979	Total Number of Agreements
Poland	—	100.0	—	71.9	86.7	755.0	1 449.4	1 438.0	128.1	4 029.1	1 330.0	5 359.1	42
German Democratic Republic	—	—	—	—	—	178.0	—	1 540.3	1 320.0	3 038.3	—	3 038.3	17
Hungary	—	—	—	—	—	6.0	4.0	86.2	148.5	241.7	3.5	245.2	43
Rumania	—	—	—	0.4	4.1	—	56.0	1.6	170.0	232.1	18.0	250.1	10
Bulgaria	—	—	—	120.0	—	18.0	50.0	—	—	188.0	—	188.0	6
Czechoslovakia	—	—	—	—	—	—	3.0	—	41.0	44.0	—	46.0	4
Total, 6 countries	—	100.0	—	192.3	90.8	957.0	1 559.4	3 066.1	1 867.6	7 773.2	1 351.5	9 124.7	122
USSR	263.0	500.0	235.0	813.0	767.0	8 773.0	4 213.3	6 608.2	594.7	22 547.2	161.0	22 928.2	80
Total, 7 countries	263.0	600.0	235.0	1 005.3	857.8	9 730.0	5 897.7	9 674.3	2 462.3	30 320.4	1 512.5	32 052.9	202
Number of Agreements Quantified													
6 countries	—	1	—	2	4	8	11	14	16	56	3	59	122
USSR	2	1	2	3	4	15	14	17	6	64	2	66	80
Total	2	2	2	5	8	23	25	31	22	120	5	125	202

(—) No agreement recorded.

a) The figures for these years are incomplete and cannot be considered as giving an indication of the total number of agreements concluded or recorded.

Source: Calculated on the base of Tables A-33. A-34. A-35. A-36. A-37. A-38. A-39 of the Annex.

equipment.[64] In this vocabulary, the term "co-operation" is solely reserved for cases where Western credits are granted not in the form of currency but as products, often giving rise to confusion if used outside narrow specialist circles.

b) The Problem of the Importance of Countertrade

The international trade statistics that are generally available do not include any assessments of countertrade in East-West economic relations. No reliable estimates of counterpurchase transactions have been found and for compensation agreements the only sources are various representative lists published on various occasions [see Annex, Tables A-33/A-39].[65]

The estimates given in Table 75 are based on lists drawn up by the Office of East-West Policy Planning, US Department of Commerce, together with some additional information. They relate to 202 signed agreements and 16 under negotiation. However, the sample is not complete. The total sum of agreements for which the value is known — about $30 billion for 1969-1977 — must be regarded as only approximate and probably considerably under estimated (by half?).

From Table 75, it can be seen that the Soviet Union leads in compensation agreements concluded since 1969. It is followed by Poland with about $5 billion worth of such agreements over the period 1969-1979. For the other CMEA countries compensation agreements play only a limited role, hardly existing in Czechoslovakia and Bulgaria. Hungary introduced them, practically speaking, only in 1976.

The leading Western countries in terms of compensation agreements (see Table 76) are Germany, France, the United States and Italy. If compensation agreements under negotiation (with the USSR in most known cases) are included, the United States is still in the lead followed by Germany (see Table 77). Information about these negotiations, however, is sketchy and the data given in Table 77 should be treated with great caution.

The share that Eastern countries' exports under compensation agreements account for in their total exports to the Western industrialised countries is difficult to quantify and, in addition, the prices on which estimates have been made have certainly escalated (particularly in the case of fuel). This has particular implications for estimates for supplies to be made over a longish period (five, ten or twenty years). The date of the agreement (pre or post-1973) is also significant.

According to the estimates of the United Nations Economic Commission for Europe, compensation agreements account for $1.5 billion of annual exports from West to East. Flows in the reverse direction are still small because the investment projects envisaged in the agreements have not yet been completed.[66] Counterpurchases, on the other hand, are thought to form a much larger share of this trade, estimates ranging from 20 to 38 per cent during recent years.[67]

As pointed out by F. L. Altmann and Hermann Clement,[68] the percentages of East-West trade quoted (in the specialised press and various publications) for compensation and counterpurchase arrangements are often inexact and there is a tendency to apply the figures for one particular contract or country to East-West trade as a whole. They therefore call for close scrutiny and great caution, a distinction being made between compensation deliveries

---

64   *Ibid.*, p. 1279.
65   A tentative list of compensation agreements among CMEA countries other than the USSR (Poland, Hungary, Rumania, the German Democratic Republic, Czechoslovakia and Bulgaria) is given by Jenelle Matheson, *et al., art. cit.*, pp. 1304-1311.
66   *Economic Survey of Europe in 1978,* Part I; *The European Economy in 1978* (Pre-publication), 10th March, 1979, p. 129.
67   *Ibid.*
68   F. L. Altmann and Hermann Clement, *op. cit.*

Table 76

Distribution of Compensation Agreements between Western Countries[a]

(Only agreements for which the value is indicated are shown)

Western Countries	With USSR		With Six Other CMEA Countries[b]	
	Number of Agreements[c]	Value of Western Exports[d] ($ million)	Number of Agreements[c]	Value of Western Exports[d] ($ million)
Austria	1 1/3	400	4	549.3
Belgium	1	10.8	1	335
Denmark	1	1.5	3 1/2	192
Finland	1	322	—	—
France	15 + 2/3	6 082.1	8 1/2	954
Germany	23 + 1/2 + 2/3	4 355.0	12	2 776.3
Italy	12	4 668.3	6 1/2	597
Japan	9	1 528.5	6	940
Netherlands	—	—	1	40
Sweden	—	—	3	357
Switzerland	—	—	—	—
United Kingdom	2 1/2 + 1/3	261	4	838
USA	3 1/2	5 299	14	1 555.2
Total	72	22 928.2	65	9 124.7

a) Only agreements whose value is known have been included in the table. For agreements concluded jointly by firms in several countries, an appropriate fraction has been attributed to each country concerned.
b) Poland, Hungary, German Democratic Republic, Rumania, Czechoslovakia, Bulgaria.
c) See a. This figure includes general agreements not yet implemented.
d) Some of the figures shown in Table 75 include the value of the overall agreement of which the compensation agreement is only a part. The absence of data on a number of major agreements that have been signed, however, more than makes up for this overestimate. Where high-low figures have been given, the average has been taken.

Source: Calculated from Tables A-33, A-34, A-35, A-36, A-37, A-38, A-39 of the Annex.

Table 77

Distribution of Compensation Agreements still under Negotiation between Western Countries[a]

(In terms of Western exports)

	Number of identified agreements	Number showing value of Western exports	Value of Western exports ($ million)	Average value of agreements ($ million)
USA	6	4	4 483.0 to 4 983.0	1 120.8 to 1 245.8
Germany	2	2	1 933.0 to 2 933.0	966.5 to 1 466.5
Japan	5	4	1 016.0 to 1 516.0	254.0 to 379.0
Italy	1	1	934.0	934.0
United Kingdom	3	1	120.0 to 220.0	120.0 to 220.0
France	2	1	300.0	300.0
Denmark	1	1	32.0	32.0
Total	20[b]	13[b]	8 818.0 to 10 866.0	678.3 to 835.8

a) Apart from the Danish agreement with Hungary, all the agreements shown in this table are being negotiated with USSR.
b) See Note a) to Table 76.

Source: Office of East-West Policy and Planning, Bureau of East-West Trade, US Department of Commerce, June 8, 1977. Calculated on the basis of Tables A-33 and A-35 of the Annex.

Table 78

Some Global Estimates of Compensation and Countertrade Deliveries by the East

Year	Share of Total Exports to the West	Comments
About 1976[1]	15-20 per cent of total exports of Western industrialised countries to the East.	"Payment in local products"
1976/1980[1]	30-40 per cent of total exports of Western industrialised countries to the East.	"Payment in local products"
1975[2]	$7 billion or 28 per cent of total trade between Eastern Europe and the industrialised West (included 20 per cent of US exports).	Estimate by US Advisory Committee on East/West Trade "Compensation Trade Arrangements"
1976[2]	30 per cent ($750 million) of the $2.5 billion of US/USSR trade turnover.	
About 1977[3]	25-30 per cent of exports from East to West.	
1976/1980[4]	38 per cent of trade, on average, between USA and USSR.	
About 1980 (forecast)[5]	38 per cent of bilateral US/USSR trade.	
About 1977[6]	20-30 per cent of Soviet trade with the West accounted for by long-term contracts based on compensation.	
About 1977[7]	25-30 per cent of East European trade with the West.	
1977[8]	$18 billion worth of deliveries by the East to the West in compensation agreements.	
1977[9]	15 per cent countertrade. A small fraction by switch arrangements, less than 10 per cent compensation.	
About 1980 (forecast)[9]	30-40 per cent of supplies from the West will be settled by compensation arrangements.	
Early 1979[10]	20-38 per cent paid for by countertrade exports from East to West.	
Early 1979[11]	12 per cent obtained in the form of counterpurchase in trade between Germany and the Soviet Union.	
1978[11]	20 per cent of Soviet trade with the West accounted for by deliveries made under compensation agreements for machinery and equipment.	
About 1978[12]	Countertrade (Gegengeschäfte) accounts for 40 per cent of East/West trade.	

Sources:
1 «Current Countertrade Policies and Practices in East-West Trade», *Business International*, November 1976, p. 4, quoted by F. L. Altmann and Hermann Clement, *Die Kompensation als Instrument im Ost-West-Handel*, Verlag G. Olzog, Munich, 1979.
2 *Ibid.*
3 J. Stankovsky, *Die Kompensationen im Ost-West-Handel*, Wiener Institut für Internationale Wirtschaftsvergleiche, March, 1978, p. 12.
4 V. Suchkov, quoted by F. L. Altmann and Hermann Clement, *op. cit.*
5 *Neue Zürcher Zeitung*, 21st June, 1978, p. 11.
6 Borissov, quoted by F. L. Altmann and Hermann Clement, *op. cit.*
7 *Economic Bulletin for Europe*, Vol. 29, p. 76.
8 *Business Week*, quoted by F. L. Altmann and Hermann Clement, *op. cit.*
9 *Doing Business with Eastern Europe*, February, 1977, p. V-75, quoted by F. L. Altmann and Hermann Clement, *op. cit.*
10 "ECE Reports Mixed Outlook for East-West Trade", *Business Eastern Europe*, March 23, 1979, p. 91.
11 "The Outlook for 1979: Where the Opportunities Lie", *Business Eastern Europe*, January 5, 1979, p. 2.
12 Stanovnik, Secretary-General of the Economic Commission for Europe, Report to the Vienna Conference (Austria) held 5th-7th March, 1979, quoted in "Ungelöste Wirtschaftsprobleme mit den sozialistischen Ländern", *Neue Zürcher Zeitung*, 16th March, 1979 (see also *Süddeutsche Zeitung* of 9th March, 1979, quoted by F. L. Altmann and Hermann Clement, *op. cit.*).

(normally scheduled for a fairly long time in the future) and counterpurchases which take place within a short time.

We have tabulated these various estimates in Table 78. In spite of their imperfections, a number of conclusions can be drawn: the share of counterpurchases is distinctly higher than that of goods supplied by way of compensation (or under compensation agreements) and the scale of these transactions in East-West trade is tending to increase over the years 1976-1980.

A point worth noting is that Eastern countries' imports under compensation agreements consist mainly of machinery equipment and turnkey plant (see Annex, Tables A-33/A-39), but they also include some highly sophisticated materials, particularly in the chemical industry.

Deliveries from Eastern countries under compensation agreements may be divided into two main groups. The first of these consists of fuels (oil and gas), timber and other basic commodities that are easy to sell in the West. These commodities are usually supplied under long-term high-value co-operation (compensation) agreements and most of them are produced with the help of Western technology. They account for a very large share of the value of countertrade deliveries. It can be seen from Table A-45 that these commodities (oil and gas, forestry products, coal and metals) will, over the next 15 years, account for some $20 billion out of the $31 billion of compensation trade identified with the USSR. For the other CMEA countries pride of place goes to compensation goods to be supplied by Poland (see Table A-34). The coal deliveries that it has been possible to quantify amount to over $1 billion and a half (certainly a very low underestimate), those of copper $1 billion and those of sulphur $670 000 000. It may be noted that deliveries by way of compensation to be made by Hungary, Rumania, Bulgaria and Czechoslovakia, (where it has been possible to quantify them) are for all intents and purposes insignificant (see Tables A-35/A-39).

The other category of supplies under counterpurchase or compensation agreements consists mainly of "soft" products that are often difficult to dispose of. It is the extension of compensation goods to these products that raises the crucial problem stressed by Mr. Brezhnev at the 25th Party Congress on 25th February, 1976: "at the moment (compensation) agreements largely relate to industries producing raw materials and intermediates but the time has now come to extend the scope of such agreements to cover the manufacturing industries and to look for new methods of co-operation in the field of production".[69]

c) **Countertrade and Compensation Agreements as a Way of Stimulating Technology Transfer to the East**

Western literature dwells on both the advantages and the disadvantages of countertrade as a way of stimulating East-West trade. For the East European countries such trade helps to improve balances of payments, penetrate Western markets, up-date technology and create Western interest in improving Eastern products.[70] Another benefit is ease of integration with Eastern countries' foreign trade planning. A disadvantage is the reluctance of Western firms to accept the inclusion of countertrade clauses. Particularly attractive Eastern goods meeting

---

69   Speech at the 25th Party Congress, quoted by Werner Beitel, *Probleme der Entwicklung des Sowjetischen Westhandels, op. cit.,* p. 68.
70   Jenelle Matheson, *et al., art. cit.,* p. 1286. Lawrence J. Brainard, takes a different view and holds that compensation agreements are a less efficient vehicle for technology transfer. The Western seller is concerned solely to supply the plant or the equipment and after that is interested solely in selling or using the products derived from it. If an unexpected bottleneck arises (technical or natural) the Western partner is not responsible. See Lawrence J. Brainard, "Financing Soviet Capital Needs in the 1980s", *The USSR in the 1980s,* NATO Colloquium 1978, *op. cit.,* p. 167.

Western quality standards would have to be produced and this could have too great an influence on Eastern countries' production planning.

For the Western countries, the main advantage of countertrade is to facilitate access to the Eastern market, particularly valuable in periods of recession. Western firms sometimes concede to Eastern insistence on countertrade when the details of the buy-back agreements are tolerable. Countertrade is also a way of securing supplies of raw materials and reducing production costs and provides "inside information" about the possible dangers of competition. On the other hand Western firms have difficulty selling consumer goods from the East. The competition among Western firms for Eastern markets is very limited. Moreover, some Eastern countries like the USSR and Poland are reluctant to include "hard" commodities as countertrade products.[71]

Are countertrade and compensation agreements really an answer to the threat of the increasing indebtedness of the East European countries to East-West trade in general and eastward technology transfer in particular? In this context a distinction must be drawn between "hard" commodities for which there is a ready market, and "soft" products for which markets are often difficult to find.

"Hard" products — energy, raw materials, etc., call for large scale contracts. Their value has been increasing at a fast rate but such increases are not infinite. More important, it is impossible to base economic co-operation solely on mutually complementary trade flows (technology in the West-East direction and energy and raw materials in the East-West direction). As Werner Beitel and Jürgen Nötzold point out with regard to Germany, mutually complementary trade would quickly come up against the barrier of Germany's limited raw materials requirements.[72] The same argument applies to other industrial products.

If the barrier is to be lowered, technology transfer has to continue even if this means reducing the West's technology lead.[73] In other words, East-West trade would have to be based on more processed and manufactured products as Mr. Brezhnev would like. But the difficulty in proceeding so, is the low quality and limited range of Eastern offerings. Improvement of this situation seems doubtful at the moment. Put differently, the East European countries that refuse to plan their national production to suit demand on their domestic markets can hardly be expected to meet foreign market demand.

It can be postulated therefore that countertrade, in its present form, can only be a make-do arrangement. It does seem that it enables the East European countries to exert pressure on the West to accept their products even if the quality of these products is not sufficiently high. It also seems, therefore, that this type of bi-lateral trade tends to reduce trade flows to the level of possibility of the less well endowed partner.[74]

The limits to countertrade or compensation as a remedy for present and future difficulties in East-West commerce caused by the mounting Eastern debt are understood by Soviet as well as Western experts. A number of alternative arrangements have been explored by both side. These are designed to obviate Soviet opposition to joint ventures. One such alternative would approximate "what would obtain under a joint ventures (e.g. with long term leases substituting for foreign ownership)"... However, the Soviet "Ministry of Foreign Trade and the Ministry of the Chemical Industry, among others, are not thought to support" this arrangement.[75] Other alternatives relate to project finance approaches. One technique is "advance payments" to be provided by groups of Western companies

---

71   Jenelle Matheson, *et al., art. cit.,* p. 1289.
72   Werner Beitel and Jürgen Nötzold, *op. cit.,* p. 49.
73   *Ibid.,* p. 50. The authors deliberately pay no account to any possible effects of such a policy from the development and strategic standpoint.
74   For more details on countertrade, see *Countertrade Practices in East-West Economic Relations, op. cit.* This source enumerates advantages and disadvantages of countertrade practice.
75   Lawrence J. Brainard, "Financing Soviet Capital Needs in the 1980s", *art. cit.,* p. 169.

in exchange for long-term export contracts. Another technique is "production payments" which would give creditors the rights to proceeds from the sale of products.[76]

All these suggestions would certainly have a positive effect and would help to prevent too great a reduction of East-West trade. But they would not seem to be a possible substitute for greater international division of labour and for a reorientation in East European countries' planning towards demand criteria (on domestic and foreign markets). The latter would seem the only way to provide a long-term answer to the difficulties of financing technology transfers to the East.

76   *Ibid.,* p. 171.

Chapter 7

EFFECT OF EAST-WEST TECHNOLOGY TRANSFER ON WESTERN ECONOMIES

1. PROBLEM OF ASSESSMENT

Any study of the effects of technology transfer on the economy should include:
— domestic price structure;
— incomes;
— distribution of incomes;
— employment;
— innovation and domestic technical progress;
— technology lag at the international level.

As recently pointed out in an American report,[1] such an undertaking is a particularly difficult task. The concepts of technology, technology changes and technology transfer are extremely ambiguous. They are extremely difficult to measure as data on product prices and quantities and production factors are either not available or inadequate, and in fact, several transfer channels are not even covered by statistical records. The American report concludes that: "In view of the state of art of research on the effects of technology transfer on US trade, production and employment, it is difficult even to make recommendations as to how to improve our knowledge of these effects".[2]

The report merely recommends data collection: on current sales and purchases of technology; on the quantity of foreign production when technology is licensed to a non-affiliated foreign recipient; and on transfers of proprietary technology not paid for by fees or royalties. The report also recommends collection of a set of qualitative as opposed to quantitative data on technology transfers.[3]

All this requires long-term research. However, general discussion on the economic advantages and disadvantages of East-West technology transfer is already under way, and at this stage, it is worthwhile listing the main problems raised, even if too little is known about them to draw definitive conclusions. These problems include:
— the economic advantages of technology transfer;
— the dangers of competition, and
— the growth of bilateralism and bartering.

---

1 *Technology Transfer: A Review of the Economic Issues, op. cit.*, pp. i-iii.
2 *Ibid.*, p. 40.
3 *Ibid.*, p. 41.

Since the technological lag and East European needs for Western technology both promote and unsettle international trade, it would be useful to dwell on the conditions needed to restore equilibrium.

2. ECONOMIC ADVANTAGES OF TECHNOLOGY TRANSFER

a) **Market Expansion and Diversification of Supply Sources**

As pointed out in Chapters 2, 3 and 6, machinery and equipment play a decisive role in the growth of East European imports from OECD countries. It is technological lag, as described in Chapter 5, that places such vast, continuous and acute requirements on the East European countries.

The problem is to know what the objective limits are to this flow of technology from West to East. Clearly, financial credit and other schemes can help to shift the limit. However, in the long run, Eastern imports of machinery and equipment can only be financed by expanding exports to match.[4]

It would seem unnecessary to point out what the industrialised countries stand to gain from exporting manufactured goods in general, and machinery and equipment in particular. Longer production runs, higher output and additional funds for Research and Development are sufficient explanation.

Nevertheless, in East-West trade, the scale of these benefits is minimised by the small percentage of total OECD country exports that goes to the CMEA countries. Although, between 1970 and 1977, machinery accounted for 31-36 per cent of these exports (see Table A-7), the share of total OECD exports that went to the East European countries during this period varied between no more than 2.9 and 4.6 per cent.[5] Their overall impact on Western economies could not have been very great.

On the basis that trade in the East-West direction will have to reach a significant level for technology transfer to have any impact worth talking about on Western economies, we need to consider, however briefly, the medium-term prospects for East European exports to the West. This has been done by Allen J. Lenz and Hedija H. Kravalis, whose main conclusions are given in the following paragraphs.[6]

Energy products were dominant in *Soviet* exports to the Western industrialised countries[7] between 1972 and 1976 when their share of total exports increased from 28 per cent to 53.2 per cent.[8] The 1976-1980 Plan foresees only a relatively moderate expansion in Soviet energy export products to the West, both because of mounting internal consumption and because of the difficulties being encountered in developing resources in Siberia. A recent CIA report forecasts that growth in Soviet oil production will come to an end in the early

---

4   Werner Beitel, *Probleme der Entwicklung des Sowjetischen Westhandels. Kreditfinanzierung, Exportstruktur, Kooperationsabkommen, op. cit.,* p. 43. See also Allen J. Lenz and Hedija H. Kravalis, "An Analysis of Recent and Potential Soviet and East European Exports to Fifteen Industrialised Western Countries", in *East European Economies Post-Helsinki — A Compendium of Papers, op. cit.,* p. 1057: "Over the long term, continued growth of East-West trade must rely on an ability of the East European countries to expand their hard currency earning exports to the West, rather than on continuing the increase in the debt that has fuelled much of the recent growth of Soviet and East European imports".

5   *Statistics of Foreign Trade,* OECD, Paris, Series A.

6   Allen J. Lenz and Hedija H. Kravalis, *art. cit.,* pp. 155-1131.

7   Austria, Belgium, Canada, Denmark, France, Germany, Italy, Japan, Luxembourg, the Netherlands, Norway, Sweden, Switzerland, United Kingdom, United States.

8   Hedija H. Kravalis, "Soviet-East European Export Potential to Western Countries", in *Issues in East-West Commercial Relations, op. cit.,* p. 172; Allen J. Lenz and Hedija H. Kravalis, *art. cit.,* p. 1062.

1980s and shows a 100 million ton drop in 1985 compared with 1980. This would have serious implications for the 44 per cent of Soviet hard currency earnings that oil exports accounted for in 1976.[9] Soviet gas exports, however, are likely to increase, particularly in exchange for pipeline ("gas for pipe"). The same applies to platinum, aluminium, diamond, chromium, nickel, timber and cotton exports which could well increase in volume and also benefit from the likely increase in world prices.[10]

The volume of Soviet manufactured goods exported, currently 4 per cent of the total, will probably be small. Quality, style and service problems are all involved here. The Soviet Union could possibly join the small number of machinery-exporting countries. However, some observers consider that Soviet leaders, though paying "lip service" to this principle, would seem to be giving priority, in the medium term, to exports of raw materials.[11]

In *Polish* exports to the Western industrialised countries, energy products and raw materials (e.g. coal, copper and sulphuric acid) are major items and could increase in volume and benefit from the probable increase in world prices.[12] However, the second largest category of Polish exports — meat and meat products — is unlikely to bring in appreciable currency earnings in future because of uncertain agricultural production and the rise in domestic demand.

Polish exports of manufactured products — tractors, ships and boats, motor vehicles and light industry products — could well benefit from the massive imports of Western technology in 1971-75.[13] Expansion here, however, will be limited by the familiar problems of quality, fashion and after-sales service and, already accepted products like shoes or clothing could well fall foul of Western import restrictions. Semi-manufactured goods could also be affected by import restrictions because Western producers of chemicals, iron and steel products and textiles are campaigning to prevent any increase in imports.[14]

The *German Democratic Republic* plans to increase its exports of manufactured goods, particularly capital goods. To what extent it can implement these plans, however, is not clear. In spite of their competitiveness, these exports are difficult to sell in the Federal Republic of Germany, where industrial standards are high. Sales are further hampered by quality, spare parts and servicing inadequacies. Besides, the German Democratic Republic's industry is operating at full capacity. Growth in production capacity will therefore depend on productivity gains and on imports of Western machinery.[15]

Czechoslovakia's export potential to the West is largely dependent on modernisation of its manufactured goods industry. This in turn depends on increased imports from the West. Czechoslovak officials, however, are cautious about accumulating deficits and would rather pay for imports with exports. Such a policy would require a sharp rise in exports to the West in order to meet requirements.[16]

Under the *Hungarian* five-year plan for 1976-1980, foreign trade is to grow faster than the economy as a whole. Trade is to shift in favour of the West, and especially to the developing countries where Hungarian manufactured goods are easier to sell than in the Western industrialised countries. Hungary hopes to develop its industrial potential with the

9   Hedija H. Kravalis, *art. cit.,* p. 170. See also Chapter 4.
10   Allen J. Lenz and Hedija H. Kravalis, *art. cit.,* p. 1075; Hedija H. Kravalis, *art. cit.,* p. 170.
11   The term "lip service" is used by Hedija H. Kravalis, *art. cit.,* p. 171. A similar idea is put forward by Werner Beitel: *Probleme der Entwicklung des Sowjetischen Westhandels. Kreditfinanzierung, Exportstruktur, Kooperationsabkommen, op. cit.,* p. 46 quoting Leonid Brezhnev's statement of 25th February, 1976: "calculations show that the various raw materials will continue to be major Soviet export items. At the same time, products from our processing industries will increase their share of exports considerably".
12   Hedija H. Kravalis, *art. cit.,* p. 171.
13   Allen J. Lenz and Hedija H. Kravalis, *art. cit.,* p. 1075.
14   Hedija H. Kravalis, *art. cit.,* p. 171.
15   Allen J. Lenz and Hedija H. Kravalis, *art. cit.,* p. 1095.
16   *Ibid.,* p. 1103.

help of imported Western technology in the form of licences, plant and equipment. It will try to sustain its food exports, representing 30 per cent of hard currency earnings, and to increase manufactured goods exports. Difficulties encountered in entering Western markets are a considerable handicap in spite of Hungarian officials' enterprising approach to marketing.[17]

Rumanian ability to increase exports to the industrialised West is uncertain.[18] Rumania is therefore trying to sustain growth in petroleum product exports and to promote light industry exports (textiles, clothing, etc.).

Bulgaria does not appear to be anxious to increase trade with the West. It plans to continue importing Western machinery, electronics and chemicals for specific projects and to extend exports to the developing countries.[19]

Lenz and Kravalis conclude that the CMEA countries' ability to expand exports to the industrialised West is dependent on three main factors:[20]

— increasing the physical volume of products available for export;
— increasing penetration of Western markets;
— the rate of Western inflation (the higher the rate the easier exporting will become).

A number of Western observers forecast an increase in raw materials prices in the early 1980s.[21] Coupled with the completion of a number of development and raw material production projects, mainly in the USSR and Poland, this increase could help East European countries to earn more from their exports to the hard currency countries.[22] There is also a possibility that the East European countries will solve certain quality problems and restructure their exports so as to get round the obstacles of Western restrictions.[23]

b) **Effects of Technology Transfer on Employment in the Western Industrialised Countries**

Unemployment is a highly sensitive issue in the West. In absolute figures, the number of unemployed has soared in all industrialised Western countries. In 1977, there were 6.9 million unemployed in the United States, 1.5 million in the United Kingdom and Italy, 1.1 million each in Japan and France and 1 060 000 (1976 figure) in Germany.[24] The percentage of workers unemployed[25] was 8.1 in Canada, 7.0 in the United States[26], 7.1 in Italy, 6.3 in the United Kingdom, 4.6 in Germany, but only 2.0 in Japan.[27]

In market economies a measure of seasonal or temporary unemployment is regarded as "normal" since, at any given time — even during periods of over-employment — part of the working population is changing jobs. This is a normal process by which production is

17   *Ibid.,* p. 1120-1121.
18   *Ibid.,* p. 1111.
19   *Ibid.,* p. 1127.
20   *Ibid.,* p. 1128. It must be kept in mind of course that the impact of these factors differs according to the level of development of a particular Western country.
21   Hedija H. Kravalis, *art. cit.,* p. 171.
22   *Ibid.,* p. 171.
23   *Ibid.,* p. 171.
24   *Monthly Bulletin of Statistics,* United Nations, New York, June, 1978 (pp. 17-20 of the French edition).
25   The percentage unemployed is the ratio of the number of unemployed workers in a given group during a given period to the total number of persons in that group, both employed and unemployed, during the same period.
26   In the United States the level of unemployment dropped from 9 per cent in early 1975 to 6 per cent in Spring 1978. "L'évolution du chômage dans les grands pays industrialisés", in *Problèmes économiques,* La Documentation française, Paris, 2nd August, 1978, pp. 19-22.
27   *Monthly Bulletin of Statistics, op. cit.,* June 1978, pp. 17-20.

adjusted to demand through the market mechanism.[28] The level of this "normal" unemployment varies with country. In the United States it is estimated at 3-4 per cent while in France 1-1.5 per cent is considered reasonable.[29] In other words, present unemployment levels in industrialised Western countries are well above "normal", giving rise to general public fears.

This "abnormal" unemployment in the West is caused by several factors. Structural unemployment is due to the fact that the geographical and qualitative structures of the demand for jobs are not in balance with those of the supply of unemployment. In France, for instance, there is an increasing shortage of labour for rough or "socially inferior" jobs and a huge surplus of job-seekers for "better status" jobs. There is also an economic, inflationary type of unemployment[30] caused: by the substitution of capital for labour which quickens as wage costs soar; by the shortfall in productive investment (because the return on capital is falling with rising wage costs) and, by the failure to increase the GDP growth rate to a level compatible with full employment, the result of the excessive wage bill.[31]

In view of the scale and urgency of the unemployment problem in the West, it is important to define the role and to assess the impact of technology transfer, especially to the East. This does not appear to have been done in detail, least of all in the context of East-West relations. As pointed out by F. L. Altmann and Hermann Clement, any structural change brought about by foreign trade affects employment, creating jobs in some sectors and reducing their number in others. East-West trade has, recently, been mainly based on credit and the balance has certainly, therefore, been tilted towards job creation in the West.[32] The repayment of loans and the inflow of goods, on the other hand, could have a negative effect on employment.

A study is needed on this subject and it should deal separately with the effects of three types of transfers:

— job creation through exports of machinery and equipment to the East;
— reduced employment levels due to competition from East European countries. This may or may not be the result of new technology sales;
— slower growth of employment if capital, which could have been used within the exporting country, is exported to the East.

An interesting fact is that the press mostly tends to deal with arguments relative to the first type of transfer: the first constitutes a good reason for exporting, and the second is quoted as a warning against policies regarded as excessively liberal in this field. In both cases, the conclusions are known in advance. The third type of transfer is seldom referred to, although the *size* of capital flows, concerning primarily delivery of plants and equipment, can easily have an effect on the domestic employment of the exporting country.[33] Since the question of competition from East European countries will be discussed later, we shall just give a few examples of policy statements which give employment as an argument in favour of technology exports to the East.

The most interesting statement — and it would be useful to see figures to back it up — is that made by Dzerman Gvishiani, Vice-President of the Soviet Committee for Science and

28   Morris Bornstein, "Unemployment in Capitalist Market Economies and Socialist Centrally Planned Economies", *The American Economic Review*, May 1978, p. 39.
29   *Ibid.*, p. 39 and Jean Marczewski, *Inflation et chômage en France*, Paris, 1977, p. 167.
30   Jean Marczewski, *op. cit.*, p. 167.
31   *Ibid.*, p. 168. It should be noted that seasonal, temporary, and even structural unemployment — and latent unemployment due to under-employment as well — also occurs in centrally-planned economies. Morris Bornstein, "Unemployment in Capitalist Market Economies and Socialist Centrally-Planned Economies", *art. cit.*, pp. 38-42.
32   F. L. Altmann and Hermann Clement, *op. cit.*
33   Jack Baranson's view on the subject is worth noting, "Technology Exports Can Hurt US", *Foreign Policy*, 1976, pp. 191-192. According to this author, American employment is at risk in key sectors such as the car, aircraft, civil electronics and chemical industries.

Technology, who said at a press conference in Vienna, when talking about the indebtedness of the East (which he believed to be temporary), that exports to the East were providing employment for 2 million persons in the West.[34]

Similar, but less sweeping, statements have also been made in the West and particularly in the United States. As already noted, a senior representative of the Department of Commerce, for instance, has said that the equivalent of 65 000 jobs had been created by exporting $1.8 billion worth of products from the United States to the Soviet Union.[35] Other spokesmen, such as Rogers Morton, United States Secretary of Commerce (August 1975), claim that the United States' strategic restrictions are directly responsible for employment losses and recommend increasing the transfer of non-strategic technology to the East European countries in order to ease the employment situation.[36] The most categoric statements relate to high technology sectors, in which the borderline between products of strategic interest and others is blurred, and attempt to influence the embargo policy in the West. The most forceful example is provided by the President of Control Data Corporation, William C. Norris, who supports elimination of the restrictions on computer sales to the East.[37] In his article, Norris states that "with more positive initiative, including moderate relaxation in export control guidelines and close co-operation between business and government, high technology exports to these countries (the Communist bloc) would grow to several billion dollars per year and create hundreds of thousands of US jobs"[38]. In the computer industry alone, it could, as already noted in Chapter 4, allow a build-up, in ten years, of an annual level of employment of 150 000 jobs.

These arguments are not shared by everyone. When Leonid Brezhnev visited Bonn in 1978, the trade union of IG Chemie, the German chemicals firm, warned against acceptance of payment by compensation in connection with the Tomsk and Tobolsk plant construction projects. Supplies of ethylene, ammonia, propylene, benzine and styrene could hurt German industry — already producing more than it could sell.[39] Sharper criticism still has come from Charles Levinson of the International Chemical Workers Federation in Geneva in his book *Vodka-Cola*.[40] His Federation recently criticised Firestone Tyres' intention to close their works in Switzerland and to make good the lost production by deliveries under joint venture agreements with East European countries.[41]

Whether these figures are correct or not, it is essential to place the problem of employment and unemployment in a general context before drawing any conclusions. Since this is too sensitive an issue and its implications too far-reaching, there is an urgent need for a detailed study on the effects of technology transfer on employment in the Western industrialised countries.

c) **Is Technology Transfer Profitable?**

In Western economies the basic criterion of decisions by business firms to export is profitability. Business secrecy makes it very difficult to obtain information on the profit earned by Western firms from technology transfer. In the absence of any detailed (and difficult) research on the subject, the only information available is vague and of doubtful

34 "Ungelöste Wirtschaftsprobleme mit den sozialistischen Ländern", *Neue Zürcher Zeitung,* 16th March, 1979.
35 Arthur Downey, then Deputy Assistant Secretary of Commerce for East-West Trade, *Détente, op. cit.,* p. 22. See also Chapter 4.
36 Rogers Morton, *op. cit.,* pp. 19 and 48-49.
37 William C. Norris, *art. cit.,* pp. 99-103. See also Chapter 4.
38 *Ibid.,* pp. 101-102. See also Chapter 4.
39 "Chemical Buy-Back: Where Will It All Lead?", *Business Eastern Europe,* 9th June, 1978, p. 178.
40 Charles Levinson, *Vodka-Cola,* Editions Stock, Paris, 1977.
41 "Chemical Buy-Back: Where Will It All Lead?", *art. cit.,* p. 178.

accuracy. One valuable source of information is the questionnaire survey taken among executives of firms exporting technology to the East.

The survey of American firms by Robert W. Clawson and William F. Kolarik, Jr., Kent University, (almost five hundred firms were interviewed), produced, for the first time, a number of representative and first-hand views on the subjects.[42] The replies to this survey on the profitability of technology transfer in US/USSR trade are summarised in Table 79.

Table 79
Profitability of United States-Soviet Trade
Percentages

Answers of Companies	For the US Business Community			For the US Company		
	Agree	Not sure/ Depends	Disagree	Agree	Not sure/ Depends	Disagree
Having high technology[a]	79.7	12.2	8.1	79.7	4.3	15.9
Not having high technology[b]	75.6	13.4	11.0	68.9	12.2	18.9
Having scientific-technical pact[c]	81.6	10.5	7.9	73.7	10.5	15.8
Not having scientific technical pact[d]	75.2	14.5	10.3	75.0	7.7	17.3

a) 74 answers concerning the US Business Community and 69 answers concerning the profitability of a given company.
b) 82 answers concerning the US Business Community and 74 answers concerning the profitability of a given company.
c) 38 answers concerning the US Business Community and 38 answers concerning the profitability of a given company.
d) 117 answers concerning the US Business Community and 104 answers concerning the profitability of a given company.

Source: Robert W. Clawson and William F. Kolarik, Jr., *art. cit.*, Tables 5 and 6.

An interesting point is that over three-quarters of the executives replying to the questionnaire regarded such exports as profitable. The percentage is higher with regard to profitability for the US business community as a whole than for the individual US companies concerned. In the latter case, there was little difference between replies from companies having and those not having a scientific/technical pact with the Soviet authorities. On the other hand, firms dealing in high technology products tended to have a more positive view of transfer profitability. The personal interviews conducted by the authors suggest an explanation: Soviet foreign trade offices often place greater priority on the acquisition of high technology, such as electronic data processing, instrumentation, etc., for which the US has a near monopoly in some cases.[43] Thus the Soviet foreign trade offices' ability to play off US suppliers against foreign competition is limited, especially in a period of economic boom, when a sellers' market situation tends to exist. US firms often have the reputation of producing the highest quality in their product sector and since the USSR is keen on "buying the best", American firms are in a good bargaining position. The prices obtained are then highly favourable.[44]

Some of the executives replying to the questionnaire and who were not sure of transfer profitability, had either lost out in their first contract, but had been promised a second, more profitable one, or were still in the negotiating stage for a first contract.[45] Negative replies were obtained from firms which had lost money on their first contract with the Soviet Union. Very often, the firms concerned had no experience with the Soviet market.[46]

42   Robert W. Clawson and William F. Kolarik, Jr., *art. cit.*
43   *Ibid.,* p. 18.
44   An executive of one such company stated that: "Our East bloc sales (including the USSR) give us our highest profit margin. They never quibble about price". *Ibid.,* pp. 18 and 19.
45   In some cases, the firm which had concluded the contract was a subsidiary which let the parent company handle the accounting. The parent company's accounting system may not include procedures for separating out and evaluating Soviet business. *Ibid.,* p. 17.
46   *Ibid.* It is of course not possible for us to judge how far these views of Clawson and Kolarik are representative of the opinion of Western businessmen.

More recently, some interesting data about the profitability of East-West trade including counterpurchase arrangements have been compiled in the survey conducted by Franz-Lothar Altmann and Hermann Clement among 694 German firms (and chambers of commerce, trade associations, etc.) in 1978.[47]

One of the questions (to which 98 replies were received) was about the way in which firms covered the extra cost of counterpurchases. 26 per cent of them did so by increasing their export prices and 7 per cent by reducing the price of the products received in compensation. 47 per cent were able to cover only part of the extra cost. The remainder (18 per cent) were unable to cover these costs in the deal itself.[48]

In the Altmann and Clement survey, an attempt is made to answer the question of whether compensation and counterpurchase contracts (*Verbundsgeschäfte*) were more profitable than ordinary contracts. Out of 95 replies on this question, only 2 per cent say that these deals are more profitable, 36 per cent consider profitability to be about the same and 59 per cent feel that they are less profitable. The 2 per cent consist of one small and one medium-sized firm. It would therefore seem that Western firms accept such deals only if they are forced to. The fact that the German firms found it difficult to include in the contract the extra costs entailed through compensation and counterpurchase arrangements comes out clearly from the replies on pricing.

In their answers to another question concerning pricing, 41 per cent of the 96 replies said that the prices in the contracts were based on world prices, 36 per cent said they matched prices prevailing in Germany and only 4 per cent claimed that prices were based on the price of similar products exported from the East to the West.[49] The chambers of commerce even said that in two-thirds of cases, prices were lower, particularly for Rumania, Bulgaria and Poland. No replies claimed that prices were higher than those in ordinary contracts.[50]

The surveys we have referred to cannot possibly give a full picture of the problem particularly since each reply is counted as one unit regardless of the value of the contract.[51] In some cases however, involving large-scale investment projects in the USSR at a cost of several billion dollars, the contracts involve financial risks associated with long-term commitments and a certain measure of interdependence. Many American firms are fearful of tying up money, technology, equipment and management potential in the USSR without any immediate return and of being at the mercy of political events. When production does finally commence, payment is made in the form of deliveries of oil, gas, etc. This implies a dependence on the USSR for products which have strategic importance.[52]

Competition among Western firms for turnkey plant sales is so keen that most are prepared to sell at any price, even at practically no profit, convinced that they will benefit in a second negotiation (which may never materialise).[53] In the case quoted by the author, the Western firm concerned found unexpected difficulties. These increased the cost of establishing the plant and thus reduced the profit margin which had initially been small.[54]

The affirmative replies about the profitability of East-West trade for the business community as a whole are not always confirmed either. For instance, Marshall Goldman and Raymond Vernon consider it a serious risk for American private corporations to sell

47 F.L. Altmann and Hermann Clement, *op. cit.*
48 *Ibid.*
49 *Ibid.*
50 *Ibid.*
51 But F. L. Altmann and Hermann Clement do make a distinction between small, medium and large firms. The size of the contracts entered into by these firms is not given in the tables they quote.
52 Franklyn D. Holzman, *International Trade Under Communism, op. cit.*, pp. 163-164.
53 "Selling Turnkey Plants to Eastern Europe: Profitability Questioned", *Business Eastern Europe*, 9th September, 1977, pp. 281-282.
54 *Ibid.*, p. 281.

know-how to Soviet partners at less than opportunity cost and also less than the total social cost to the United States.[55] The reasons are quite simple. By the time a firm has developed a unique technology, a large proportion of its total costs will already have been incurred. The cost of supplying this technology to a new buyer constitutes only a small percentage of the firm's total costs. In other words, the marginal cost of these additional sales is much smaller than the total cost of production and sales. From the firm's point of view, however, an additional sale is worthwhile since the selling price is higher than the marginal cost.[56]

The national interest is also involved when R & D has been largely financed from public funds and when the product may have military uses.

To combat these undesirable effects, Marshall Goldman and Raymond Vernon propose that a government agency be set up with the following functions: to set prices representing a net social gain for the United States, and to take various measures to reduce the undesirable effects connected with such transactions, including refusal of Eximbank credits. They also suggest setting a limit, say $5 million, above which sales of technology would be subject to review by the Federal authorities.[57]

3. THE DANGERS OF COMPETITION

a) **Developments in Imports from the East in Competition-sensitive Sectors**

The expansion in exports of Western technology to the East during the first half of the seventies raises the problem of the possible impact of these exports on the Western economies. In other words will this not increase the economic potential of the East European countries and enable them to compete with the Western countries in sectors where competition is already relatively keen in the West?

The first question is to define those sectors which are sensitive to competition from East European countries. A study just published by the US Department of Commerce makes its criterion the taking of legal action against imports from the East. The definition is admittedly arbitrary and used simply for lack of a more direct yardstick.[58]

The list adopted excludes raw materials, agricultural produce and pharmaceutical industries. Sectors are classified as "very sensitive" and "moderately sensitive" (see Table 80), depending on whether the problem involved is of a general nature already familiar to the industrialised countries (textiles, clothing, steel and footwear) or relates to sectors referred to only occasionally or presenting certain risks.

Table 80, which sums up the results of this survey, shows that in spite of very steep growth in Western imports in these sectors, the share of the Centrally Planned countries in the Western industrialised countries' total imports has hardly changed — with a few exceptions — for the products concerned. Textile fibres and a number of chemicals are the only sectors showing any steep increase in that share. The textile fibre concerned is cotton from the USSR. 60 per cent of the chemicals concerned also come from the USSR. But this figure includes uranium, Soviet exports of which tripled in 1977.[59]

The study also includes a more detailed investigation of the steel and clothing sectors. For steel, the conclusion is that growth in exports to the West amounted to only 34 per cent

55 Marshall I. Goldman, "US Policies on Technology Sales to the USSR", in *East-West Technological Co-operation*, NATO Colloquium, 1976, *op. cit.*, p. 111.
56 *Ibid.*
57 *Ibid.*, p. 117.
58 Karen Taylor and Deborah Lamb, "Communist Exports to the West in Import-sensitive Sectors", in: *Issues in East-West Commercial Relations, op. cit.*, pp. 129-131.
59 *Ibid.*, pp. 132-133.

Table 80

Centrally Planned Economies Exports to Industrialized West in Import Sensitive Sectors as Percent of World Exports to Industrialized West, 1973-1977[a]

SITC and description	CPE exports to IW as a percent of world exports to IW					1973-77 exports to IW, percent increase	Principal CPE suppliers in 1977 (millions of dollars)
	1973	1974	1975	1976	1977		
I. MODERATELY OR POTENTIALLY SENSITIVE SECTORS							
26 Textile fibers	9.1	9.3	10.5	10.1	11.6	33.5	USSR (512.2); People's Republic of China (278); other CPE's (82.8).
51 Chemical elements and compounds (organic and inorganic chemicals)	3.5	4.0	3.9	3.9	5.1	205.8	USSR (570.7); Hungary (77.7); Poland (67.2); Czechoslovakia (65.5); other CPE's (175.6).
56 Manufactured fertilizer	9.1	6.6	7.7	8.0	6.9	56.5	USSR (50.1); German Democratic Republic (45.4); Poland (22.8); Rumania (19.2); other CPE's (25.3).
58 Plastic materials	0.7	0.6	0.6	0.7	0.1	125.3	Czechoslovakia (16.5); German Democratic Republic (15.6); USSR (15.1); other CPE's (21.5).
59 Chemical materials not elsewhere specified	3.3	3.9	2.6	2.2	2.1	10.6	People's Republic of China (53.9); Poland (22.1); other CPE's (26.9).
69 Manufactures of metal, not elsewhere specified	1.8	1.8	1.7	1.7	1.7	64.0	Poland (58.6); Yugoslavia (48.0); other CPE's (100.1).
72 Electrical equipment and electronic products	1.6	1.3	1.3	1.3	1.3	78.0	Yugoslavia (166.6); Hungary (82.6); German Democratic Republic (66.8); Poland (61.3); other CPE's (110.4).
73 Transport equipment	1.3	1.5	1.0	1.2	1.1	68.3	Poland (313.0); USSR (148.7); Yugoslavia (148.0); other CPE's (169.4).
Total for all moderately sensitive sectors	2.3	2.2	2.1	2.1	2.3	76.9	USSR (1 341.7); Poland (588.7); Yugoslavia (426.8); People's Republic of China (288.5); other CPE's (930.8).
II. HIGHLY SENSITIVE SECTORS							
65 Textile yarns and fabric products	3.7	4.0	4.2	4.4	4.2	58.1	People's Republic of China (355.5); Czechoslovakia (104.5); Poland (80.1); other CPE's (277.2).
67 Iron and steel products	4.1	3.2	3.0	3.6	3.6	33.7	Czechoslovakia (208.2); Poland (134.6); Rumania (120.5); other CPE's (396.1).
84 Clothing	7.7	8.3	7.9	7.4	7.3	81.6	Yugoslavia (432.1); Hungary (218.7); Rumania (204.9); People's Republic of China

Total for all highly sensitive sectors	4.9	4.7	4.8	5.0	5.0	62.7	Czechoslovakia (47.4); other CPE's (39.8). Yugoslavia (630.6); People's Republic of China (578.9); Poland (460.4); Rumania (459.3); Czechoslovakia (455.4); other CPE's (785.8).
Total for all sensitive sectors—both moderately and highly sensitive	3.1	3.0	3.0	3.0	3.1	69.8	USSR (1 493.9); Yugoslavia (1 057.4); Poland (1 019.1); People's Republic of China (967.4); other CPE's (2 482.1).

a) Centrally Planned Economies include: Bulgaria, Czechoslovakia, German Democratic Republic, Hungary, People's Republic of China, Poland, Rumania, USSR, Yugoslavia.
Industrialized West include: Austria, Belgium, Canada, Denmark, Germany, France, Italy, Japan, Luxembourg, Netherlands, Norway, Sweden, Switzerland, United Kingdom, United States.

Sources: Karen Taylor and Deborah Lamb, "Communist Exports to the West in Import Sensitive Sectors", in: Issues in East-West Commercial Relations, op. cit., p. 132.

283

between 1973 and 1977 whereas total exports from the Planned Economy countries to the West went up by 58 per cent over the same period. The authors conclude that the countries concerned have made no attempt to use their iron and steel products as a way of earning hard currency.[60] As regards clothing, the rate at which Western imports increased between 1973 and 1977 (82 per cent) was higher than the total increase in imports (58 per cent). However, their share of total imports dropped from 7.7 per cent in 1973 to 7.3 per cent in 1977.[61] Some products, however, did increase their share: women's and children's underclothes (10.8 per cent), handkerchiefs (14.8 per cent), men's and boys' outer garments (10.1 per cent) and fur clothing (10.7 per cent).[62]

The study's general conclusion is that exports from the Centrally Planned countries were not focused on the sensitive sectors and that their pattern did not vary very much from that of world exports. In the absence of specific data, however, on the pattern of Western imports from the developing countries (South Korea, Taiwan, Hong Kong, India and Pakistan), this conclusion may be premature.[63] In the clothing sector, the labour-intensive nature of this industry in the West makes it more exposed to penetration by the Centrally Planned countries and there are relatively more products that could, potentially, disrupt Western markets.[64] As regards the steel industry, this is still in process of "internationalisation" and it is only since the mid-1970s that surplus capacity at world level has threatened national industries. Currently, however, exports from the Centrally Planned countries do not according to this study, present any serious problems for the Western industrialised countries.[65] The study therefore concludes that although some specific sectors of products may encounter problems from time to time, exports from the Centrally Planned countries have not been disruptive or particularly significant in competition-sensitive sectors.[66]

b) **Competition between Western Firms for Deals with CMEA Countries**

In East-West trade, the two negotiating partners are often unequally matched: a privately owned firm versus a State enterprise which often takes advantage of its monopoly or monopsony (*bargain asymmetry issues or whipsawing*). The only way a Western firm can negotiate with an East European country is through its foreign trade monopoly. The East European organisation can however, take advantage of the competition between Western firms. Since collusion between competing American and European firms is prohibited by anti-trust (or anti-cartel) legislation, the East European buyer is able to force down prices.

The fierce competition between Western firms for turnkey contracts is often referred to in the press and headlines such as "race to the death" for turnkey plant orders even appear.[67] The same technology is often available from several different Western firms. For instance, one analyst believes that if General Motors had turned down the Polish Polmot offer, Mercedes-Benz or Volvo would have taken it up and would have been able to meet the same requirements.[68]

An indication of the concern felt regarding the practice of playing off Western firms against one another is the fact that the condemnation of monopolistic practices and

60   *Ibid.*, p. 136.
61   *Ibid.*, p. 144.
62   *Ibid.*, p. 145.
63   *Ibid.*, pp. 144 and 155.
64   *Ibid.*, p. 156.
65   *Ibid.*, p. 156.
66   *Ibid.*, p. 156.
67   "Selling Turnkey Plants to Eastern Europe: Profitability Questioned", *Business Eastern Europe, art. cit.*, p. 282.
68   Jack Baranson, "Technology Exports Can Hurt US", *art. cit.*, 1976, p. 190.

whipsawing of Western firms was proposed for negotiation at the Belgrade Conference of 1978.[69] The latter was intended as a continuation of the Helsinki talks with a view to improving East-West trade relations.[70]

c) **Views of Company Executives on the Dangers of East European Competition on Western Markets**

Western firms often wonder about the dangers of competition resulting from their licensing agreements and technology exports to East European countries.[71] Unfortunately, there are very few surveys on Western firms' views on the subject.

Table 81

Fear For Setting Up the USSR As a Future Competitor

Percentages

Answers of Compagnies	In the industrialised countries			In the lesser developed countries		
	Agree	Not sure/ Depends	Disagree	Agree	Not sure/ Depends	Disagree
All companies investigated[a]	31.0[f]	9.0	61.0[g]	40.0[h]	15.0	45.0[i]
Companies having high technology[b]	29.3	9.3	61.3	34.7	14.7	50.7
Companies not having high technology[c]	31.8	7.1	61.2	44.0	16.7	39.3
Companies having scientific technical pact[d]	25.6	7.7	66.7	28.2	10.3	61.5
Companies not having scientific technical pact[e]	31.7	9.2	59.2	42.9	17.6	39.5

a) 163 answers concerning industrialised countries and 162 answers concerning lesser developed countries.
b) 75 answers concerning industrialised countries as well as lesser developed countries.
c) 85 answers concerning industrialised countries and 84 answers concerning lesser developed countries.
d) 39 answers concerning industrialised countries as well as lesser developed countries.
e) 120 answers concerning industrialised countries and 119 answers concerning lesser developed countries.
f) 6 agree strongly and 25 not strongly.
g) 25 disagree strongly and 36 not strongly.
h) 11 agree strongly and 29 not strongly.
i) 15 disagree strongly and 30 not strongly.

Source: Robert W. Clawson and William F. Kolarik, Jr., *art. cit.*, Tables 4, 5, 6.

Tables 81 and 82 show the results obtained by Robert W. Clawson and William F. Kolarik, Jr., in this field. Table 81 shows that about two-thirds of the American company executives who replied to the questionnaire were not worried about Soviet competition in Western industrialised countries. The percentages are roughly the same for firms with and without high technology and for those with and without a scientific-technical agreements. The executives are far more concerned about competition in developing countries, especially in firms with no high technology and with no scientific and technological co-operation agreements. The most optimistic replies came, in fact, from companies which had concluded such agreements.

Clawson and Kolarik's survey also shows the difference in attitudes reflecting the form of the technology transfer: licensing or product sales (see Table 82).

69 *Report to the Congress of the United States on Implementation of the Final Act of the Conference on Security and Co-operation in Europe: Findings and Recommendations Two Years After Helsinki, op. cit.,* p. 77.
70 *Ibid.*
71 David Winter, "A Lawyer's View of the Management of the Transfer of Technology within Industrial Co-operation with Particular References to East-West Trade", in *Proceedings of the UN/ECE Seminar on the Management of the Transfer of Technology within Industrial Co-operation, op. cit.,* pp. 181-182. David Winter, incidentally, takes the view that these fears are unduly magnified.

The results suggest that American businessmen are becoming less willing to sell know-how to the Soviet Union in the form of licences. Some 68.2 per cent of those who replied stated that they feared competition in industrialised countries following technology sales in the form of licensing, while a slightly lower percentage (64.4 per cent) feared competition in the developing countries. The attitude towards product sales incorporating advanced technology is completely different. Here competition is apparently not feared.

Table 82

Attitudes towards USSR as a Future Business Rival of US Companies

Percentages

		In Developed Nations			In Less Developed Countries		
		Yes	Not Sure	No	Yes	Not Sure	No
Entire Sample: Fear USSR as Competitor[a]		30.7	8.6	60.7	40.1	15.4	44.4
Do you favour *licensing* your firm's latest technology to the USSR?[b]	Yes	29.5	38.5	49.5	33.9	36.4	52.9
	Not Sure	2.3	0	1.1	1.7	0	1.5
	No	68.2	61.5	49.5	64.4	63.6	45.6
Number of answers		(44)	(13)	(93)	(59)	(22)	(68)
Do you favour *product sales* to the USSR incorporating your company's latest technology?[b]	Yes	80.4	84.6	96.6	85.2	91.3	95.2
	Not Sure	4.3	7.7	0	4.9	0	0
	No	15.2	7.7	3.4	9.8	8.7	4.8
Number of answers		(46)	(13)	(89)	(61)	(23)	(63)

a) Percentaged horizontally.
b) Percentaged vertically.

Source: Robert W. Clawson and William F. Kolarik, Jr., *art. cit.*, Tables 7 and 8.

One explanation for this attitude toward licensing is the complexity and cost of negotiating licensing agreements with the Soviet Union.[72] Some of the firms entering the field for the first time had had disappointing experiences, word of which had spread to firms with similar production profiles. Some American firms which had advanced technology but sold "stable" products (i.e. which were not outdated too quickly through technical progress) such as cars, chemicals and textiles, thought that their goods could be easily sold by the USSR on Western markets. The *Lada* agreement negotiated in Switzerland is an often-quoted example. Other firms were more sceptical of the USSR's ability to reproduce and sell their technology. Still others were fully aware of Soviet ability to compete on Western markets, but adopted a dynamic approach on the grounds that products would quickly become outdated because of technological advances (e.g. electronics, computers, numerically controlled machine-tools, etc.). In these sectors, performance quality, not price, is the critical factor in acquisition decisions, and Soviet dumping is regarded as unlikely.[73]

American businessmen's relatively optimistic view of competition from East European countries does not appear to be shared by all Western partners. According to a survey by the Institute of Foreign Trade and Overseas Economy, Hamburg University, 55 per cent of the 40 companies interviewed were afraid of competition on East European markets and 50 per cent were worried about not being able to control the quality of goods manufactured under

72   Robert W. Clawson and William F. Kolarik, Jr., *art. cit.*, p. 20.
73   *Ibid.*, p. 23.

co-operation agreement.[74] As a result, these firms tended to sell relatively outdated technology to the East.[75]

The German firms' pessimism is confirmed by the survey made by F. L. Altmann and Hermann Clement.[76]

Out of 125 replies to questionnaires, 57.6 per cent felt that domestic sales were affected by compensation or countertrade imports (*Verbundsgeschäfte*). The authors however do not think it is safe to conclude from this that there is any real market disruption (*Marktstörung*) or intensive (*lästige*) competition. They believe that the firms consulted had not yet had enough experience in this field.[77] Still, these German firms are all more or less agreed (73.7 per cent take this view) that specific branches of industry suffer fierce competition as a result of compensation or countertrade imports (*Verbundsgeschäfte*). Examples quoted are textile products, certain tools and machinery, the electrical industry, clothing, and consumer goods in general. The difficulties stem mainly from the lower prices asked for products supplied under compensation or countertrade agreements.[78]

d) Possible Implications of Competition

As a result of increasing production capacities and the debt, the result of imports of Western technology, East European countries will necessarily have to increase their sales on the world market.

There are very few in-depth investigations of East European countries' ability to compete with Western countries. One such study discusses the future impact of competition from the USSR in the United States.[79] After analysing the almost 200 contracts concluded between private American firms and the Soviet Union between 1970 and 1974, the authors of the study attempt to identify the sectors in which significant competition might occur. In fact, the study is confined to the four sectors in which technical progress is particularly pronounced: semiconductors, civil aviation, construction machinery and equipment, and synthetic fibres.

The findings of this research indicate that there is little chance of strong Soviet competition in the field of semi conductors or in civil aviation. In the field of semi-conductors, swift technical progress, sharply fluctuating sales and rapidly falling prices make it unlikely that the Soviet Union will venture to compete with the Western countries. Its success in this field is even less likely. The same situation exists in the field of civil aviation, where maintenance, supply of components and after-sales service have been particularly deficient in the USSR.

The Soviet Union may make a few attempts to compete with the West in the sector of construction machinery and equipment and in synthetic fibres.

The general conclusions of the research done by Herbert Levine and his associates indicate that the chances of the Soviet Union competing with the Western countries during the years 1975 to 1985 in the fields necessitating advanced design and production technology are fairly limited ("it is highly unlikely"). In the authors' view, this applies specifically to the sectors involving high technology and those dominated by rapid technical progress. The

---

74 "Kehrseiten des Technologietransfers. Klagen deutscher Osthandelsfirmen", *Neue Zürcher Zeitung*, 1st and 2nd January, 1978.
75 *Ibid.*
76 F. L. Altmann and Hermann Clement, *op. cit.*
77 *Ibid.*
78 *Ibid.*
79 Herbert S. Levine, M. Mark Earle, Charles H. Movitt, Anne R. Lieberman, *Transfer of US Technology to the Soviet Union: Impact on US Commercial Interests, op. cit.* The conclusions are summarised in *Impact on US Commercial Interests of Technology Transfer to the USSR*, External Research Study, Department of State, 6th August, 1976.

assimilation of high technology is a complex process and the Soviets have had little success in this in the past. Even in the case where the Soviet Union may acquire turnkey plants in sectors where technical progress is very rapid, it would not maintain its competitive position given the prevailing technological level. To remain competitive, the USSR must itself contribute more to the development of its own technological progress.

In general, according to Herbert Levine and his colleagues, the prospects for Soviet competition depend essentially on future strategy with regard to foreign trade. Will the Soviets maintain their traditional policy of treating foreign trade as a means of making good short-term deficits, or will they aim at an increasing degree of interrelatedness with the world economy?

Fuller integration of the Eastern countries into the world economy poses the problem of competition in different terms. Western firms may fear competition from Eastern Europe in fields, other than high technology, which require very considerable funds. It is understandable that in this context Western entrepreneurs should wonder about the long-term effects of technology exports and the possibility of their "encouraging" the eventual emergence of competitors.

Totally opposite views have also been expressed on the subject. According to William C. Norris, President of Control Data Corporation, there is virtually no technology today that is exclusive to the United States. If the latter do not help East European countries to meet their know-how needs, other countries will. The United States would then lose in three ways: the value of the know-how; suffer product competition in world markets; and would not get know-how in return that could be the basis for greater competitive thrust for United States industry on world markets.[80]

Fred Bucy strongly opposes such arguments, comparing them to those used by gun-runners to salve their consciences. He thinks that the United States could prevent nations from being whipsawed into selling high technology. They might even be able to change the pattern of the Soviet offers to provide truly mutual competitive advantages.[81]

Fred Bucy is especially opposed to the opinion that selling technological know-how to East European countries is not likely to increase competition. In his statement, he stressed that[82] even obsolete technology in which East European countries showed little interest might lead to effective competition. Every high-technology company had a "corporate memory" of know-how and technology. It is a data bank containing technical background accumulated through years of experience. Much of it was never documented, but most of it was recorded in the memories of experienced engineers and embodied in manufacturing equipment. Even when selling obsolete know-how, therefore, the United States contributed invaluable background information to the potential competitor, giving it a priceless advantage in both time and cost.[83]

Bucy also criticised the arguments used by those who advocated sales of technological know-how on the assumption that the CMEA countries did not have the infrastructure necessary to support it. "The people concerned had short memories, viz. the development of Sputnik. It would be deluding oneself to expect the Soviet Union, once the technology is provided, to become a "hungry market" for Western technology (e.g. in electronics). This argument ignored a simple thrust of Soviet industrial policy: to make the East as self-sufficient as possible. The Soviet Union was not likely to favour Western manufacturers as long as any Eastern bloc manufacturer had essentially the same product for sale. Lastly, the argument that selling manufacturing know-how would generate enough revenue to sustain a high level of Western investment in R & D, was no more convincing. The only way

80 William C. Norris, *art. cit.*, p. 102.
81 Statement made at the *East-West Technological Trade Symposium*, sponsored by the US Department of Commerce, 19th November 1975, Washington, D.C., p. 40.
82 *Ibid.*, pp. 39-41.
83 *Ibid.*, p. 39.

to generate the continuing flow of revenue necessary to support R & D in manufacturing was by obtaining market share. Lump-sum payments, however large, lack the staying power to finance on-going research. Royalty payments and license agreements maintain the flow over a longer time, but they can rarely be made large enough to fund adequate R & D. And payment in products is shaky, at best. The only lasting solution was free competitive access to Eastern bloc markets".[84]

e) The Problem of Unfair Competition (Dumping)

The differences of conception concerning economic systems in general and international trade in particular are reflected in the mutual accusations of *dumping*, brought by the market-economy countries, and *discrimination*, brought by the socialist countries. In fact, these charges are not strictly on the same level. While the market-economy countries complain that the socialist countries do not play the game by their *own rules*, the socialist countries complain that they are subjected — quite tentatively, incidentally — to certain quantitative restrictions and tariff preferences which they themselves apply in a generalised manner and which are *radically bound up with other, more severe measures* in their system of imperative planning.

The currently accepted definition of dumping is the sale of a product abroad at a price below that applying at the same time and in the same conditions on the domestic market.[85] The definition applied by the GATT is more restrictive in that it includes two other conditions: the concept of "normal" value and the material injury caused to one of the contracting parties or the resulting curbs on the development of local industry.[86] Owing to the difficulty of defining what is the "normal" value for a socialist country, some Western countries like Australia, Sweden and the United States have introduced another criterion of dumping, namely the market price value in a third, non-socialist country with similar production costs.[87]

In fact, all these complaints simply amount to an attempt to extend the anti-trust law to the international context. Moreover, the trust in question here would be the socialist State itself. The market-economy countries are thus locked into a certain contradiction: having condemned interference with competition on the part of their own firms, they implicitly accept the same interference on the part of the State firms in the Eastern countries by the very fact of trading with them.

The monopoly exercised by the socialist State over all internal or external transactions is in no doubt. The socialist foreign trading enterprise is endowed with a national monopoly for a specific category of exports and imports. It does not have to engage in separate transactions for a given product, and losses and profits offset each other. The foreign trade monopoly and authoritarian pricing completely isolate domestic prices from foreign prices. The socialist trading organisation can also use its monopoly position to apply different prices to each partner in accordance with demand elasticity. The "losses" incurred in certain cases can easily be offset by gains in others where bargaining power is stronger. Moreover, a total loss may easily result from the provisions of the plan and be covered by the State budget.[88]

In any case, as pointed out by Fred Bucy, any market which the CMEA countries might temporarily grant the United States would be enjoyed only at their sufference. The moment a

---

84   *Ibid.,* pp. 40-41.
85   Gottfried Haberler, *The Theory of International Trade,* London, 1936, p. 296.
86   General Agreement on Tariffs and Trade (GATT), Article VI, paragraph 1.
87   J. Wilczynski, *The Economics and Politics of East-West Trade, op. cit.,* p. 173.
88   *Ibid.,* pp. 169-170.

CMEA country chose to capture a market, either within or outside the CMEA, it could penetrate it deeply on the basis of price alone.[89]

With imports of Western technology acquiring greater importance in East-West trade, the East European countries' export possibilities and hence prices are bound to be affected. The most usual complaints of dumping by the socialist countries relate to the prices at which products are offered on Western markets. J. Wilczynski, who has assembled a large number of such complaints,[90] shows that during the years 1957 to 1968 the products of the socialist countries were often sold at well below the general level of prices applying in the West, the gap varying considerably from product to product and country to country. For raw materials (zinc, aluminium, tin, titanium) the price gap ranged from 5 to 20 per cent, for oil as high as 40 per cent, for textile goods and footwear from 70 to 75 per cent, for machine-tools up to 40 per cent, for chemicals up to 75 per cent and for watches, cameras and musical instruments from 50 to 75 per cent. The markdown was smaller in the case of exports with an essential element of built-in technology, such as turnkey plants, optical instruments and so on, and generally did not exceed 20 per cent.

More recently an attempt has been made to calculate the reduced rates at which products from the East are sold in the West. The reduction is 5-25 per cent for Soviet products, 8-17 per cent for Rumanian products, 4-12 per cent for products from the German Democratic Republic, 4-13 per cent for Czechoslovakian products, 10-24 per cent for Bulgarian products and 8-25 per cent for Polish products. No Hungarian products have been sold at reduced prices.[91]

Examples of lower prices by East European countries are not lacking and a whole list could be made. The Soviet Union, for example, has been reported as selling low-density polyethylene (LDPE) at about DM.1 per kilogramme compared with the DM.1.30-1.40 charged by local German producers. Total Soviet sales are said to be 20-30 thousand tons per year.[92]

It is not easy, however, to prove that CMEA-produced chemicals are being sold below cost. Assembling the evidence takes a great deal of time whereas the harm suffered is immediate. Several methods have been devised to work out a trigger price based on Western producers' costs; selling at lower than this price would warrant anti-dumping action.[93]

The Commission of the European Communities in Brussels is said to favour this formula but is reluctant to apply a standard rule to all discount-priced chemicals. Apparnetly it feels that the chemical industry in the EEC has somewhat inflated the harm it has suffered. Its investigations on dumping will probably continue but only in certain specific cases.[94]

Dumping by way of export subsidies was also widely practised by the market-econonmy countries in their trade with the socialist countries. J. Wilczynski cites in detail the gaps between domestic wheat prices in Australia and the prices offered to the socialist partners, and the amount of the subsidies necessitated by the policy.[95] Subsidies for wheat sales were also applied by Canada, France and the United States. There are plenty of instances of unfair competition among the Western countries in their trade with socialist countries. Great Britain, for example, has complained that the other European countries generally offer prices 5 to 25 per cent lower than British prices.[96]

---

89   Fred Bucy, Statement made at the *East-West Technological Trade Symposium, op. cit.,* p. 41.
90   J. Wilczynski, *The Economics and Politics of East-West Trade, op. cit.,* pp. 149-153.
91   "Discount Rates for Sale of Eastern Production in the West", *Eastern Europe Reports,* 1976, No. 19.
92   "Chemical Buy-Back: Where Will It All Lead?", *Business Eastern Europe,* 9th June, 1978, p. 178.
93   "Chemicals in the East Explode West", *The Economist,* 10th February, 1979, p. 85.
94   *Ibid.*
95   J. Wilczynski, *The Economics and Politics of East-West Trade, op. cit.,* pp. 142-145.
96   *The Economist,* 10th September, 1966, p. 1053; cited by J. Wilczynski, *op. cit.,* p. 145.

In addition to price markdowns, there is the practice of offering disguised advantages such as the best credit terms (no cash down-payment or only a small one, low interest rates, longer repayment periods), convenient methods of payment (in non-convertible local currency, barter) and other facilities such as training of local technicians or premiums and favours granted to employees.[97]

Recently complaints about East European dumping concern the Soviet and other East European shipping lines' policy of granting 30 to 40 per cent discounts and reductions on most European and African routes.[98] According to the journal *Seatrade*, (quoted by *The Economist*, London) ships belonging to the Soviet Union, Poland, the German Democratic Republic and two other CMEA countries controlled 35 per cent of the freight market between Northern Europe and the Mediterranean, 25 per cent of that between Northern Europe and the Western seaboard of Latin America and more than 20 per cent of the traffic between the Gulf of Mexico and the Mediterranean. In the Far East, Fesco, the Soviet shipping operator has already acquired 12 per cent of the market between Japan and the West Coast of the United States.[99] Some observers think that all major international routes are under threat from Soviet competition.[100] The same is true of German Democratic Republic's shipping.[101]

The official American view is that this penetration of the market by the Soviet Fleet is attributable to unethical freight rate practice.[102] The policies have even been confirmed by Guzhenko, the Soviet minister for merchant shipping: "During the summer talks between Soviet Merchant Marine Minister Guzhenko and Edmund Dell, British Secretary of State for Trade in London, Mr. Guzhenko admitted that Soviet lines were undercutting liner Conference rates. He said, however, that the Soviet Union wanted to join the North Atlantic and Pacific Conferences and did not intend to act in any way to deepen differences. The major problems of the East African and other lines were also discussed".[103]

In addition to this discounting practice there are also fears in the West with regard to the composition and planned expansion of the Soviet Fleet. It would seem that the Soviet Union has its eye on the more profitable trades rather than what is necessary under bilateral trade agreements. It has been said that forecasts for Soviet shipbuilding to 1985 do not seem to match future Soviet Merchant Marine requirements.[104] The Soviet Fleet is expected to continue to expand at the higher rates observed in the past, thus reducing or eliminating altogether the technological gap in this field in most categories.[105]

97   *Ibid.*, pp. 153-154.
98   "Le pavillon soviétique bat les records en matière de dumping", *L'Echo de la bourse*, Brussels, 6th September, 1977, quoted in *Problèmes économiques*, La Documentation française, Paris, 14th December, 1977, pp. 24-25.
99   "Russia's Merchant Fleet", *The Economist*, 18th June, 1977, London, as quoted by *Problèmes économiques*, La Documentation française, Paris, 14th December, 1977, p. 23.
100   Philippe Poirier d'Ange d'Orsay, "Pour notre flotte de commerce: une menace venue de l'Est", *Le Figaro*, Paris, 10th-11th April, 1976, p. 5.
101   A mine detection unit built in the United Kingdom was recently shipped to Hamburg by a German Democratic Republic's freighter. German Democratic Republic's operators offer rates 30, 40 or even 50 per cent below those of Western lines. "Les pratiques de dumping des pavillons de l'Est", *L'Echo de la bourse*, Brussels, 20th August, 1977, quoted in *Problèmes économiques*, La Documentation française, Paris, 14th December, 1977, pp. 25-26. Denmark, United Kingdom and Australia have recently called for immediate sanctions and a common long-term policy to fight against the Soviets' extremely low freight rates. A. Tirapolsky and B. Despinay, *art. cit.*, p. 43.
102   John P. Hardt, "Maritime Development Involving the Soviet Union, the United States and the West", in: *Issues in East-West Commercial Relations, op. cit.*, p. 252.
103   *Ibid.*, p. 252.
104   *Ibid.*, p. 253.
105   According to Timofei Guzhenko, the intention in the Soviet 1976-1980 five-year Plan is to increase the fleet by 4.6 million deadweight tons. Mr. Guzhenko calls this growth "slow and steady": John P. Hardt, "Maritime Development Involving the Soviet Union, the United States and the West", *art. cit.*, p. 253.

According to one Western expert, the East European countries are able to offer cheaper rates because of the accounting methods which they employ. Soviet shipping operators do not have to pay for the amortisation of their vessels (whereas in the West such payments amount to approximately 35 to 50 per cent of operating costs).[106] Soviet ships do not have to pay for insurance (the State meets the bill) and wage costs are much lower: $30 per week for a Soviet seaman compared to $50 per week in the West.[107]

Transfer of Western technology is partly responsible for the boom in Soviet shipping and the competition it is offering. By 1980, the Soviet Union intends to increase its container and roll-on and roll-off fleet from 14 to 40 ships.[108] With this in mind, it has been acquiring Western technology in several ways: importing equipment from Western countries, direct purchase of new vessels (CMEA countries as a whole have bought almost 20 per cent of the ships now being built in the world),[109] taking advantage of the machinery and equipment purchased by other CMEA countries from the West and then exported to the USSR in vessels built in East European shipyards. This often applies to the Polish vessels sold to the Soviet Union, and the Poles have been heard to complain that they have to pay for this machinery in hard currency while they themselves are paid in transferable roubles.

To the criticisms from the West, the leaders of the socialist countries make rather dilatory responses. "We have limited quantities for export. Then any country can oppose dumping by means of a simple administrative decision", replies the Polish Deputy Minister for Foreign Trade, Stanislaw Dlugosz.[110] "We are founder members of the GATT and as such we observe the anti-dumping articles", replies the Czechoslovak Deputy Minister for Foreign Trade, Jaroslav Jakubec.[111]

The Soviet leaders' attitude towards complaints of freight-rate dumping is rigid: "Lower transport costs, which are due to the benefits enjoyed by the Soviet merchant shipping in our economic system, enable Soviet operators to allow discounts bringing their rates below those of the West. These benefits include lower operating costs than in the West..."[112]

Western countries do in fact have ways of fighting dumping practices by taking measures under the GATT, under bilateral agreements or through national legislation. A special study would be needed to describe these measures, and this would fall well outside the scope of the possible impact of technological transfer on Western economies. The GATT as negotiated in 1947, did not include any provisions directly concerned with imports from non-market economies.[113] Protection measures were then regarded as exceptional and included the general escape clause, anti-dumping and countervailing duties applicable in the event of market disruption, where this was likely to cause considerable injury.

106   "Russia's Merchant Fleet", *art. cit.*

107   *Ibid.* The figures quoted here are somewhat problematic and should be viewed as being approximate. Since Soviet employees are paid in rubles, comparisons with Western salaries are difficult to make. It is assumed that *The Economist* used the official Soviet exchange rate, and it would have been better had the source compared the "price" a Western ship-owner must pay when he charters an East European crew. There are some Swedish shipping companies which hire Polish crews for industrial fishing and in these instances the wage differences are greater than the figures quoted here. The purchasing power of the ruble is actually not that different from the official exchange rate. In 1972, it was 6.33 FF for a ruble while the exchange rate was 6.09 FF for a ruble. See Anita Tiraspolsky, "Le pouvoir d'achat du rouble en 1972", *Revue de l'Est*, Paris, janvier 1974, p. 102.

108   *Ibid.*

109   *Ibid.,* p. 24.

110   Guy Schwartz, "Les sept moyens de coopérer avec l'Est", *Usine nouvelle,* Paris, June 1976, reprinted in: *Problèmes économiques,* La Documentation française, Paris, 1st September, 1976, p. 25.

111   *Ibid.*

112   "Le pavillon soviétique bat les records en matière de dumping", *L'Echo de la bourse, art. cit.,* in: *Problèmes économiques,* La Documentation française, Paris, 14th December, 1977, p. 24.

113   Karen C. Taylor, "Import Protection and East-West Trade: A Survey of Industrialized Country Practices", in *East European Economies Post-Helsinki — A Compendium of Papers, op. cit.,* p. 1135.

Current Western restrictions on imports from the East are somewhat varied. There is the refusal by the United States to give to the Soviet Union and to some of the CMEA countries Most Favoured Nation Treatment and the quota system applied by the European Community. In addition, CMEA products may be subject to "safeguard action", "anti-dumping" investigations and applications for compensatory customs tariffs. In the United States market disruption proceedings may be taken against them.[114]

The measures applied by the European Economic Communities include common rules for imports from State-trading countries (including an import surveillance system), anti-dumping regulations and partial control of the unilateral measures taken by the member States. Any legal or material person can lodge a complaint against the dumping, but very few countervailing duty actions have been taken.[115] In 1975, the quantitative restrictions established by the Community against State-trading countries included 1 570 different restrictions, of which 269 were against Czechoslovakia, 247 against Rumania, 238 against Poland and only 105 against the USSR. Any further quantitative restrictions will have to be established using a special procedure under Community rules.[116]

Various reports on complaints submitted to the Community concerning market distortion have appeared in the press. For example, the European Council of Chemical Manufacturers has recently filed a complaint with the European Economic Commission concerning imports from CMEA countries at prices "based on political criteria", that are a threat to the Western petro-chemicals market, and listing 60 of the most vulnerable products.[117] Other complaints to the EEC have also been reported.[118]

It is significant that many of these complaints relate to products for which the contribution of Western technology has recently been substantial. For instance, Great Britain has lodged complaints against exports of motor cars from Poland, Czechoslovakia, the German Democratic Republic and the USSR. These complaints essentially concern the Soviet-produced *Lada*. Such high-technology products, as equipment for plastics manufacture, electrical engines, polyester fibres, thermostats and electronic alarm clocks have also been the subject of complaints by EEC members.[119, 120]

4. TECHNOLOGY TRANSFERS AND RESUMPTION OF BILATERALISM IN EAST-WEST TRADE

a) East-West Trade and its Rules

In the market-economy countries the system of trade and international payments is regulated by certain rules established by international agreements and institutions. One of the most important rules is that whereby trade is conducted by a multitude of private firms, each trying to maximise its profit. In the absence of government intervention, private firms

---

114   Karen C. Taylor and Deborah Lamb, *art. cit.*, p. 128. On controls on American imports, see: Karen C. Taylor, "A Summary of US Laws Applying to Imports of Communist Products", in *Issues in East-West Commercial Relations, op. cit.*, pp. 173-175.

115   Karen C. Taylor, "Import Protection and East-West Trade: A Survey of Industrialized Country Practices", *art. cit.*, pp. 1156-1158.

116   *Ibid.*, p. 1159.

117   "Comecon Imports Under Scrutiny", *Financial Times,* London, 27th February, 1978.

118   A. Tiraspolsky and B. Despinay, *art. cit.*, pp. 44-46.

119   *Ibid.*

120   The Soviets have estimated that foreign currency earnings from sales of passenger motor cars manufactured at the Togliatti plant amount to the cost of the equipment imported. A. Dostal', "Vneshnyaya Torgovlya i kredit" (External Trade and Credit), *Ekonomicheskaya Gazeta,* No. 31, 1977, p. 21.

will try to sell their products at lower cost on the domestic market and to import the cheaper products from abroad. Domestic and international prices are determined essentially by supply and demand, and trade tends to equalise them (not taking into account transport and distribution costs, and customs tariffs). Inasmuch as prices determine the amount and direction of trade, there are mutual advantages for the trade partners.

The socialist countries, on the other hand, are not necessarily guided in their foreign trade decisions by domestic/foreign price differentials.[121] Their domestic prices are fixed by the government, do not reflect equilibrium of supply and demand, and the distribution of products for which there is excessive demand is rationed. The State trade enterprises are not necessarily guided by economic considerations: ideological, political and military motives often play a decisive role. For this reason, a private exporter to a socialist country may be unable to compete for a segment of the socialist market to which he might reasonably aspire as a result of his efficiency and prices.

This problem was well understood before the war, and a clause requiring the observance of commercial and financial considerations was inserted in the Soviet-British trade agreement of 1930 at Britain's insistence. This clause has subsequently been written into many trade pacts, even though there was no possibility of implementing it.[122]

CMEA countries have always operated exchange controls and used multiple rates of exchange. Nevertheless, when trading with Western industrialised countries, debts were settled promptly in hard currency. This was also the procedure used by the Soviet Union in the period between the two World Wars.

With the significant purchases of Western technology and grain by CMEA countries since the early 1970s and the increase in their convertible currency debt, this policy is now changing. The CMEA countries are now constantly pressing for a larger proportion of countertrade in their agreements with Western countries. In other words, they are "exporting" the bilateralism currently prevailing in the CMEA.

Table 75 gives the results of various calculations regarding the scale of East-West compensation and counterpurchase deliveries. Counterpurchase deliveries alone seem to have amounted to $10 billion for the period 1976-1980[123] and current estimates (see Table 78) forecast that compensation and counterpurchase deliveries will account for 30-40 per cent of exports from East to West in 1980.

In fact, compensation trade is beginning to develop outside the East European markets. According to estimates, about 20 per cent of world trade is now subject to various restrictions, and Olivier Long, Director-General of the GATT, believes that this is menacing the world trade system.[124] In February 1978, Mr. René Monory, French Minister for Industry, said that the question of limiting compensation trade in Europe would soon have to be tackled.[125] All this naturally raises the problem of the present and future economic implications of compensation trade in East/West trade.

121   On the difference between internal and external prices in the USSR see the following articles from the *Revue de l'Est*, Paris, No. 1, 1974: Keith Bush, "Les prix de détail à Moscou et dans quatre villes occidentales en novembre 1971", Anita Tiraspolsky, "Le pouvoir d'achat du rouble en 1972", Alexander Woroniak, "Le problème de la conversion du rouble en dollar".

122   The Soviet-British trade agreement of 1930 was denounced by Great Britain, which considered that State trading and the most-favoured nation clause are fundamentally irreconcilable. But this clause was reinserted in the 1934 agreement between the two countries on the grounds that it was the only legal guarantee, however imperfect. J. Wilczynski, *The Economics of Politics of East-West Trade*, *op. cit.*, pp. 131-132.

123   Quoted from: Economic Commission for Europe, "The Outlook for 1979: Where the Opportunities Lie", *Business Eastern Europe*, January 5, 1979, p. 2.

124   "Sicherung der Unternehmensautonomie. Warnung Olivier Long vor protektionistischen Welthandelspraktiken", *Neue Zürcher Zeitung*, 12th May, 1978, p. 13.

125   "Les ventes d'usines "clés en main": Les risques d'un essor dans la confusion", *La Croix*, Paris, 10th February, 1978.

b) Economic Effects of Compensation Trade

The simplest indicator of the pressure applied by East European countries to make barter a feature of trade with the industrialised Western countries is the share of Western products which is to be paid for in kind. Variation in this share depends mainly on whether the Western exports include plant or other items.

For the Soviet Union, buy-back commitments generally do not exceed 20-30 per cent of total Western sales, but in some cases, the Soviet partners insist on 100 per cent, or even 120 per cent to incorporate payments for loans included in the deal.[126] If the Western firms do not agree, the Soviet side will accept "mixed" payment (e.g. 70 per cent buy-back and 30 per cent counterpurchasing) or even 100 per cent counterpurchasing although in that case the figure would be increased — to 110 per cent, for instance. The Soviets are also prepared to pay for only about 5 to 10 per cent of their imports, other than turnkey plants, through countertrade. However, they have recently increased this to 20 to 30 per cent. The restriction did not apply to imports less than $1 million.[127]

Poland has recently started restricting convertible currency imports to self-repaying projects, i.e., those where foreign credit can entirely be paid by the return on the concerned investment over a given period. This type of credit totalled $3.6 billion between 1971 and 1976.[128] For other currency imports Poland is asking for a 25 to 30 per cent countertrade share, but Western exporters fear that this may rise to 50 to 60 per cent by the 1980s.[129]

Bulgaria and Rumania want particularly high buy-back payments levels: 100 per cent in both cases.[130, 131] The German Democratic Republic is more moderate (it used to ask for 20 per cent but has now doubled this to 40 per cent). But these are firm requirements and there is no possibility for protracted bargaining.[132] Yugoslav firms ask for 30 per cent countertrade for capital imports and 100 per cent for other types.[133]

The primary difficulty facing Western firms in buy-back deals is how to dispose of the products. Their first concern is to try to obtain products which they themselves can use,[134] include in their foreign sales,[135] or sell easily.[136] In some cases, Western firms negotiate

---

126 "Developments in Countertrade: Part VIII – the Soviet Union", *Business Eastern Europe,* 5th August, 1977, pp. 244-245.

127 *Ibid.,* p. 245. On "mixed payment" see also "Financing Deals with EE(VI): The Soviet Union", *Business Eastern Europe,* July 28, 1978, p. 236.

128 "Poland Limits Investments to Self-Repaying Credits", *Business Eastern Europe,* 22nd July, 1977, p. 228.

129 "Developments in Countertrade: Part VI – Poland", *Business Eastern Europe,* 15th July, 1977, p. 220.

130 Swedish negotiators are reported to have turned down a Bulgarian request for 200 per cent countertrade: "How a Swedish Company Handles Counterpurchase Deals", *Business Eastern Europe,* 25th February, 1977, p. 57.

131 The Rumanians are said to start bargaining on the basis of 100 per cent countertrade but are prepared, in the end, to accept 30 per cent. "Developments in Countertrade: Part V – the GDR", *Business Eastern Europe,* 8th July, 1977, p. 213.

132 *Ibid.,* p. 213.

133 "Counterpurchase Pressure to Continue in Yugoslavia", *Business Eastern Europe,* 4th March, 1977, p. 70.

134 "How Daimler-Benz Handles Countertrade Commitments", *Business Eastern Europe,* 14th January, 1977, p. 9.

135 One Swedish firm was able to build part of the buy-back equipment from Poland into a plant it was delivering, thus boosting its profits. "How a Swedish Company Handles Counterpurchase Deals", *art. cit.,* pp. 57-58.

136 Countries like the German Democratic Republic and Czechoslovakia have products that are easier to sell in the West and are therefore less intent on countertrading. For example, in the knowledge that their products are easier to sell, the Czechs occasionally ask for 150 per cent buy-back but such deals may be more attractive to Western firms than others at only 30 per cent. In both countries, however, most of the goods in buy-back deals are already earmarked. In the chemicals and textiles

(Cont'd on next page)

tied sales agreements to dispose of the East European products acquired in this way.[137]

Western firms have to undertake full scale market research projects for these East European products, involving time-consuming and difficult work,[138] and encompassing both product type and quality.[139] Very often, the outcome of this market research is not entirely satisfactory and Western firms are forced to take products which they will have to sell in turn. Examples of such purchases abound. German firms often receive felt slippers, starch, rabbit skins, canned vegetables, onions or electrical motors in exchange for the goods they have sold to Eastern Europe. For instance, Krupp was sent a large batch of tights among the goods delivered in exchange for a DM.240 million order for two plants. Other West European countries come up against similar problems. Some years ago Fiat became known as the biggest egg merchant in Europe.[140]

Certain dealers in London, Amsterdam, Vienna, Paris, Lausanne or Milan, etc., specialise in selling products off-loaded onto Western firms under buy-back agreements. Some of the firms which regularly trade with Eastern Europe, such as Daimler-Benz, have set up their own subsidiaries for this purpose. The products are often sold at a discount,[141] thus creating competition for other Western firms,[142] and the end-result is that the marketing network that the East European countries ought to have in Western industrialised countries is replaced by Western firms.[143]

Re-selling is not the only complication in countertrading. The search for suitable products in Eastern Europe is hampered by differences in the administrative responsibilities of the East European Foreign Trading Organisations,[144] the various requirements regarding

(Cont'd note 136)

industries countertrade share has fallen to 10 and 15 per cent, respectively. Some products from the German Democratic Republic and Czechoslovakia are already being sold in the West through the countries' own agencies, to which Western firms have to pay special commission. "Developments in Countertrade: Part V – the GDR", *art. cit.*, pp. 213-214; "Developments in Countertrade - Part VII – Czechoslovakia", *Business Eastern Europe*, 22nd July, 1977, pp. 228-230.

137   "Counterpurchases Trouble West German Machine Builders", *Business Eastern Europe*, 29th July, 1977, p. 233.

138   See the Claus Welcker Interview in "Counterpurchases Trouble West German Machine Builders", *art. cit.*, p. 233. He mentions difficulties in obtaining products in the case of Czechoslovakia, Poland and the German Democratic Republic, *Business Eastern Europe*, *op. cit.*, 8th July, 1977, pp. 213-214; 15th July, 1977, *op. cit.*, pp. 220-221; *op. cit.*, 22nd July, 1977, p. 229. Representatives of the company set up by Daimler-Benz to sell products obtained in exchange for passenger motor cars (50 per cent of these products come from East European countries) regularly tour Eastern Europe to find products that could sell in the West; "How Daimler-Benz Handles Countertrade Commitments", *art. cit.*, p. 9.

139   Western firms often prefer to carry out their own quality control on the products they buy under countertrade agreements. See "How Daimler-Benz Handles Countertrade Commitments", *art. cit.*, p. 10. Robert E. Weygand, "International Trade Without Money", *Harvard Business Review*, November-December 1977, as reproduced in *Problèmes économiques*, La Documentation française, Paris, 14th June, 1978, p. 17. "Building Countertrade into Long-range Planning", *Business Eastern Europe*, 11th February, 1977, pp. 42-43.

140   Peter Stoltz, "Machines contre tomates. Les échanges avec les pays de l'Est", Revue hebdomadaire de presse, *Economies des pays communistes*, NATO, 14th March, 1978, p. 26. As translated in *Kölner Stadt-Anzeiger*, 18th February, 1978.

141   As the Poles often insist on over-inflated prices for countertrade products (especially for machinery), these can only be sold at a discount of 15-25 per cent. "Developments in Countertrade: Part VI, Poland", *art. cit.*, p. 220.

142   It has been reported that German wine merchants are facing competition from cut-price East European red wine obtained by a German exporter in exchange for machinery, "Counterpurchases Trouble West German Machine Builders", *art. cit.*, p. 234.

143   *Ibid.*, p. 234.

144   "Developments in Countertrade: Part VI – Poland", *art. cit.*, pp. 220-221, mentions complications arising from the various spheres of responsibility of the Polish foreign trade organisations.

contractual penalties for non-compliance with countertrade commitments,[145] and the often inflexible delivery times[146] which are not always met.[147]

It is difficult to estimate the importance of countertrade. Buy-back goods are often basic commodities such as oil, coal, wood, etc., and therefore easy to sell in the West.[148] The difficulties lie with the manufactured goods supplied under compensation agreements and counterpurchase products. This type of trade is often criticised in the West. Mr. Hartwig, President of the West German Federation of Wholesale and Foreign Trade has said that "East European countries regularly supply goods that are not wanted at the least suitable moment in time".[149] Others take the view that these agreements represent a step backwards since they constitute a return to the trading practices of the Middle Ages.[150]

In spite of these criticisms, it is generally agreed that countertrade is likely to develop in the near future. A Swedish company involved in East-West trade considers that "buy-back payments will become the main form of East-West trade in the next few years".[151] Any firm willing to countertrade will have an advantage over its competitors.[152]

Other Western authors have made similar predictions: Robert E. Weygand writes that the current crisis and difficulties encountered in maintaining their balance of payments are inducing many countries to continue with or introduce the procedure of paying for their imports in kind.[153] Another Western observer, Everett M. Jacobs, goes a step further in a report to the NATO Colloquium of January 1978: "because they cost so little to the Soviet Union in real terms, compensation agreements will be pushed by the Soviet planners and will doubtless become more important in East-West trade in the near future.[154]

5. NEW TRENDS IN THE FORMS AND CONTENT OF EAST-WEST TRADE AND THE PROBLEM OF RESTORING EQUILIBRIUM

There is a surprising conflict in attitudes in the West towards the development of East-West trade: on the one hand countertrade is said to be unprofitable and on the other it is expected to increase.

The real problem would seem to be that of whether Eastern and Western economies will develop in complementary fashion. Will present forms of trade help to reorganise structures on both sides in terms of the broader division of labour, or even "internationalisation of the production process"?

The new forms of co-operation described in Chapter 3 give the impression that technology transfer fosters greater international co-operation through the various agreements for co-production, technical assistance, and so on. It might therefore be thought that East-West trade is likely to become more intensive.

145   Regarding contractual penalties, see for example: "Developments in Countertrade: Part VII – Czechoslovakia", *art. cit.*, p. 230; "How a Swedish Company Handles Counterpurchase Deals", *art. cit.*, p. 58; "Counterpurchases Trouble West German Machine Builders", *art. cit.*, p. 224.
146   "Developments in Countertrade: Part V – the GDR", *art. cit.*, p. 214.
147   "Kehrseiten des Technologietransfers: Klagen deutscher Osthandelsfirmen", *Neue Zürcher Zeitung*, 1-2 January, 1978, p. 9.
148   See Chapter 6 above. Payment in easily sold commodities accounts for over 2/3 of the total in compensation contracts with the USSR (see Table A-45).
149   Peter Stoltz, *art. cit.*, p. 27.
150   *Ibid.*, p. 27.
151   "How A Swedish Company Handles Counterpurchase Deals", *art. cit.*, p. 58.
152   *Ibid.*, p. 58.
153   Robert E. Weygand, *art. cit.*, p. 19.
154   Everett M. Jacobs, "The Global Impact of Foreign Trade on Soviet Growth", in: *The USSR in the 1980s. Economic Growth and the Role of Foreign Trade, op. cit.*, p. 205.

The real issue, however, is whether current forms of trade are helping to fit Eastern and Western economic structures into one another.[155] While opinions on this issue differ, it could be maintained that present trends in East-West trade, particularly the growth of barter, bilateral and buy-back deals, will not be conducive to increased trade nor to the complementary development of Western and Eastern economies. Increased bilateral countertrade could mean that the East European economies will be "protected" from the necessity to adjust themselves to Western demand. Even if the East European countries were prepared, in their National Economic Plan, to adapt their economy so as to facilitate East-West trade, an argument can be made that their systems would still be too rigid and influenced too much by non-economic considerations of power, thereby negating the effect of such adaption. In other words, there might still be a tendency for Western buyers to be offered products "not allocated" under the plan so that deficits can be made up. It can also be argued that plant purchases (under compensation agreements — even at 100 per cent) do not necessarily guarantee that the best sectors, from the standpoint of exports, become "open" to the West. On the whole, it can be said that the future development of East-West trade will be greatly affected by the existing technological gap.

---

155   Werner Beitel, *Probleme der Entwicklung des Sowjetischen Westhandels, op. cit.,* pp. 49 *et seq.*

CONCLUSION

1. The Scale of Technology Flows between East and West

The first objective of the study was to quantify technology flows in East-West trade.

However, it has to be admitted that there is no way of giving a complete answer to this question from the statistical sources available. For the "measurable" share of technology, i.e. that incorporated in products, we can, generally speaking, refer only to machinery and licensing. Assessing the transfer of know-how can only be done indirectly, by studying the nature of co-operation agreements. Disaggregation of statistics by product using the standard classification (SITC) is a time-consuming exercise and, even if taken as far as the five-digit headings, does not always reveal research density, or the presence of high technology. What is more, technology is changing and developing all the time so that the same statistical heading may cover products of differing sophistication.

Nevertheless, in spite of these statistical shortcomings, the main features of East-West technology transfer are clearly visible.

The direction of technology flows is mainly from West to East. Between 1971 and 1974 "technology-intensive" products accounted, on average, for 46 per cent of all Eastern Europe's imports from the West [1] but only 15 per cent of exports in the opposite direction. This gap has widened further during recent years, the corresponding averages for the period 1976-77 being 49 and 12 per cent. [2]

If machinery trade is taken as a criterion of technology flow, the dominant direction is also from West to East. Between 1970 and 1976, the machinery share of CMEA countries' imports from the OECD countries varied between 31 and 35 per cent, while that of OECD countries' exports was about 7-9 per cent. [3] A point worth noting is that machinery imports to CMEA countries from the industrialised West are not necessarily high technology products. For the Soviet Union the share of high technology [4] in imports from the industrialised West ranged from 14 to 17 per cent between 1972 and 1976 and for the other European CMEA countries between 11 and 12 per cent. [5]

CMEA machinery imports from the West increased at a very rapid rate between 1970 and 1975 (9 to 10 times), even faster than imports in general for which the factor was 7.5. [6] Extremely high increases in machinery imports were recorded in 1974 and 1975 (35 and

---

1 See Table 1. United Nations figure based on ten industrialised Western countries.
2 See Table 2, United Nations figure.
3 See Table A-7 of the Annex.
4 For the definition of "high technology", see Table A-2. The term includes machinery and also certain instruments (SITC category 8), so that the percentages quoted cannot be directly compared to those for the share of machinery, but the error entailed is slight.
5 See Table 3.
6 Calculated from the figures in Table 9.

55 per cent, respectively. In 1976, CMEA machinery imports levelled off and in 1977 increased by only 10 per cent.[7] If allowance is made for price escalation (see Table 62) a reduction of machinery imports occurred in 1977 and 1978. Expansion of the kind experienced between 1970 and 1975 is, at the moment, hardly to be expected.

It has to be admitted that measuring technology transfer in the form of know-how with any accuracy is practically impossible. But there is little doubt, and this is confirmed by Western researchers, that East-West industrial co-operation agreements are a powerful medium for this type of transfer.[8]

The only way to analyse industrial co-operation agreements is by reference to surveys. These merely list contracts without necessarily specifying their value, except in the case of turnkey plants and compensation agreements. Here contract values are given,[9] enabling the relative importance of a particular trade flow to be assessed. The USSR has by far the largest contracts, as confirmed by its share (over 50%) of total machinery imports from industrialised countries.[10] The main lesson to be learnt from the surveys, whose results are summarised in Chapter 3, concerns the link between the various forms of transfer and the way in which know-how is passed on. Know-how transfer is particularly effective through sales of plant and joint ventures. The rising number of industrial co-operation contracts in recent years is impressive. Their attraction for the East seems to consist not only in the technology to be acquired but also in the forms of payment (credit and/or countertrade) so that it is difficult to tell which of the two is more implicit in the term "co-operation".

2. East-West Relations and Policies Regarding Technology Transfer

Technology transfer is far too important an aspect of East-West relations to be studied without reference to policies of both parties. However, it is very difficult to establish the connection between policy and actual transfers and it therefore seemed preferable to consider the problem as a whole rather than to look for any particular impact.

In recent years, East-West relations have developed within a political framework which is more conducive to the development of closer commercial ties. The policy of détente has, for instance, helped to make technology transfer more acceptable and less controversial than it was during such periods as the Cold War. Arguments for closer trading ties between East and West have also been bolstered by Eastern support of the notion of the international division of labour — though as is pointed out in Chapter 4, Eastern support of this notion is greatly nuanced.

Chapter 4 deals in part with the East European countries' policy and also discusses changing technology needs, believed to be the prime force in policy decisions. Little is known about the arguments which may have been used in the East — either for or against extending trade with the West — but the policy has been fairly obvious. In the beginning, the Soviet Union and later all the CMEA countries aimed at self-sufficiency. Subsequently, after the sixties, East European leaders started to move increasingly towards the international division of labour, although this policy was not allowed to act as an obstacle to integration among CMEA countries where there was also a demand for even closer internal ties.

Our study does not reveal any link between Western or Eastern policies and the trends in technology transfer observed in recent years. During the first half of the seventies, both seemed to be developing along parallel directions: policies were becoming more favourable and trade was expanding. More recently, however, in spite of policies favouring transfer economic difficulties have hindered the expansion of technology transfer.

7 See Table 8.
8 In this connection, see Carl H. McMillan's. "East-West. Industrial Cooperation", *art. cit.*
9 See Table 30 and Tables A-32 and A-39.
10 See Tables 8 and 9.

3. **The East's Need for Western Technology and the Impact of Transfers on East European Economies**

This study clearly shows that several factors make the East European countries' need for Western technology pressing and vital.

Since the Second World War, the slowdown in the growth of the active population and the depletion of easily accessible energy sources and raw materials have reduced growth rates and the productivity factors of production in Eastern Europe. Only investment based on modern technology can reverse the trend.

National efforts to innovate and introduce new technology have proved inadequate as the administrative planning system is not very conducive to innovation. A further difficulty is the high cost of modern technology if Research and Development activities were conducted in all sectors. A technological lag, favouring the West, has occurred in practically all major sectors (see Chapter 5). The lag is evident both at the general scientific level and at the level of application of technology in production. As regards science in general, the lag compared with the United States may be measured in terms of the number of Nobel prizes, quote frequency, or simply the number of scientific publications. Technology lag in the application of technology in production is clearly demonstrated in the pattern of East-West trade.

It must be borne in mind however, that the technological lag is not an indispensable condition for the appearance and increase in foreign technology needs. Its only effect is to dictate a certain type of trade which is usually found between industrialised and developing countries. The level of economic maturity reached by the East European countries does not warrant such a situation, and the fact that trade patterns are actually tending towards this model is largely due to the East's acute technology needs. However, unless combined with a policy aimed at strict self-sufficiency, modernising the economy in the East European countries is bound to heighten the need for Western technology.

Current policies in Eastern Europe encourage the use of Western technology. The pressure for innovation is such that Western technology has to be chosen for those sectors where national performance is inadequate. Technical progress is not the only aim in using Western technology, however. Constant calls are also made to compensate for the often unexpected weaknesses in the national production system.

These weaknesses, whose immediate cause is the "tightness" of the national plans, are also partly due to the lack of national savings. Western technology is generally imported on credit and its contribution is therefore all the more valuable. But this also explains why most of the machines imported from the West are the ordinary models — not the latest inventions.[11]

While the East European countries' acute need for Western technology is beyond question, the impact of Western technology on these countries is very much a matter of debate. In Chapter 5, the only publications quoted relate to the Soviet Union. However, Western technology also reaches the latter via the other CMEA countries, directly or incorporated in the goods they make. While we know very little about the effects of the latter kind of technology transfer to the East, we also have much to learn even about the impact of the direct transfer of technology on the Soviet economy.

The first point made in the publications we have referred to in Chapter 5 is the small share that equipment exports from the West account for in total Soviet investment. However, in such case studies as the mineral fertilizers industry, the impact is clearly considerable. The same conclusion emerges from general econometric studies.

The opposite conclusion, however, is reached in Sutton's historical survey and by studies of the diffusion and assimilation of new technology in the Soviet Union. Innovation difficulties under the Soviet system are blamed for the persistence of the technological gap in

---

11 See Table A-46 of the Annex.

spite of massive imports of Western technology. This is often the line of argument used by Western advocates of a substantial increase in technology exports to the East: "since in any case the Soviets are incapable of diffusing new technology or innovating themselves, why worry, why refuse profitable sales of high technology?".

The analysis of the technological gap between the USSR and the West, as summed up in Chapter 5, should give the propounders of this argument cause for caution. In several sectors, — e.g. high voltage power transmission, numerically-controlled machine-tools, jet engines, etc., — Soviet performance is quite remarkable and even if Soviet products do not achieve the same technical standard as that of the West, (e.g. passenger motor cars) the results achieved could well have considerable effects on national production, and even on international competitiveness. A good example is the success of *Lada* cars, based on Fiat models. In other words, the impact of Western technology must be studied very carefully sector by sector — almost technology by technology — and hasty generalisations avoided.[12]

4. Effect of Economic Factors on Transfers

While economic factors undoubtedly have considerable effects on East-West technology transfer, such effects are not easy to identify or quantify.

Transfer is greatly affected by the immediate economic situation but seems to be less sensitive to price fluctuations. Favourable trends in the terms of trade for the East European countries between 1970 and 1975 were certainly accompanied by an increase in machinery purchase from the West (see Table 60). Nevertheless, a further improvement in the terms of trade in 1976 and 1977 did not prevent their volume from levelling off (see Tables 8 and 62). In the same way, the worsening terms of East-West trade in machinery for the CMEA countries over the period 1971-1974 and for the whole of the period 1968/69 to 1976/77 seems to have had little influence on the steep growth between 1970 and 1975 or on the recent standstill.[13]

Even a brief analysis of East-West transfer of technology (incorporated in machinery) shows that it depends very much on payment possibilities. The deficit in the balance of payments is almost entirely due to that in the trade balance, which is approximately equal to the East's total machinery purchases (see Tables 66 and 67). The latter correlation is hardly surprising in view of the way in which Western sales of equipment are financed. In other words, it was Western credit, in various forms, which enabled machinery and equipment sales to the East to rise as they did between 1970 and 1975.

Again, it was the East European countries' mounting debt and their desire to keep the deficits in their trade balances with hard currency countries within limits that was directly responsible for failure of machinery purchases to increase from 1976 onwards.

We may therefore conclude that future growth in technology sales to the East will depend on one of two factors: continuation of Western credit and/or an increase in East European sales on the Western market.

The first condition is easy to fulfil and Western firms seem anxious to grant credit terms (as this is made very easy through government guarantees). However, in view of the present and future debt situation, credit expansion poses obvious problems. The USSR is obviously in a more favourable position, but it would be very unwise to rely on the "umbrella" theory.

---

12 In some quarters it is thought that Western controls on high technology should be relaxed on the grounds that, come what may, armament production receives priority in the Soviet Union, which avoids any dependence on foreign technology in this sector. See William C. Norris, *art. cit.*, p. 102. In other words, controls are unnecessary (or should be minimal) since the Soviet Union is able anyhow to meet all its own military requirements. However, even if this were true up to now there is no reason why it should continue to be so in the future. What is more the statement seems to go too far and calls for factual proof.

13 See Table 61 as compared to Table 62.

Namely, that the Soviet Union would shoulder the debts of defaulting CMEA countries. In the long run, regardless of the rate at which prices increase in the West, loans must inevitably become free gifts, if East European countries are unable to find enough goods to export.

If there is to be an increase in technology flows to the East it is essential that Eastern exports to the West increase as well. This is hampered by difficulties on both sides. In the West, the possibility of competition is treated most seriously, and the distinction made by East European countries between internal and external prices breeds suspicion of unfair competition. We have tried to collect claims of evidence of such competition which does exist, but it is difficult to determine to what extent. Western firms, however, are highly sensitive to this problem.

Even if Western opposition to competition were overcome, the problem of the content of Eastern European exports would still remain. Here, distinctions need to be drawn between raw materials, energy and manufactured products.

For raw materials, possibilities of increasing exports from Eastern Europe are limited, both by the amount of capital necessary to increase production and for non-energy raw-materials, the capacity of the Western markets to absorb them.

In the case of energy, the present shortage should, in theory, put the Soviet Union in a favourable position because of its vast proved and probable reserves. But expanding production, particularly that of oil, demands massive capital investment, advanced technology and time. The energy conservation measures now being introduced in the Eastern European countries show that their governments are not very optimistic in the short term. Western capital and high technology could play a decisive role in this context. But this would call for programmes on a different scale and this report is not the place for a discussion on their advisibility.[14]

For the time being, therefore, there is only one answer and that is for the East to increase its exports of manufactured goods. This depends on availabilities and product quality. These aspects are governed by many factors, such as the East European countries' economic policy, their willingness to accept a more widely-based international division of labour, etc. But the success of this policy would depend very much on the ability of the East to innovate and on a continued inflow of Western technology and its subsequent diffusion and assimilation. In short, future Western technology flow *will depend, in part, on the way in which it is used*. If planned economies prove incapable of using it, then no solution that is found will give more than temporary relief.

5. Impact of Transfer on the Western Economies

It is difficult to measure the impact of Eastward technology transfer on the Western economies since East-West trade is only a very small part of total Western foreign trade (2.9 to 4.6 per cent for OECD countries between 1970 and 1976) and its effects are overshadowed by other economic factors.

The most tangible result of technology transfer to the East appears to be the potential for increased supplies of energy, raw materials and intermediate products. Trading profits in general are practically impossible to quantify.

The benefits of this trade may be assessed at national or company level. According to surveys among Western firms, technology transfer to the East appears to be profitable. Nevertheless, there are firms which reflect at length about selling their latest technology and which are afraid of competition, especially in the developing countries. Profitability at

---

14 See Chapter 4. One of the most important conclusions drawn there would seem to be that expansion in East-West technology transfer goes beyond the realms of technology and economics. The impression is even gained that the political and strategic consequences — hardly touched upon in this study — often take first place.

company level is somewhat distorted because of the credit guarantees granted by Governments. Competition may also threaten firms other than those selling technology.

Surveys of Western firms often relate to compensation agreements in which Western exports consist mainly of machinery and equipment. A survey among German firms underlines the existence of additional costs which can only partly be covered. These compensation agreements are nevertheless accepted or even sought after by Western firms. The main reason seems to be, that despite disadvantages, compensation agreements do enable Western firms to "do business" — support production.

Unfortunately, the benefits of technology transfer at the national level have not yet been studied. This is a difficult problem because the national interest cannot be measured solely in terms of profit or growth. More is known about the impact of transfer on international trading systems than about the economic benefits. Because of the worsening debt situation, East European countries are leaning increasingly towards bilateral and countertrade agreements, and many developing countries imitate them as soon as they encounter financial difficulties.

In practice, it seems that the bigger the debt, the greater is the use of bartering in technology transfer from West to East. As has been noted, bilateralism can, at worst, only serve to prevent technology transfer from shrinking too fast and suddenly drying up. At best, it can only maintain trade at the present level. Because it protects the Eastern countries from making hard adjustments to Western demand, it can be said to slow down, even hinder, growth in trade.

ANNEX

SITC Index	Product Group	SITC Index	Product Group
071.3	Coffee essences, extracts	* 861.7	Medical instruments
122.2	Cigarettes	* 861.9	Measuring and controlling, scientific instruments
* 231.2	Synthetic rubber		
431.2	Hydrogenated oils, fat	862.4	Photo film
* 512.86	Sulphonamides	* 891.1	Sound recorders
* 513.2	Chemical elements n.e.s.	899.97	Vacuum flasks
* 513.5	Metallic oxide for paint	* 266.3	Regenerated fibre to spin
* 515.1	Radio-active elements	711.3	Steam engines, turbines
* 531.01	Synthetic organic dye	711.81	Water engines and turbines
533.3	Prepared paints, enamels, lacquers, etc.	* 714.2	Calculating and accounting machines
* 541.63	Sera, vaccines	* 715.1	Metal cutting tools
* 541.3	Antibiotics	715.23	Welding machinery
* 599.75	Anti-knock preparations	* 718.22	Type setting machinery
551.2	Synthetic perfume and flavour materials	718.3	Food processing machines
		718.5	Mineral crushing etc., and glass working machines
* 571.1	Prepared explosives		
581.31	Vulcanized fibres	719.15	Refrigerating equipment, non-domestic
629.4	Rubber belting		
655.5	Elastic fabrics and trimmings of elastic	719.32	Fork lift trucks
			Packaging and filling machines
675.03	Alloy steel, hoop and strip	719.62	Ball, roller bearings
678.2	Tubes and pipes of iron and steel, seamless	* 719.7	Telecommunications equipments
		* 724.9	n.e.s.
691.2	Finished structural parts and structures of aluminium		X-ray apparatus
		* 736.2	Batteries and cells
695.24	Tools for use in hand or machine	* 729.11	Transistors, valves, etc.
* 695.26	Carbide tool tips, etc.	* 729.3	Electric measuring and control
698.11	Locks and keys	* 729.5	equipment
* 861.1	Optical elements		Special motor vehicles
* 861.5	Cinema cameras	732.4	

* Research Intensive Product Groups *within* the Research Intensive Industries.

Source: Gaps in Technology — Analytical Report, OECD, Paris, 1970, Book III, Appendice 2, pp. 231-232.

High Technology

SITC heading or subheading	Groups of Products	SITC heading or subheading	Groups of Products
71142	Jet and gas turbines for aircraft	72911	Primary batteries and cells
7117	Nuclear reactors	7293	Tubes, transistors, photocells, etc.
7142	Calculating machines (including electronic computers)	72952	Electrical measuring and control instruments
7143	Statistical machines (punch card or tape)	7297	Electron and proton accelerators
71492	Parts of office machinery (including computer parts)	7299	Electrical machinery, n.e.s. (including electromagnets, traffic control equipment, signalling apparatus, etc.)
7151	Machine tools for metal		
71852	Glass-working machinery	7341	Aircraft, heavier than air
7192	Pumps and centrifuges	73492	Aircraft parts
71952	Machine tools for wood, plastic, etc.	7351	Warships
		73592	Special purpose vessels (including submersible vessels)
71954	Parts and accessories for machine tools	8611	Optical elements
71992	Cocks, valves, etc.	8613	Optical instruments
7249	Telecommunications equipment (excluding TV and radio receivers)	86161	Image projectors
		8619	Measuring and control instruments, n.e.s.

Some items which might well contain high technology products have not been included in this list:

SITC		SITC	
7111	Steam-generating boilers	8614	Photographic cameras
71181	Water turbines	8624	Photographic plates, film, etc.
71822	Type making and setting machinery	8641	Watches
		8642	Clocks
71994	Metal-plastic joints (gaskets)		
726	Electromedical and X-ray apparatus		

Source : *Quantification of Western Exports of High Technology Products to the Communist Countries*, Office of East-West Policy and Planning, Bureau of East-West Trade, 17th October, 1977, pp. 4 and 14.

Table A-3

Foreign Trade of East European Countries and the Soviet Union by Commodity Groups 1969-1975

	Exports				Imports			
	Current value in million US dollars 1975	Annual average percentage change 1969-1973	Percentage change over previous year 1974	1975	Current value in million US dollars 1975	Annual average percentage change 1969-1973	Percentage change over previous year 1974	1975
Bulgaria								
Machinery and equipment	2 001	20.2	19.4	24.8	2 352	8.8	22.4	26.9
Raw materials and semi-finished products	810	3.6	45.2	4.2	2 584	9.7	32.8	29.5
Foodstuffs and raw materials for food production	1 521	9.5	3.7	30.8	349	2.6	111.2	−0.4
Consumer goods	499	5.1	2.4	20.5	285	7.9	38.8	7.5
Total	4 831	1.1	16.3	22.1	5 570	8.8	32.3	24.8
Czechoslovakia								
Machinery and equipment	4 127	10.6	8.6	20.3	3 270	14.0	20.0	14.3
Raw materials and semi-finished products	2 616	10.8	28.6	12.8	4 234	8.8	29.3	18.4
Foodstuffs and raw materials for food production	369	7.4	53.2	−4.4	851	4.9	4.6	3.5
Consumer goods	1 575	9.3	12.0	28.7	701	10.4	29.6	18.1
Total	8 687	10.3	16.7	18.0	9 056	10.1	22.8	15.3
German Democratic Republic								
Machinery and equipment	5 114	11.2	11.8	18.4	3 455	17.4	−1.7	35.7
Raw materials and semi-finished products	2 788	11.3	26.1	15.3	5 585	13.5	43.1	15.0
Foodstuffs and raw materials for food production	612	10.9	21.4	11.7	1 595	7.0	29.7	1.7
Consumer goods	1 574	7.2	13.6	7.7	655	19.9	2.3	−3.0
Total	10 088	10.5	16.3	15.3	11 290	13.9	22.8	17.0
Hungary								
Machinery and equipment	2 385	15.3	13.4	23.7	2 239	12.2	37.2	27.2
Raw materials and semi-finished products	1 215	14.9	12.0	0.1	4 063	11.5	37.2	21.4
Foodstuffs and raw materials for food production	1 325	17.8	12.2	3.2	449	11.4	49.6	−7.6
Consumer goods	1 205	11.4	7.6	10.8	480	16.4	19.4	19.9
Total	6 130	14.9	11.6	11.2	7 231	12.0	36.8	20.6
Poland								
Machinery and equipment	3 998	14.4	23.4	29.9	4 689	21.5	25.0	16.1
Raw materials and semi-finished products	3 813	12.0	15.1	25.4	6 181	16.6	44.7	27.1
Foodstuffs and raw materials for food production	914	12.6	8.6	−4.2	1 080	15.1	2.8	8.2
Consumer goods	1 564	13.9	19.1	25.3	595	20.0	20.5	1.4

Machinery and equipment	1 352	19.3	9.8	34.9	1 854	10.2	18.5	5.9
Raw materials and semi-finished products	2 255	12.6	56.9	6.9	2 896	14.4	66.4	4.9
Foodstuffs and raw materials for food production	873	14.4	27.6	−11.8	391	24.4	110.2	−9.5
Consumer goods	861	22.2	9.7	11.4	201	4.2	22.7	−0.2
Total	5 341	16.1	30.4	9.6	5 342	12.5	46.8	3.9
Soviet Union								
Machinery and equipment	5 959	10.8	15.6	12.9	11 991	11.3	14.5	48.2
Raw materials and semi-finished products	23 392	11.9	33.5	20.1	10 646	12.9	46.9	18.8
Foodstuffs and raw materials for food production	1 530	−2.2	66.4	−21.7	8 135	22.2	2.6	90.5
Consumer goods	988	12.9	26.9	23.9	4 598	7.9	11.3	26.1
Total	31 869	10.6	31.2	15.9	35 370	12.9	21.2	41.6
Total Eastern Europe								
Machinery and equipment	18 977	13.4	14.1	23.6	17 859	14.9	18.4	20.6
Raw materials and semi-finished products	13 497	11.4	37.1	13.8	25 543	12.4	41.8	19.3
Foodstuffs and raw materials for food production	5 614	12.5	15.2	5.4	4 715	8.8	34.2	1.4
Consumer goods	7 278	10.9	12.0	17.3	2 917	14.9	18.4	6.7
Total	45 366	12.3	20.1	17.1	51 034	13.0	30.6	17.0
Total Eastern Europe and the Soviet Union								
Machinery and equipment	24 936	12.7	14.5	20.7	29 850	13.5	16.9	30.6
Raw materials and semi-finished products	36 889	11.7	34.8	17.8	36 189	12.5	43.3	19.1
Foodstuffs and raw materials for food production	7 144	8.3	26.0	−2.2	12 850	15.1	16.6	44.8
Consumer goods	8 266	11.0	13.6	18.1	7 515	10.5	14.2	18.0
Total	77 235	11.5	24.6	16.6	86 404	13.0	26.9	26.2

Source: National statistics, as quoted in *Economic Bulletin for Europe*, Vol. 28, *op. cit.*, 1976, pp. 58-60.

Table A-3a

Foreign Trade of East European Countries and the Soviet Union by Commodity Group — 1972-1978

	Exports					Imports				
	Value in million US dollars	Annual percentage change				Value in million US dollars	Annual percentage change			
	1978	1972-1976 (average)	1976	1977	1978	1978	1972-1976 (average)	1976	1977	1978
Bulgaria										
Machinery and equipment	3 660	22.9	18.3	25.5	15.3	3 143	16.3	3.4	7.2	12.9
Raw materials and semi-finished products	1 265	16.6	16.3	13.9	10.3	3 877	18.1	6.7	19.2	10.4
Foodstuffs and raw materials for food products	1 909	9.7	11.3	3.1	2.4	343	26.0	-10.6	-15.6	22.0
Consumer goods	739	7.7	5.9	16.7	12.2	316	9.9	-0.7	3.2	1.3
Total	7 573	15.3	14.5	15.8	10.7	7 679	17.0	3.8	11.5	11.5
Czechoslovakia										
Machinery and equipment	5 645	11.9	16.7	14.4	12.7	4 588	16.6	9.5	20.5	11.0
Raw materials and semi-finished products	2 828	12.0	6.3	8.6	3.6	5 150	15.1	10.0	9.8	7.4
Foodstuffs and raw materials for food products	395	7.2	-7.6	3.4	26.2	949	6.7	23.3	6.9	-7.5
Consumer goods	1 804	11.2	13.0	10.7	5.9	716	6.2	1.3	1.9	12.8
Total	10 672	11.6	11.8	11.7	9.4	11 403	14.2	104	12.9	7.7
German Democratic Republic										
Machinery and equipment	7 680	12.9	13.7	10.4	16.1	5 215	15.4	20.3	14.9	3.8
Raw materials and semi-finished products	3 463	18.3	13.3	1.9	3.4	7 362	20.0	18.6	9.1	0.7
Foodstuffs and raw materials for food products	726	19.2	22.6	-15.7	17.8	1 979	10.8	11.7	-3.9	-3.1
Consumer goods	2 094	5.2	4.0	12.4	8.2	782	17.5	-3.3	4.1	12.7
Total	13 963	13.2	12.6	5.8	10.3	15 338	17.0	16.9	8.6	1.7
Hungary										
Machinery and equipment	2 289	17.8	6.2	14.1	3.7	2 853	9.3	13.6	17.7	17.9
Raw materials and semi-finished products	1 755	12.9	13.8	19.2	0.3	4 460	13.3	-6.5	14.1	12.0
Foodstuffs and raw materials for food products	1 367	9.4	-8.4	16.1	-5.3	515	4.9	-6.6	24.7	-11.3
Consumer goods	1 354	8.1	-0.3	5.7	3.8	630	11.4	0.4	17.1	18.8

310

Machinery and equipment	5 820	18.4	..	15.6	10.3	5 874	25.8	..	2.4	6.3
Raw materials and semi-finished products	4 368	21.5	..	5.5	9.3	6 964	23.6	..	6.1	3.4
Foodstuffs and raw materials for good product	1 088	13.6	..	5.2	1.6	1 533	19.2	..	5.2	8.4
Consumer goods	2 195	16.8	..	16.5	13.9	966	16.1	..	20.2	1.6
Total	13 471	18.8	7.1	11.3	9.8	15 337	23.3	10.6	5.4	4.9
Rumania										
Machinery and equipment	2 105	21.6	16.6	19.1	12.1	3 034	12.6	4.6	33.0	17.7
Raw materials and semi-finished products	2 952	20.0	14.1	3.2	11.4	4 254	24.4	18.3	6.9	16.1
Foodstuffs and raw materials for food products	1 065	16.8	13.1	34.3	−19.6	580	22.5	33.2	−3.5	15.5
Consumer goods	1 295	16.6	16.9	16.1	11.0	305	10.2	4.0	31.6	10.9
Total	7 417	19.2	14.9	14.4	5.6	8 173	19.1	14.1	15.1	16.5
Soviet Union										
Machinery and equipment	9 874	14.9	21.0	15.0	11.8	20 500	22.1	15.5	9.9	26.6
Raw materials and semi-finished products	37 834	20.2	18.5	19.9	6.8	13 179	18.7	1.1	4.4	10.0
Foodstuffs and raw materials for food products	1 108	−6.0	−27.1	22.6	−23.9	9 371	30.8	6.9	−4.4	6.0
Consumer goods	1 562	18.5	12.8	6.8	23.1	5 759	9.9	5.3	7.2	5.0
Total	50 378	17.7	16.6	4.7	7.3	48 809	20.3	7.8	4.7	14.8
Total Eastern Europe										
Machinery and equipment	27 199	15.7	14.5	15.4	13.0	24 707	16.9	16.2	14.2	9.9
Raw materials and semi-finished products	16 631	17.6	11.1	7.2	6.5	32 067	19.2	16.1	10.0	7.2
Foodstuffs and raw materials for food products	6 550	12.3	9.9	11.0	0.0	5 899	12.5	3.5	3.1	1.3
Consumer goods	9 481	10.6	10.7	13.0	9.2	3 715	14.0	9.6	12.6	9.8
Total	59 861	14.9	—	11.6	8.4	66 388	17.3	13.4	10.5	7.8
Total Eastern Europe and the Soviet Union										
Machinery and equipment	37 073	15.6	12.2	15.0	12.6	45 207	19.0	13.6	12.3	17.8
Raw materials and semi-finished products	54 465	19.5	10.4	15.9	6.8	45 246	19.1	7.8	8.5	8.0
Foodstuffs and raw materials for food products	7 658	7.7	16.5	13.2	−3.3	15 270	21.5	10.9	−1.3	4.1
Consumer goods	11 043	11.5	6.5	12.2	11.2	9 474	12.2	5.8	9.2	7.0
Total	110 239	16.1	10.3	15.1	7.9	115 197	18.7	9.3	8.1	10.8

Source: Economic Bulletin for Europe, op. cit., Vol. 30. No. 1. Pre-publication. Table 2.3 (for 1976); *Economic Bulletin for Europe*, Vol. 31. No. 1. Pre-publication. Table 2.3 (for 1972-1976, 1977 and 1978).

Table A-4

Foreign Trade of Eastern Europe
Millions of dollars, f.o.b.

Exports

Destination	Year						Annual average growth rate (per cent)					
	1960	1965	1970	1974	1975	1978	1960-1965	1965-1970	1970-1975	1974-1975	1977	1978
World	13 187	19 939	30 893	66 522	74 728	110 239	8.6	9.2	19.3	12.3	15.1	7.9
of which:												
Developed market economy countries	2 616	4 052	7 266	19 721	19 138	30 222	9.1	12.4	21.5	−3.0	10.0	5.2
Developing countries	1 132	2 772	4 262	11 083	12 188	16 329	19.6	9.0	23.4	10.0	30.8	6.5
Socialist countries	9 439	13 115	19 365	35 718	43 402	60 772	6.8	8.1	17.5	21.5	14.4	..

Imports[a]

Origin	Year						Annual average growth rate (per cent)					
	1960	1965	1970	1974	1975	1978	1960-1965	1965-1970	1970-1975	1974-1975	1977	1978
World	13 390	19 652	30 176	68 727	84 570	115 197	8.0	9.0	23.0	23.0	8.1	10.8
of which:												
Developed market economy countries	2 878	4 390	7 800	24 781	30 201	38 978	8.8	12.2	31.0	21.9	1.3	8.5
Developing countries	1 277	2 437	3 493	9 161	11 161	11 948	13.8	7.5	26.0	21.8	20.2	2.6
Socialist countries	9 235	12 825	18 883	34 785	43 208	62 219	6.8	8.0	18.0	24.2	..	..

Turnover[a]

Destination/Origin	Year						Annual average growth rate (per cent)					
	1960	1965	1970	1974	1975	1978	1960-1965	1965-1970	1970-1975	1974-1975	1977	1978
World	26 577	39 591	61 069	135 249	159 298	225 436	8.3	9.1	21.0	17.8	..	..
of which:												
Developed market economy countries	5 494	8 442	15 066	44 502	49 339	69 200	9.0	12.3	27.0	10.9	..	..
Developing countries	2 409	5 209	7 755	20 244	23 349	28 277	16.7	8.3	25.0	15.3	..	..
Socialist countries	18 674	25 940	38 248	70 503	86 610	122 991	6.8	8.1	17.8	22.8	..	..

a) Hungary c.i.f.

Sources: *Economic Bulletin for Europe, op. cit.,* Vol. 31, No. 1, Pre-publication, Table 2.2. "Trade Relations among Countries having Different Economic and Social Systems", UNCTAD, TD/B/615/Add.1, 13th September, 1976.

Table A-5
Growth of Imports of OECD Countries from East European Countries by Major Commodity[a]
In million US dollars

Major SITC Categories		1961	1965	1970	1971	1972	1973	1974	1975	1976	1977
						I. Total Eastern Europe					
Food and beverages (0+1+4)	Value	633	819	1 123	1 208	1 457	1 907	2 080	2 093	1 932	1 990
	Index	56	73	100	108	130	170	185	186	172	177
Crude materials (2)	Value	537	949	1 293	1 290	1 458	2 217	2 994	2 746	3 047	3 310
	Index	42	73	100	100	113	171	232	212	235	256
Mineral fuels (3)	Value	412	745	1 229	1 533	1 625	2 337	5 083	6 455	8 165	8 745
	Index	34	61	100	125	132	190	414	525	664	711
Chemicals (5)	Value	125	187	322	357	403	552	965	927	1 104	1 523
	Index	39	58	100	111	126	171	300	288	339	473
Manufactured goods (6+8+9)	Value	375	943	1 643	1 773	2 315	3 417	4 321	3 891	4 964	5 603
	Index	23	57	100	108	142	210	266	237	302	341
Machinery and transport equipment (7)	Value	120	205	464	588	731	1 060	1 147	1 516	1 809	1 927
	Index	26	39	100	127	158	228	247	327	390	415
Total (0 to 9)	Value	2 204	3 848	6 073	6 749	7 990	11 490	16 589	17 627	21 022	23 090
	Index	36	63	100	111	132	190	273	290	346	380
						II. USSR					
Food and beverages (0+1+4)	Value	176	126	177	223	167	174	329	305	224	235
	Index	100	71	100	126	94	98	186	172	127	133
Crude materials (2)	Value	33	570	836	857	965	1 507	2 134	1 966	2 170	2 372
	Index	40	68	100	102	115	180	255	235	260	284
Mineral fuels (3)	Value	247	509	851	1 066	1 097	1 660	3 590	4 582	6 102	6 852
	Index	29	50	100	125	129	196	422	538	717	805
Chemicals (5)	Value	30	41	79	86	96	151	320	322	389	733
	Index	38	52	100	109	122	191	405	408	489	928
Manufactured goods	Value	114	404	517	473	649	1 115	1 301	938	1 415	1 563

Table A-6

Growth of Exports of OECD Countries to East European Countries by Major Commodity[a]

In million US dollars

Major SITC Categories		1961	1965	1970	1971	1972	1973	1974	1975	1976	1977
							I. Total Eastern Europe				
Food and beverages (0+1+4)	Value	311	745	705	860	1 579	2 613	1 964	3 458	4 463	3 318
	Index	44	106	100	122	224	371	279	490	633	471
Crude materials (2)	Value	217	385	542	499	700	1 146	1 303	1 240	1 494	1 699
	Index	40	71	100	92	129	211	240	229	276	313
Mineral fuels (3)	Value	3	13	85	71	86	115	157	201	207	228
	Index	2	15	100	84	101	135	185	236	244	268
Chemicals (5)	Value	191	413	749	832	1 018	1 347	2 703	2 839	2 914	3 395
	Index	26	55	100	111	136	180	361	379	389	453
Manufactured goods (6+8+9)	Value	729	890	2 090	2 342	2 854	4 289	7 262	8 731	8 893	8 728
	Index	35	61	100	112	138	206	347	418	425	418
Machinery and transport equipment (7)	Value	689	1 022	2 218	2 300	3 112	4 415	5 980	9 351	9 457	10 283
	Index	31	46	100	104	140	199	269	422	426	463
Total (0 to 9)	Value	2 141	3 468	6 388	6 904	9 349	13 924	19 368	25 819	27 427	27 655
	Index	34	54	100	108	146	218	303	404	429	433
							II. USSR				
Food and beverages (0+1+4)	Value	69	309	233	282	908	1 539	769	2 158	2 637	1 908
	Index	30	133	100	121	390	661	330	926	1 131	819
Crude materials (2)	Value	55	112	172	146	217	381	388	358	533	614
	Index	32	65	100	85	126	222	226	208	310	357
Mineral fuels (3)	Value	—	1	5	7	7	9	23	38	39	51
	Index	—	20	100	140	140	180	460	760	780	1 020
Chemicals (5)	Value	39	139	255	276	304	326	776	967	920	1 225
	Index	15	53	100	108	119	128	304	379	361	480
Manufactured goods	Value	257	323	947	1 027	1 252	1 770	3 227	4 400	4 717	4 563

(6+8+9)	Index	22	78	100	92	126	217	253	181	274	302
Machinery and transport equipment (7)	Value	20	48	93	135	183	267	238	372	462	371
	Index	22	52	100	145	197	287	256	400	497	399
Total (0 to 9)	Value	920	1 749	2 552	2 841	3 158	4 874	7 912	8 484	10 761	12 128
	Index	36	68.5	100	111	124	191	310	332	427	475

III. Other Eastern European Countries

Food and beverages (0+1+4)	Value	458	693	945	984	1 290	1 733	1 750	1 788	1 710	1 754
	Index	26	73	100	104	137	183	185	189	181	186
Crude materials (2)	Value	204	328	456	433	493	710	860	780	876	938
	Index	45	71	100	95	109	156	189	171	192	205
Mineral fuels (3)	Value	164	236	378	467	528	677	1 493	1 874	2 064	1 893
	Index	43	62	100	123	140	179	395	496	546	501
Chemicals (5)	Value	95	146	243	271	307	401	645	605	715	790
	Index	33	60	100	112	126	165	265	249	290	325
Manufactured goods (6+8+9)	Value	262	539	1 126	1 280	1 666	2 302	3 020	2 953	3 549	4 040
	Index	23	48	100	115	150	207	272	262	315	359
Machinery and transport equipment (7)	Value	100	157	371	453	548	793	909	1 144	1 347	1 566
	Index	27	42	100	122	148	214	245	308	363	419
Total (0 to 9)	Value	1 284	2 099	3 521	3 908	4 833	6 616	8 677	9 143	10 260	10 962
	Index	36	60	100	111	138	188	246	260	291	311

a) Index numbers based on 1970.

Sources: Statistics of Foreign Trade, Series C, OECD, Paris, for 1976 and 1977: for other years, OECD Statistical data.

	Index	35	45	100	88	117	168	225	445	478	523
Total (0 to 9)	Value	781	1 346	2 639	2 640	3 895	5 755	7 491	12 498	13 755	13 736
	Index	30	51	100	100	148	218	284	474	521	520

III. Other Eastern European Countries

Food and beverages (0+1+4)	Value	243	437	471	578	671	1 073	1 195	1 299	1 826	1 410
	Index	52	93	100	123	142	228	254	276	388	299
Crude materials (2)	Value	163	273	370	353	483	765	915	881	961	1 085
	Index	44	74	100	95	131	207	247	238	260	293
Mineral fuels (3)	Value	3	12	80	64	79	106	134	162	168	177
	Index	4	15	100	80	99	133	168	203	210	221
Chemicals (5)	Value	151	273	494	556	715	1 021	1 927	1 872	1 994	2 170
	Index	31	55	100	113	145	207	390	379	404	439
Manufactured goods (6+8+9)	Value	472	567	1 143	1 315	1 603	2 519	4 036	4 331	4 176	4 165
	Index	41	50	100	115	141	220	353	379	365	364
Machinery and transport equipment (7)	Value	328	559	1 190	1 398	1 905	2 686	3 671	4 775	4 548	4 908
	Index	28	46	100	117	160	225	308	401	382	412
Total (0 to 9)	Value	1 360	2 121	3 749	4 263	5 454	8 169	11 877	13 321	13 672	13 919
	Index	36	57	100	114	146	218	317	355	365	371

a) Index numbers based on 1970.

Sources: *Statistics of Foreign Trade*. Series C. OECD. Paris, for 1976 and 1977: for other years. OECD Statistical data.

Commodity Pattern of Trade of OECD Countries with the Seven East European Countries

	1970	1971	1972	1973	1974	1975	1976	1977
				A.	Imports			
Food and beverages (0 + 1 + 4)	18.5	17.9	18.2	16.6	12.5	11.9	9.2	8.6
Crude materials (2)	21.3	19.1	18.2	19.3	18.0	15.6	14.5	14.3
Mineral fuels (3)	20.2	22.7	20.3	20.3	30.6	36.6	38.9	37.9
Chemicals (5)	5.3	5.3	5.0	4.8	5.8	5.3	5.2	6.6
Manufactured goods (6 + 8 + 9)	27.1	26.3	29.0	29.7	26.0	22.1	23.6	24.3
Machinery and transport equipment (7)	7.6	8.7	9.1	9.2	6.9	8.6	8.6	8.3
Total (0 to 9)	100	100	100	100	100	100	100	100
				B.	Exports			
Food and beverages (0 + 1 + 4)	11.0	12.5	16.9	18.8	10.1	13.3	16.3	12.0
Crude materials (2)	8.5	7.2	7.5	8.2	6.8	4.8	5.5	6.1
Mineral fuels (3)	1.3	1.0	0.9	0.8	0.8	0.8	0.8	0.8
Chemicals (5)	11.7	12.1	10.9	9.7	14.0	11.0	10.5	12.3
Manufactured goods (6 + 8 + 9)	32.7	33.9	30.5	30.8	37.5	33.8	32.5	31.6
Machinery and transport equipment (7)	34.7	33.3	33.3	31.7	30.9	36.2	34.5	37.2
Total (0 to 9)	100	100	100	100	100	100	100	100

Source: Statistics of Foreign Trade, Series A. OECD. Paris.

Table A-8

Trade Balance of OECD Countries with the Seven East European Countries

Million US $

	1961	1965	1970	1971	1972	1973	1974	1975	1976	1977
Food and beverages (0 + 1 + 4)	−322	−74	−418	−348	122	706	−116	1 365	2 531	1 328
Crude materials (2)	−320	−564	−751	−791	−758	1 071	−1 691	−1 506	−1 553	−1 611
Mineral fuels (3)	−409	−732	−1 144	−1 462	−1 539	−2 222	−4 926	−6 254	−7 958	−8 517
Chemicals (5)	66	225	427	475	615	795	1 738	1 912	1 810	1 872
Manufactured goods (6 + 8 + 9)	354	−53	447	569	539	872	2 941	4 840	3 929	3 125
Machinery and transport equipment (7)	569	817	1 754	1 712	2 381	3 355	4 833	7 836	7 648	8 356
Total (0 to 9)	63	−380	315	155	1 359	2 434	2 779	8 192	6 407	4 553

Sources: Statistics of Foreign Trade. Series C. OECD. Paris. for 1976 and 1977: for other years. OECD Statistical data. Calculated from Tables A-5 and A-6.

317

Table A-9

Commodity Structure of Trade with Developing Countries of Eastern Europe

Percentage share in total exports or imports[a]

	SITC	1965	1970	1971	1972	1973	1974
Exports to Eastern Europe							
Food, beverages and tobacco	0+1	43.2	39.7	34.8	32.9	37.0	37.2
Crude materials (excluding fuels) oils and fats	2+4	32.7	30.7	28.7	24.8	23.5	23.0
Mineral fuels and related materials	3	0.2	1.5	4.2	8.0	9.3	12.2
Chemicals and manufactured goods	5+6+8	17.1	22.9	26.6	30.0	24.8	23.1
Machinery and transport equipment	7	6.8	5.2	5.7	4.3	5.4	4.5
Total	0-8	100.0	100.0	100.0	100.0	100.0	100.0
Imports from Eastern Europe							
Food, beverages and tobacco	0+1	9.2	10.1	11.7	9.5	13.8	16.3
Crude materials (excluding fuels) oils and fats	2+4	5.3	6.5	7.1	8.5	7.8	9.7
Mineral fuels and related materials	3	9.8	7.0	7.7	8.4	8.7	13.2
Chemicals and manufactured goods	5+6+8	33.1	31.2	30.7	31.5	28.4	30.1
Machinery and transport equipment	7	42.6	45.2	42.8	42.2	41.3	30.7
Total	0-8	100.0	100.0	100.0	100.0	100.0	100.0

a) SITC 0-8.

Source: "Trade Relations among Countries having Different Economic and Social Systems", UNCTAD, TD/B/615/Add. 1, 13th September, 1976.

Table A-10

Commodity Structure of Trade of the USSR with Developing Countries

Percentage share in total exports or imports[a]

	SITC	1973	1974	1975
Exports to developing countries				
Food, beverages and tobacco	0+1	6.9	7.9	7.4
Crude materials (excluding fuels) oils and fats	2+4	10.8	14.2	11.0
Mineral fuels and related materials	3	16.2	24.5	28.4
Chemicals and manufactured goods	5+6+8	18.8	19.7	16.3
Machinery and transport equipment	7	47.3	33.7	36.9
Total	0-8	100.0	100.0	100.0
Imports from developing countries				
Food, beverages and tobacco	0+1	36.2	37.9	53.0
Crude materials (excluding fuels) oils and fats	2+4	23.9	24.3	13.8
Mineral fuels and related materials	3	13.0	13.1	12.0
Chemicals and manufactured goods	5+6+8	23.2	20.8	16.5
Machinery and transport equipment	7	3.7	3.9	4.7
Total	0-8	100.0	100.0	100.0

a) SITC 0-8.

Source: "Trade Relations among Countries having Different Economic and Social Systems", UNCTAD, TD/B/615/Add.1, 13th September, 1976.

Note: The data cover trade only with those developing countries for which the foreign trade statistics of the USSR provide a commodity breakdown. Figures exclude trade with countries for which there is no commodity breakdown.

Table A-11

Structure of Exports of Machinery and Transport Equipment from OECD Countries to the Seven East European Countries

A = Value in million US $
B = Percentage share of total

		1961	1965	1970	1971	1972	1973	1974	1975	1976	1977
SECTION 7 Machinery and Transport Equipment	A	689	1 022	2 218	2 300	3 112	4 415	5 980	9 351	9 457	10 283
	B	100	100	100	100	100	100	100	100	100	100
Index 1961 = 100		100	148.3	321.9	333.8	451.7	640.7	867.9	1 357.2	1 371.7	1 491.1
DIVISION 71 Machinery other than Electric	A	521	590	1 482	1 576	2 269	3 363	4 228	6 295	6 565	7 334
	B	75.6	57.7	66.8	68.5	72.9	76.2	70.7	67.3	69.5	71.4
712 Agricultural machinery and implements	A	3	8	40	35	31	85	101	268	277	113
	B	0.4	0.8	1.8	1.5	1.0	1.9	1.7	2.9	2.9	1.1
715 Metalworking machinery	A	53	70	293	251	426	649	841	1 072	1 163	1 329
	B	7.7	6.8	13.2	10.9	13.7	14.7	14.1	11.5	12.3	12.9
718 Machines for special industries	A	108	77	186	191	231	328	470	695	690	698
	B	15.7	7.5	8.4	8.3	7.4	7.4	7.9	7.4	7.3	6.8
719 Machinery and appliances (other than electrical) and machine parts n.e.s.	A	259	325	687	814	1 160	1 696	2 111	2 352	3 524	4 126
	B	37.6	31.8	31.0	35.4	37.3	38.4	35.3	25.2	37.3	40.2
DIVISION 72 Electrical Machinery Apparatus and Appliances	A	121	160	380	390	487	638	954	1 352	1 317	1 651
	B	17.6	15.7	17.1	17.0	15.6	14.5	16.0	14.5	13.9	16.1
722 Electric power machinery and apparatus for making or breaking electrical circuits (switchgear, etc.)	A	47	42	71	88	104	129	199	278	345	426
	B	6.8	4.1	3.2	3.8	3.3	2.9	3.3	3.0	3.7	4.1
729 Other electrical machinery and apparatus	A	44	70	201	199	250	339	491	724	612	818
	B	6.4	6.8	9.1	8.7	8.0	7.7	8.2	7.7	6.5	8.0
DIVISION 73 Transport Equipment	A	47	272	357	334	357	414	799	1 705	1 570	1 289
	B	6.8	26.6	16.1	14.5	11.5	9.4	13.3	18.2	16.6	12.5
732 Road motor vehicles	A	13	31	113	126	148	221	294	733	662	494
	B	1.9	3.0	5.1	5.5	4.8	5.0	4.9	7.8	7.0	4.8
735 Ships and boats	A	22	218	209	144	96	145	379	706	621	630
	B	3.2	21.3	9.4	6.3	3.1	3.3	6.3	7.5	6.6	6.1

Sources: Statistics of Foreign Trade. Series C. OECD. Paris. for 1976 and 1977: for other years. OECD Statistical data.

319

Table A-12

Structure of Imports of Machinery and Transport Equipment to OECD Countries from the Seven East European Countries

A = Value in million US $
B = Percentage share of total

		1961	1965	1970	1971	1972	1973	1974	1975	1976	1977
SECTION 7 Machinery and Transport Equipment	A	120	205	464	588	731	1 060	1 147	1 516	1 809	1 927
	B	100	100	100	100	100	100	100	100	100	100
Index 1961 = 100		100	170.8	386.6	490	609.1	883.3	955.8	1 263.3	1 507.5	1 605
DIVISION 71 Machinery other than Electric	A	69	98	256	296	332	451	566	752	795	793
	B	57.5	47.8	55.2	50.3	45.4	42.5	49.3	49.6	43.9	41.1
712 Agricultural machinery and implements	A	6	12	19	23	41	64	86	138	145	159
	B	5.0	5.9	4.1	3.9	5.6	6.0	7.5	9.1	8.0	8.2
715 Metalworking machinery	A	25	33	93	105	97	125	168	204	190	184
	B	20.8	16.1	20.0	17.9	13.3	11.8	14.6	13.5	10.5	9.5
718 Machines for special industries	A	8	13	26	34	34	35	41	62	71	69
	B	7.3	6.3	5.6	5.8	4.7	3.3	3.6	4.1	3.9	3.6
719 Machinery and appliances (other than electrical) and machine parts n.e.s.	A	9	19	71	81	102	148	171	184	195	205
	B	7.5	9.3	15.3	13.8	14.0	14.0	14.9	12.1	10.8	10.6
DIVISION 72 Electrical Machinery Apparatus and Appliances	A	18	33	89	100	143	221	258	299	340	403
	B	15.0	16.1	19.2	17.0	19.6	20.8	22.5	19.7	18.8	20.9
722 Electric power machinery and apparatus for making or breaking electrical circuits (switchgear, etc.)	A	5	9	30	32	42	74	93	109	118	129
	B	4.2	4.4	6.5	5.4	5.7	7.0	8.1	7.2	6.5	6.7
729 Other electrical machinery and apparatus	A	8	11	29	33	51	74	79	85	105	137
	B	7.3	5.4	6.3	5.5	7.0	7.0	6.9	5.6	5.8	7.1
DIVISION 73 Transport Equipment	A	33	74	119	192	257	389	323	465	674	732
	B	27.5	36.1	25.6	32.7	35.2	36.7	28.2	30.7	37.2	38.0
732 Road motor vehicles	A	23	42	44	53	107	169	145	230	279	322
	B	19.2	20.5	9.5	9.0	14.6	15.9	12.6	15.2	15.4	16.7
735 Ships and boats	A	6	19	60	120	117	182	132	174	325	336
	B	5.0	9.3	12.9	20.4	16.0	17.2	11.5	11.5	18.0	17.4

Sources: *Statistics of Foreign Trade*, Series C. OECD, Paris, for 1976 and 1977; for other years, OECD Statistical data.

Table A-13

Structure of Exports of Machinery and Transport Equipment from OECD Countries to the USSR

A = Value in million US $
B = Percentage share of total

		1961	1965	1970	1971	1972	1973	1974	1975	1976	1977
SECTION 7 Machinery and Transport Equipment	A	361	462	1 028	903	1 207	1 729	2 309	4 576	4 909	5 375
	B	100	100	100	100	100	100	100	100	100	100
Index 1961 = 100		100	128	284.8	250.1	334.3	478.9	639.6	1 267.6	1 359.8	1 488
DIVISION 71 Machinery other than Electric	A	283	224	648	617	938	1 401	1 741	3 121	3 474	3 905
	B	78.4	48.5	63.0	68.3	77.7	81.0	75.4	68.2	70.8	72.7
711 Power generating machinery	A	..	12	13	17	27	31	45	126	137	
	B	..	2.6	1.3	1.9	2.2	1.8	1.9	2.8	2.8	
712 Agricultural machinery and implements	A	—	1	30	18	5	32	17	142	181	191
	B		—	2.9	2.0	0.4	1.9	0.7	3.1	3.7	3.4
714 Office machines	A	—	23	24	30	32	35	28	56	67	
	B	—	5.0	2.3	3.3	2.7	2.0	1.2	1.2	1.4	
715 Metalworking machinery	A	26	24	192	143	255	375	506	634	658	744
	B	7.2	5.2	18.7	15.8	21.1	21.7	23.1	13.9	13.4	13.8
717 Textile and leather machinery	A	—	29	29	18	8	63	96	212	175	
	B	—	6.3	2.8	2.0	0.7	3.6	4.2	4.6	3.6	
718 Machines for special industries	A	69	28	77	67	72	114	174	295	328	287
	B	19.1	6.1	7.5	7.4	6.0	6.6	7.5	6.4	6.7	5.3
719 Machinery and appliances (other than electrical) and machine parts n.e.s.	A	140	134	283	324	502	751	876	1 657	1 929	2 363
	B	38.8	29.0	27.5	35.9	41.6	43.4	37.9	36.2	39.3	44.0
DIVISION 72 Electrical Machinery Apparatus and Appliances	A	51	47	163	135	162	193	326	553	493	738
	B	14.1	10.2	15.9	15.0	13.4	11.1	14.1	12.1	10.0	13.7
722 Electric power machinery and apparatus for making or breaking electrical circuits (switchgear, etc.)	A	18	9	18	23	23	25	45	73	99	168
	B	5.0	1.9	1.8	2.5	1.9	1.4	1.9	1.6	2.0	3.1
723 Equipment for distributing electricity	A	—	6	25	23	23	27	60	91	79	
	B	—	1.2	2.4	2.5	1.9	1.6	2.6	2.0	1.6	
724 Telecommunications apparatus	A	—	4	18	8	12	17	22	25	30	
	B	—	0.9	1.7	0.9	1.0	1.0	1.0	0.5	0.6	
726 Electric apparatus for medical purpose	A	—	3	4	6	7	9	12	22	25	
	B	—	0.6	0.4	0.7	0.6	0.5	0.5	0.5	0.5	
729 Other electrical machinery	A	19	17	97	73	95	113	183	335	252	384

Table A-14

Structure of Imports of Machinery and Transport Equipment to OECD Countries from the USSR

A = Value in million US $ B = Percentage share of total		1961	1965	1970	1971	1972	1973	1974	1975	1976	1977
SECTION 7 Machinery and Transport Equipment	A	20	48	93	135	183	267	238	372	462	371
	B	100	100	100	100	100	100	100	100	100	100
Index 1961 = 100		100	240	465	675	915	1 335	1 190	1 850	2 310	1 855
DIVISION 71 Machinery other than Electric	A	11	20	51	75	95	126	114	166	194	131
	B	55.0	41.7	54.8	55.6	51.9	47.2	47.9	44.6	42.0	35.3
712 Agricultural machinery and implements	A	1	2	3	3	6	11	15	24	28	21
	B	5.0	4.2	3.2	2.2	3.3	4.1	6.3	6.5	6.1	6.2
715 Metalworking machinery	A	3	5	19	28	33	34	36	58	52	43
	B	15.0	10.4	20.4	20.7	18.0	12.7	15.1	15.6	11.3	11.6
718 Machines for special industries	A	2	5	6	14	11	7	8	14	20	11
	B	10.0	10.4	6.5	10.4	6.0	2.6	3.4	3.8	4.3	2.9
719 Machinery and appliances (other than electrical) and machine parts n.e.s.	A	2	7	16	22	34	60	39	43	45	26
	B	10.0	14.6	17.2	16.3	18.6	22.5	16.4	11.5	9.7	7.0
DIVISION 72 Electrical Machinery Apparatus and Appliances	A	3	5	12	11	29	34	36	41	47	46
	B	15.0	10.4	12.9	8.1	15.8	12.7	15.1	11.0	10.2	12.4
722 Electric power machinery and apparatus for making or breaking electrical circuits (switchgear, etc.)	A	—	1	3	4	7	7	7	13	18	11
	B	—	2.0	3.2	3.0	3.8	2.6	2.9	3.5	3.9	2.9
729 Other electrical machinery and apparatus	A	2	1	4	3	12	13	14	13	14	20
	B	10.0	2.0	4.3	2.2	6.6	4.9	5.9	3.5	3.0	5.4
DIVISION 73 Transport Equipment	A	6	22	30	49	58	107	89	165	221	194
	B	30.0	45.8	32.3	36.3	31.7	40.0	37.4	44.4	47.8	52.3
732 Road motor vehicles	A	3	11	9	13	30	58	63	114	142	160
	B	15.0	22.9	9.7	9.6	16.4	21.7	26.5	30.6	30.7	43.1
735 Ships and boats	A	3	7	19	33	16	41	15	38	70	21
	B	15.2	14.6	20.4	24.4	8.7	15.4	6.3	10.2	15.2	5.7

Sources: Statistics of Foreign Trade. Series C. OECD. Paris. for 1976 and 1977: for other years. OECD Statistical data.

		6.0	5.7	9.4	8.1	7.9	6.5	.	7.3	5.1	7.1
DIVISION 73 Transport Equipment	A	27	192	218	151	108	136	242	902	942	731
	B	7.5	41.6	21.2	16.7	8.9	7.9	10.5	19.7	19.2	13.6
731 Railway vehicles	A	—	0.1	6	12	12	19	25	45	33	
	B	—	0.0	0.6	1.3	1.0	1.1	1.1	1.0	0.7	
732 Road motor vehicles	A	2	4	30	37	44	26	49	389	375	153
	B	0.6	0.9	2.9	4.1	3.6	1.5	2.1	8.5	7.6	2.8
733 Road vehicles other than motor vehicles	A	—	0.1	0.3	0.3	0.8	3	5	16	21	
	B	—	0.0	0.0	0.0	0.1	0.2	0.2	0.3	0.4	
735 Ships and boats	A	18	188	181	102	51	86	162	448	512	535
	B	5.0	40.6	17.6	11.3	4.2	5.0	7.0	9.8	10.4	10.0

Sources: *Statistics of Foreign Trade*: "Trade by Commodities". Series C. OECD. Paris. for 1976 and 1977: for other years, OECD Statistical data.

Table A-15

Structure of Trade between the German Democratic Republic and the Federal Republic of Germany between 1965 and 1976

	Annual Average		1975	1976	1977	1978
	1965-1970	1971-1975				

A. Exports from the Federal Republic of Germany to the German Democratic Republic

	1965-1970	1971-1975	1975	1976	1977	1978
Total Exports DM million	1 844	3 203	3 921	4 269	4 343	4 524
of which:						
Machinery, equipment and electrical goods (%)	22.9	23.8	22.8	28.9	30.3	31.0
of which:						
Machinery (%)	16.5	17.5	15.5	20.2	21.7	21.4

B. Imports by the Federal Republic of Germany from the German Democratic Republic

	1965-1970	1971-1975	1975	1976	1977	1978
Total Imports, DM million	1 541	2 791	3 342	3 877	3 961	3 900
of which:						
Machinery, equipment and electrical goods (%)	13.7	10.7	9.9	10.5	11.0	10.5
of which:						
Machinery (%)	5.9	3.7	2.9	3.5	2.9	3.0

Sources: DIW Wochenbericht 9-10/77 (West Berlin); *DIW Wochenbericht* 20/78, pp. 200-201; *DIW Wochenbericht* 10/79, pp. 112-113-114, 8 March, 1979.

Table A-16

Structure of Exports of Machinery and Transport Equipment from OECD Countries to the German Democratic Republic

A = Value in million US $
B = Percentage share of total

		1961	1965	1970	1971	1972	1973	1974	1975	1976	1977
SECTION 7 Machinery and Transport Equipment	A	22	56	159	185	220	183	220	336	371	306
	B	100	100	100	100	100	100	100	100	100	100
Index 1961 = 100		100	255	722	841	1 000	832	1 000	1 527	1 691	1 386
DIVISION 71 Machinery other than Electric	A	10	31	118	119	102	144	150	193	190	183
	B	45.5	55.4	74.2	64.3	46.4	78.7	68.2	57.4	51.1	60.0
712 Agricultural machinery and implements	A	—1	1	1	1	1	3	6	4	4	1
	B	—	1.8	0.6	0.5	0.4	0.5	1.4	1.8	1.1	1.3
715 Metalworking machinery	A	1	2	10	8	10	9	8	14	13	19
	B	4.5	3.6	6.3	4.3	4.5	4.9	3.6	4.2	3.5	6.2
718 Machines for special industries	A	3	5	14	10	10	15	20	28	34	29
	B	13.6	8.9	8.8	5.4	4.5	8.2	9.1	8.3	9.1	9.5
719 Machinery and appliances (other than electrical) and machine parts n.e.s.	A	4	16	65	71	62	73	72	96	105	98
	B	18.2	28.6	40.9	38.4	28.2	39.9	32.7	28.6	28.2	32.1
DIVISION 72 Electrical Machinery Apparatus and Appliances	A	10	9	18	23	25	23	28	35	36	41
	B	45.5	16.1	11.3	12.4	11.4	12.6	12.7	10.4	9.7	13.4
722 Electric power machinery and apparatus for making or breaking electrical circuits (switchgear, etc.)	A	5	2	5	8	8	8	9	17	14	14
	B	22.7	3.6	3.1	4.3	3.6	4.4	4.1	5.1	3.8	4.6
729 Other electrical machinery and apparatus	A	2	3	9	10	8	9	12	11	15	18
	B	9.1	5.4	5.7	5.4	3.6	4.9	5.5	3.3	4.0	5.9
DIVISION 73 Transport Equipment	A	2	16	22	43	93	17	41	109	145	81
	B	9.1	28.6	13.8	23.2	42.3	9.3	18.6	32.4	39.0	26.6
732 Road motor vehicles	A	1	1	3	3	5	3	7	6	12	21
	B	4.5	1.8	1.9	1.6	2.3	1.6	3.5	1.8	3.2	6.9
735 Ships and boats	A	—	12	11	5	5	13	31	28	9	48
	B	—	21.4	6.9	2.7	2.3	7.1	15.9	8.3	2.4	15.7

Sources: Statistics of Foreign Trade. Series C. OECD. Paris. for 1976 and 1977: for other years. OECD Statistical data.

324

Table A-17

Structure of Imports of Machinery and Transport Equipment by OECD Countries from the German Democratic Republic

		1961	1965	1970	1971	1972	1973	1974	1975	1976	1977
A = Value in million US $											
B = Percentage share of total											
SECTION 7											
Machinery and Transport Equipment	A	30	56	93	120	130	171	218	229	257	278
	B	100	100	100	100	100	100	100	100	100	100
Index 1961 = 100		100	186.7	310	400	433	570	727	763	857	926
DIVISION 71											
Machinery other than Electric	A	21	25	54	56	54	68	101	96	108	111
	B	70.0	41.7	58.1	46.7	41.5	39.8	46.3	41.9	42.0	41.4
712 Agricultural machinery and implements	A	1	1	1	1	2	2	4	5	5	9
	B	3.3	1.8	1.1	0.8	1.5	1.2	1.8	2.2	1.9	3.4
715 Metalworking machinery	A	5	6	17	16	16	20	31	31	29	28
	B	16.7	10.7	18.3	13.3	12.3	11.7	14.2	13.5	11.3	10.0
718 Machines for special industries	A	4	5	10	9	9	11	15	15	20	23
	B	13.3	8.9	10.8	7.5	6.9	6.4	6.9	6.6	7.8	8.2
719 Machinery and appliances (other than electrical) and machine parts n.e.s.	A	3	4	11	12	14	18	30	25	28	32
	B	10.0	7.1	11.8	10	10.8	10.5	13.8	10.9	10.9	11.5
DIVISION 72											
Electrical Machinery Apparatus and Appliances	A	5	13	28	29	34	56	63	75	77	95
	B	16.7	23.2	30.1	24.2	26.2	32.7	28.9	32.8	30.0	34.2
722 Electric power machinery and apparatus for making or breaking electrical circuits (switchgear, etc.)	A	2	3	7	8	10	21	24	28	27	32
	B	6.7	5.4	7.5	6.7	7.7	12.3	11.0	12.2	10.5	11.5
729 Other electrical machinery and apparatus	A	2	3	7	7	10	14	14	15	17	20
	B	6.7	5.4	7.5	5.8	7.7	7.9	6.4	6.6	6.6	7.2
DIVISION 73											
Transport Equipment	A	4	18	11	35	42	47	55	58	73	68
	B	13.3	32.1	11.8	29.2	32.3	27.5	25.2	25.3	30.1	24.5
732 Road motor vehicles	A	2	7	2	3	6	9	10	7	9	11
	B	6.7	12.5	2.2	2.5	4.6	5.3	4.6	3.1	3.5	4.0
735 Ships and boats	A	1	9	5	27	29	35	39	37	48	46
	B	3.3	16.1	5.4	22.5	22.3	20.5	17.9	16.2	18.7	16.5

Sources: Statistics of Foreign Trade. Series C. OECD. Paris, for 1976 and 1977; for other years. OECD Statistical data.

325

Table A-18

Structure of Exports of Machinery and Transport Equipment from OECD Countries to Czechoslovakia

A = Value in million US $
B = Percentage share of total

		1961	1965	1970	1971	1972	1973	1974	1975	1976	1977
SECTION 7											
Machinery and Transport Equipment	A	60	110	284	296	314	430	566	689	707	789
	B	100	100	100	100	100	100	100	100	100	100
Index 1961 = 100		100	183	473	493	523	716	943	1 148	1 178	1 315
DIVISION 71											
Machinery other than Electric	A	43	81	222	222	232	323	418	515	517	593
	B	71.7	73.6	78.2	75.0	73.9	75.1	73.9	74.7	73.1	75.2
712 Agricultural machinery and implements	A	—	2	4	4	4	5	7	19	10	14
	B	—	1.8	1.4	1.4	1.3	1.2	1.2	2.8	1.4	1.8
715 Metalworking machinery	A	7	12	19	17	24	32	38	65	84	66
	B	16.7	10.9	6.7	5.7	7.6	7.4	6.7	9.4	11.9	8.4
718 Machines for special industries	A	6	11	36	41	47	46	56	70	74	75
	B	10	10.0	12.7	13.9	15.0	10.7	10.0	10.2	10.5	9.5
719 Machinery and appliances (other than electrical) and machine parts n.e.s.	A	22	41	99	112	103	161	225	253	251	322
	B	36.7	37.2	34.9	37.8	32.8	37.4	39.8	36.7	35.5	40.8
DIVISION 72											
Electrical Machinery Apparatus and Appliances	A	13	21	42	53	57	77	109	130	146	149
	B	21.7	19.0	14.8	17.9	18.2	17.9	19.3	18.9	20.7	18.9
722 Electric power machinery and apparatus for making or breaking electrical circuits (switchgear, etc.)	A	2	3	6	9	7	13	20	18	25	25
	B	3.3	2.7	2.1	3.0	2.2	3.0	3.5	2.6	3.5	3.2
729 Other electrical machinery and apparatus	A	9	14	22	30	30	43	60	81	82	79
	B	15.0	12.7	7.7	10.1	9.6	10.0	10.6	4.8	11.6	10.0
DIVISION 73											
Transport Equipment	A	4	8	20	21	25	30	39	45	44	47
	B	6.7	9.1	7.0	7.1	8.0	7.0	6.9	6.5	6.2	6.0
732 Road motor vehicles	A	4	7	18	20	24	26	30	33	29	39
	B	6.7	6.4	6.3	6.8	7.6	6.0	5.3	4.8	4.1	4.9
735 Ships and boats	A	—	—	—	—	—	—	—	—	—	—
	B	—	—	—	—	—	—	—	—	—	—

Sources: Statistics of Foreign Trade. Series C, OECD, Paris, for 1976 and 1977; for other years, OECD Statistical data.

Table A-19

Structure of Exports of Machinery and Transport Equipment to OECD Countries from Czechoslovakia

A = Value in million US $ B = Percentage share of total		1961	1965	1970	1971	1972	1973	1974	1975	1976	1977
SECTION 7 Machinery and Transport Equipment	A	52	69	135	145	176	236	234	287	273	303
	B	100	100	100	100	100	100	100	100	100	100
Index 1961 = 100		100	133	260	279	338	454	450	552	525	582
DIVISION 71 Machinery other than Electric	A	28	37	91	96	100	126	147	179	168	174
	B	53.8	53.6	67.4	66.2	56.8	53.4	62.8	62.4	61.5	57.4
712 Agricultural machinery and implements	A	3	8	11	11	20	26	31	48	46	53
	B	5.8	11.6	8.1	7.6	11.4	11.0	13.2	16.7	16.8	17.5
715 Metalworking machinery	A	12	14	35	37	28	36	45	49	42	46
	B	23.1	20.3	25.9	25.5	15.9	15.3	19.2	17.1	15.4	15.2
718 Machines for special industries	A	2	3	6	7	7	9	10	12	12	13
	B	3.8	4.3	4.4	4.8	4.0	3.8	4.3	4.2	4.4	4.3
719 Machinery and appliances (other than electrical) and machine parts n.e.s.	A	3	4	20	20	22	26	31	35	37	37
	B	5.8	5.8	14.8	13.8	12.5	11.0	13.2	12.2	13.6	12.2
DIVISION 72 Electrical Machinery Apparatus and Appliances	A	5	5	14	16	22	35	38	39	45	57
	B	9.6	7.2	10.4	11.0	12.5	14.8	16.2	13.6	16.5	18.8
722 Electric power machinery and apparatus for making or breaking electrical circuits (switchgear, etc.)	A	2	2	6	7	7	13	17	17	21	25
	B	3.8	2.9	4.4	4.8	4.0	5.5	7.3	5.9	7.7	8.3
729 Other electrical machinery and apparatus	A	1	2	5	5	7	10	10	12	13	18
	B	1.9	2.9	3.7	3.4	4.0	4.2	4.3	4.2	4.8	5.9
DIVISION 73 Transport Equipment	A	18	27	31	33	54	75	49	68	61	72
	B	34.6	39.1	23.0	22.8	30.7	31.8	20.9	23.7	22.3	23.7
732 Road motor vehicles	A	16	23	24	26	43	64	36	56	50	61
	B	30.8	33.3	17.8	17.9	24.4	27.1	15.4	19.5	18.3	20.1
735 Ships and boats	A	—	—	3	2	5	4	4	2	1	2
	B	—	—	2.2	1.4	2.8	1.7	1.7	0.7	0.4	0.7

Sources: Statistics of Foreign Trade. Series C. OECD. Paris, for 1976 and 1977; for other years. OECD Statistical data.

Table A-20

Czechoslovakia

Structure of Exports to the Non-Socialist Countries and Imports from the Non-Socialist Countries 1973 to 1977

Per cent

	Czechoslovakia exports					Czechoslovakia imports				
	1973	1974	1975	1976	1977	1973	1974	1975	1976	1977
SECTION 7 Machinery and Transport Equipment	100	100	100	100	100	100	100	100	100	100
71 Machinery	65.0	71.0	72.1	71.7	68.2	83.6	80.0	78.6	80.2	80.7
of which:										
712 Agricultural machinery and implements	8.5	9.2	9.6	11.7	11.4	—	—	—	—	—
715 Metalworking machinery	10.8	12.5	14.3	11.7		—	—	—	8.6	6.9
714 Office machines	—	—	—	—	—	5.3	4.9	5.9	3.9	4.7
717 Textile and leather machinery	6.6	7.5	8.2	8.3	6.8	8.0	7.9	6.4	5.6	5.6
718 Machines for special industries	4.1	4.4	4.4	7.1	8.5	12.8	12.6	12.7	12.9	11.6
719 Other machines, appliances, machines parts	34.0	36.6	35.5	32.6	29.8	45.2	45.1	41.9	46.3	48.7
72 Electrical machinery, apparatus appliances	14.0	14.7	12.5	13.5	17.6	12.7	14.8	16.2	15.1	14.5
73 Transport equipment	21.0	14.3	15.4	14.8	14.1	3.7	5.2	5.2	4.7	3.9

Source: Czechoslovak Yearbook of Foreign Trade 1975, pp. 90 and 91; 1976, pp. 93, 94, 106, 107; 1977, pp. 90, 91, 103, 104; 1978, pp. 86, 87, 99, 100. Published by the Czechoslovak Chamber of Commerce.

Table A-21

Structure of Exports of Machinery and Transport Equipment from OECD Countries to Poland

A = Value in million US $
B = Percentage share of total

		1961	1965	1970	1971	1972	1973	1974	1975	1976	1977
SECTION 7											
Machinery and Transport Equipment	A	106	113	241	283	577	1 090	1 528	2 084	1 990	1 865
	B	100	100	100	100	100	100	100	100	100	100
Index 1961 = 100		100	106	225	264	539	935	1 428	1 948	1 860	1 772
DIVISION 71											
Machinery other than Electric	A	79	81	162	178	422	801	1 060	1 378	1 379	1 304
	B	74.5	71.7	67.2	63.2	73.1	73.5	69.4	66.1	69.3	70.0
712 Agricultural machinery and implements	A	—	1	1	1	6	18	21	51	42	17
	B	—	—	—	—	1.0	16.5	1.4	2.4	2.1	0.9
715 Metalworking machinery	A	6	11	25	24	75	127	175	213	260	242
	B	5.6	9.7	10.4	8.5	13.0	11.7	11.5	10.2	13.1	13.0
718 Machines for special industries	A	10	13	18	20	47	133	143	182	156	181
	B	9.3	11.5	7.5	7.1	8.1	12.2	9.4	8.7	7.8	9.7
719 Machinery and appliances (other than electrical) and machine parts n.e.s.	A	32	40	70	85	179	336	509	690	724	678
	B	29.9	35.4	29.0	30.0	31.0	30.8	33.3	33.1	36.7	36.4
DIVISION 72											
Electrical Machinery and Appliances	A	23	25	47	51	98	163	222	305	350	339
	B	21.7	22.1	19.5	18.0	17.0	15.0	14.5	14.6	17.6	18.2
722 Electric power machinery and apparatus for making or breaking electrical circuits (switchgear, etc.)	A	13	9	16	16	28	42	60	89	127	118
	B	12.1	8.0	6.6	5.7	4.8	3.9	3.9	4.3	6.4	6.3
729 Other electrical machinery and apparatus	A	5	10	22	25	49	88	114	139	136	154
	B	4.7	8.8	9.1	8.8	8.5	8.1	7.5	6.7	6.8	8.3
DIVISION 73											
Transport Equipment	A	4	7	32	53	58	126	246	401	261	222
	B	3.8	6.8	13.3	18.7	10.0	11.6	16.1	19.2	13.1	11.9
732 Road motor vehicles	A	2	4	15	16	25	88	109	156	137	158
	B	1.9	3.5	6.2	5.7	4.3	8.1	7.1	7.5	6.9	8.5
735 Ships and boats	A	1	2	16	35	30	32	124	214	96	45
	B	0.9	1.8	6.6	12.4	5.2	2.9	8.1	10.3	4.8	2.4

Sources: Statistics of Foreign Trade. Series C. OECD. Paris. for 1976 and 1977: for other years. OECD Statistical data.

Table A-22

Structure of Imports of Machinery and Transport Equipment to OECD Countries from Poland

A = Value in million US $
B = Percentage share of total

		1961	1965	1970	1971	1972	1973	1974	1975	1976	1977
SECTION 7 Machinery and Transport Equipment	A	9	16	62	93	137	221	241	364	481	604
	B	100	100	100	100	100	100	100	100	100	100
Index 1961 = 100		100	177	688	1 033	1 522	2 456	2 678	4 044	5 344	6 711
DIVISION 71 Machinery other than Electric	A	4	9	27	31	40	65	114	189	179	208
	B	44.4	56.3	43.5	33.3	29.2	29.4	47.3	51.9	37.2	34.4
712 Agricultural machinery and implements	A	—	—	1	2	2	7	14	27	21	30
	B	—	—	1.6	2.2	1.5	3.2	5.8	7.4	4.4	5.0
715 Metalworking machinery	A	2	4	12	12	12	21	—	37	35	35
	B	22.2	25.0	19.4	12.9	8.8	9.5	14.9	10.2	7.3	5.8
718 Machines for special industries	A	—	—	1	2	4	4	5	19	15	17
	B	—	—	1.6	2.2	2.9	1.8	2.1	5.2	3.1	2.8
719 Machinery and appliances (other than electrical) and machine parts n.e.s.	A	1	3	10	13	15	20	34	37	37	48
	B	11.1	18.8	16.1	14.0	10.9	9.0	14.1	10.2	7.7	7.9
DIVISION 72 Electrical Machinery Apparatus and Appliances	A	1	2	9	13	16	29	39	49	59	69
	B	11.1	12.5	14.5	14.0	11.7	13.1	16.2	13.5	12.3	11.4
722 Electric power machinery and apparatus for making or breaking electrical circuits (switchgear, etc.)	A	—	1	4	5	7	13	19	20	22	24
	B	—	6.2	6.5	5.4	5.1	5.9	7.9	5.5	4.6	4.0
729 Other electrical machinery and apparatus	A	1	1	3	5	5	10	11	11	13	20
	B	11.1	4.4	4.8	5.4	3.6	4.5	4.6	3.0	2.7	3.3
DIVISION 73 Transport Equipment	A	4	5	26	49	81	127	88	126	243	327
	B	44.4	31.2	41.9	52.7	59.1	57.5	36.5	34.6	50.5	54.1
732 Road motor vehicles	A	1	—	3	4	16	21	15	30	35	46
	B	11.1	—	4.8	4.3	7.2	9.5	6.2	8.2	7.3	7.6
735 Ships and boats	A	1	2	20	43	60	100	63	86	196	265
	B	11.0	18.8	32.2	46.2	43.8	45.2	26.1	23.6	40.7	43.9

Sources: *Statistics of Foreign Trade.* Series C. OECD, Paris. for 1976 and 1977; for other years, OECD Statistical data.

Table A-23

Poland

Proportion of Exports and Imports of Machinery and Transport Equipment in Poland's Trade with the Non-Socialist Countries

	1971	1972	1973	1974	1975	1976	1977	1978
					Polish Exports			
Total Exports, Machinery and Transport Equipment millions zl/dev.	6 012	7 076	8 234	10 085	13 055	14 819	16 827	19 103
of which:								
Non-Socialist Countries	907	1 187	1 351	1 783	2 658	3 068	3 583	4 341
Per cent share	15.1	16.8	16.4	17.7	20.4	20.7	21.3	22.7
of which:								
Industrialised countries	7.9	9.9	11.5	10.0	12.4	13.1	13.7	13.7
Developing countries	7.2	6.9	4.9	7.7	8.0	7.6	7.6	9.0
					Polish Imports			
Total Exports, Machinery and Transport Equipment millions zl/dev.	5 569	7 598	10 641	13 400	15 712	17 999	18 061	18 936
of which:								
Non-Socialist Countries	1 236	2 465	4 129	5 892	7 887	8 086	7 058	6 800
Per cent share	22.2	32.4	38.8	44.0	50.2	44.9	39.1	35.9
of which:								
Industrialised countries	22.2	32.3	38.5	43.8	50.1	44.8	39.0	35.9
Developing countries	—	0.1	0.3	0.2	0.1	0.1	0.1	—

Source: Rocznik Statystyczny Handlu Zagranicznego (Statistical Yearbook of Foreign Trade). 1976. GUS Warsaw. pp. 48-49 : 1977. GUS Warsaw. pp. 46-47-48 : 1978. GUS Warsaw. pp. 46-48 : 1978. GUS Warsaw. pp. 46-48.

Table A-24
Structure of Exports of Machinery and Transport Equipment from OECD Countries to Hungary

A = Value in million US $
B = Percentage share of total

		1961	1965	1970	1971	1972	1973	1974	1975	1976	1977
SECTION 7 Machinery and Transport Equipment	A	36	67	140	196	235	280	408	471	535	725
	B	100	100	100	100	100	100	100	100	100	100
Index 1961 = 100		100	186	389	544	653	778	1 133	1 308	1 486	2 014
DIVISION 71 Machinery other than Electric	A	23	42	89	132	172	198	289	325	370	505
	B	63.9	62.7	63.6	67.3	73.2	70.7	70.8	69.1	69.2	69.7
712 Agricultural machinery and implements	A	—	—	3	9	14	26	49	38	33	47
	B	—	—	2.1	4.6	6.0	9.3	12.0	8.1	6.2	6.5
715 Metalworking machinery	A	4	6	5	8	10	10	14	21	33	63
	B	11.1	9.0	3.6	4.1	4.3	3.6	3.4	4.5	6.2	8.7
718 Machines for special industries	A	4	7	13	15	16	18	31	35	40	65
	B	11.1	10.4	9.3	7.7	6.8	6.4	7.6	7.4	7.5	9.0
719 Machinery and appliances (other than electrical) and machine parts n.e.s.	A	12	19	43	64	92	103	132	162	195	254
	B	33.3	28.4	30.7	32.7	39.1	36.8	32.4	34.5	36.4	35.0
DIVISION 72 Electrical Machinery Apparatus and Appliances	A	10	16	33	41	44	59	87	112	115	157
	B	27.8	23.9	23.6	20.9	18.7	21.1	21.3	23.8	21.5	21.7
722 Electric power machinery and apparatus for making or breaking electrical circuits (switchgear, etc.)	A	3	4	6	9	10	14	22	32	34	42
	B	8.3	6.0	4.3	4.6	4.3	5.0	5.4	6.8	6.4	5.8
729 Other electrical machinery and apparatus	A	5	8	16	20	22	28	42	53	49	74
	B	13.9	11.9	11.4	10.2	9.4	10.0	10.3	11.1	9.2	10.2
DIVISION 73 Transport Equipment	A	3	9	19	24	19	23	33	35	51	64
	B	8.3	13.4	13.6	12.2	8.1	8.2	8.1	7.4	9.5	8.8
732 Road motor vehicles	A	2	3	16	20	14	18	27	27	39	54
	B	5.6	4.5	11.4	10.2	6.0	6.4	6.6	5.7	7.3	7.4
735 Ships and boats	A	—	—	1	1	—	—	—	—	—	1
	B	—	—	0.7	0.5	—	—	—	—	—	0.1

Sources: Statistics of Foreign Trade, Series C, OECD, Paris, for 1976 and 1977; for other years, OECD Statistical data.

Table A-25

Structure of Imports of Machinery and Transport Equipment to OECD Countries from Hungary

A = Value in million US $
B = Percentage share of total

		1961	1965	1970	1971	1972	1973	1974	1975	1976	1977
SECTION 7											
Machinery and Transport Equipment	A	8	12	34	40	46	73	95	116	141	169
	B	100	100	100	100	100	100	100	100	100	100
Index 1961 = 100		100	150	425	500	575	912	1 187	1 450	1 762	2 112
DIVISION 71											
Machinery other than Electric	A	3	4	13	14	12	20	30	40	42	52
	B	38.0	33.3	38.2	35.0	26.1	27.4	31.6	34.5	29.8	30.8
712 Agricultural machinery and implements	A	—	—	1	1	1	—	1	2	3	4
	B	—	—	2.9	2.5	2.2	—	1.1	1.7	2.1	2.4
715 Metalworking machinery	A	2	2	4	4	3	4	8	10	9	11
	B	25.0	16.7	11.8	10.0	6.5	5.5	8.4	8.6	6.4	6.5
718 Machines for special industries	A	—	—	—	1	1	1	1	2	2	3
	B	—	—	—	2.5	2.2	1.4	1.1	1.7	1.4	1.8
719 Machinery and appliances (other than electrical) and machine parts n.e.s.	A	—	1	6	7	7	11	17	18	21	24
	B	—	0.8	17.6	17.5	15.2	15.1	17.9	15.5	14.9	14.2
DIVISION 72											
Electrical Machinery Apparatus and Appliances	A	4	6	17	21	27	45	54	65	80	93
	B	50.0	50.0	50.0	52.5	58.7	61.6	56.8	56.0	56.7	55.0
722 Electric power machinery and apparatus for making or breaking electrical circuits (switchgear, etc.)	A	—	1	3	4	6	8	13	16	17	16
	B	—	8.3	8.8	10.0	13.0	11.0	13.7	13.8	12.1	9.5
729 Other electrical machinery and apparatus	A	3	4	10	12	14	23	25	28	38	46
	B	38.0	33.3	29.4	30.0	30.4	31.5	26.3	24.1	27.0	27.3
DIVISION 73											
Transport Equipment	A	1	2	4	5	6	9	11	12	19	24
	B	12.5	16.7	11.8	12.5	13.0	12.3	11.6	10.3	13.5	13.9
732 Road motor vehicles	A	—	—	2	2	3	3	5	8	12	13
	B	—	—	5.9	5.0	6.5	4.1	5.3	6.9	8.5	7.9
735 Ships and boats	A	—	—	—	—	1	2	2	1	1	2
	B	—	—	—	—	2.2	2.6	2.1	0.9	0.7	1.2

Sources: Statistics of Foreign Trade. Series C. OECD. Paris. for 1976 and 1977 : for other years. OECD Statistical data.

333

Table A-25a

Structure of Imports of Machinery and Transport Equipment to OECD Countries from Rumania

A = Value in million US $
B = Percentage share of total

		1961	1965	1970	1971	1972	1973	1974	1975	1976	1977
SECTION 7 Machinery and Transport Equipment Index 1970 = 100	A	—	—	30	30	41	70	89	114	152	158
	B	100	100	100	100	100	100	100	100	100	100
		—	—	100	100	136.6	233.3	296.6	380	506.6	526
DIVISION 71 Machinery other than Electric	A	—	—	13	19	24	36	43	61	79	89
	B	—	—	43.3	63.3	58.5	51.0	48.0	53.4	52.6	56.3
712 Agricultural machinery and implements	A	—	—	1	6	10	18	20	32	41	41
	B	—	—	3.3	20.0	24.4	25.7	22.4	28.1	27.0	25.9
715 Metalworking machinery	A	—	—	4	6	4	6	7	11	15	12
	B	—	—	13.3	20.0	9.8	8.6	7.9	9.6	9.9	7.6
718 Machines for special industries	A	—	—	2	—	1	2	1	2	2	2
	B	—	—	6.6	—	2.4	2.9	1.1	1.8	1.3	1.2
719 Machinery and appliances (other than electrical) and machine parts n.e.s.	A	—	—	5	5	7	7	11	13	15	27
	B	—	—	16.6	16.6	17.0	10.0	12.3	11.4	10.0	17.0
DIVISION 72 Electrical Machinery Apparatus and Appliances	A	—	—	5	4	7	11	15	18	16	26
	B	—	—	16.6	13.3	17.0	15.0	16.6	15.9	10.5	16.4
722 Electric power machinery and apparatus for making or breaking electrical circuits (switchgear, etc.)	A	—	—	3	3	4	5	6	8	6	13
	B	—	—	10.0	10.0	9.8	7.1	6.7	7.0	3.9	8.2
729 Other electrical machinery and apparatus	A	—	—	1	1	2	3	3	4	7	7
	B	—	—	3.3	3.3	4.8	4.3	3.3	3.5	4.6	4.4
DIVISION 73 Transport Equipment	A	—	—	12	8	10	24	32	35	56	44
	B	—	—	40.0	24.0	24.4	34.0	35.8	30.7	36.8	27.8
732 Road motor vehicles	A	—	—	4	4	8	16	16	15	31	28
	B	—	—	13.3	13.3	19.5	22.9	18.0	13.2	20.4	17.7
735 Ships and boats	A	—	—	7	2	—	1	9	11	9	—
	B	—	—	23.3	6.6	—	1.4	10.0	9.6	5.9	—

Sources: Statistics of Foreign Trade, Series C, OECD, Paris, for 1976 and 1977; for other years, OECD Statistical data.

Table A-25b

Structure of Exports of Machinery and Transport Equipment from OECD Countries to Rumania

A = Value in million US $
B = Percentage share of total

		1961	1965	1970	1971	1972	1973	1974	1975	1976	1977
SECTION 7 Machinery and Transport Equipment	A	93	142	266	319	446	554	720	696	534	883
	B	100	100	100	100	100	100	100	100	100	100
Index 1970 = 100		100	153	286	343	480	596	774	748	574	948
DIVISION 71 Machinery other than Electric	A	79	91	171	221	332	399	423	433	340	606
	B	84.9	64.0	64.3	69.4	74.4	72.0	58.8	62.2	63.7	68.6
712 Agricultural machinery and implements	A	1	2	1	1	1	1	2	1	2	2
	B	1.1	1.4	0.4	0.3	0.2	0.1	0.3	0.1	0.3	0.2
715 Metalworking machinery	A	8	14	37	39	46	85	79	94	82	167
	B	8.6	9.8	13.9	12.2	10.3	15.3	11.0	13.5	15.4	18.9
718 Machines for special industries	A	17	7	22	27	30	32	31	40	23	48
	B	18.3	4.9	8.3	8.5	6.7	5.8	4.3	5.7	4.3	5.4
719 Machinery and appliances (other than electrical) and machine parts n.e.s.	A	46	50	79	107	181	218	217	224	160	270
	B	49.5	35.2	29.7	33.5	40.5	39.4	30.1	32.2	30.0	30.6
DIVISION 72 Electrical Machinery Apparatus and Appliances	A	11	34	60	63	76	95	130	149	115	161
	B	11.8	23.0	22.6	19.7	17.0	17.1	18.1	21.4	21.5	18.1
722 Electric power machinery and apparatus for making or breaking electrical circuits (switchgear, etc.)	A	4	11	16	17	23	25	37	39	33	44
	B	4.3	7.7	6.0	5.3	5.2	4.5	5.1	5.6	6.2	5.0
729 Other electrical machinery and apparatus	A	4	14	28	29	36	45	54	69	52	79
	B	4.3	9.8	10.5	9.2	8.1	8.1	7.5	9.9	9.7	9.0
DIVISION 73 Transport Equipment	A	3	18	34	35	38	60	167	114	78	116
	B	3.3	3.1	12.8	10.9	8.5	10.8	23.2	16.4	14.6	13.2
732 Road motor vehicles	A	2	10	22	25	31	53	58	59	35	45
	B	2.2	7.0	8.3	7.8	7.0	9.6	8.1	8.5	6.6	5.1
735 Ships and boats	A	—	4	—	—	1	—	47	15	4	—
	B	—	2.8	—	—	0.2	—	6.5	2.2	0.7	—

Sources: Statistics of Foreign Trade. Series C. OECD. Paris. for 1976 and 1977: for other years. OECD Statistical data.

335

Table A-25c

Structure of Imports of Machinery and Transport Equipment to OECD Countries from Bulgaria

A = Value in million US $
B = Percentage share of total

		1961	1965	1970	1971	1972	1973	1974	1975	1976	1977
SECTION 7 Machinery and Transport Equipment	A	—	—	17	24	19	22	31	34	42	44
	B	100	100	100	100	100	100	100	100	100	100
Index 1970 = 100		—	—	100	141	112	129	182	200	247	258
DIVISION 71 Machinery other than Electric	A	—	—	6	6	6	11	18	22	25	24
	B	—	—	35.3	25.0	31.6	50.0	58.1	64.7	59.5	54.5
712 Agricultural machinery and implements	A	—	—	—	—	—	—	—	—	—	—
	B	—	—	—	—	—	—	—	—	—	—
715 Metalworking machinery	A	—	—	2	2	2	4	5	7	9	9
	B	—	—	11.8	8.3	10.5	18.2	16.1	20.6	21.4	20.5
718 Machines for special industries	A	—	—	—	—	—	—	—	—	—	—
	B	—	—	—	—	—	—	—	—	—	—
719 Machinery and appliances (other than electrical) and machine parts n.e.s.	A	—	—	3	2	3	5	9	12	13	11
	B	—	—	17.6	8.3	15.8	22.7	29.0	35.3	31.0	25.0
DIVISION 72 Electrical Machinery Apparatus and Appliances	A	—	—	5	5	7	12	13	11	17	17
	B	—	—	29.4	20.8	36.8	50.0	41.9	32.4	40.0	38.6
722 Electric power machinery and apparatus for making or breaking electrical circuits (switchgear, etc.)	A	—	—	3	2	3	8	9	7	8	8
	B	—	—	17.6	8.3	15.8	36.4	29.0	20.6	19.0	18.2
729 Other electrical machinery and apparatus	A	—	—	1	1	1	2	2	2	3	6
	B	—	—	5.9	4.2	5.8	9.0	6.5	5.9	7.0	13.5
DIVISION 73 Transport Equipment	A	—	—	6	13	5	—	—	1	1	3
	B	—	—	35.3	54.2	26.3	—	—	2.9	0.5	6.8
732 Road motor vehicles	A	—	—	—	—	—	—	—	1	—	3
	B	—	—	—	—	—	—	—	2.9	—	6.8
735 Ships and boats	A	—	—	6	13	5	—	—	—	—	—
	B	—	—	35.3	54.2	26.3	—	—	—	—	—

Sources: Statistics of Foreign Trade. Series C. OECD. Paris. for 1976 and 1977: for other years. OECD Statistical data.

336

Table A-25d

Structure of Exports of Machinery and Transport Equipment from OECD Countries to Bulgaria

A = Value in million US $ B = Percentage share of total		1961	1965	1970	1971	1972	1973	1974	1975	1976	1977
SECTION 7 Machinery and Transport Equipment	A	11	72	102	119	112	148	230	498	405	340
	B	100	100	100	100	100	100	100	100	100	100
Index 1970 = 100		100	654	927	1 081	1 018	1 345	2 090	4 527	3 681	3 081
DIVISION 71 Machinery other than Electric	A	5	40	73	87	71	97	147	330	295	242
	B	45.5	55.6	71.6	73.1	63.4	65.5	63.8	66.3	72.8	71.4
712 Agricultural machinery and implements	A	—	1	1	—	1	2	4	12	5	10
	B	—	1.4	1.0	—	0.9	1.4	1.7	2.4	1.2	2.9
715 Metalworking machinery	A	—	2	5	13	7	11	21	32	34	29
	B	—	2.8	4.9	10.9	6.2	7.4	9.1	6.4	8.4	8.6
718 Machines for special industries	A	—	5	6	11	8	14	15	44	35	14
	B	—	6.9	5.9	9.2	7.1	9.5	6.5	8.8	8.6	4.1
719 Machinery and appliances (other than electrical) and machine parts n.e.s.	A	3	27	48	51	42	54	80	170'	160	142
	B	27.3	37.5	47.1	42.9	59.2	36.5	34.8	34.1	39.5	41.9
DIVISION 72 Electrical Machinery Apparatus and Appliances	A	3	9	17	25	24	27	53	68	61	71
	B	27.3	12.5	16.7	21.0	21.4	18.2	23.0	13.7	15.1	20.9
722 Electric power machinery and apparatus for making or breaking electrical circuits (switchgear, etc.)	A	1	3	5	6	4	4	7	11	12	15
	B	9.1	4.2	4.9	5.0	3.6	2.7	3.0	2.2	3.0	4.4
729 Other electrical machinery and apparatus	A	1	4	7	10	11	13	27	36	27	29
	B	9.1	5.5	6.9	8.4	9.8	8.8	11.7	7.2	6.7	8.6
DIVISION 73 Transport Equipment	A	3	23	12	7	17	24	31	100	49	27
	B	27.2	31.9	11.8	5.9	15.2	16.2	13.4	20.0	12.1	7.9
732 Road motor vehicles	A	—	3	9	6	7	8	13	63	36	24
	B	—	4.2	8.8	5.0	6.2	5.4	5.7	12.7	8.9	7.1
735 Ships and boats	A	—	—	—	—	9	14	13	—	—	—
	B	—	—	—	—	8.0	9.5	5.7	—	—	—

Sources: Statistics of Foreign Trade. Series C. OECD. Paris. for 1976 and 1977: for other years. OECD Statistical data.

Table A-26

**Types of Contractual Arrangements Included in Different Definitions
of East-West Industrial (inter firm) Cooperation** [a]

1. Sale of equipment for complete production systems, or "turnkey" plant sales (usually including technical assistance).

2. Licensing of patents, copyrights and production know-how.

3. Franchising of trademarks and marketing know-how.

4. Licensing or franchising with provision for market sharing and quality control.

5. Co-operative sourcing: long-term agreement for purchases and sales between partners, especially in the form of exchanges of industrial raw materials and intermediate products.

6. Sub-contracting: contractual agreement for provision of production services, for a short-term and on the basis of existing capabilities, but often to design specifications furnished by the contractor. Some or all components also frequently supplied by contractor.

7. Sale of plant, equipment and/or technology (1-3 above) with provision for complete or partial payment in resulting or related products.

8. Production contracting: contractual agreement for production on a continuing basis, to partner specifications, of intermediate or final goods to be incorporated into the partner's product or to be marketed by him. In contrast to sub-contracting, production-contracting usually is on the basis of a partially transferred production capability, in the form of capital equipment and/or technology (on basis of a license or technical assistance contract).

9. Co-production: mutual agreement to narrow specialization and exchange components so that each partner may produce and market the same end product in his respective market area. Usually on the basis of some shared technology.

10. Product specialization: mutual agreement to narrow the range of end products produced by each partner and then to exchange them so that each commands a full line in his respective market area. In contrast to co-operative sourcing, product specialization involves adjustment in existing product lines.

11. Co-marketing: agreement to divide market areas for some product(s) and/or to assume responsibilities for marketing and servicing each other's product(s) in respective areas. Joint marketing in third markets may be included. Does not in practice stand alone, unless in the form of a joint marketing company (i.e., combined with 14 below). Otherwise combined with various forms, especially 4, 8, 9, 10.

12. Project co-operation: joint tendering for development projects in third countries.

13. Joint Research and Development: joint planning, and the coordinated implementation, of R & D programs, with provision for joint commercial rights to all product or process technology developed under the agreement.

14. Any of the above in the framework of a specially formed mixed company or joint venture between the partner firms (on the basis of joint equity participation, profit and risk sharing, joint management).

---

a) The terminology used here is not standard, and the types are variously designated in the literature. For example, the term "production-sharing" is sometimes used to designate all or some of types 8 through 10. For more extensive definition, discussion and illustration of the many types of arrangement included in this table, but using somewhat different terminology, the interested reader is referred to *Rapport analytique sur la coopération industrielle entre les pays de la CEE,* Economic Commission for Europe, United Nations, Geneva, 1973, and St. Charles, D.P., "East-West Business Arrangements: Typology", in Carl H. McMillan (ed.), *Changing Perspectives in East Commerce,* Heath-Lexington, Lexington, Mass., 1974.

Source: Carl H. McMillan, "East-West Industrial Co-operation" in *East European Economies Post Helsinki. op. cit.,* p. 1182.

Table A-27

Categories of East-West Industrial Co-operation and their Definition:
a Comparison of Schemes used by Paul Marer and Joseph C. Miller and by the United Nations

Category and Definition Used	ECE Definition
1. Scientific-technical agreement (STA) are signed between a US corporation and the Soviet State Committee for Science and Technology (or its counterpart in an EE country) to explore what technology or know-how might be exchanged subsequently, on a commercial basis. While STAs are not commercial contracts, they are of interest because they identify industries and projects where the Eastern partner intends to obtain Western technology and where US firms have both an interest and a capability to compete.	Category not included.
2. Know-how (K) is transferred under a technical assistance contract. Only if such a contract was signed on its own, rather than as a component of a license or equipment sale or more complex forms of IC, was it included as a separate category. *a*	Category not included.
3. Licensing (L), regardless of the means of payment. L is not shown separately if it occurred as part of a more complex transaction, such as turnkey, co-production or joint venture, to avoid double-counting the number of agreements or projects. "L-direct" records transactions by a US-based firm directly: "L-indirect" records transactions carried out by a European, Canadian, or Japanese subsidiary or affiliate. *b*	Licensing (L) with payment in resultant products.
4. Turnkey (TK) includes all contracts where the supplier has significant on-site installation or supervision responsibilities, namely: prime contractor; contractor (US firm is one of several contractors but agreement is directly with the host country); and subcontractor (supplies machinery and equipment under subcontract and has on-site installation/training responsibilities).	Turnkey (supply of complete plants or production lines) with payment in resultant products (Subcontractors are not included).
5. Subcontracting (SC) in the Eastern country, with the Western partner usually providing the technical know-how and sometimes the machinery, equipment, and parts.	Same.
6. Co-production (CP) each partner specializes either in the production of parts of a product assembled by one or both partners or in the production of a limited number of finished products exchanged to complete each partner's range of products.	Same.
7. Joint ventures (JV) co-ownership of capital, co-management, and sharing of risk and profit, if the JV is located on the territory of the Eastern partner.	Same, plus including joint marketing ventures typically located in the West which market products of the Eastern country.
8. Category not included (no known US cases).	Joint Tendering or Projects (JTP) supply of complete plants or production lines to a third Western party.
9. Reverse Licensing (RL): US firms is importing a license. This category is not part of the total.	Category not included.

a) It is useful to distinguish know-how (1) to make: (2) to operate and maintain: and (3) to apply. Whereas a license and a patent normally cover what can be put on paper, know-how is less readily rendered on paper, rather it must be demonstrated.
b) This study shows separately licensing with payment in cash, in resultant product, and payment not speficified.
Source: Paul Marer, Joseph C. Miller, "US Participation in East-West Industrial Co-operation Agreements" in: *Journal of International Business Studies,* Fall-Winter, 1977, pp. 28-29.

Table A-28

**Companies having Co-operation Agreements with State Committee
for Science and Technology of the USSR**

Country/Company	1974 ranks in *Fortune* 300 largest non-US industrials	Country/Company	1974 rank in *Fortune* 300 largest non-US industrial
Germany[a]		Netherlands[a]	
Schering AG	283	Synres Nederland Sigma	—
Werkzeugmaschinen-		Vereinigde Machinefabrieken	295
Fabrik Gildemeister	—	AKZO NV	—
Ruhrkohle AG	33		
Krupp	51	Switzerland[a]	
Lurgi-Gesellschaften	—	Durisol AG	—
Otto Wolf AG	—		
Robert Bosch	63	Belgium[a]	
Daimler-Benz	16	Picanol	—
AEG-Telefunken	30	Marconi Ltd.	—
Kimsch	—	Dunlop-Pirelli	46
Thyssen-Roehenwerke	8	Lucas Industries	182
Bayer	15	Rank Xerox	174
Siemens	13	Rolls Royce[b]	—
Hoechst	10	Beecham Group	188
Degussa	126	Shell Oil[b]	1
Henkel	108		
BASF	9	Sweden[a]	
Hemscheidt	—	Volvo	80
		LBK Producer	—
Austria[a]		Sandvik	256
Schoeller Bleckmann Stahlwerke	—		
Voest	90	Japan[a]	
Manfred Swarovski GmbH	—	Mitsubishi	—
		Mitsui	214
Italy[a]		Tokyo Boeki	—
Pirelli-Dunlop	46	Mayekawa	—
Snia Viscosa	237	Teijin Co.	—
Metenco	—	C. Itoh Co.	180
Liquichimica	—		
Finmeccanica	—	Finland[a]	
Montecatini Edison	—	W. Rosenlev	—
ENI	—		
		France[a]	
Canada[a]		Moet Hennessy	—
Polysar Ltd.	—		
Canadian Broadcasting Co.	—		

Country/Company	1974 rank in *Fortune* 300 largest non-US companies	Country/Company	1974 rank in *Fortune* 300 largest non-US companies
United States		United States (cont'd)	
Abbott Laboratories	249	International Harvester	26
Allis-Chalmers	158	International Paper	56
American Can Co.	67	Kaiser Industries	186
American Home Products	92	Litton Industries	53
Armco Steel	51	Lockheed	49
Arthur Andersen	—	R. J. Reynolds Industries	48
Bechtel Corp.	—	Rohm & Haas	196
Bendix Corp.	77	Stanford Research Institute	—
Boeing Co.	39	Singer Co.	66
Bristol-Myers	125	Sperry Rand	70
Brown & Root	—	Standard Oil of Indiana	13
Burroughs Corp.	134	Union Oil Products	—
Coca-Cola	74	Union Carbide	22
Colgate-Palmolive	69	Varian Associates	492
Control Data Corp.	187	Louis Berger, Inc.	—
Corning International	190	McKinsey & Co.	—
Deere & Co.	75	Monsanto	43
Dresser Industries	146	Norton Simon	123
FMC Corp.	91	Occidental Petroleum	20
General Electric	8	Pepsico Inc.	—
General Dynamics	98	Pfizer International Inc.	130
Gould Inc.	259	Philip Morris	57
Gulf Oil	7	Raymond Loewy	—
H. H. Robertson Co.	—	Reichold Chemicals	347
Hewlett-Packard	225	Revlon International	291
ITT Corporation	10	Phillips Petroleum	25
Industrial Nucleonics	—		

a) List derived from published sources, therefore not exhaustive.
b) These could be subsidiaries since the author groups these firms with the Belgian firms.
Source: Theriot, Lawrence H., "US Governmental and Private Industry Co-operation with the Soviet Union in the Field of Science and Technology", in: *The Soviet Economy in a New Perspective, op. cit.*, pp. 763, 764, 766.

Table A-29

Examples of Licences sold by Western Firms to CMEA Countries

Buying CMEA Country	Description of the Technology	Western Licensor [a]	
Bulgaria	Transistor radiograms	Remap (Fr)	
	Automatic switches	Sace (I)	
	Grey cast iron	Mechanite (UK)	
	Marine diesel engines	Sulzer (Swi)	
	Container handlers	Rubery Owen (UK)	
Czechoslovakia	Molded footwear	British Rubber (UK)	
	Desulphurization of gas	Integral (A)	
	Direction indicator switches	Lucas (UK)	
	High-tensility nylon thread	Teijin (Ja)	
	Ethylene	Toyo Engineering (Ja)	
	Polyethylene	Union Carbide (US)	
	Waste incineration	Vereinigte Kesselwerke (FRG)	
Hungary	Insulation for electrical rotary machines	Alsthom (Fr)	
	Equipment for welding spiral pipes	Friedrich Kochs (FRG)	
	Boilers	Merloni (I)	
	Automatic control elements	Bosch (FRG)	
	High tension switches	Brown, Boveri (Swi)	
	Shock absorbers	Girling (UK)	
	Hypodermic needles	Hampden Industries (US)	
	Concrete pipes	Rocla Industries (Aul)	
	Automatic machine-tools	San Giorgio (I)	
	Licence for the production	Semperit (A)	1976
	Licences, know-how base concentrates for chemicals	Franz von Furtenbach (A)	1977
	Licences, parts, materials for making boilers	Chaffoteaux et Maury (Fr)	1976
	Licence, know-how to make bus engines (11 000 p.a.)	MAN (FRG)	1975
	Licence, know-how to make colour TV	Standard-Elektrik Lorenz AG (FRG)	1975
	Licence to make automatic washing machines	AEG (FRG)	1977
	Licence and equipment for making windows from synthetic materials (PVC)	Kömmerling (FRG)	1977
	Licence and semi-finished chemicals	Terosen (FRG)	1977
	Licenses, know-how. Gas welding pistols	AGA Svetsprodukter (Swe)	1976
	Licence and know-how to make truck axles	Vauxhall Motors (UK)	1975
	Licences for hospital equipment	Dent, Hellyer (UK)	1976
	Licences, equipment for making tractors	Steiger (US)	1974
Poland	Buses	Berliet (Fr)	
	High-pressure safety valves	Bopp and Reuther (FRG)	
	Lubricating oils	British Petroleum (UK)	
	Small electric motors	Hitachi (Ja)	
	Razor blades	Gillette (US)	
	Synthetic yarns	Marubeni-Ida (Ja)	
	Nylon	Toray Industries (Ja)	
	Shoe prod. licences, know-how	Alsa Schuhbedarf (FRG)	1976
	Licences, know-how for automotive industry trucks	Steyer-Daimler-Puch (A)	1975

Buying CMEA Country	Description of the Technology	Western Licensor [a]	
Poland (*cont'd*)	Licence for making gear boxes for rotary plugs	Huard (Fr)	1977
	Licences for prod. of concrete mixers and containers	Stetter (FRG)	1977
	Licences to produce relays for railway signal boxes	(Swe)	1975
	Licences, know-how for making electric typewriters	(Swe)	1976
	Licence for making construction equipment	Clark Equipment (US)	1972
	Licences for making tractors and accessories	International Harvester (US)	1974
	Licences, equipment for making semi-conductors, rectifiers	Westinghouse (US)	1974
	Licences, machines for making medical equipment	General Electric (US)	1976
Rumania	Semi-conductors	Compagnie générale de TSF (Fr)	
	Transformers for TV sets	Philips (N)	
USSR (not distributed by industrial branches)	Porous acetylene bottles	L'Air Liquide (Fr)	
	Axis-blower for nuclear power stations	A.G. Kühnle, Kopp & Kausch (FRG)	
	Chemical treatment of steel strips	Anchem Products (UK)	
	Furnaces for sulphur burning	Chemibau Zieren (FRG)	
	Numerically controlled machine-tools	Fujitsu (Ja)	
	Modular switches	Isostat (Fr)	
	Motor vehicle brakes	Knorr-Bremsen (FRG)	
	Machine-tool heads	Line (Fr)	
	Resistors and equipment for their manufacture	Précis (Fr)	
	Coating of metal sheets for motor vehicles	Pro Finish Metals (US)	
	Prefabricated houses	Tchersmachiner (Swe)	
	Electro-hydraulic cranes	Xegglound and Sioner (Swe)	

(*Cont'd on next page*)

Buying CMEA Country	Description of the Technology	Western Licensor [a]
USSR (distributed by industrial sectors)		*Announced*

Automotive

	Togliattigrad automotive plant – Positork automatic ignition device	DBA (Fr) 1/76

Business Equipment

	Electric typewriters	Olympia Werke (FRG) (announced July 1974)

Chemicals and Petrochemicals

	Aromatics	Arco Chemical (US) 11/72
	Chloropropene monomer on butadiene base	BP Chemicals International (UK) 3/73
	Reinforced plastic foil	Ewald Dörken (FRG) 8/73
	Alpha calcium-sulphate semihydrate refining	Gebr. Giulini (FRG) 9/74
	High solid latex	International Synthetic Rubber (UK) 3/73
	Acetic acid	Lummus Co. and Monsanto 12/73
	Automatic zinc-removing devices used in electrolysis	Montedison (I) 12/72
	Isocyanate processing	Upjohn Co. (US) 10/72
	200 cm. reactor for production of suspension PVC	Chemische Werke Huels AG (FRG) 4/75
	"Pattex" contact glue	Henkel & Co. (FRG) 5/75
	Polymerization agent Liladox, a percarbonic acid derivative	Kemanord (Swe) 7/75
	"Betanal", a herbicide for turnip and beet fields	Schering AG (FRG) 5/75
	Porous material for acetylene bottles	L'Air liquide (Fr) 7/76
	Synthesized standard gases	Seitetsu Kagaku Kogyo (Ja) 3/77

Construction

	Roadbuilding and paving equipment	CMI Corp. (US) 10/76

Consumer Goods

	Stainless steel razors	Wilkinson Sword (UK) 8/73
	Padlocks and mortise locks	Wärtsilä (Fin) 8/76
	Photoflash cubes	Bellmann (FRG) 11/76

Electrical Equipment

	Air preheaters for power stations	Kraftanlagen Heidelberg (FRG) 2/73
	Axial bellows for power static cauldrons	Kühnle, Kopp & Kausch (FRG) 8/72
	Cassette magnet head	Wolfgang Bogen (FRG) 5/74
	High-voltage powerline insulation materials	General Cable (US) 2/77

Buying CMEA Country	Description of the Technology	Western Licensor [a]	

USSR (*cont'd*) | | *Announced* |

Electronics

| Automatic line for reed relays | Wm. Günther (FRG) | 7/77 |
| Thermistors plant | Murata Manufacturing (Ja) | 8/77 |

Food Products and Tobacco

| Marlboro cigarettes | Philip Morris (US) | 2/77 |

Household Equipment

| Phonograph cabinets | Berlin Consult (FRG) | 1/74 |
| Electric stoves | Merloni SpA (I) | 9/73 |

Iron and Steel

Conversion coating of cold rolled steel strips	Amchem Products (UK)	9/72
Direct reduction process to be used in Kursk furnace	Midrex Corp. (US)	4/75
Steel structure manufacturing plant	Blohm & Voss (FRG)	1/77

Machine-Tools

Wedge presses and related transport equipment	Eumuco (FRG)	12/73
Abrasive material	Norton (US)	1/73
Universal presses	Aïda Engineering (Ja)	7/74

Materials-Handling Equipment

| Containers | Renault Industries Equipements et Techniques (Fr) | 12/73 |

Medical Equipment

| Disposable plastic medical goods | Portex (UK) | 8/77 |

Metalworking

| Aluminium wire | W. C. Heraeus (FRG) | 8/77 |

Mining and Metallurgy

| Aluminium casting; manufacture of equipment | Péchiney Ugine Kuhlmann (Fr) | 11/76 |

Packaging

| Nylon film production plant | Kohjin (Ja) | 6/76 |

Petroleum and Gas

Ethyl-benzene	Universal Oil Products	1/74
Gas dessiccation Orenburg natural gas complex	Davy Power Gas (FRG)	3/76
Oil drilling platform	Armco International (US)	7/76

(*Cont'd on next page*)

Buying CMEA Country	Description of the Technology	Western Licensor[a]	
USSR (*cont'd*)			*Announced*
	Offshore exploitation of gas and oil, including blowout preventers, preventer control devices, Sea King and Marine Riser systems	Seitetsu Kagaku (Ja)	5/77
	Printing		
	Two-web offset presses	Maschinenfabrik Augsburg-Nürenberg (FRG)	9/74
	Pulp and Paper		
	Know-how and equipment for production of "Super Perga" paper	Greaker Industrier (No)	5/75
	Rubber		
	Butadiene-type poly-chloroprene rubber	DuPont de Nemours (US)	8/74
	Shipping and Shipbuilding		
	Pipe-sealing technology	Chuetsu-Waukesha (Ja)	6/77
	Textiles, Clothing and Leather		
	Yield-increasing raw wool scouring	Sover SA (Be)	1/74
	Clothing factory	McIntosh Confectie (N)	1/77
	Corset tulle	Gold-Zack Werke (FRG)	8/77

a) *Country abbreviations:* A: Austria, Aul: Australia, Be: Belgium, Fin: Finland, Fr: France, FRG: Federal Republic of Germany, I: Italy, Ja: Japan, N: Netherlands, No: Norway, Swe: Sweden, Swi: Switzerland, UK: United Kingdom, US: United States.

Sources:
— "Doing Business with Eastern Europe", *Business International,* October 1975.
— J. Wilczynski, *Technology in Comecon,* MacMillan, London and Basingstoke, 1974, p. 303.
— J. Wilczynski, "Licences in the West-East-West Transfer of Technology", *Journal of World Trade Law,* March-April, 1977, p. 133.
— Office of East-West Policy and Planning, Bureau of East-West Trade, US Department of Commerce, 8th June, 1977.

Note of Bureau of East-West Trade: Although information on these transactions has been taken from published sources, the Bureau cannot vouch for its accuracy.

**Examples of Licences sold by the COMECON Countries
to Western Firms**

The Selling Comecon Country	Description of the Technology	Western Licensee Firm [a]
German Democratic Republic	Steel bar faggoting machines	Ataka and Co. (Ja)
Bulgaria	Automatic reeling and placing of spinning spools	Carelli Industriali Tessili (I)
	Electrolytic refining of copper	Inspiration Consolidated Co. (US)
	Perfected process for producing yoghurt	Milifoma (FRG)
	Protection of graphite electrodes in steel production	British Steel Corp. (UK)
Czechoslovakia	Automatic textile-winding machines	Ensju (Ja)
	Spindleless spinning machines	Daiwa Spinning (Ja)
	Production of electric ovens	Horn (FRG)
	Soft contact lenses	Bausch and Lomb (US)
	Skin protection varnish	Albus (Sp)
	Spindleless spinning machines	Nuova San Giorgio (I)
	Vertical forging presses	Kurimoto (Ja)
Hungary	Manufacture of equipment for condensing air	Mitsubishi Heavy Industries (Ja)
	Manufacture of small rechargeable battery cells	William Old (UK)
	Method of water purification	Ebara Infilco (Ja)
	Rust prevention process	Teccomex (Sp)
	Proteins from grasses	Alfa-Laval (Swe)
	Substitute body tissues	MGA Technology (US)
Poland	Manufacture of extract of the smoking-house smoke	Hercules Powder (US)
	Method of forging crankshafts	Sulzer (Swi)
	Forging of crankshafts	Endo Ironworks (Ja)
	Carousel furnaces	Creusot-Loire (Fr)
	Automatic safety winches	Dusterloch (FRG)
USSR	Automatic loading of pulpwood	J. M. Voith (A)
	Casting of aluminium ingots	Kaiser Aluminium Chemical Corp. (US)
	Construction of blast furnaces	Andco (Can)
	Improved methods of steel making	Ashmore, Benson, Pease & Co. (UK)
	Manufacture of a new type of metalcutting machine	Demag (FRG)
	Manufacturing of a pulsed waterflow gauge used in mining	Joy Manufacturing Co. (US)
	Manufacture of specialised mining machines	Sociedad Metallurgica Duro Felguera (Sp)
	Production of double-walled plastic tubes by extrusion	Anger Plastik Verarbeitungsmaschinen Gesellschaft (A)
	Synthetic acids from paraffin	Adzina Moto (Ja)
	Needle-cutting technique	Amtel (US)
	Steel and alloys	Avesta (Swe)
	Anti-cancer drug	Bristol-Myers (US)
	Cooling of blast furnaces	Broken Hill (Aul)
	Thin-walled tubing	Carpenter Technology (US)

(Cont'd on next page)

Table 30 *(cont'd)*

The Selling Comecon Country	Description of the Technology	Western Licensee Firm [a]
USSR *(cont'd)*	Continuous welding electrodes	Chemetron (US)
	Particle accelerators	Energy Science (US)
	Tube cold rolling mills	Innocenti (US)
	Polycarbonates	Montedison (I)
	Gas-permeating membrane	Rhône-Poulenc (Fr)
	High pressure polyethylene	Salzgitter (FRG)
	Rotary printing machines	Schnellpressen Fabrik (FRG)
	Aluminium from alunite	Southwire (US)
	Pneumatic transporter system	Sumitomo (Ja)
	Chemical disposal of waste	Toyo Engineering (Ja)
	Surgical instruments	US Surgical (US)

a) *Country abbreviations:* A : Austria, Aul : Australia, Can : Canada, Fr : France, FRG : Federal Republic of Germany, I : Italy, Ja : Japan, Sp : Spain, Swe : Sweden, Swi : Switzerland, UK : United Kingdom, US : United States.

Source: J. Wilczynski, *Technology in Comecon, op. cit.,* p. 309 ; and J. Wilczynski, "Licences in the West-East-West Transfer of Technology", *art. cit.,* p. 133.

Table A-31

Recent Co-operation Agreements with the USSR

Western Partner	Soviet Partner	Purpose of Agreement	Announced
Agriculture			
Gi è Gi SAS (Italy)	Tractorexport	Production and marketing cattle-feeding complexes.	1/74
Elanco (US)	Ministry of Agriculture	Joint tests of antibiotics for livestock breeding; exchange of results.	10/75
E. I. DuPont de Nemours (US)	SCST*	Research, production and application of agricultural chemicals.	8/77
Agricultural Equipment			
Robert Bosch (Germany)	SCST	Outfitting vehicles incl. tractors and agricultural machines, hydraulics and pneumatics, TV technology.	12/73
Verenigde Machinefabrieken (Netherlands)	SCST	Production of turbine blades, rotary screen printing machines, milk and products, offal processing in slaughterhouses.	12/73
International Harvester (US)	Soyuzselchoztechnika	Testing corn-tilling machinery.	7/74
International Harvester (US)	SCST	R & D; manufacture of tractors and other farm and construction machinery.	10/75
Aircraft			
Boeing (US)	SCST	R & D; helicopter and passenger aircraft construction, air traffic control.	6/74
Rockwell International (US)	Aviaexport	Rockwell equipment assembled into Soviet planes for sale in the US.	9/73
Lucas Aerospace Ltd.	SCST	Hydraulic and pneumatic aircraft control systems, onboard power-generating systems and switch gear, engine monitoring systems.	5/75
Automotive			
Daimler-Benz (Germany)	SCST	Car production; R & D on safety, urban transport, pollution.	3/73
Fiat SpA (Italy)		Expansion of Togliattigrad car works (negotiating).	
Marposs Finike Italiana		Automobile construction and fine mechanics, ball bearings; sales in third markets.	2/75
Volvo (Sweden)	SCST	Production and R & D: combustion engines, electric motors.	6/74
Daimler-Benz (Germany)	SCST	Automotive manufacturing.	10/76
Fiat (Italy)	SCST	Renewed agreement to expand auto output and manufacture farm and building industry vehicles.	11/76

* SCST = State Committee for Science and Technology.

(Cont'd on next page)

Table A-31 (cont'd)

Western Partner	Soviet Partner	Purpose of Agreement	Announced
Business Equipment			
Rank Xerox (UK)	SCST	R & D: copying and duplicating machines.	6/73
Chemicals and Petrochemicals			
Agfa-Gevaert (Germany/Belgium)		R & D: photochemistry.	4/73
BASF (Germany)	SCST	R & D: plastic materials, dyestuffs, fuels, mineral oil additives.	8/72
Bayer (Germany)	SCST	Inorganic pigments, synthetic rubber and raw materials, lacquers, environmental protection.	6/73
ENI (Italy)	Ministry of Chemical Industry	Construction of chemical plants against Soviet payments in kind.	3/73
Klimsch & Co. (Germany)	SCST	R & D and modernization of photographic reproduction equipment and technology.	12/73
Liquichimica (Italy)	Ministries of Oil Refining and Petro-chemicals and Chemicals	R & D: paraffins, olefins and derivatives; industrial biosynthesis; lubricating oil additives.	7/74
Monsanto (US)	SCST	R & D: use of computers in chemical industry; development of rubber compound products.	11/73
Montedison (Italy)	Foreign Trade Ministry	Construction of chemical plants against Soviet payments in products.	8/73
Montedison (Italy)	Soyuzchimexport	Barter: Italian base chemicals for Soviet chemical intermediates.	6/74
Occidental Petroleum (US)		Construction of fertilizer plant.	4/73
Pressindustria (Italy)	SCST	Chemical and food industries, pharmacology and cosmetics.	5/74
Reichhold Chemie (Austria)	SCST	Production of lacquers, resins and related raw materials.	3/73**
		Production of synthetic polymeric materials, wood chemicals extraction.	11/74
Schenectady Chemicals (US)	Ministry of Electrotechnical Industry	Enamel and varnish wire insulating techniques (protocol).	2/75
Snia Viscosa (Italy)	SCST	Artificial and synthetic fibers, environmental protection.	2/74**
Synres Nederland, Sigma (Netherlands); DeSoto (US)	R & D Institute of the Paint Industry	Know-how on composition and use of paints in extreme climates.	5/74
Rohm & Haas (US)	SCST	R & D in plastics, herbicides, pesticides and petroleum additives.	5/75
Montedison SpA (Italy)	Licensintorg	Joint R & D of polycarbonates to serve as possible alloy, glass and ceramic substitutes.	5/75

	Refining and Petrochemical Industry		
Hempel's Marine Paints (Denmark)	SCST	R & D; production of varnishes and paints; anti-corrosive coatings for ocean-going vessels.	1/76
Krauss-Maffei (Austria)	SCST	Design of plastics processing equipment.	6/76
Worthington Pumps Italiana	SCST	Joint production of pumps for chemical and petrochemical industries.	6/76
	Ministry of Chemical and Petroleum Machine Building	Joint R & D of protein and chemical pumping equipment; exchange of experts and information.	6/76
Imperial Chemical Industries (UK)	SCST	Exchange of information on crop protection, chemicals, health products, chlorinated rubber, fibers, fertilizers, paints.	6/76
L'Oréal (France)	SCST	Supplies of chemical products.	7/76
Vianova Kunstharz (Austria)	SCST	Joint R & D; exchange of experience and experts in production and application of synthetic resins, varnishes, paints, enamels.	8/76
Maschinenfabrik A. Gentil (Germany)	All-Union Research Institute of Hydro-machinery Construction and Technology	Joint development of pumps for chemical industry.	10/76
Mitsui Group (Japan)	SCST	Scientific and technical cooperation in chemical and petrochemical industries and electrical engineering.	11/76
Chemie Linz (Austria)	SCST	Cooperation and joint R & D in plastics, pharmaceuticals, glues; exchange of experts and test results.	12/76
Essochem Impex (Belgium)	SCST	Petrochemical products.	4/77
Ballestra (Italy)	SCST	Production of technical and medical white oils, sulphonates, detergents.	7/77
Computers			
Burroughs (US)	SCST	Computer technology, inc. education, design, programming and application.	8/74
Control Data (US)	SCST	R & D: computer technology.	2/73
	SCST	R & D: computers for transportation, medicine and education.	11/73
Sperry Rand (US)	SCST	Computer application in air traffic control and other areas, also the manufacture of farming machines.	6/74

(Cont'd on next page)

Table A-31 (*cont'd*)

Western Partner	Soviet Partner	Purpose of Agreement	Announced
Construction			
Bechtel (US)	SCST	Control and organization of planning and building large-scale industrial units.	7/73
Manfred Swarovski GmbH** (Austria)	SCST	Co-operation in traffic safety.	7/75
H. H. Robertson (US)	SCST	Exchange of information on rubber materials and dyes for structural coatings.	8/75
Durisol (Switzerland)	SCST	Exchange of experiences in using lightweight building materials.	8/75
Metecno (Italy)	SCST	Uses of prefab. panels in industrial and civil construction.	3/76
Stetter (Germany)	SCST; Gosstroy	R & D; production of concrete mixing and setting equipment.	9/76
Otis Elevator (US)	Ministry of Construction, Road and Municipal Machine-building		11/76
Consumer Goods			
Loewy Raymond-William Snaith (US); Cie de l'Esthétique industrielle (France)	All-Union Research Institute of Technical Aesthetics	Industrial and consumer design for cars, boats, household appliances, motorcycles.	1/74
Revlon (US)	SCST	Joint R & D; exchange of staff for cosmetics, perfumes, fragrances, pharmaceuticals.	2/76
Electrical Equipment			
General Electric (US)	SCST	Power generation, transmission and utilization.	1/73
Matsushita Electric Industrial Co. (Japan)	SCST	Electrical and electronic equipment.	10/74
Siemens; Kraftwerk-Union (Germany)	Ministry of Electrical Engineering	Computer testing, turbogenerators, water jackets, anticorrosion measures.	7/73
General Electric's UK subsidiary	SCST	Development and engineering of turbogenerators, heavy machinery, and AC & DC transmission systems.	5/75
Tokyo Boeki (Japan)	SCST	Scientific-technical co-operation and exchange of information on electrical technology.	7/75
Maekawa Shoji (Japan)	SCST	Refrigeration; compressors.	9/75
Gould (US)	SCST; Ministry of Electrotechnical Industry	Condensers, electro-motors, electronic tools, storage batteries, powder metallurgy.	12/75
Alsthom (France)	Energomashexport	Hydro-accumulating power stations.	6/76

Siemens (Germany)	SCST	Electrical engineering.	10/76
AEG-Telefunken (Germany)	SCST	Agreement renewed for joint production of household appliances.	2/77
Hitachi (Japan)	Energomashexport	Construction and equipping of power stations in third countries.	8/77
Jungner (Sweden)	Ministry of Electro-technical Industry	Development of locomotive batteries.	9/77
Electronics			
Arthur Andersen (US)		Application of control systems in industry, production calculations (protocol).	5/73
Hewlett-Packard (US)	SCST	Medical electronics, mini computers, measuring instruments.	6/73
Nippon Electric (Japan)	SCST	R & D: communications and electronic technology; household electrical appliances.	8/74
Olivetti SpA (Italy)**	SCST	Electronics and automation of production and management.	7/75
Wolfgang Bogen (Germany)	SCST	Exchange information and experts in recording techniques.	1/76
Marconi (UK)	SCST	Joint R & D on radio and TV equipment.	2/76
AEG-Telefunken (Germany)	SCST	Electronics; radio engineering; precision instruments.	10/76
CBS (US)	State Committee for Radio Broadcasting and Television	Exchange programs and expertise; develop technical cooperation.	10/76
Cameca (France)	Burevestnik	Jointly develop and produce microelectronic screen microscopes.	4/77
Strömberg (Finland)	SCST	Computer control of cement production.	7/77
Nippon Electric Co. (Japan)	SCST	Communications; electronic technology.	7/77
W. C. Heraeus (Germany)	Ministry of Electronics Industry	Joint production of diascripters.	8/77
Jungner (Sweden)	Ministry of the Electrotechnical Industry	Development of locomotive batteries.	9/77

** Renewal.

(Cont'd on next page)

353

Table A-31 (*cont'd*)

Western Partner	Soviet Partner	Purpose of Agreement	Announced
Environment			
Silvani Anticendi (Italy)	SCST	Firefighting technology.	10/75
Food			
Coca-Cola (US)	SCST	R & D in agricultural and food industry, including new foodstuffs with higher nutritional value; production of soft drinks; land reclamation methods.	7/74
Pepsico (US)	Soyuzplodimport	Pepsi-Cola marketing in USSR against buyback of Soviet liquors.	12/72
Hennessy (France)	SCST	Food, wine, cosmetics.	2/76
Rieber & Son (Norway)	Ministry of Fish Industry	Fish processing.	3/77
Seagram (US)	SCST	Food processing.	4/77
Gervais-Danone (France)	SCST	Joint R & D; exchange of information in food, production, packaging, storage, marketing.	7/77
Household Equipment			
Kymi Kymmene (Finland)	SCST; Ministry of the Construction Materials Industry	Plumbing and bathroom fixtures.	2/77
Iron and Steel			
Fried. Krupp (Germany)	SCST	R & D: continuous casting methods.	2/73
Lurgi-Gesellschaften (Germany)	SCST	Construction of metallurgical and chemical plants.	2/74
Otto Wolff (Germany)	SCST	Production of rolling mills for third markets.	2/74
Perusyhtyma Oy, Lemminkäinen Oy, Finnbothia (Finland)	Prommashimport	Mining iron ore against counterdeliveries of iron ore pellets.	12/73
Schoeller-Bleckmann Stahlwerke (Austria)	SCST	R & D: rapid machining of high alloyed steel.	7/73
Vöest (Austria)	SCST	R & D: converter steel technology.	7/72
Vöest-Alpine (Austria)	Soyuzpromexport	Austrian imports of bituminous coal and iron ore.	6/74

Company	Soviet Organisation	Description	Date
Assn. of French Machine-Tool Builders	Ministry of Machine Building and Instrument Making Industry	Automatic production lines for car spare parts; design of manufacturing equipment for furniture, wood packaging, saws.	7/72
Elliot Co. (US)	Energomashexport	Production of compressors.	12/73
Georg Fischer (Switzerland)	SCST	Engineering, casting techniques.	7/73
Werkzeugmaschinen-Fabrik Gildemeister (Germany)	SCST	R & D: machine tools, metalworking machines.	4/73
Sandvik (Sweden)	SCST	Joint R & D in metal-cutting tools, super-hard materials in metal processing.	12/75
Gildemeister (Germany)	SCST	Machine tools.	10/76
Liebherr (Germany)	Ministry for Machine Tools and Tool Building Industry	Grinding machines.	4/77

Machinery

Company	Soviet Organisation	Description	Date
Mitsubishi (Japan)	SCST	R & D: new machinery and equipment.	4/74
Silvani Anticendi SpA (Italy)	SCST	Scientific-technological co-operation in fire-fighting.	7/75
Ishikawajima-Harima Heavy Industries (Japan)	Ministry of Heavy, Power and Transport Machinebuilding	Design of rolling mills, furnace equipment, multipurpose cranes.	7/76
Fried. Krupp (Germany)	SCST	Heavy machinebuilding.	10/76

Materials Handling

Company	Soviet Organisation	Description	Date
Interpool Ltd (UK), jointly with Associated Container Transportation (Australia), Nippon Yusen Kabushiki Kaisha and Yamashita Shinnihon Steamship (both Japan)	Sovfracht	Leasing containers.	2/73
Strick Co. (US)	Sovinflot	Leasing refrigerated containers.	8/74

(Cont'd on next page)

Table A-31 (cont'd)

Western Partner	Soviet Partner	Purpose of Agreement	Announced
Medical Equipment			
Siemens (Germany)	Ministry of Health	R & D: Medical equipment, computerized diagnosis.	7/74
LKB Produkter (Sweden)	SCST	Clinical biochemical instruments.	6/76
Metalworking			
Occidental Petroleum (US)	Various FTOs	Barter: Metal-processing against Soviet raw nickel.	12/72
Degussa Deutsche Gold- und Silberscheideanstalt (Germany)	SCST	Scientific-technical co-operation in metalworking, participation in various Soviet industrial projects, including transfer of know-how and technology.	7/75
Vereinigte Edelstahlwerke (Austria)	Paton Institute, Kiev	Technical application of electroslag remelting process.	10/75
Mining			
Joy Manufacturing (US)	Ministry of Coal Industry	Machine building for coal and mining industries (protocol).	7/72
Kaiser Resources (Canada)	Ministry of Coal Industry; SCST	Hydraulic and open-cut strip mining.	8/74
Ruhrkohle (Germany)	Ministry of Coal Industry; SCST	Design of mining equipment, safety systems; processing and preparation of residual products.	5/73
Hermann Hemscheidt, Maschinenfabrik (Germany)	SCST; Ministry of Coal Industry	Joint R & D of machinery for working thin and medium coal seams.	11/75
Ruhr Kohle (Germany)	SCST	Coal equipment and technology.	10/76
Multisector			
AEG-Telefunken (Germany)	SCST	Data processing, energy, tools, communication, transportation.	8/72
AKZO (Netherlands)	SCST	R & D: yarns, fibers, chemicals, pharmaceuticals, plastic and rubber products.	8/72
Armco Steel Corp. (US)	SCST	Ferrous metallurgy, offshore drilling equipment.	1/74
Bendix Corp. (US)	SCST	Automotive, aerospace and electronics products, scientific instruments, automation and machine tool products.	11/74
Food Machinery Corp. (US)	SCST	Agricultural, food industry, packaging, oil industry and materials-handling equipment.	12/73

356

Company	Partner	Description	Date
	Coal, Oil Extracting, Chemical, Oil Refining and Petrochemicals	petroleum and derivatives; production of chemicals and synthetic fuels.	
Industrial Nucleonics (US)	SCST	Automation and control in pulp and paper industry; steel, rubber and plastics manufacture.	5/74
Kaiser Steel (US)	SCST	Aluminium production, special steel, large-diameter pipes, off-shore drilling platforms, coal hydroextracting technologies.	2/74
Linde AG (Germany)	SCST	R & D: cryogenic techniques, sewage treatment, natural gas refining, chemical equipment.	5/74
Lockheed Aircraft (US)	SCST	Aviation industry, navigation systems and apparatus, special machine-tool building, cross-country vehicles, minicomputers, mineral exploration, medical electronic systems.	2/74
Norton Simon (US)	SCST	Technology for cosmetics, soft drinks, production of baby food, food and packaging industry.	10/74
Occidental Petroleum (US)	SCST	Extracting and processing oil and gas; agricultural fertilizers and chemicals; metalworking, metalcoating, projecting and building of hotels, utilization of solid wastes.	7/72
Philip Morris (US)	SCST	Paper-packing materials, chemical technology, tobacco-leaf cultivation, cigarette manufacture.	4/74
Singer Co. (US)	SCST	Calculators, training equipment, navigation equipment, consumer electrical products.	10/73
Stanford Research Institute (US)	SCST	Scientific, technological and economic activities.	10/73
Universal Oil Products (US)	SCST; Ministry of Oil Refining and Petrochemicals	Petroleum refining, organic chemicals and plastics technology (protocol).	12/74
Allis-Chalmers (US)	SCST; Ministries of Heavy, Power and Transport Engineering; Ferrous Metallurgy; and Nonferrous Metallurgy	Exchange of know-how in engineering and metallurgy.	4/75
Colgate-Palmolive (US)	Ministry of Chemical Industry	Medicine, sports goods and detergents.	5/75
G. L. Rexroth GmbH, member of Mannesmann Group (Germany)		Co-operation in manufacture of control systems, especially for hydraulic excavators.	5/75

(Cont'd on next page)

357

Table A-31 *(cont'd)*

Western Partner	Soviet Partner	Purpose of Agreement	Announced
Multisector *(cont'd)*			
Loewy Raymond International (US)	SCST	Joint design of packaging, industrial interiors, e.g. shopping centers and hotels.	7/75
Nokia Oy (Finland)	SCST	Scientific-technical co-operation in electrical engineering, electronics rubber and paper production.	5/75
Rolls-Royce Ltd (UK)	SCST	R & D in industrial engines and aerospace technology.	5/75
Shell (UK)	SCST	Agrochemicals and oil drilling	5/75
Finmeccanica (Italy)	SCST	R & D: production of electrical and power-generating equipment; automobile manufacture.	10/75
Cie générale d'électricité (France)	SCST	Energy; petrochemical equipment; machine-tools.	10/75
Union Carbide (US)	SCST	Chemicals; metallurgy; electric welding; environmental protection.	10/75
Corning Glass (US)	SCST	Glass; glass ceramics; electronics; biochemistry.	12/75
Ente Partecipazioni e Finanziamento Industria Manifatturiera (Italy)	SCST	Exchange of information and specialists; joint R & D in aluminium, shipbuilding, food processing, fishing industries.	6/76
C. Itoh (Japan)	SCST	Electrical and electronic engineering; chemistry.	7/76
Airco (US)	SCST	Manufacture of welding equipment; shipbuilding; medical equipment; refrigeration.	11/76
Babcock & Wilcox (US)	SCST	Joint research.	5/77
Plessey (UK)	SCST	Renewed agreement in aviation, telecommunications, data processing, semiconductors, hydraulic and pneumatic systems.	6/77
Mitsubishi (Japan)	SCST	Extension of agreement in machinebuilding, chemicals, petrochemicals, ferrous metallurgy, shipbuilding, engine manufacture.	7/77
Nonferrous Metals			
Klöckner-Humboldt-Deutz (Germany)	Licensintorg	Preparation and extraction of nonferrous metals.	3/74
Minemet (France)	Mekanobrabotka Institute	R & D: flotation units.	2/73
Tréfimétaux (France)	Gipros-Wetmetobrabotka Institute	R & D: high-resistant copper alloys.	2/73
Nuclear			
Gulf Oil Corp. (US)	SCST	Nuclear energy, atomic power.	10/74

Company	Soviet partner	Subject	Date
American Can (US)	Ministry of Engineering for Light and Food Industries	Container and packaging technology.	1/73
	SCST	Container and packaging technology.	7/74
Petroleum and Gas			
Brown & Root (US)	SCST	Engineering and construction of gas; oil transportation methods.	6/73
Cooper Industries (US)	Licensintorg	Natural gas exploitation.	9/72
Dresser Industries (US)	SCST	Geophysics research in oil and gas extraction.	6/74
ENI (Italy)	Ministry of Gas Industry	Opening and exploiting gas deposits.	4/72
Japanese consortium		Developing and exploiting oil and natural gas deposits on continental shelf of Sakhalin peninsula.	2/75
Occidental Petroleum (US)	SCST	Oil and gas exploration.	7/72
Petroleum Services, Subsidiary of Dresser Industries (US)	SCST	Private agreement involving only the subsidiary; oil and gas extraction.	10/73
Worthington Pump (US)	Ministry of Chemical and Oil-Machine Building	Consultation and R & D on oil pipeline technology (protocol).	12/74
Phillips Petroleum (US)	Ministry of Oil Extraction Industry	Joint prospecting and exploitation of oil deposits; recovering oil as secondary crude from other industrial processes.	9/75
Standard Oil of Indiana (US)	SCST	Oil prospecting and extraction; oil refining.	9/75
British Petroleum (UK)	SCST	Exchange of information on lubricants, bitumen, other petroleum by products; pollution control; synthetic proteins.	6/76
Liquichimica (Italy)	Ministry of Oil Extraction Industry	Joint R & D in oil-exploiting technology.	12/76
Essochem Impex (Belgium)	Ministry of Oil Refining and Petrochemical Industry	Lube oil additives.	4/77
Cameron Iron Works (US)	SCST	Exchange of specialists and information in petrochemical engineering.	5/77
Pharmaceuticals			
American Home Products (US)	SCST	R & D: pharmaceuticals and medical instruments.	9/74
Beecham Group (UK)	SCST	Exchange of documentation; R & D on antibiotics.	2/73
Schering AG (Germany)	SCST	R & D.	9/73

(Cont'd on next page)

359

Table A-31 (cont'd)

Western Partner	Soviet Partner	Purpose of Agreement	Announced
Pharmaceuticals (cont'd)			
Pfizer Inc. (US)	SCST	Exchange of scientific-technical information, joint R & D of pharmaceuticals, particularly veterinary and botanic genetics.	6/75
Revlon Inc. (US)	Soyuzchimexport, Raznoexport	Joint development and marketing of "Epas" perfume.	5/75
Rosenlew Oy (Finland)	SCST	Scientific-technical co-operation in microbiological products.	7/75
Abbott Laboratories (US)	SCST	Joint R & D in infant nutrition, drugs.	11/75
	Ministry of Health; Ministry of the Meat and Dairy Industry	Execution of research.	—
Bristol Myers (US)		Exchange of information in antibiotics, nonnarcotic analgesics, chimotherapy in cancer research.	9/76
Precision Instruments			
Lip (France)		Production of quartz-based wrist watch components.	11/72
Rank Taylor Hobson (UK)	State Committee for Standards	Development of surface and optical metrology equipment for auto, machine tool, aircraft and other industries.	11/74
Varian Associates (US)	SCST	High energy particle accelerators for scientific, industrial and medical application, analytic and measuring instruments, vacuum apparatus and components.	6/74
LKB-Produkter (Sweden)	SCST	R & D, design and production of scientific instruments.	1/76
Disa Elektronik (Denmark)	SCST	Production of scientific instruments.	8/77
Pulp and Paper			
International Paper (US)	Ministry of Pulp and Paper	R & D in paper industry and on plant construction.	6/74
Parsons & Whittemore (France)	Prommashimport	Construction of pulp plant against counter-deliveries of pulp.	8/73
Printing			
BASF (Germany)	SCST	Expansion of agreement to include photopolymeric printing plates.	8/76
Railway			
Plasser & Theurer (Austria)	SCST	Rail laying, maintenance, and track tamping	7/73

360

Continental Gummiwerke (Germany)	SCST	9/74	Know-how for rubber mixing factory.
Dunlop-Pirelli (UK/Italy)	SCST	11/72	R & D: car tire.
Services			
McKinsey & Co. (US)	SCST	9/74	Management, technology, science.
Volvo Fritid (Sweden)	SCST	6/75	Scientific and technical co-operation.
NCR (US)	SCST	10/76	Automation technology for retail stores, restaurants.
Rank Taylor Hobson (UK)	Gosstandort	5/77	Research into surface measurement and evaluation.
Seibu (Japan)	Soviet government authorities	5/77	Joint research into retailing techniques.
Shipping and Shipbuilding			
Svensk Varvs Industri Föreningen (Sweden)	SCST	5/76	Modernization of equipment and engines.
Wärtsila (Finland)	Ministry of Shipbuilding Industry	1/77	Scientific technical co-operation; increased trade in ships and ship equipment.
Bos Kalis Westminster Group (Netherlands)	SCST; Ministry of Maritime Fleet; Ministry of River Fleet	3/77	Study of underwater formations; development of dredging methods.
Wärtsila (Finland)	Ministry of Shipbuilding Industry; SCST	7/77	Design of ship structures; standardization of ship engines.
Telecommunications			
ITT Corp. (US)	SCST	7/73	Telecommunications, electronic components, consumer goods, scientific and technical information publication.
		12/73	US-USSR satellite communications link.
Textiles			
Agache (France)	Techmashexport	5/74	Textiles and fashion goods.
Karl Mayer (Germany)	SCST	1/75	Design and manufacture of equipment for textile industry.
Picanol (Belgium)	SCST	2/73	Looms.
MacIntosh Confectie (Netherlands)	Soviet government authorities	1/77	Clothing factory; commercial co-operation.
Teijin (Japan)	SCST	7/77	Production and marketing of synthetic fibers.

(Cont'd on next page)

361

Table A-31 (cont'd)

Western Partner	Soviet Partner	Purpose of Agreement	Announced
Wood and Products			
International Paper (US)	SCST	Scientific and technical co-operation in pulp and paper industry	1/75
Nihon Chip Boeki, head of 30 Japanese-firm consortium		Woodworking equipment against Soviet wood chips.	8/72
Valmet (Finland)	Ministry of Pulp and Paper Industry	Paper-making machinery.	9/77
Canadian Institute of the Pulp and Paper Industry	Soviet Pulp and Paper Scientific and Technical Information Institute	New methods of pulp cooking; transportation and storage of sawed timber.	10/77
Miscellaneous Equipment			
Nummela (Finland)	SCST	Joint development of fire extinguishers for trucks.	8/77

Source: "Doing Business with Eastern Europe", *Business International*, October 1975 and November 1977, USSR, APX 9.2.

Table A-32

Major Western Exports of Technologically Advanced Industrial Equipment and Plants to the Comecon Countries in the Early 1970s

Description and Approximate Value of the Contract	Western Exporter [a] and the Importing Comecon Country
Acrylic fibre and tyre plants ($70 m.)	Courtaulds (UK) – USSR
Aluminium panels plant ($22 m.)	Ataka & Co. (J) – USSR
Ammonia plant ($75 m.)	Toyo Engineering (J) – German Democratic Republic
Ammonia plants ($110 m.)	Toyo Engineering (J) – USSR
Automated production lines for textiles ($22 m.)	Industriewerke Karlsruhe Augsburg (FRG) – USSR
Automatic transfer lines for a truck plant ($39 m.)	Renault (Fr) – USSR
Ball-bearings plant ($15 m.)	Nippon Seiko (J) – Bulgaria
Earth-moving equipment ($28 m.)	Caterpillar Mitsubishi (J) – USSR
Ethylene plant ($30 m.)	Selas Nederland (UK) – Bulgaria
Ethylene plant ($23 m.)	Toyo Engineering (J) – Czechoslovakia
Ethylene plant ($40 m.)	Voëst (A) – German Democratic Republic
Fertilizer plant ($64 m.)	Creusot-Loire (Fr) – Poland
Foundry for heavy trucks ($440 m.)	Swindell Dressler (US) – USSR
Gas pipeline construction plant ($20 m.)	Cooper Industries (US) – USSR
Gas treatment plant ($53 m.)	Cie centrale d'études industrielles (Fr) – USSR
Gearbox manufacturing plant ($35 m.)	Liebherr-Verzahn-Technik (FRG) – USSR
High-density polyethylene plant ($29 m.)	Constructors John Brown (UK) – Czechoslovakia
Iron-ore pelletizing plant ($12 m.)	Ashmore, Benson, Pease & Co. (UK) – USSR
Machinery and equipment for woodworking plants ($25 m.)	Teamwood International (FRG) – USSR
Metallurgical plants ($19 m.)	Gute Hoffnungshütte (FRG) – Rumania
Microwave radio line equipment ($11 m.)	A/S NERA (No) – Bulgaria
Nylon plant ($80 m.)	SNIA Viscosa (I) – USSR
Nylon plant ($50 m.)	Toray Industries (J) – Poland
Oil refinery ($100 m.)	SNAM Progetti (I) – Poland
Olefin plant ($37 m.)	Linde (FRG) – Hungary
Plant for electrolytic zinc coating of steel ($21 m.)	Anchem Products (US) – USSR
Polyester and nylon spinning complex ($250 m.)	Toray Industries (J) – USSR
Polyester plants ($100 m.)	Constructors John Brown (UK) – USSR
Polyester tyre plant ($17 m.)	Krupp (FRG) – USSR
Polyethylene plant ($56 m.)	Simon Carves (UK) – USSR
Polypropylene plant ($18 m.)	Itoh & Co. (J) – Czechoslovakia
Polyurethane plant ($28 m.)	ENSA (Fr) – USSR
Polyvinyl chloride floor coverings plant ($12 m.)	Pegulan Werke (FRG) – Poland
Synthetic fibre plants ($120 m.)	Salzgitter (FRG) – USSR
Woodworking production lines ($45 m.)	Bisonwerke, Baehre & Greten (FRG) – USSR

a) A = Austria, Fr = France, FRG = Federal Republic of Germany, I = Italy, J = Japan, No = Norway, UK = United Kingdom, US = United States.

Source: Compiled from daily and periodical literature published in the Comecon and Western countries by J. Wilczynski, *Technology in Comecon*, op. cit., pp. 320-321.

Table A-33
Identified Soviet Compensation Agreements with the West

Western Country	Western Supplier	Year Signed	Type of Soviet Import	Value of Soviet Imports (Million US$)	Type of Soviet Export	First Year of Soviet Deliveries	Value of Soviet Exports (Million US$) 1975-80	1981-85	Remarks
					Forestry				
Japan	KS Industries	1969	Forestry handling equipment	163	Timber products	1969-74	n. a.	n. a.	
Japan	Japan Chip Trading Company	1971	Wood chip plant	45	Wood chips and pulp	1972-81	145	50	
Japan	KS Industries	1974	Forestry handling equipment	(525) 500-550	Timber products	1975-79	1 000	n. a.	
France	Parsons Whitmore	1973	Pulp-paper complex	60	Wood pulp	1977	34	50	Finland and Sweden also participated in transaction.
Japan	Mitsubishi Heavy Industry Ltd. – Mitsubishi Corp.	1976	Plant for making ground pulp	5.5	Ground pulp	1978	n. a.	n. a.	USSR to deliver undisclosed percentage production capacity.
Finland	Enso-Gutzeit	1976	55 000 ton/yr. double cardboard and special boards	n. a.	1.5 million cu.m. of timber and hardwood per year	1976	n. a.	n. a.	1976-80.
					Chemicals				
Germany	Salzgitter	1972	Polyethylene plant capacity: 120 000 t/yr.	63	Polyethylene	1972	170 (1971-1983)		USSR contracted to export between 150 000 and 250 000 tons/yr. of polyethylene to Germany.
Germany	Salzgitter	1973	Polyethylene plant capacity: 240 000 t/yr.	87	Polyethylene	n. a.	225 (1978-1986)		250/350 000 tons of low density polyethylene in total.
France	Litwin S.A.	1973	Styrene/Polystyrene	120	Polystyrene	1978	50	60	
Italy	Montedison	1973 (preliminary agreement)	11 Chemical plants	(500)*	Ammonia	1978	287 (1978-1987) (ammonia) 78 (1976-1985) (urea)		5 plants ordered so far; contracts signed for one urea and one acrylonitrile plant in 1974; two freon plants in 1975; one urea plant in 1976. * figure is based on the value of the 7 plants ordered so far: the total value of all 11 plants is given as $800 million.
United Kingdom/	John Brown, Union Carbide	1974	Polyethylene plant	68	High density Polyethylene	n. a.	n. a.	n. a.	

Country	Company	Facility	Date	Value	Products	Date	Value	Value	Remarks
	Subs. od Baxter Lab.)								
France	Creusot-Loire	Ammonia Plants (4)	1974	220	Ammonia	1979-80	100**	225**	At least 150 000 mt/y to be shipped to France over 10 year period. Most to be sold in France and Europe. 40 000 mt/y of ammonia to Rhône-Poulenc. ** refers to shipments during 1975-80 only; approx. 300 000 tonnes of ammonia will be shipped each year during the 10-year life of the agreement.
Italy	ENI	Chemical Plants (6)	1974 (preliminary agreement)	(135) (110-165)*	Chemical products	n. a.	n. a.	n. a.	*figure is based on the value of the 1 plant ordered so far: the total value of all 6 plants is given as $670-1 000 million.
United States	Occidental Petroleum	Ammonia plants (2); fertilizer storage and handling facilities; ammonia pipeline	1974	5 000	Ammonia, urea, potash	1978	5 000 (1978-1987)		10 year agreement; part of 20 year agreement. 2.1 million t/yr of ammonia for first 10 years; 2.5 million t/yr of ammonia for next 10 years; 1 million t/yr of urea and same of potash for 20 years.
Italy	Tecnimont	Polypropylene	1975	(115) a 100-130	Possibly chemical intermediates	n. a.	n. a.	n. a.	
Italy	ENI	Urea Plants	1975	170	Possibly urea, ammonia	n. a.	n. a.	n. a.	
Italy	Pessindustria	Surface-active detergent plant	1975	8.3	Monoethylene glycol, organic chemicals, surface active agents	n. a.	n. a.	n. a.	
Germany/ United Kingdom	Klöckner Group & Davy Powergas GmbH	Phthalic anhydride plant	1976	50	Chemical products including phthalic anhydride, fumaric acid and urea	1981 a	47 over 10 yr. period (1981-90)		Mixed counterpurchase and buyback.
France	Rhône-Poulenc S.A.	Chemical Plant	1976 (preliminary agreement)	1 300 b	Petroleum products and ammonia	1980	34.5 (1981-1990)		Sov. prods. to Rhône-Poulenc for in-house use. Only $400 m. in actual contracts so far. Latest is 1977 Speichim $97 m. contract. Total value is scheduled to rise to $1 000 million in terms of the French export value; 30 000 t/yr. of methanol as part payment for this deal over 10 years.
France	Technip S.A.	2 Petrochemical facilities	1976	501.1	Petroleum (m.$400.8) Resultant prod. (m.$100.3)	1980	501.1	n. a.	

(Cont'd on next page)

Table A-33 (cont'd)

Western Country	Western Supplier	Year Signed	Type of Soviet Import	Value of Soviet Imports (Million US$)	Type of Soviet Export	First Year of Soviet Deliveries	Value of Soviet Exports (Million US$)	Remarks
					Chemicals (cont'd)			
Germany/United Kingdom	Davy Powergas (UK) ICI, Klöckner (Germany)	1977	2 Methanol Plants	250	Methanol	1981-1991	350[a] n.a.	Contract in dollars not sterling. ICI and Klöckner to purchase methanol approx. 20% of plants prod. or 300 000 mt/yr.
Germany	Klöckner	1975	PVC Plant	40	PVC	1977	n.a.	10 000 t.p.a. for 10 years of vinyl chloride from 1977, 8 000 t.p.a. for 10 years of PVC from 1980.
United States	Lummus-Monsanto	1975	Acetic Acid Plant	200	Acetic Acid	n.a.	n.a.	
Germany	Krupp-Koppers GmbH and Friedrich Uhde GmbH	1976	Dimethyl terephthalate complex and addition to an existing polyester fibre plant	124.7	Raw cotton and base materials of aromatics	n.a.	n.a.	Possible $1.1 billion private credits. Hermes guarantee of credit facilities. Soviet goods mainly to German importers.
Germany	Hoechst	1976	De-emulsifiers for oil exploration, synthetic fibres and mat. textile dyes, and pigments, plant protection, pharm. preparations	130[a]	n. a.	n. a.	n. a.	Soviets now requesting 100% countertrade for deals over $50 m.
Germany	Salzgitter	1976	Ethylene oxide plant	100	Resultant chemical product	1979		
Germany	Hoechst-Uhde-Wacker	1974	VCM plant capacity: 270 000 T/yr.	47	VCM	1976	66 (1976-1979)	40 000/50 000 tons of VCM over 4 years.
Germany	Klöckner-Hills	1974	PVC plant capacity: 250 000 t/yr.	71	VCM and PVC		33	10 000 t/yr. of VCM over 10 years. But deliveries not started yet. Also PVC buy-back.
Germany	Klöckner-Hills	1974	PVC plant capacity: 250 000 t/yr.	71	PVC and VCM	1978	54 (1978-1987)	8 000 t/yr. of PVC over 10 years (rest paid in VCM).
Germany	Krupp, Korf, Salzgitter, Siemens, Demag	1975	Production for pelleting of iron ore	1,000		1976	450 1,000	
Germany	Salzgitter	1976	Ethylene oxide plant	100	MEG	1979	100 (1979-1988)	27 000 t/yr. of MEG over 10 years.
Germany	Krupp-Koppers	1976	Dimethylterephthalate	75	Methanol	1981	100 (1981-1990)	Methanol to be used by Dynamit

Country	Contractor	Year	Plant description		Product	Year		Products/Remarks
		1978	60 000 and 120 000 t/yr. capacity	125		1981	150 (1981-1990)	products.
Germany	Uhde-Hoechst	1977	Polyester staple fibre plant: capacity 35 000 t/yr.	70	Methanol	1981	50 (1980-1989) (without urea)	Methanol as buy-back (with DMT and paraxylene).
Germany	Klöckner-Davy Power Gaz	1977	Phtalic anhydride plant plus maleic anhydri-plant: capacity: 60 000 t/yr. and 3 000 t/yr.	50	Phtalic anhydride and maleic anhydride	1980	70 (1980-1989)	5 000 t/yr. of phtalic anhydride and 3 000 t/yr. of maleic anhydride (plus urea buy-back).
United Kingdom/United States	John Brown Contractors/Union Carbide	1974	Polyethylene plant: capacity: 200 000 t/yr. high-density polyethylene	40	High-density polyethylene	1980	162 (1983-1993)	10 000 t/yr. of high-density polyethylene.
United Kingdom/United States	John Brown Contractors/Union Carbide	1977	Polyethylene plant: capacity: 200 000 t/yr. high-density polyethylene	90	Polyethylene	1983	224	Total of 240 000 tons of low-density polyethylene probably over 10 years.
Italy	Snia Viscosa	—	Caprolactam plant: capacity: 80 000 t/yr.	180	Caprolactam	1976	115 (1979-1988)	28 000 t/yr. of caprolactam over 8 years.
Italy	ENI	1975	Chemical plants	1,000				(for 1975-1980).
Italy	ENI	1975	2 plants	(170)		1979		
Italy	Snamprogetti/Anic	1975	3 urea plants capacity: 1 500 t/day each	200	Ammonia			100 000 t/yr. of ammonia over 10 years.
Italy	Montedison	1975	Chemical plants	500	Acrylonitrile	1977	175 250	
Japan	Marubeni	1975	Extension of plant to 75 000 t/yr. of acrylonitrile	10		1978	30 (1978-1982)	10 000 t/yr. of acrylonitrile over 5 years.
Japan	Mitsui/Toyo	1976	4 ammonia plants; capacity 1 360 t/day each	90	Ammonia and urea	1977	240 (1977-1987)	100 000 ton/yr. of ammonia and 100 000 ton/yr. of urea for 10 years.
France	Litwin	1974	Chemical plants	100		1977	50 60	
France	Creusot-Loire	1975	Chemical plants	220	Pipe	1979	100 225	
France	Creusot-Loire	1975	Ammonia	200				
France	Technip S.A.	1976	2 aromatic complexes; capacity: 125 000 t/yr. of benzene, 165 000 t/yr. of orthoxylene, 165 000 t/yr. of paraxylene		Orthoxylene, paraxylene, benzene, diesel oil, naphta	1980	950 (1980-1989) and (1981-1990)	20 000 t/yr. of orthoxylene, 20 000 t/yr. of paraxylene, 50 000 t/yr of benzene; also diesel oil and naphta over 10 years.
France	Rhône-Poulenc	1976	Chemical plant	1,203				See also remarks for Rhône-Poulenc Deal, in this Table.

(Cont'd on next page)

Table A-33 (cont'd)

Western Country	Western Supplier	Year Signed	Type of Soviet Import	Value of Soviet Imports (Million US$)	Type of Soviet Export	First Year of Soviet Deliveries	Value of Soviet Exports (Million US$)	Remarks
Chemicals (cont'd)								
United Kingdom/ Germany/France	Woodal-Duckham, TBA-Bishop, Klöckner, INA	1979	Glass fibre plant	36	Methanol or ethylene glycol	—	—	For commission at Polotsk in 1982. Part of a 950 million credit line negotiated in 1975.
Natural Resource Development								
Austria	Voest, Oemv	1969	Large-diam. pipe	100	Natural gas	1969	900 / 1000	Similar agreements signed in 1974 and 1975.
Finland	n. a.	1970	Pipe	n. a.	Natural gas	1974	n. a.	
Germany	Ruhrgas, Mannesmann Export AG	1970, 1972, 1974	Large-diam. pipe	1 500	Natural gas	1974	2 800 / 4 700	Three separate deals.
Italy	ENI	1971	Large-diam. pipe	190	Natural gas	1974	1 200 / 3 200	
France	Gaz de France	1972	Gas field equipment	250	Natural gas	1976	700 / 1462	
France	CMP (Incl. a subs. of International Systems and Control Corp.)	1974	Filter separators and gas compressor stations	26	Natural gas	n. a.	n. a.	
Italy	Finsider	1974	Large-diam. pipe	1 500	Scrap metal, coal, iron ore	1975	n. a. / (1000)*	*based on 65% coverage of Soviet import cost by Soviet exports. (Credit terms and costs are unknown and hence excluded)
Japan	Southern Yakutsk Coal Development Corp., Ltd.	1974	Coal development equipment	450	Coal	1979	80 / 860	Soviets have requested additional credits $150 million – some Japanese Eximbank credits used.
Japan	Sakhalin Oil Development Co., Ltd. (SODECO)	1975	Oil exploration equipment	(200) 150-250	Oil and gas	n. a.	n. a.	Japan will get half of output over 20 years. Financing between SODECO and Soviet bank for foreign trade.
France/ Austria/ Germany	Unspecified firms	1976	Large-diam. pipe and equipment	900 a	Natural gas	1981	n. a.	Part of triangular deal with Iran with participation of French and Austrian interest.
Finland	n. a.	1976	Construction of a mining complex (Kostamush)	n. a.	Iron pellet	n. a.	n. a.	Series of counterdeliveries over unspecified time.
Other								
United States	Airco, Inc.	1971	1 500 ton/yr. steel mill	n. a.	Chromium ore	n. a.	15 (for 1976)	This agreement is being extended to include welding equipment for super conduc. materials

Country	Company	Year	Project	Value	Product	Year	Value	n.a.	Notes
Finland	Rauma-Repola Oy	1977	3 Tankers	n. a.	Ten 3 500 HP engines	1977-79	n. a.	n. a.	materials and petroleum prod.
Germany	Olympia-Werke	n. a. (1977)	Electric typewriter plant	40	Typewriters	1977	n. a.	n. a.	
United States	Philip-Morris (via Eur. sub)	1977	Tobacco production machinery	n. a.	Tobacco	1978	n. a.	n. .a	Reciprocal value agreement.
Germany	Glahe International	1976	Exhibition hall built with German labour and materials	31.1	n. a.	n. a.	n. a.	n. a.	Glahe will charge rentals to West. users for a period of 10 years after which hall will revert to USSR.
Germany	Ruterbau of Hanover-Langenhagen	1977	Air terminal at Moscow Airport	93.2	Energy and construction material	n. a.	n. a.	n. a.	
United States	Pepsi-Cola	1974	Equipment, soft drink concentrates	n. a.	Vodka	n. a.	n. a.	n. a.	In 1976, Pepsi Cola announced it will supply equip. for add. cola plants. Equip. del. to start in mid-1977. Payment made in Vodka.
France	Péchiney-Ugine-Kuhlmann	1976	Alumina refinery	250	Aluminium	1979	n. a.	n. a.	USSR to sell back 50 000 tons of aluminium products from a mill in Siberia, also being constructed by Pechiney over 8 1/2 years.
Finland	Kalmet	1976	Floating hotels; timber carriers; roll on/roll off vessels; ships	322	12 000 ton capacity floating dock	n. a.	n. a.	n. a.	
United Kingdom	Dumbee-Combex-Marx	1976	Toy moulds	n. a.	Toys	n. a.	n. a.	n. a.	10 year agreement. Soviet toys to be sold outside CMEA.
United States	FMC	1977	Tomato paste factory	n. a.	Tomato paste	n. a.	n. a.	n. a.	
Denmark	Danish-Co-operative Union	1977	Clothing and canned hams	1.5	Oil prod.	n. a.	1.5	n. a.	
France	Péchiney-Ugine Kuhlmann Group	1976	Aluminum complex (refinery)	(800) 600-1000	Aluminium	n. a.	n. a.	n. a.	Agreement provides for the counterdelivery of 50 000 tonnes of aluminium products from another plan over 8 1/2 years.
Japan	Yokohama, Aiwa, Tairiku Trading Companies	1976	Tires, steel rope, consumer goods	40	Timber, seafood	1976	60	n. a.	

(Cont'd on next page)

Table A-33 (cont'd)

Western Country	Western Supplier	Year Signed	Type of Soviet Import	Value of Soviet Imports (Million US$)	Type of Soviet Export	First Year of Deliveries	Value of Soviet Exports (Million US$)	Remarks
Potential Agreements with the West								
Japan/United States	US Japanese firms in competition	U.N.	Pulp-paper plants	1 000-2 000	Wood products	Early 1980s	n.a.	n.a.
United Kingdom	Price & Pierce, Ltd.	U.N.	Wood processing plant	120-220	Forestry products	n.a.	n.a.	n.a.
United States/Japan	Tokyo Gas, El Paso Co., Occidental Petroleum	U.N.	Development of natural gas deposits in Eastern Siberia	100	Liquified natural gas	1980s	n.a.	n.a.
United States	Tenneco Inc., Texas Eastern, Brown & Root Inc.	U.N.	Development of natural gas deposits in Western Siberia – North Star project	3 000ᵃ	Liquified natural gas	1980s	n.a.	n.a.
France	Parfums Christian Dior	U.N.	Know-how for cosmetic production and packaging	n.a.	Unspecified items for use in cosmetics		n.a.	n.a.
France	Creusot-Loire	U.N.	300 000 ton/yr. steel pipe manufacturing plant	300	Unspecified related equipment		n.a.	n.a.
United States	Bendix Corp.	U.N.	Construction of spark plug factory (50-75 mill. units/yr.)	n.a.	Spark plugs	1980s	n.a.	Production: 50-75 million plugs, of which 25 per cent to be sold in West by Bendix.
United States/Italy/Germany	Lummus, Monsanto, Phillips Petroleum, Montedison	U.N.	Petrochemical complex	2 800ᵃ	Plant output	Mid-1985s	n.a.	US: olefins; Italy: polyolefins; Germany: aromatics likely specialised.
Japan	Nippon Electric Co.	U.N.	Electronic equip.	n.a.	n.a.	n.a.	n.a.	Nippon willing to accept 30% counter-purchase of Mash-Promintorg products.
Japan	Consortium	U.N.	4 pressurized water reactors and equipment	432	Enriched uranium	n.a.	n.a.	n.a.
United Kingdom	Alfred Dunhill, Ltd.	U.N.	Cigarette factory	n.a.	Cigarettes	n.a.	n.a.	20-30% of value of contract to be taken in product.; Dunhill willing to take Soviet cigarettes for one-third of its

Country	Company		Facility	Value	Product				Comments
					concentrate				counterpurchase.
United States	Colgate-Palmolive Co.	U.N.	Powdered laundry det. plant	n. a.	Powdered laundry detergent	n. a.	n. a.	n. a.	Partial compensation in kind; also counterpurchase of other products.
Japan	Taiyo Fishery Co.	U.N.	Refrig. plants fish-net mfg. plants, canneries	34	Alaska pollack whale meat and other fish		450		
Germany	German Consortium (preliminary agreement)	U.N.	Steel Complex	1 000-2 000 a	Iron ore pellets	1979	450	1 000	Initial contract and agreement signed in 1975, for cash only. Second stage may bring total value of Soviet imports to at least $2 billion and will include compensation in result product.

n. a. = not available.
U.N. = Under negotiation.
a) Estimated.
b) Value for whole deal including several contracts. *Business Eastern Europe*, 23rd February, 1979, p. 63.

Sources: Office of East-West Policy and Planning, Bureau of East-West Trade. US Departement of Commerce 8.6.77. Franz Lothar Altmann, Hermann Clement, *Die Kompensation als Instrument im Ost-West Handel, op. cit.*

Note by the Bureau of East-West Trade: Although information on these transactions has been taken from published sources, the Bureau cannot vouch for its accuracy.

371

(Cont'd on next page)

Table A-34

Identified Polish Compensation Transactions with the West

Western Country	Western Supplier	Year Signed	Type of E.E. Import	Value of E.E. Imports (Million of US $)	Type of E.E. Export	Value of E.E. Exports (Million of US $)	Remarks
United Kingdom	Massey Fergusson	1974	Equipment for Ursus Trac. Plant	350	Diesel engines and tractors	n. a.	$257.9 m credit backed by EGCD Deliveries for next 5 years.
Austria	Chemië-Linz-Voest-Alpine	1974	Melanine Plant (resin)	43	Part of plant output	n. a.	
Austria	Voest-Alpine	1975	Steel Products	287	Coal	348 over 9-year period	Austrian banks provide $287 m finance Polish purchases over 5-year period.
Belgium	n. a.	1975	Coal Mining Equipment	335	1.5 million tons of coal/yr. for 10-15 yrs.	335	Poland receives $335 m in credit.
France	n. a.	1975	Equipment	n. a.	Coking coal	n. a.	
Sweden	Byggnads AB Gunnar Haellstroem	1974	Prefab. bldg. elements plant	9	Prefab. houses	9	
United States	Westinghouse	1974	License, equipment for semi-conductors and rectifiers	10	Semi-conductors, rectifiers	n. a.	License fees will cost 30-50% of production.
United States	Clark Equip.	1972	License for manuf. of construction equipment	n. a.	Axles	n. a.	
Austria	Steyr-Daimler-Puch	1975	Licenses, trucks, and know-how	161	Diesel engines and truck parts	285	Counterdelivery will cover 1980.
France	Westinghouse CII	1974	IR 15-80 Computer System	10	Computer installations	n. a.	
United Kingdom	Petrocarbon Developments, Ltd.	1975	PVC and chlorine Plants	400	PVC	n. a.	Japanese firm also involved.
Sweden	Stansab-Electronic	1975	Equip. for computer-monitor systems	15	n. a.	15	

372

Country	Firm	Year	Item	Value	Description		Notes
United States	FMC	1973	300 mt powdered pectin plant	2.3	n. a.	n. a.	n. a.
United States	Waterbury Farrell	1973	Steel Rolling mill	4.4	Surface-grinding machinery	n. a.	n. a.
United States	Waterbury Farrell	1975	Brass and copper strip mill	55.4	Misc. products, incl. copper and brass prod.	n. a.	n. a.
Japan	n. a.	1970	Industrial Plants	100	2-3 million tons of coal per yr. for 10 years	(Estimate: $300 million for 10 years)	$100 m in private Japanese credits.
United States	International Harvester	1974	Licensing for manuf. of tractors and accessories	n. a.	Tractors, access.	n. a.	Bumar will sell in CMEA; IH will sell elsewhere.
Denmark	F. L. Smidth	1973	Cement Plants (2)	80	1.25 million tons of coal/yr. for 7-8 yrs.	n. a.	Purchases partly financed by Danish Export Credit Council.
Germany	Siemens AG; Kabel- und Metallwerke Gutehoffnungshütte AG, Norddeütsche Raffinerie, Metallgesellschaft AG	1976	Equipment for Expansion of Poland's Copper Industry	125	Deliveries of 40,000 tons/yr of unfabricated copper with a limited commitment to buy semifinished products such as cathodes and wire bar.	804-984 over 12-years	12-year agreement. Hermes credit guarantees extended.
France	Creusot-Loire	1976	Equip., technology for fert. plant	360	Fertilizer	n. a.	Joint marketing in third country.
France	Produits chimiques Ugine-Kuhlmann	1976	n. a.	n. a.	Sulphur	n. a.	France and Poland will co-operate in chemical production. 4 yr. agreement.

(Cont'd on next page)

373

Table A-34 (cont'd)

Western Country	Western Supplier	Year Signed	Type of E.E. Import	Value of E.E. Imports (Million of US $)	Type of E.E. Export	Value of E.E. Exports (Million of US $)	Remarks
Japan	Mitsui Ship-building and Engineering Co.	1976	Know-how and equipment for a chemical equip. mfg. plant	3	Chemical Equipment	n. a.	Part of output will be sold in Japan.
Switzerland	Emil Haefely & Cie AG	1976	High-Voltage generator	n. a.	Elec. energy	n. a.	
Sweden	n. a.	1974	Construction of oil refinery w/cap. of 10 m ton/year of crude oil	333	Resultant products to Sweden	n. a.	To export 50% production to Sweden.
Sweden	ASEA	1977	Electrically-driven industrial robots	n. a.	Electronic & automation systems	n. a.	Total value of agreement is $6 m.
Sweden	n. a.	1976	Licence, know-how & parts for production of elec. typewriters	n. a.	Counter deliveries of parts and components to Sweden	n. a.	
Sweden	n. a.	1975	Licence to pro. relay devices for RR signal boxes	n. a.	Counter deliveries of relay devices to Swedish firms	n. a.	
United Kingdom	Cementation International	1977	Airline terminal complex (Warsaw)	85	Construction by Pol. firm on CI's contracts w/ 3rd world nations; CI to pur. construction material from Poland	35 in construction at 3rd World sites; 4.5 m/yr. purchases by CI	ECGD guarantees extended for $47 m loan.
United States	Katy Industries	1976	Machinery and working programs for shoe production	n. a.	Shoes	n. a.	Five yr. agreement; some part of production to be sold in US and Canada.
Germany	ALSO-Schuhbedorf GmbH	1976	Lic. & know-how	n. a.	Ladies' Shoes	n. a.	5.5 m pairs/yr. to be mktd. in

Country	Company	Year			French Operations		
United States	General Electric	1976	Lic. & machines for prod. of medical equipment	n. a.	Joint output; particularly, electrocardiogram meters	n. a.	being expanded to include heavy trucks
Italy	n. a.	1976	Equipment	150	Coal	150	$150 m Italian credit for Polish purchases.
Germany	Krupp-led Consortium	1976	Coal Gasification plants (Framework Agreement)[a]	800	Ammonia, Urea and methanol	n. a.	Result prods. to be marketed by joint German/Polish Company.
Netherlands	Hoogovens Ijmuiden BV	1977	Steel structures	40	Coking Coal	400	CP 10-yr. agreements. 750,000 t/yr. of Polish Coal exported to Netherlands tied to $85 m loan for expansion of coal industry. Dutch government guarantee extended.
France	Rhône-Poulenc Institut français du pétrole	1975	Chemical Products, textile fibers	14 (annually)	Sulphur	n. a.	A 10 year agreement.
Germany	Alfred Hempel KG	1977	Machine and equipment to modernize corundum works	3.1	Corundum products	10.2 (20% of plant output)	Alfred Hempel will handle all market corundum products on exclusive basis in West until 1990s.

(Cont'd on next page)

375

Table A-34 (*cont'd*)

Western Country	Western Supplier	Year Signed	Type of E.E. Import	Value of E.E. Imports (Million of US$)	Type of E.E. Export	Value of E.E. Exports (Million of US$)	Remarks
France	Huard	1977	License to produce gearboxes for rotary plows	n. a.	Gearboxes for rotary plows	n. a.	200 gearboxes to Huard this year. Most of next year's output will be assembled into land-rollers for use in Poland.
Germany	Stetter	1977	License for prod. of concrete mixer components and concrete containers	n. a.	Concrete mixer components and concrete containers	8.4 in 1978	Part of a co-operation agreement signed in mid-1960s.
Germany	Krebs-Klöckner	1975	Soda ash plant	182			250,000 t/yr. soda ash to be marketed by Poland.
United States	Occidental Petroleum	1978	Phosphate	1 330		20 years (1978-1997) 670	Swap deal: phosphate rock in exchange for 500,000 t/yr. of sulphur.

n. a.: Not available.
a) First contract worth $63 million for equipment to develop a large coal-mining operation in Upper Silesia was expected to be signed shortly after 4th May 1977.

Sources: Office of East-West Policy and Planning, Bureau of East-West Trade, US Department of Commerce, 8th June, 1977. Franz-Lothar Altmann, Hermann Clement, *Die Kompensation als Instrument im Ost-West Handel, op. cit.*, 1979.

Note of Bureau of East-West Trade: Although information on these transactions has been taken from published sources, the Bureau cannot vouch for its accuracy.

Identified Hungarian Compensation Transactions with the West

Western Country	Western Supplier	Year Signed	Type of E. E. Import	Value of E. E. Imports (Million of US $)	Type of E. E. Export	Value of E. E. Exports (Million of US $)	Remarks
United States	Steiger	1974	Licens. and equip. for manuf. of tractors	n. a.	Tractor axles	n. a.	
United States	Steiger	1976	Technology and components for tractor manuf.	80	Tractor axles	20	$100 million 2-way trade over 5 yrs. (Ext. 1974 agreement).
United States	Corning Glass	1975	Know-how, blue-prints	n. a.	Blood gas analyzers	n. a.	Joint venture. Corning to market 40 per cent of output.
Germany	Hildebrand	1974	Parquet floor plants (4)	6	Finished Parquet	n. a.	No information on delivery dates.
United States	Katy Industries	1976	Equip. Manage., designs, know-how to manuf. women's shoes	3.2	Women's shoes	66	
United Kingdom	Vauxhall	1975	Lic. and know-how for truck axles.	n. a.	Truck axles	15	
United Kingdom	Dent and Hellyer, Ltd.	1976	Lic. for manuf. of hospital equip.	n. a.	Sterilizing equip.	n. a.	$4.8 million total trade by 1980.
Austria	Semperit	1976	Lic. for tyre production	n. a.	Tyres	n. a.	
Germany	Burghard & Weber	1975	Lic. and know-how for multispindle drilling machine production	n. a.	Machine-tools	n. a.	Majority of prod. capacity to go to Germany.
Austria	Robert Lanschwart	1976	Lic. and know-how for computer component prod.; computer components	n. a.	Computer components	n. a.	Total prod. value of $14.59 million in Austria and Hungary prod. over 5 yrs; joint sales in develop. countries.
Germany	Gildemeister	1976	Lic. and know-how for prod. of medium and large universal turning lathes	n. a.	Machine-tools	4.1-6.2 (1977-81)	

(Cont'd on next page)

Table A-35 *(cont'd)*

Western Country	Western Supplier	Year Signed	Type of E.E. Import	Value of E.E. Imports (Million of US $)	Type of E.E. Export	Value of E.E. Exports (Million of US $)	Remarks
France	Sté. Prorea/Sorice C. G. Buettner	1977	Sodium tripolyphosphate processing equip. (natrium polyphosphate)	5.01	Various industrial products (autoclaves elec. household appliances)	3	Part of Tisza chemical works.
Austria	Franz von Furtenback	1977	Lic. and know-how with information up-dates; base concentrates during initial stages	n. a.	Related finished products	n. a.	
United States	Levi Strauss	1977	Material machinery, patterns	0.4	Levis	n. a.	5 yr. agreement; Levi will take 60 per cent of production.
United States	Colgate/ Palmolive	1975	n. a.	1	Paste-tubes (partial payment)	n. a.	
Denmark	Regnecentralen	1977	Electronic data processing system	44	Unspecified commodities (comprise large portion of payment)	n. a.	
Germany	Barton	1976	Know-how for prod. panelled flooring	n. a.	Panelled flooring	n. a.	Similar to 1974 Hildebrand agreement.
France	Chaffoteaux et Maury	1976	Lic. for boilers; parts and materials	n. a.	Boiler components and gas apparatus	n. a.	Hungary will have Socialist and Scandinavian market rights; 7 yr. agreement.
Sweden	AGA Svetsprodukter	1976	Lic. and know-how	n. a.	Gas welding pistols	0.3369	6 yr. agreement.
Germany	MAN	1975	Lic. and know-how for prod. of MAN bus engines at rate of 11 000/yr.	n. a.	9 000 Hungarian built MAN engines	n. a.	
Belgium	Marreau-Vervaeke	1976	Equip. and Tech. for prod. of fiber plate	n. a.	35 000 cu. m. of fiber plate over 5 yrs.	n. a.	
Germany	Standard Elektrik Lorenz AG	1975	Lic. and know-how for colour TV prod.	n. a.	Colour TVs and components	8 295/yr	Duration not specified; coproduction agreement.
Switzerland	Ghelfi AG	1976	Tech. and know-how for environmental equip.	n. a.	Resultant products	n. a.	Joint-production; Nikex to have Hungarian

sorting equip.

Country	Partner	Year	Description		Product		Remarks
Italy	Montedison	1977	Synthetic raw mat. and organic and inorganic chemicals	80	Olefins and aromatic	28	4 yr. agreement.
Italy	Snia Viscosa	1977	Organic chemical industry products	8	Aromatic compounds and petrochemical mat.	8-9	4 yr. agreement.
Italy	EFIM	1977	Design a 100 000 t/yr. aluminium smelter	n. a.	Aluminium	n. a.	Plant to be operational by 1984.
Japan	Tokai Metals and Toyo Menka	1977	Aluminium foil rolling mill	4	Aluminium foil	2	Operative by end 1978. 6 yr. agreement.
Germany	Gartemann und Hollman	1977	Bag-making machines and parts	2.6	Machine tools	0.39	
Switzerland	Sechy Co.	1977	Welding-machine accessories	n. a.	Welding machines (100-150/yr.)	n. a.	
Austria	Hodry Metall-warenfabrik R. Hoppe	1977	Know-how for prod. of furniture hinges and fittings	n. a.	Furniture hinges and fittings	n. a.	Expected annual turnover $1.8-2.1 million. 10 yr. agreement.
France/ Belgium	Bekoto Petersime Pvba	1977	Know-how for prod. egg-collecting and incubator vehicles	n. a.	Egg-collecting and incubator vehicles	n. a.	5 yr. agreement.
Germany	Terosen	1977	Licensing, 10 semi-finished chemicals	n. a.	Finished chemicals	n. a.	Renewed 1971 agreement, anti-corrosion paints for car chassis, cooling liquids, cleaning agents produced.
Denmark	Consortium led by Atlas A/S	Under Negoti-ation	Pig slaughter house and canning plant	32	n. a.	n. a.	Under negotiation as of June 1977.
United States	Universal Machinery Equipment Co.	1977	Know-how and elec. equip. for electrical furnace production	n. a.	Remaining Parts	n. a.	Joint product US company to market in West.
Germany	Kömmerling GmbH	1977	Lic. for window prod. from synthetic mat.; extruders and process tools	n. a.	PVC	n. a.	

(Cont'd on next page)

379

Table A-35 *(cont'd)*

Western Country	Western Supplier	Year Signed	Type of E.E. Import	Value of E.E. Imports (Million of US $)	Type of E.E. Export	Value of E.E. Exports (Million of US $)	Remarks
United Kingdom	Nova Jersey Knit, Ltd.	1977	Mach. and yarn for knitwear factory	n. a.	Knitwear garments	n. a.	Expansion of 5 yr. co-operation agreement on prod. of jersey cloth.
United States	Hesston	1977	Harvesters (80) and hay-handling systems (12)	4.5	Heads and gearboxes for Hesston's harvesters	4.5	Hesston's deliveries completed by 1980. Hungary deliveries completed by 1982.
Belgium	CRC Chemicals Europe	1977	Packaging for Hungary brake-cleaning medium	n. a.	Concentrates of car-grooming compound and protective oil	n. a.	Hungary will turn out finished product in spray tubes.
Germany	Ahlmann and Co.	1977	Bathtub enameling machines	n. a.	Bathtubs	n. a.	Ahlmann has agreed to purchase most of the 70 000/yr. exports over first 3 yrs. to cover 75 per cent of export contract value.
Germany	AEG	1977	Lic. for auto. washing machine productions	n. a.	Washing machines	n. a.	AEG has agreed to purchase small portion of 20-25 000 units/yrs. output.
Netherlands	Naarden	1979	Concentrates for production of tonic water	n. a.	n. a.	n. a.	
United States	Coca-Cola and Pepsi-Cola	1979	Concentrate and processing and cooling equipment	3.5	n. a.	n. a.	

n. a.: Not available.

Source: Office of East-West Policy and Planning, Bureau of East-West Trade, US Department of Commerce, 8th June, 1977. *Business Eastern Europe*, 27th April, 1979, p. 136.

Table A-36

Identified German Democratic Republic Compensation Transactions with the West

Western Country	Western Supplier	Year Signed	Type of E.E. Import	Value of E.E. Imports (Million of US $)	Type of E.E. Export	Value of E.E. Exports (Million of US $)	Remarks
France	Arbel Industries S.A. and Société franco-belge de matériel	1974	Rail Wagons	178	Materials for Wagons	23	Delivery starts 1975, ends 1977.
Germany	Salamander	1976	Shoes	14-16	n. a.	n. a.	In past, Salamander has accepted part payment in hosiery and furs.
United States	Dow Chemical	1976	Chemicals	n. a.	Metalworking Products, plastics and chemicals	n. a.	10-year umbrella agreement.
Germany	Friedrich Uhde	176	PVC complex, pheripheral equipment and infrastructure work	451	PVC and soda lye	30% of total value (over 8-10 year period)	"Treuarbeit" guarantee of 90% of the loan.
Italy	Montedison	1976	n. a.	n. a.	Chemicals	n. a.	Exchange to take place over 1976-1980 period.
Italy	Danieli Group and ASEA (Sweden)	1977	Steel Mill (melting shop)	240	Output of plant and machinery & metallurgical products	240	$180 m in Italian credits extended.
Denmark/France	Hoeggaard et Schultz and Kampsax/CFEM	1976	Rolling mill	72	Construction work	24-28	30% counter-purchase.
Germany	Hoechst	1976	Chemical complex for prod. of chlorine, caustic soda, chlorinvynyl monomer, PVC	n. a.	Resultant products	n. a.	
Austria	Chemie Linz	1976	Pesticides and herbicides agents and fertilizers	58.38	Potassium salt and special chemicals	58.38	
Austria	ADS Anker – Data System	1977	Cash registers	n. a.	Office machines	n. a.	Contract initially runs 1977-80. Total value $15.3 m.
Austria	Vereinigte Edelstahlwerke	1977	Fine steel products	n. a.	Potash fertilizers	n. a.	Fertilizer deliveries through 1980.

(Cont'd on next page)

Table A-36 (cont'd)

Western Country	Western Supplier	Year Signed	Type of E.E. Import	Value of E.E. Imports (Million of US $)	Type of E.E. Export	Value of E.E. Exports (Million of US $)	Remarks
Germany	Uhde/Hoechst	1976	Complex of 4 plants, including one for producing caustic soda	472ᵃ	Resultant products (caustic soda)	Total for 1980-1987: 32	30 000 t/yr. of caustic soda for 8 years.
Germany	Uhde/Hoechst	1976	PVC plant in complex	472ᵃ	PVC	Total for 1980-1987: 80	15 000 t/yr. of PVC.
Germany	Dow Chemical	1977	Propylene oxyde	250	Propylene	Total for 1979-1988: 85	50 000 t/yr. of propylene.
United Kingdom	Catalytic	1977	Chlorine plant	n. a.	Caustic soda	Total for 5 yrs.: 11.5	20 000 t/yr. of caustic soda for 5 years.
Japan	Mitsui	1977	Benzene plant (part of aromatic complex)	360ᵇ	Benzene, butadiene, butylene	Total for 1983-1988: 150ᵇ	165 t. of benzene agreed over 6 years and 10-20 000 t/yr. of butadiene and butylene.
Japan	Toyo Engineering Kobe Steel, Hitachi, Mitsubishi	1977	Chemical plant	450	n. a.	90	Deliveries up to 1985.

n. a.: Not available.
a) Value for whole deal including several contracts.
b) Value for whole deal, including contracts other than for the benzene plant.

Sources: Office of East-West Policy and Planning, Bureau of East-West Trade, US Department of Commerce, 8th June, 1977. Franz-Lothar Altmann, Hermann Clement, *Die Kompensation als Instrument im Ost-West Handel, op. cit.*

Table A-37

Identified Czechoslovak Compensation Transactions with the West

Western Country	Western Supplier	Year Signed	Type of E.E. Import	Value of E.E. Imports (Million of US$)	Type of E.E. Export	Value of E.E. Exports (Million of US$)	Remarks
United Kingdom	International Computers, Ltd.	1975	Computers	3	n. a.	2	
United Kingdom	Schweppes	1976 (under negot.)	Soft drinks	n. a.	Cash, soft drinks tomatoes, etc.	n. a.	
Italy	Montedison	1977	Chemical and plastic products	41	Chemical and plastic products	n. a.	
Japan	Chisso	1972	Polypropylene plant	n. a.	Polypropylene	5 yearly	Expansion of 1975 agreement; new orders $41.32 m. 8 000 t/yr. of polypropylene sold through C. Itoh.

n. a.: Not available.

Sources: Office of East-West Policy and Planning, Bureau of East-West Trade, US Department of Commerce, 8th June, 1977. Franz-Lothar Altmann, Hermann Clement, *Die Kompensation als Instrument im Ost-West Handel, op. cit.*

Note of the Bureau of East-West Trade: Although information on these transactions has been taken from published sources, the Bureau cannot vouch for its accuracy.

383

Table A-38

Identified Rumanian Compensation Transactions with the West

Western Country	Western Supplier	Year Signed	Type of E.E. Import	Value of E.E. Imports (Million of US$)	Type of E.E. Export	Value of E.E. Exports (Million of US$)	Remarks
United States	Lipe Rollway	1975	Roller bearing plant	56	Bearings	31 (over 10-year period)	Marketing in WE and the United States.
United States	Control Data Corporation	1973	Equipment, know-how for manufacture of computer peripherals	n. a.	Printers and card-readers	n. a.	Joint venture: marketing in WE.
Germany	Censor Industrial Handling Systems	1976	Roller bearing equipment	1.6	Roller bearings	n. a.	50% of contract value to be repaid in roller bearings.
United States	Delaval	1972	Pumps and centrifuge equipment	0.4	n. a.	n. a.	
United States	General Gulf Atomic (Gulf Oil)	1973	Fuel components and assembly of nuclear reactor	4.1	n. a.	4.1	Will cover 10-year period.
United Kingdom Germany	Brush Electrical VFW-Fokker	1974 1977	Locomotive components Short haul jetliner and components	n. a. n. a.	Locomotives Jetliners	n. a. n. a.	Joint venture: total value $425-855 million. Production up to 1985(?). 50 jetliners to be used in Rumania, other 50 marketed abroad.
Italy	Fiat SpA	1977	Joint production of earth-moving machines	n. a.	Joint production of earth-moving machines	n. a.	Production to start in 1980 in Italy and Rumania. Annual two-way trade expected to total $12 million. Agreement still subject to official approval as of 18.2.1977.
Italy	De Nora	1978	2 chlorine plants 85 000 t/yr.	18			
France	Citroën	1977	Design, engineering, parts and materials for assembly of cars	170	Cars	Half of annual output	Coproduction. Joint marketing in CMEA and WE.

n. a.: Not available.

Sources: Office of East-West Policy and Planning, Bureau of East-West Trade, US Department of Commerce, 8th June, 1977. Franz-Lothar Altmann, Hermann Clement, Die Kompensation als Instrument im Ost-West Handel, op. cit.
Note of Bureau of East-West Trade: Although information on these transactions has been taken from published sources, the Bureau cannot vouch for its accuracy.

Table A-39

Identified Bulgarian Compensation Transactions with the West

Western Country	Western Supplier	Year Signed	Type of E.E. Import	Value of E.E. Imports (Million of US$)	Type of E.E. Export	Value of E.E. Exports (Million of US$)	Remarks
France	Technip	1975	Ethylene Plant	50	Handl. and Hoist Mach.; Eng. Goods, Petro-Chem. Prods.	n. a.	
Japan	Komatsu, Ltd.	1975	Bulldozers, loaders, and scrapers. A service shop and a spare parts stock	n. a.	n. a.	n. a.	Part of the undisclosed contract value will be paid in counter-deliveries.
Italy	n. a.	n. a.	Cannery	n. a.	n. a.	n. a.	
United States/ United Kingdom	G. M., Vauxhall Motors	1976	Heavy-duty trucks	n. a.	Fork-lift carts and trucks	n. a.	Initially G. M. will use Bulgarian fork-lifts in own plants, but other mktg. possibilities later on.
Germany	Klöckner/Hills	1974	PVC plant: 120 000 t/yr, capacity	18	PVC	25	About 5 000 t/yr. of PVC for 7-8 years.
Germany	Lurgi	1976	Polypropylene plant: 80 000 t/yr. capacity	n. a.	Benzene and other products	n. a.	Benzene quoted among other products taken as buy-back.
France/Italy	Technip/ Technipetrol	1972	Ethylene plant: capacity 250 000 t/yr.	120	Ethylene	Total 1979-1983: 10	6 000 t/yr. of MEG over 4/5 years (buy-back comprises other chemicals and equipment).

n. a.: Not available.

Sources: Office of East-West Policy and Planning, Bureau of East-West Trade, US Department of Commerce, 8th June, 1977. Franz-Lothar Altmann, Hermann Clement, *Die Kompensation als Instrument im Ost-West Handel, op. cit.*

Note of the Bureau of East-West Trade: Although information on these transactions has been taken from published sources, the Bureau cannot vouch for its accuracy.

Table A-40

Some Macro-Economic Measurements Relating to the Level and Rate of "Technical Progress" in the USSR

Author(s) and date of publication	Years to which estimates refer	Purpose(s) of measurement	Technique(s) used	Results
F. Seton (1958)[5]	1950-55	Sources of Soviet economic growth.	Cobb-Douglas production function — "dynamic version" where elasticities do not add up to unity.	Continuing high rates of Soviet economic growth 1950-55 are mainly the result of increasing efficiency (technological, administrative or both) rather than capital accumulation or labour influx. Technical progress proceeded at over 6% per annum[12].
A. Bergson (1963)[3]	1950-58	Dynamic efficiency (growth in total factor productivity) of Soviet economy compared with selected Western countries.	Production function with the following variables: NNP valued at 1937 prices and "given year" prices, labour, reproducible fixed capital and livestock herds.	Annual average rate of growth of total factor productivity in USSR = 2.7% (unadjusted for intersectoral shifts in employment or changes in inventories and assuming a 20% annual net return for capital, inventories and livestock). Equivalent figure for USA = 1.7%.
B. Balassa (1964)[2]	1950-58	Dynamic efficiency.	As above but out put measured in terms of GNP; capital weighted in terms of share of GNP including an allowance for depreciation to make it consistent with MIT estimates for Western countries.[4]	Average annual rate of growth of total factor productivity for economy as a whole: Japan = 4.1% (1951-59); Federal Republic of Germany = 3.4% (1950-59); USSR = 3.3% (1950-58); USA = 2.6% (1948-57); UK = 0.7% (1949-59). Soviet performance can be largely explained by rapid accumulation and reallocation of labour from agriculture to industry.
J. S. Berliner (1964)[1]	1960	Static efficiency (total factor productivity) of Soviet and American economies.	Kendrick-type production function (homogeneous of degree one: $P = \alpha L + \beta K$) and Cobb-Douglas ($P = \beta L^\alpha K^{1-\alpha}$). No adjustment made for relative qualities of land, labour and capital.	"Relative efficiency of the USSR ranges from 36-39% of USA if the USA used Soviet inputs and if outputs were valued at Soviet prices, to 87-98% of the USA if the USSR used American inputs and if outputs were valued at American prices." Due to absence of qualitative adjustments, the estimates measure relative efficiency of total economies rather than efficiency of pure systems of economic organisation.
S. H. Cohn (1966)[6]	1950-58 and 1958-64	Sources of Soviet economic growth.	"Simplified" Cobb-Douglas production function with GNP as the output variable.	Annual average rates of increase: L K Residual Production 1950-58 1.2 8.3 3.7 7.1

	L + λ	Residual	Production
	4.8	2.2	7.0

Author	Period	Method	Commentary
... Solov'ev (1966)[7]	economic growth...	Specific... by Diriynes and Kurz". Measure of output excludes non-productive sectors (c.f. Cohn). Inputs include working capital, consumer durables, agricultural land and mineral resources.	This estimate of "dynamic efficiency» corresponds closely to Cohn's estimate for a similar period; as with Cohn's estimate, the rate of growth of inputs declined less rapidly than factor productivity. The authors further suggest that 1.5% of the residual (2.2%) is accounted for by "embodied and disembodied technical progress".
A. Tolkachev (1966)[9]	1950-64	Sources of Soviet economic growth.	Cobb-Douglas production function: $Q = 1.186 \times X^{0.155} \times Y^{0.845}$ where Q = net output, X = man hours, Y = production funds.	The equation reveals that a 1% increase in production funds leads to 0.845% growth in net output whereas a 1% growth in man hours leads to only a 0.155% growth (N.B. The 1.186 is not the "residual": it is an adjustive constant).
A. Bergson (1968)[10]	(a) 1960 (b) 1950-62	(a) Static efficiency. (b) Dynamic efficiency: Soviet economy compared with major Western countries.	Implied Cobb-Douglas production function. Employment variable adjusted for educational level and sex distribution.	(a) Static efficiency of USSR economy in 1960 was between 1/3 and 1/2 that of USA and well behind leading West European countries with the exception of Italy. (b) Dynamic efficiency of USSR economy 1950-62 superior to that of USA and Britain but inferior to that of other West European countries especially Italy.
M. L. Weitzman (1970)[11]	1950-69	Sources of Soviet economic growth.	CES, Hicks neutral production function.	Soviet decline in rate of economic growth throughout the period especially after late 1950s. This is explained primarily by diminishing returns to capital rather than a declining "residual"; Weitzman estimates that "technical change" has occurred at a rate of 2% per annum, throughout the period, which is "respectable" by world standards. This result is challenged by E. R. Brubaker (1971).[13]

387

(Cont'd on next page)

Table A-40 (cont'd)

Author(s) and date of publication	Years to which estimates refer	Purpose(s) of measurement	Technique(s) used	Results
E. R. Brubaker (1972)[15]	1960	Static efficiency of *particular* sectors of selected West European countries and USSR.	Systematic comparison of data on *output per man hour* compiled by Bergson (1968) *op. cit.*, Cohn (1966) *op. cit.*, and Maddison (1965).[16]	There are wide variations in the results, especially between estimates using American price weights. The general picture is that in 1960 the Soviet economy, overall, appears to have been "unambiguously less efficient," than Germany, France and on a par with Italy. This aggregate shortfall is primarily due to the low output per man hour in commerce and other services; other sectors, such as industry, agriculture and transport and communications perform relatively well according to this criterion.
A. Bergson (1973)[14]	1970-75-80	Future sources of Soviet economic growth.	Cobb-Douglas production function.	The 1970-75 plan envisages an annual average rate of growth of material income of 6.7%. This can only be obtained by considerable increases in productivity, given that even in official figures there is a planned decline in the rate of growth of capital investment (6.7% for 1970-75 c/p 7.5% for 1967-70). Employment is most unlikely to grow at an increasing rate. Thus a new model of economic growth is implied, founded on sharp increases in efficiency and technical progress. Assuming a continued decline in the rate of growth of capital stock, in order to fulfil the plan the average annual rate of increase of total factor productivity would need to be 3.0% (c.f. actual rate of 1.7% in 1950-58 and 0.7% in 1958-67).
J. Hocke, O. Kyn and H. J. Wagener (1973)[17]	1950-70	Growth of factor productivity of Soviet *industry* (i.e., dynamic efficiency).	Six variants of Coob-Douglas production function (three assuming constant returns to scale and three allowing for increasing or decreasing returns to scale).	Average rate of "technological change" in Soviet industry was between 3-5.5%. "This is neither extremely low nor extremely high ... [and] ... indicates a good overall performance." Rate of change may have been greater at the beginning of the period (4-7.3%) than at the end (3-3.7%); thus there has been an average annual decline in the rate of technological change of 0.09%-0.18%.
P. Desai (1974)[18]	1955-71	Comparative rates of "technical change" in seven branches of So-	Cobb-Douglas and CES production function with constant and variable returns to scale with the	Soviet heavy industry appears to be more progressive technologically than light industry. Rates of "technical change" for the period, by branches of heavy industry,

388

	4.89%
Machinery and metalworking	2.80-3.36%
Construction materials	1.69-3.15%
Electric power	1.74%
Ferrous metals	

gross output was the dependent
variable and raw materials were
explicitly considered (c/p Weitzman).

Sources:
1 J. S. Berliner, *American Economic Review*, May 1964, pp. 480-9.
2 B. Balassa, *Ibid*. pp. 490-505.
3 A. Bergson, in (Eds), A. Bergson and S. Kuznets, *Economic Trends in the Soviet Union*, Cambridge, Mass., 1963, pp. 1-37.
4 E. D. Domar, *Review of Economics and Statistics*, February 1964.
5 F. Seton, *American Economic Review*, May 1959, pp. 1-14.
6 S. H. Cohn, in *New Directions in the Soviet Economy*, Joint Committee of US Congress, Washington, D.C., 1966, pp. 99-132.
7 B. N. Mikhalevskii and Yu. P. Solov'ev, *Ekonomika i Matematicheskie Metody*, 1966, No. 6, pp. 823-40. Comment by F. Denton in *Soviet Studies*, 1967-68, No. 4, pp. 501-9.
8 P. Dhrymes and M. Kurz, *Econometrica*, 1964, No. 3, p. 32.
9 A. Tolkachev, *Planovoe Khozyaisvo*, 1966, No. 6, pp. 1-9.
10 A. Bergson, *Planning and Productivity under Soviet Socialism*, New York, 1968.
11 M. L. Weitzman, *American Economic Review*, September 1970, pp. 676-92.
12 See also P. J. D. Wiles, *The Prediction of Communist Economic Performance*, Cambridge, 1971, p. 337.
13 E. R. Brubaker, *American Economic Review*, Vol. LXII, 1972, pp. 675-8.
14 A. Bergson, *Problems of Communism*, March-April 1974, pp. 1-9.
15 E. R. Brubaker, *Soviet Studies*, 1971-1972, No. 3, pp. 435-49.
16 A. Maddison, *Banca Nazionale Del Lavoro Quarterly Review*, March 1965.
17 J. Hocke, O. Kyn and H. J. Wagener, *Forschungsbericht*, 1973, Osteuropa-Institut, Munich, pp. 1-44.
18 P. Desai, *Technical Change. Factor Elasticity of Substitution and Returns to Scale in Branches of Soviet Industry in the Postwar Period*, manuscript of article prepared at Institute of International Studies, University of California, Berkeley, April 1974.
   Quoted from: Ronald Amann, «Some Approaches to the Comparative Assessment of Soviet Technology: its Level and Rate of Development", in: *The Technological Level of Soviet Industry*, *op. cit.*, pp. 16-19.

389

Table A-41

East-West Long-term Commercial Agreements in Force at the end of Septembre 1975
Period covered and, in brackets, date signed

	Bulgaria	Czechoslovakia	German Democratic Republic	Poland	Rumania	USSR	Hungary
Benelux	1970-74 (13.5.70)	1972-74 (6.4.72)		1971-74 (25.11.72)	1970-74 (8.12.70)	1971-74 (14.7.71)	1971-74 (13.7.71)
Germany	1970-74 (12.2.71)	1970-74 (17.12.70)		1970-74 (15.10.70)	1970-74 (22.12.69)	1972-74 (1.7.72)	1970-74 (27.10.70)
Denmark	1971-75 (31.8.70)			1971-75 (3.12.70)	1971-75 (21.8.70)	1970-75 (24.10.69)	
Finland	1973-77 (21.12.72)	n.s. (19.9.74)	1973-78 (20.6.73)	1971-75 (4.12.70)	1971-75 (28.9.70)	1976-80 (12.9.74)	1969-73[1] (22.10.68)
	n.s.[2]					1971-75 (26.3.69)	n.s.[2] (2.5.74)
France	1970-74 (23.1.70)	1970-74 (23.2.70)		1970-74 (23.12.69)	1970-74 (9.1.70)	1970-74 (26.5.69)	1970-74 (5.1.70)
Greece	1975-79[3] (28.11.75)	1964-67[3] (22.7.64)		1972-75[4] (18.1.73)	1971-75 (15.12.70)		
United Kingdom	1970-75 (27.4.70)	1972-74 (27.6.72)		1971-74 (21.4.71)	1972-74 (15.6.72)	1969-75 (3.6.69)	1972-74 (21.3.72)
Ireland	1970-74 (23.4.70)	1972-74 (14.12.72)			1971-74 (20.7.71)	1973-74[5] (28.12.73)	
Iceland		1971-76[3] (12.10.72)	1973-77 (6.2.73)	1970-74 (12.9.69)	1972-77 (16.6.72)	1972-75 (2.11.71)	1970-75 (19.5.70)
Italy	1970-74 (21.1.70)	1970-74 (14.11.69)		1970-74 (18.2.70)	1970-74 (4.12.69)	1970-74 (15.1.70)	1970-74 (15.11.69)
Yugoslavia	1971-75 (10.12.70)	1971-75 (26.1.71)	1971-75 (19.2.71)	1971-75 (13.11.70)	1971-75 (12.10.70)	1971-75 (10.2.71)	1971-75 (4.2.71)

Austria	1974-84[6] (28.6.73)	1973-77[3] (7.11.72)	1973-77 (30.8.73)	1972-76 (9.9.71)	1971-75 (24.9.70)	1971-75 (5.8.70)	1973-78 (11.11.72)
Portugal	1975-80[3] (11.2.75)		1975-79 (25.1.75)				1975-80[3] (23.1.75)
Sweden	1972-76 (14.9.72)	1973-77 (30.3.73)	1974-78 (26.7.73)	1973-77 (25.10.72)	1973-77 (10.5.73)	1971-75 (8.10.70)	1974-78 (5.12.73)
Switzerland	1973-77[6] (23.11.72)	1971-75[6] (7.5.71)		1973-76[6] (25.6.73)	1973-78[6] (13.12.72)		1974-78[6] (30.10.73)
Spain	1971-75[7] (3.6.71)	1972-76[3] (5.10.71)	(1.4.74)	1970-74[7] (3.4.70)	1971-75[7] (5.3.71)	1972-75 (15.9.72)	1971-75[7] (18.11.70)
Turkey	1975-76[3] (29.8.75)	[5]	[5]	[5]	1971-73 (27.10.70)	[5]	1974-75[3] (12.11.74)
United States				[8]	[8]	1973-75[8,9] (18.10.72)	
Cyprus	1970-74 (24.2.70)	1973-78 (6.6.73)	1968-74 (6.12.68)	1962-78 (16.3.62)	1962-75 (19.6.62)	1966-77 (22.12.61)	1962-75 (6.3.62)

n.s.: Period not specified.
1 Long-term agreement usually extended by annual protocol.
2 Agreement on reciprocal removal of obstacles to trade.
3 Period automatically extended if agreement not terminated.
4 From 1st October, 1972 to 30 th September, 1975.
5 Commercial agreements or protocols extended each year.
6 Long term agreements on trade relations. At the end of the ten-year period, the agreement is automatically extended for one year failing notice of termination during the preceding six months.
7 Agreement on commercial movement of goods or persons by land or by sea and on industrial co-operation.
8 Protocols in force on various forms of trade relations.
9 Agreement not yet in force.

Source: F. Levcik and J. Stankovsky, Industrielle Kooperation zwischen Ost und West, op. cit., pp. 154-155.

Table A-42

East-West Agreements on Economic, Industrial and Technical Co-operation in Force at end of September 1975

Period and, in brackets, date signed

	Bulgaria	Czechoslovakia	German Democratic Republic	Poland	Rumania	USSR	Hungary
Belgium and Luxembourg	n.d.[1] (14.6.66) 1974-84 (3.74)	n.s. (10.9.75)	1974-84 (31.8.74)	1973-83 (22.11.73)	n.s. (16.9.68)	10 years (27.9.74)	1975-85[2] (6.10.75)
Germany	1974-84 (19.7.74)	[3]	1975-85 (11.11.74)	1974-84 (1.11.74)	n.d. (3.8.67) 1973-83 (29.6.73)	1973-83 (19.5.73)	1974-84[4] (11.11.74)
Denmark	n.s. (22.4.75) 1975 (22.4.75)	n.s. (9.11.70)	1974-83 (21.2.74)	n.s. (20.11.74) 1971-75 (3.12.70)	n.s. (29.8.67)	n.s. (17.7.70)	n.s. (20.10.69)
Finland	1970-74 (14.11.69) 1975-85 (12.8.74)	1971-75[2] (1.3.71)	1973-83 (20.6.73)	1974-84 (30.1.74)	1970-74 (25.9.69)	1971-81 (20.4.71)	1974-84[2] (2.9.74)
France	n.s.[1] (10.7.68) 1974-84 (11.74)	n.s.[5] (23.2.70) nouv. accord paraphé	1973-83 (19.7.73) 1975-80 (11.7.75)	1972-82 (5.10.72) 1975-80 (20.6.75)	n.s. (2.2.67)	1971-81 (27.10.71)	1974-84 (25.11.74)
United Kingdom	1974-84 (5.74)	n.s.[5] (26.3.68) 1972-77 (8.9.72)	1973-83 (18.12.73)	6 (10.10.67) 1973-82 (20.3.73)	6 (9.3.67) 1972-77 (15.6.72)	6 (19.1.68) 1974-84 (6.5.74)	n.s. (24.3.72)
Italy	n.s.[1] (20.9.66) 1974-84 (27.5.74)	n.s. (30.4.70)	1973-83 (18.4.73)	n.s.[1] (14.7.65) 1974-84 (17.1.74)	n.s.[1] (6.9.65) 1973-83 (22.5.73)	1974-84 (25.2.74) 1974-84 (25.7.74)	n.s.[1] (1.12.65) 1974-84 (25.5.74)
Yugoslavia	n.s. (16.7.64)			n.s. (2.2.58)	n.s. (13.1.67)	n.s. (27.2.69) 1971-76 (27.1.71)	n.s. (27.3.63)

	(11.12.74)	paraphé	(12.6.74)	(22.8.67) 1974-84 (2.7.74)	(14.5.75)	(15.7.75)	(18.7.75)
Norway	n.s. (22.9.70)		1974-84 (12.74)	1973-82 (28.11.72)	n.s. (29.11.68)	n.s. (19.5.72)	n.s. (17.1.70)
Austria	1974-84[1] (28.6.73)	n.d.[1] (12.9.71)	1974-84 (7.12.74)	1974-84 (6.9.73)	n.d. (20.2.68) 1974-84 (18.7.74)	n.d. (24.5.68) 1973-83 (1.2.73)	n.d. (15.11.68)
Portugal	(2.75)			1975-80 (14.5.75)			1975-80 (23.1.75)
Sweden	n.s. (26.5.70) 1975 (19.2.75)	n.s.[7] (13.10.71)		1975-85 (5.6.75)	n.s. (9.4.68)	n.s. (12.1.70) 1975 (25.4.75)	n.s. (12.5.69)
Switzerland	1973-77[8] (23.11.72)	1971-75[8] (7.5.71)	1975-80[2] (27.6.75)	1973-78[8] (25.6.73)	1973-78[8] (13.12.72)		1974-78[8] (30.10.73)
Spain	1971-75[9] (3.6.71)	1972-76[9] (5.10.71)		1975-85 (3.6.74)	1971-75[9] (5.3.71)	1972-75 (15.9.72)	1971-75[9] (18.11.70)
Turkey	1975-80[2] (13.9.75)						
United States*	1979-81 (9.2.79)					1974-84 (29.6.74)	

n.s.: Period not specified.
1 Agreements on economic and industrial co-operation.
2 Automatic extension.
3 Single long-term agreement on trade and industrial co-operation.
4 Agreement on economic and technical co-operation.
5 Agreement on economic co-operation.
6 Five-year agreement (may be extended) in the field of applied science and technology.
7 Agreement on economic and technical co-operation.
8 Agreements on trade relations.
9 Agreement on trade, movement of goods and persons by land and sea, scientific and technical co-operation.
* Luther J. Carter, "United States and Bulgaria to Cooperate in Research", *Science*, March 10, 1978, p. 1051.

Source: F. Levcik and J. Stankovsky, *Industrielle Kooperation zwischen Ost und West, op. cit.*, pp. 158-159.

Table A-43

Dollar Unit Values of Machinery Exported to and Imported from Eastern Europe and the Soviet Union

SITC Sub-groups		Total of 10 Western countries				Germany			
		Exports		Imports		Exports		Imports	
		1965-1968	1971-1974	1965-1968	1971-1974	1965	1974	1965	1974
Engines (internal combustion)	7115	3 666	4 778	2 070	2 391	4 830	8 817	4 138	2 528
Tractors	7125	1 483	2 421	663	875	1 215	4 281	834	796
Machine tools	7151	3 773	5 902	1 467	1 667	4 207	9 386	1 339	2 197
Metalwork machinery	7152	2 193	4 006	665	429	2 032	6 030	321	729
Textile machinery	7171	3 790	6 457	2 823	5 209	3 614	9 458	6 827	5 767
Printing machinery	7182	4 912	8 292	2 789	3 911	5 186	10 738	1 270	5 833
Construction and mining machinery n.e.s.	7184	1 693	2 649	984	1 232	1 677	3 361	581	1 826
Mineral, glass machinery	7185	2 770	3 872	930	1 014	2 292	4 279	738	932
Heating and cooling equipment	7191	2 471	3 451	986	1 004	1 881	6 444	521	1 086
Pumps and centrifuges	7192	3 832	5 552	1 696	1 813	3 231	8 941	933	1 878
Handling equipment	7193	1 869	2 943	990	1 016	1 561	5 943	774	701
Powered tools	7195	3 019	4 950	2 038	2 007	3 307	7 760	1 309	1 496
Ball bearings	7197	2 432	4 048	1 810	2 689	2 568	5 883	1 874	3 424
Appliances	7198	3 574	5 884	1 789	2 077	3 726	7 803	856	1 018
Parts and accessories	7199	3 225	3 282	752	1 098	3 125	5 713	394	1 148
Electric power machinery	7221	2 151	3 719	1 220	1 081	3 003	7 341	836	1 255
Electrical apparatus	7222	4 542	7 816	1 960	3 284	8 058	25 647	6 250	5 416
Wire and cable	7231	808	1 150	712	845	1 561	1 817	945	1 428
Telecommunications equipment	7249	14 045	17 928	7 777	7 689	17 433	47 037	5 466	12 884
Domestic electrical equipment	7250	1 649	1 833	1 308	1 469	3 474	4 928	1 461	1 619
Valves, tubes, transistors	7293	17 420	48 669	6 569	13 240	1 763	12 471	1 518	1 752
Electrical measuring instruments	7295	18 801	35 970	8 020	10 877	24 525	43 662	12 778	19 096
Electrical machinery n.e.s.	7299	2 426	3 359	479	1 045	2 567	4 574	3 380	1 097
Cars	7321	1 301	1 340	942	1 255	1 334	2 997	884	1 509
Lorries and trucks	7323	1 641	2 530	915	1 288	1 731	2 694	799	1 759
Chassis, bodies	7328	1 326	2 136	1 349	1 708	3 189	4 039	2 042	1 742
Ships and boats	7350	624	2 823	551	1 231				
Arithmetic average		4 126	7 324	2 011	2 720	4 373	10 079	2 272	3 112

Table A-43 (continued)

SITC Sub-groups	Italy				France			
	Exports		Imports		Exports		Imports	
	1965	1974	1965	1974	1965	1974	1965	1974
7115 Engines (internal combustion)	3 175	4 420	856	3 147	4 541	6 536	2 334	3 068
7125 Tractors	1 173	2 055	602	957	1 154	6 079	695	864
7151 Machine tools	4 351	5 983	1 399	1 443	2 591	6 460	1 585	1 717
7152 Metalwork machinery	2 115	4 593	1 030	1 891	1 215	7 844	239	136
7171 Textile machinery	3 582	7 177	2 510	3 735	3 747	6 355	4 860	3 652
7182 Printing machinery	3 685	5 229	2 570	2 759	3 707	10 751	2 514	3 777
7184 Construction and mining machinery n.e.s.	739	3 254	764	620	1 798	2 377	663	878
7185 Mineral, glass machinery	3 825	3 865	397	727	2 316	3 618	434	1 747
7191 Heating and cooling equipment	2 311	2 818	..	1 352	2 901	3 433	3 302	2 536
7192 Pumps and centrifuges	2 247	6 696	3 994	2 708	2 875	4 487	2 516	2 323
7193 Handling equipment	2 049	3 482	1 107	1 075	2 152	3 980	1 203	1 498
7195 Powered tools	2 569	6 284	1 874	1 666	7 408	8 840	1 943	3 502
7197 Ball bearings	2 260	5 372	1 893	3 089	5 736	6 006	1 481	2 514
7198 Appliances	2 603	6 427	..	1 415	4 564	5 871	3 805	1 782
7199 Parts and accessories	2 480	3 005	568	842	3 527	3 187	274	619
7221 Electric power machinery	3 170	4 272	873	1 112	2 514	5 876	1 597	1 020
7222 Electrical apparatus	5 791	11 773	5 663	4 940	8 276	16 539	793	7 847
7231 Wire and cable	764	1 999	886	2 524	1 288	1 785	523	1 585
7249 Telecommunications equipment	26 110	23 028	5 447	5 672	14 947	21 190	6 007	4 646
7250 Domestic electrical equipment	1 926	1 829	1 188	2 588	3 975	3 552	1 637	1 220
7293 Valves, tubes, transistors	11 624	10 103	1 473	1 045	50 725	56 686	15 005	31 656
7295 Electrical measuring instruments	21 794	28 551	17 561	6 951	21 195	4 913	2 528	4 527
7299 Electrical machinery n.e.s.	1 923	3 236	386	879	1 443	1 097	872	1 186
7321 Cars	1 334	1 873	888	1 111	2 491	2 843	1 087	1 196
7323 Lorries and trucks	1 539	2 902	1 594	1 083	3 005	2 364	1 298	2 047
7328 Chassis, bodies	637	1 948	1 624	1 157	..	..	..	..
7350 Ships and boats	299	728	1 421	4 590	691	11 923	894	1 239
Arithmetic average	4 288	6 033	2 202	3 744	6 183	8 284	2 311	3 415

Source: United Nations. Trade Statistics from the *Economic Bulletin for Europe*. Vol. 28. New York, 1976. p. 135.

395

Table A-44
Various Estimates of USSR Gold Sales (1965-1977)

Year	Quantity (tonnes)				Value (US$ Million)				Addendum: Price in US$ per fine ounce	
	ISB Annual Reports[1]	Variant A[2]	Variant B[3]	Variant C[4]	ISB Annual Reports[5]	Variant A[2]	Variant B[3]	Variant D[6]	Official rate	Free market
1965	489	500			550	560		310	35	
1966	—	—			—			—	35	
1967	—	14			—	16		—	35	
1968	—	11			—	12		—	35	
1969	5	—			6.1	—		—	35	37.9
1970	50	—			58.2	—		—	35	36.2
1971	90	—			116.9	—		—	35	40.4
1972	200	150			370.0	300		325	38	57.5
1973	330	300			1 340.0	1 000		800	42.22	126.3
1974	150				800.5				—	166.0
1975	150			150	713.7				—	148.0
1976[7]	350			300[9]	—	1 400[8]	1 000 to 1 200	—	—	90 to 120[10]
1977[11]	—	—	330 to 370		—		1 700	—		

1 International Settlement Bank — sales by "Communist Countries" not simply USSR. Years 1974-1976, 47th Annual Report Basle, 13th June, 1977.
2 Non-official estimates prepared for the United States Government and not previously published.
3 Chase Manhatten Bank: 1976, Werner Beitel, *op. cit.* p. 17: 1977, *Neue Zürcher Zeitung,* 21st June, 1978, p. 11.
4 Non-official estimates, unpublished and said to be those of the South African Government.
5 At the official price up to 1968 and the free market price thereafter (col. 10).
6 Richard Rockingham-Gill "The Effects of the Increase in the Price of Gold and Minerals on the USSR's Monetary Reserves", *Problèmes économiques;* Paris, 11 December 1974. Translation of an article published in *Osteuropa-Wirtschaft,* April 1974.
7 The 1976 level was maintained during the first months of 1977.
8 *Soviet Economic Problems and Prospects, op. cit.,* p. 23.
9 Figure expected for 1976.
10 Average price at the International Monetary Fund's two first gold auctions.
11 Annual production (date not given) of approximately 440 tons. Report of Chamber of Mines of South Africa, September 1978, given in *Problèmes économiques.* Paris, 18 July, 1979, p. 26.

Table A-45
Soviet Union Countertrade Commitments under Compensation Agreements broken down by Commodity Group
Soviet exports in $ million

	1976-1980	1981-1985	Other Periods[a]	Total	Number of Agreements Considered[e]
	1	2	3	4	5
Forestry products	1 279	100	. .	1 379	3
Oil and gas	6 102.6	10 162	. .	16 264.6	7
Coal	80	860	400[c]	1 340[c]	3
Metals[bd]	28	. .	600[c]	628[c]	3
Chemicals	1 222	1 665	8 528	11 415	28
Total	8 711.6	12 787	9 528	31 027.1	44

a) Agreements for different periods from those in the headings to Columns 1 and 2 (e.g. 1979-1983, 1978-1986, etc.).
b) Scrap metal, chromium, iron ore (and a small contract for $10.85 million for textile raw materials and oil products). See Table A-33.
c) It has been estimated on an arbitrary basis that coal accounts for $400 million out of the $1 000 million in the contract signed with Finsider, the Italian firm, in 1974 for compensation deliveries of "scrap-metal, coal and iron ore".
d) $600 million of the Finsider contract, see Note c).
e) Only contracts whose value has been given are considered. For all known contracts, see Table A-33. The Finsider contract (see Note c) has been counted twice.
Source: Based on Table A-33.

USSR High Technology and Total Machinery Imports from OECD Countries
In million US dollars

	1972	1974	1975	1976
Machinery and transport equipment (SITC Section 7) (from OECD countries)[1]	1 207	2 309	4 576	4 909
High technology imports[2] *of which:*	582.4	1 036.2	1 583.5	1 627.1
Machine tools for working metals (7151)	. .	448.8	550.3	576.9
Cocks, valves, etc. (71992)	. .	107.8	283.1	252.6
Pumps and centrifuges (7192)	. .	107.8	167.4	245.0
Electrical machinery, n.e.s. (7299)	. .	119.7	214.3	139.6
Other electrical measuring and controlling instruments (72952)	. .	53.4	60.2	69.1
Share of high technology in total machinery imports from OECD countries[3]	48.3%	44.9%	34.6%	33.1%

1   *Statistics of Foreign Trade*, OECD, Paris.
2   *Quantification of Western Exports of High Technology Products to the Communist Countries*, Office of East-West Policy and Planning, Bureau of East-West Trade, October 17, 1977, pp. 7 and 9.
3   This is an approximate figure as trade statistics for machinery (SITC Section 7) and high technology do not cover exactly the same ground.
   Trade in machinery covers SITC Section 7 only, whereas high technology also includes optical elements, optical instruments, image projectors and measuring and control instruments (SITC Groups 8611, 8613, 86161 and 8619) but Soviet imports of these four product groups totalled only $47.3 million in 1974, 71.1 million in 1975 and 77.8 million in 1976, or 4.6, 4.5 and 4.8 per cent respectively of the USSR's total high technology imports from Western industrialised countries.
   The term "Western industrialised countries" is more restrictive than "OECD countries" as it does not include: Australia, Finland, Greece, Ireland, Iceland, New Zealand, Portugal, Spain, Turkey.
   Nevertheless, the calculation errors are slight and to some extent cancel one another out.

Table A-47

Shares of Individuals Countries in Total Sales of Machinery and Transport Equipment in General and High Technology by Western Industrialised Countries to the USSR in 1974 and 1976

	1974		1976	
	Machinery[a]	High Technology[b]	Machinery[a]	High Technology[b]
United States	10.3	*13.3*	15.1	12.7
Canada	0.1	0.1	0.6	0.6
Japan	10.2	8.6	17.8	13.9
Belgium	0.9	*1.3*	0.4	*0.9*
France	14.8	*15.0*	12.4	10.9
Germany (including West Berlin)	30.5	*34.0*	27.0	*34.5*
Netherlands	2.5	*4.0*	1.2	*2.4*
Italy	8.6	7.6	9.6	8.4
Austria	4.1	2.0	1.9	*2.4*
Norway	0.7	0.1	0.4	0.4
Sweden	8.2	4.0	4.4	3.0
Switzerland	5.4	*6.1*	4.2	*6.2*
United Kingdom	3.1	*3.5*	4.1	3.3
Denmark	1.0	0.5	0.8	0.5
Total	100.5[c]	100.0	100.4[c]	100.0

a)   *Vneshnyaya Torgovlya SSSR v 1974 godu* (USSR Foreign Trade in 1974); *Vneshnyaya Torgovlya SSSR v 1976 godu* (USSR Foreign Trade in 1976). Moscow, 1977. Machinery and transport equipment.
b)   Percentages taken from Table 7.
c)   Figures do not add up to 100 per cent because of rounding off.
Italics indicate that the country's share of high technology exports to the USSR is higher than its share of exports of machinery and transport equipment in general.

SITC Sub-groups		United States	Canada	Japan	Belgium Luxembourg	France	Germany
7142	Calculating machines (including electronic computers)	39	—	4 773	—	5	
7143	Statistical machines (punch card or tape)	—	—	12	4	26	
71492	Parts of office machinery (including computer parts)	969	—	255	—	—	
7151	Machine-tools for metal	1 003	—	14 648	710	4 117	18
71852	Glass-working machinery	170	—	70	4 619	2 954	
7192	Pumps and centrifuges	37 226	256	8 861	199	17 033	16
71952	Machine-tools for wood, plastics, etc.	—	279	—	618	241	24
71954	Parts and accessories for machine-tools	2 764	—	6 255	575	6 780	8
71992	Cocks, valves, etc.	44	—	5 168	11	59 389	19
7249	Telecommunications equipment (excludig TV and radio receivers)	1 765	135	5 619	21	2 231	4
72911	Primary batteries and cells	1	—	29	—	38	
7293	Tubes, transistors, photocells, etc.	47	4	431	5	1 725	
72952	Electrical measuring and control instruments	6 929	29	4 570	4	9 120	14
7299	Electrical machinery	13 260	12	6 808	2 169	14 103	59
7341	Aircraft heavier than air	439	—	—	—	—	
73492	Aircraft parts	115	6	—	1	—	
7359	Special purpose vessels	—	—	—	—	—	
8611	Optical elements	46	—	109	—	45	
8613	Optical instruments	—	—	2 198	11	98	1
86161	Image projectors	14	—	66	1	546	
8619	Measuring and control instruments	2 851	9	3 517	4	4 798	11
	Total specified	67 682	730	63 389	8 952	122 649	181
	Total calculated by Department of Commerce[b]	137 581	729	88 559	13 685	155 501	352

a) *Statistics of Foreign Trade.* Series C. OECD. Paris. 1974.
b) See Table 7.
c) No OECD total published. Calculated by adding up the figures published for each country.
d) Total for the 14 countries listed in the statistics of the Department of Commerce and included in this table.

m the Major Industrialised Countries in 1974[a]

dollars

Netherlands	Italy	Austria	Norway	Sweden	Switzerland	United Kingdom	Denmark	OECD Total	Total calculated by Dept. of Commerce
—	724	18	—	618	448	135	163	7 337	7 281
4	16	—	—	1	1	21	1	341[c]	14 539
—	—	4	—	45	2	1 305	71	2 652	4 064
60	6 860	1 438	—	2 226	3 231	1 206	1	131 622[c]	448 796
—	22	—	—	1 773	—	813	—	11 102	11 139
1 502	4 369	407	234	14 503	2 970	2 453	672	114 326	107 823
12	1 422	137	—	2 758	505	139	—	53 351	30 440
40	2 647	121	102	737	1 515	1 375	7	32 904	31 901
128	20 308	1 961	—	528	549	323	125	108 409	107 840
7	435	166	—	477	40	790	25	20 833	16 808
—	—	1	—	—	—	38	—	107	107
—	123	13	—	78	21	97	7	3 222	3 225
1 286	2 648	1 129	129	—	2 896	6 753	3 398	54 178	53 409
5 601	2 457	688	175	7 145	699	7 241	3	120 436	119 734
—	—	—	—	—	—	—	—	439	439
—	2	—	—	—	—	116	—	242	242
30 834	—	—	—	48	—	—	13	41 900	30 937
—	—	1	—	1	214	8	1	871	873
—	—	179	5	—	379	204	—	4 319	4 322
17	19	—	—	—	244	17	1	926	927
272	2 921	212	3	2 875	7 046	4 868	341	44 268	41 237
39 763	44 973	6 475	648	33 813	20 760	28 946	4 829	753 785	1 036 123
41 214	78 713	20 448	1 316	41 197	63 511	36 391	4 919	1 036 208[d]	1 036 208

Table A-49

**Main Types of High Technology Exported to the Soviet Union
by the Major Industrialised Countries in 1974**

	Share in total high technology exports from the 14 countries to the USSR	Share of the USSR in exports of some types of high technology [a]	
United States	13.3	Parts of office machinery (71492)	36.5
		Pumps and centrifuges (7192)	32.6
		Aircraft, heavier than air (7341)	100.0
		Aircraft parts (73492)	47.5
Canada	0.1	Aircraft parts (73492)	2.5
Japan	8.6	Calculating machines (including electronic computers (7142)	65.1
		Optical instruments (8613)	50.9
Belgium/Luxembourg	1.3	Glass-working machinery (71852)	41.6
		Electrical machinery (7299)	1.8
France	15.0	Image projectors (86161)	59.0
		Cocks, valves, etc. (71992)	54.7
		Primary batteries and cells (72911)	35.5
		Tubes, transistors, photocells, etc. (7293)	53.5
		Glass-working machinery (71852)	26.6
		Parts and accessories for machine tools (71954)	20.6
Germany (including West Berlin)	34.0	Optical elements (8611)	51.2
		Electrical machinery (7299)	49.2
		Machine-tools for wood, plastics, etc. (71952)	45.5
The Netherlands	4.0	Electrical machinery (7299)	4.7
		Special purpose vessels (7359)	73.6
Italy	7.6	Cocks, valves, etc. (71992)	18.7
		Calculating machines (including electronic computers) (7142)	9.9
		Parts and accessories for machine tools (71954)	8.0
Austria	2.0	Optical instruments (8613)	4.1
		Electrical measuring and control instruments (72952)	2.1
Norway	0.1	Pumps and centrifuges (7192)	0.2
		Electrical measuring and control instruments (72952)	0.2
Sweden	4.0	Calculating machines (including electronic computers) (7142)	8.4
		Glass-working machinery (71852)	16.0
		Pumps and centrifuges (7192)	12.7
		Machine tools for wood plastics, etc. (71952)	5.2
		Electrical machinery (7299)	5.9
		Measuring and control instruments (8619)	6.5
Switzerland	6.1	Calculating machines (including electronic computers) (7142)	6.1
		Optical elements (8611)	24.6
		Image projectors (86161)	26.3
		Measuring and control instruments (8619)	15.9

(Cont'd on next page)

400

Share in total high technology exports from the 14 countries to the USSR		Share of the USSR in exports of some types of high technology [a]	
United Kingdom	3.5	Statistical machines (punch card or tape) (7143)	6.2
		Parts of office machinery (including computer parts) (71492)	49.2
		Parts and accessories for machine-tools (71954)	4.8
		Telecommunications equipment (7249)	3.8
		Primary batteries and cells (72911)	35.5
		Electrical measuring and control instruments (72952)	12.5
		Electrical machinery (7299)	6.0
		Aircraft parts (73492)	47.9
		Optical instruments (8613)	4.7
		Measuring and control instruments (8619)	11.0
Denmark	0.5	Calculating machines (including electronic computers) (7142)	2.2
		Parts of office machinery (71492)	2.7
		Pumps and centrifuges (7192)	0.6
		Electrical measuring and control instruments (72952)	6.3
		Measuring and control instruments (8619)	0.8

a) Machine-tools for metal have not been included because there is too big a difference between OECD export figures and the total indicated by the Department of Commerce.

Source: Calculated from Annex Table A-48.

SELECTED BIBLIOGRAPHY

In order to structure the literature and to avoid undue duplication (though some was unavoidable) the authors have attempted to group this selected bibliography by broad subject matter. This grouping does *not* follow the individual chapter headings. As is evident from the analysis, the individual chapters reflect broad categories of issues and problems. The literature in the field is, for the most part, general and does not necessarily address directly these issues and problems. However, the literature can be roughly "classified". The headings chosen for the bibliography reflect one attempt to do so.

I. GENERAL STUDIES ABOUT EAST-WEST TRADE
(including studies about East-West
trade of individual countries)

A Background Study on East-West Trade, prepared for the Committee on Foreign Relations, United States Senate, 89th Congress, 1st Session, Legislative Reference Service of the Library of Congress, US Government Printing Office, Washington, D.C., April 1965.

Alder, Karlsson G., *The Political Economy of East-West-South Co-operation,* Studien über Wirtschafts und Systemvergleiche Hg/F/Nemschak, Band 7, Springer Verlag, Wien-New York, 1976.

Altmann, F. L., Clement, H., *Die Kompensation als Instrument im Ost-West Handel,* Verlag G. Olzog, München, 1979.

American-Soviet Trade, A Joint Seminar on the Organisational and Legal Aspects, Moscow, December, 1975, US Department of Commerce, Washington, D.C., September 1976.

A New Look at the Trade Policy Toward the Communist Bloc, Sub-Committee on Foreign Policy, Joint Economic Committee, US Government Printing Office, Washington, D.C., 1961.

Background Materials Relating to the US-USSR Commercial Agreements, Committee on Finance, US Senate, 93rd Congress, 2nd Session, US Government Printing Office, Washington, D.C., 2nd April, 1974.

Balcerowicz, Leszek; Bozyk, Pawel; Ladyka, Stanislaw; Misala, Jozef; Zimny, Zbigniew, *La coopération économique des pays de l'Est avec la Suisse,* Institut universitaire de hautes études internationales, Genève, Sijthoff, Leiden, 1976.

Baumer, Max, and Jacobsen, Hanns-Dieter, "CMEA and the World Economy: Institutional Concepts", *East European Economies Post-Helsinki — A Compendium of Papers,* Joint Economic Committee, Congress of the United States, 95th Congress, 1st Session, August 25, 1977, US Government Printing Office, Washington, D.C., 1977.

Beitel, Werner, *Probleme der Entwicklung des Sowjetischen Westhandels. Kreditfinanzierung. Exportstruktur. Kooperationsabkommen,* Stiftung Wissenschaft und Politik, Ebenhausen, September 1977.

Beitel, Werner, Nötzold Jürgen. *Entwicklungstendenzen und Perspektiven der Wirtschaftsbeziehungen der Bundesrepublik Deutschland mit der Sowjetunion,* Stiftung Wissenschaft und Politik, Ebenhausen, März 1978.

Beitel, Werner, Nötzold, Jürgen, *Deutsch-Sowjetische Wirtschaftsbeziehungen in der Zeit der Weimarer Republik,* Stiftung Wissenschaft und Politik, Ebenhausen, März 1977.

Beitel, Werner, Nötzold, Jürgen, *Zur Problematik der Wirtschaftlichen Kooperation mit der UdSSR, Erfahrungen aus den Wirtschaftlichen Beziehungen zwischen der Weimarer Republik und der Sowjetunion,* Stiftung Wissenschaft und Politik, Ebenhausen, Oktober 1976.

Bethenhagen, Jochem; Lodahl, Maria, Machowski, Heinrich, *Entwicklung und Struktur der Osteinfuhren der EWG. Eine Analyse unter besonderer Berücksichtigung der gemeinsamen Importpolitik,* reprint from *Vierteljahrshefte zur Wirtschaftsforschung,* Heft 2, 1977.

Blumenthal, W. Michael, excerpts from an address in Moscow, December 6, 1978, at the opening meeting of the USSR Trade and Economic Council, following the conclusions on December 5, 1978 of the 7th Session of the Soviet-American Commission on Trade and Economic Relations.

Bogomolov, O. T.; Shmelov, N. P.; Fufryansky, N. A.; Borisov, B. A.; Seifulin, F. A., (ed), *All-European Economic Co-operation,* Soviet Committee for European Security, USSR Academy of Sciences, Moscow, 1973.

Bot, Bernard R., "EEC-CMEA: Is a Meaningful Relationship Possible?", *Common Market Law Review,* Vol. 13, No. 3, August 1976.

Brada, Josef, (ed), *Quantitative and Analytical Studies in East-West Economic Relations,* International Development Research Centre, Indiana University, Bloomington, Indiana, 1976.

Brada, Josef C., Wipf, Larry J., "The Export Performance of East European Nations in Western Markets", *Weltwirtschaftliches Archiv, Review of World Economics,* Zeitschrift des Instituts für Weltwirtschaft Kiel, Heft 1, Band 111, 1975.

Brainard, Lawrence J., *Eastern Europe's New Five-Year Plans, The Outlook for Intra-CMEA and East-West Trade,* VIII AAASS Convention, St. Louis, 7th October, 1976.

Brougher, Jack, "USSR Foreign Trade: A Greater Role for Trade with the West", *Soviet Economy in a New Perspective, A Compendium of Papers,* Joint Economic Committee, Congress of the United States, 94th Congress, 2nd Session, US Government Printing Office, Washington, D.C., October 14, 1976.

Brown, Alan A., and Neuberger, Egon, (ed), *International Trade and Central Planning— An Analysis of Economic Interactions,* University of California Press, Berkeley and Los Angeles, 1968.

Communist States and Developing Countries: Aid and Trade in 1974, Special Report, The Department of State, Bureau of Public Affairs, Office of Media Services, Washington, D.C., February 1976.

Cviic, Christopher, "Comecon and East-West Trade Policies", *Comecon: Progress and Prospects,* Colloquium, 1977, Directorate of Economic Affairs, NATO, Brussels, March 16-18, 1977.

Developments in Countertrade: Part V: "The GDR", *Business Eastern Europe,* July 8, 1977. Part VI: "Poland", *Business Eastern Europe,* July 15, 1977. Part VII: "Czechoslovakia", *Business Eastern Europe,* July 22, 1977. Part VII: "The Soviet Union", *Business Eastern Europe,* August 5, 1978.

Die Entwicklung der Wirtschaftlichen Ost-West Beziehungen als Problem der Westeuropäischen und Atlantischen Gemeinschaft, Stiftung Wissenschaft und Politik, Ebenhausen, September 1975.

Die Entwicklung der Beziehungen zwischen EG und RGW vor der KSZE Folgekonferenz in Belgrad, Stiftung Wissenschaft und Politik, Ebenhausen, März 1977.

Die Internationalisierung der Produktion und ihre Bedeutung für die Ost-West-Wirtschaftsbeziehungen vom Handel zur Kooperation, Stiftung Wissenschaft und Politik, Ebenhausen, May 1976.

East European Economies Post-Helsinki—A Compendium of Papers, Joint Economic Committee, Congress of the United States, 95th Congress, 1st Session, US Government Printing Office, Washington, D.C., August 25, 1977.

"East-West Economic Co-operation", *L'Est,* Quaderni CESES, Milano, No. 9, 1978.

"East-West Trade", *Law and Contemporary Problems;* Part I, No. 3; Part 2, No. 4, School of Law, Duke University, Summer and Autumn 1972.

ECE and East-West Trade Promotion, Committee on the Development of Trade, Economic Commission for Europe, United Nations, Geneva, 1975.

Economic Relations between East and West: Prospects and Problems, a tripartite report by fifteen experts from the European Community, Japan and North America, The Brookings Institution, Washington, D.C., 1978.

Ehrhardt, Carl A., "EEC and CMEA Tediously Nearing Each Other", *Aussen Politik* (English edition), Vol. 28, 2/77, 2nd Quarter, 1977.

Ericson, Paul, Farell, John, "Soviet Trade and Payments with the West", *Soviet Economy in a New Perspective—A Compendium of Papers,* Joint Economic Committee, Congress of the United States, 94th Congress, 2nd Session, US Government Printing Office, Washington, D.C., 14th October, 1976.

Etude et prospection des marchés dans le commerce Est-Ouest, Comité pour le développement du commerce, Commission économique pour l'Europe, Nations Unies, Genève, TRADE/INF. 4, 1977.

"Evolution récente du commerce européen, évolution du commerce et des paiementes entre l'Est et l'Ouest", 1965-1974, *Economic Bulletin for Europe,* Vol. 28, United Nations, New York, 1976.

Examen des travaux récents sur les projections du commerce intrarégional jusqu'en 1990, Economic and social Council, Economic Commission for Europe, United Nations, Geneva, TRADE/AC.11/R.2, 18 mai 1979.

Goldman, Marshall I., *Détente and Dollars, Doing Business with the Soviets,* Basic Books, Inc., Publishers, New York, 1975.

Goldman, Marshall I., "Autarchy or Integration — the USSR and the World Economy", *Soviet Economy in a New Perspective — A Compendium of Papers,* Joint Economic Committee, Congress of the United States, 94th Congress, 2nd Session, US Government Printing Office, Washington, D.C., October 14, 1976.

Gvishiani, Jermen M., *Adress at the International Industrial Conference,* Stanford Research Institute and the Conference Board, 21st September, 1973.

Haberler, Gottfried, *The Theory of International Trade,* London, 1936.

Hannigan, J. B., and McMillan, C. H., *The Participation of Canadian Firms in East-West Trade: A Statistical Profile,* East-West Commercial Relations Series, Institute of Soviet and East European Studies, Carleton University, Ottawa, Research Report, No. 11, June 1979.

405

Hanson, Philip, *USSR: Foreign Trade Implications of the 1976-1980 Plan,* EIU Special Report No. 36, The Economist Intelligence Unit, Ltd., London, October 1976.

Hardt, John P., "Les relations commerciales soviétiques et l'évolution politique en URSS", *Revue d'études comparatives Est-Ouest,* Volume 6, No. 2, Editions du Centre national de la recherche scientifique, Paris, 1975.

Hewett, Edward A., "Recent Developments in East-West European Economic Relations and Their Implications for US-East European Economic Relations", *East European Economies Post-Helsinki – A Compendium of Papers,* Joint Economic Committee, Congress of the United States, 95th Congress, 1st Session, US Government Printing Office, Washington, D.C., August 25, 1977.

"Hitch in EEC Trade Talks with Comecon", *The Times,* November 27, 1978.

Holzman, Franklyn D., *Foreign Trade Under Central Planning,* Harvard University Press, Cambridge, Mass., 1974.

Holzman, Franklyn D., *International Trade under Communism, Politics and Economics,* Basic Books Inc., New York, 1976.

Holzman, Franklyn D., "Some Systemic Factors Contributing to the Convertible Currency Shortages of Centrally Planned Economies", *Current Issues in East-West Trade and Payments,* Vol. 69, No. 2, American Economic Association, May 1979.

Hoya, Thomas W., "The Changing Nature of East-West Trade", *Columbia Journal of Transnational Law,* Vol. 12, No. 1, 1973.

International Economics – Comparisons and Interdependences. edited by Friedrich Levcik, Festschrift für Franz Nemschak, Springer-Verlag, Wien, New York, 1978.

Internationale Wirtschaftsorganisationen und Ost-West-Kooperation, Stiftung Wissenschaft und Politik, Ebenhausen, June 1979.

Iskra, W., "RGW und EWG: Möglichkeiten der Zusammenarbeit", *Probleme des Friedens und des Sozialismus,* Vol. 19, No. 6, 1976.

Issues in East-West Commercial Relations – A Compendium of Papers, Submitted to the Joint Economic Committee, Congress of the United States, 95th Congress, 2nd Session, US Government Printing Office, Washington, D.C., January 12, 1979.

Jonquières, Guy de, "EEC and Comecon Fail to Agree on Basis for Talks", *Financial Times,* November 27, 1978.

Kaysen, Carl, *Review of US-USSR Exchanges and Relations, Executive Summary,* National Academy of Sciences, Washington, D.C., n.d.

Knirsch, Peter, *Interdependence in East-West Economic Relations,* paper prepared for the Marshall Plan Commemoration Conference, OECD, Paris, June 2-3, 1977.

Knirsch, Peter, «Comecon and the Developing Nations», *Comecon and Prospects,* Colloquium 1977, Directorate of Economic Affairs, NATO, Brussels, March 16-18, 1977.

Knirsch, Peter, *Osteuropa und die neue Weltwirtschaftsordnung,* Separatdruck aus Sozialwissenschaftliche Studien des Schweizerischen Instituts für Auslandforschung, Band 6 (Neue Folge), *Umstrittene Weltwirtschaftsordnung,* Herausgegeben von Prof. Dr. Daniel Frei.

Kostecki, M. M., *East-West Trade and the GATT System,* the Macmillan Press, for the Trade Policy Research Centre, London, 1979.

Kravalis, Hedija H., "Soviet-East European Export Potential to Western Countries", *Issues in East-West Commercial Relations – A Compendium of Papers,* Submitted to the Joint Economic Committee, Congress of the United States, 95th Congress, 2nd Session, US Government Printing Office, Washington, D.C., January 12, 1979.

Lange-Prollius, Horst, *Praxis des Ostwesthandels. Die Wirtschaftsbeziehungen 1977-1990,* Mit Beiträgen von Erich Kissner und Helmuth Bohunovsky, Econ. Verlag, Düsseldorf-Wien, 1977.

Lavigne, Marie, "Comecon-CEE. Une lente partie d'échecs", *Le Figaro,* Paris, 10-11 April, 1976.

Lavigne, Marie, "Compte rendu de l'ouvrage de Friedrich et Jan Stankovsky", *Revue d'études comparatives Est-Ouest,* No. 1, Editions du Centre national de la recherche scientifique, Paris, 1978.

Lavigne, Marie, *Les relations économiques Est-Ouest,* Presses universitaires de France, Paris, 1979.

Lebahn, Axel, "RGW und EG – Faktoren des Ost-West-Handels", *Aussenpolitik,* No. 2, 1978.

Lebahn, Axel, *Sozialistische Wirtschaftsintegration und Ost-West-Handel im sowjetischen Recht,* Duncker und Humblot, Berlin-München, 1976.

Lebahn, Axel, "Neuentwicklungen der Geschäftstätigkeit und Rechtsgrundlagen der internationalen Comecon-Banken IBEC und IIB", *Recht der Internationalen Wirtschaft,* Heidelberg, Januar 1979.

Lenz, Allen J., Kravalis, Hedija H., "An Analysis of Recent and Potential Soviet and East European Exports to Fifteen-Industrial Western Countries", *East European Economies Post-Helsinki – A Compendium of Papers,* Joint Economic Committee, Congress of the United States, 95th Congress, 1st Session, August 25, 1977, US Government Printing Office, Washington, D.C., 1977.

"Les relations commerciales de la France avec les pays de l'Europe de l'Est à commerce d'Etat", Conseil économique et social, Session de 1976, Séance du 8 juin 1976, *Journal officiel,* Paris, 22 octobre 1976.

Levcik, Friedrich, Stankovsky, Jan, *Industrielle Kooperation zwischen Ost und West – Studien über Wirtschaft und Systemvergleiche,* Band 8, Springer Verlag, Wien, New York, 1977.

Levinson, Charles, *Vodka Cola,* Editions Stock, Paris, 1977.

Loeber, Dietrich André, *East-West Trade,* A Sourcebook on the International Economic Relations of Socialist Countries and Their Legal Aspects. Vols I-IV, Oceana Publications, Inc. Dobbs Ferry, New York, 1976-1977: Vol. I, 1976; Vol. II, 1976; Vol. III, 1977; Vol. IV, 1977.

Marer, Paul, "Soviet Economic Policy in Eastern Europe", *Reorientation and Commercial Relations of the Economies of Eastern Europe – A Compendium of Papers,* Joint Economic Committee, Congress of the United States, 93rd Congress, 2nd Session, US Government Printing Office, Washington, D.C., August 16, 1974.

Marer, Paul, Statement in *US Policy Towards Eastern Europe,* Hearings Before the Sub-Committee on Europe and the Middle East, Committee on International Relations, US House of Representatives, 95th Congress, 2nd Session, September 7 and 12, 1978, Washington, D.C., 1979.

Masu, Stefan, *La coopération économique des pays de l'Est avec la Suisse: la Roumanie,* Institut universitaire des hautes études internationales, Genève, 1975.

McMillan, Carl H., *Canada's Postwar Economic Relations with the USSR: An Appraisal,* East-West Commercial Relations Series, Institute of Soviet and East European Studies, Carleton University, Ottawa, Research Paper No. 10, January 1979.

McMillan, Carl, H., (ed.), *Changing Perspectives in East-West Commerce,* Carleton University, Lexington Books, Toronto, London, 1974.

McMillan, Carl H., *The International Organisation of Inter-Firm Cooperation,* (With special reference to East-West relationships), IEA Conference, Dresden, 29th June-3rd July, 1976.

Morton, Rogers, *The United States Role in East-West Trade, Problems and Prospects,* US Department of Commerce, August 1975.

Nove, Alec, "East-West Trade: Problems, Prospects, Issues", *The Washington Papers,* Vol. VI, No. 53, The Center for Strategic and International Studies, Georgetown University, Washington, D.C., Sage Publications, Beverly Hills/London, 1978.

Nykryn, Jaroslaw, Karel, Stefan, *La coopération économique des pays de l'Est avec la Suisse,* Institut universitaire des hautes études internationales, Genève, Sijthoff, Leiden, 1976.

Osthandel In Der Krise, Herausgegeben von Stefan Graf Bethlen, Berichte und Studien der Hanns-seidel-Stiftung e.V.Müchen, Günter Olzog Verlag, München-Wien, 1976.

Patolichev, N., "Thirty Years of Co-operation in Foreign Trade", *Foreign Trade,* No. 4, USSR Ministry of Foreign Trade, (English translation), 1979.

Peter, Paul, *La coopération économique des pays de l'Est avec la Suisse: Hungary,* Institut universitaire des hautes études internationales, Genève, Sijthoff, 1976.

Peterson, Peter G., *US-Soviet Commercial Relationships in a New Era,* Department of Commerce, Washington, D.C., August 1972.

Petrov, Latchezar, *La coopération économique de la Suisse avec les pays de l'Est: la Bulgarie,* Institut universitaire des hautes études internationales, Genève, 1975.

Pisar, Samuel, "Co-existence and Commerce", *Guidelines for Transactions between East and West,* Allen Lane, The Penguin Press, 1970.

Pisar, Samuel, *Transactions entre l'Est et l'Ouest. Le cadre commercial et juridique,* Dunod, Paris, 1972.

Project Interdependence: US and World Energy Outlook Through 1990, A Report printed at the request of John D. Dingell, Chairman, Sub-Committee on Energy and Power, Committee on Interstate and Foreign Commerce, United States House of Representatives and Henry M. Jackson, Chairman, Committee on Energy and Natural Resources, Ernest F. Hollings, Vice Chairman, The National Ocean Policy Study of the Committee on Commerce, Science and Transportation, United States Senate, by the Congressional Research Service, Library of Congress, 95th Congress, 1st Session, US Government Printing Office, Washington, D.C., November 1977.

Reorientation and Commercial Relations of the Economies of Eastern Europe – A Compendium of Papers, Joint Economic Committee, Congress of the United States, 93rd Congress, 2nd Session, US Government Printing Office, Washington, D.C., August 16, 1974.

Report to the United States Senate of the Senate Delegation on Parliamentary Exchange with the Soviet Union, Senate, US Government Printing Office, Washington, D.C., 1979.

Revue des tendances, des politiques et des problèmes récents du commerce intrarégional, note du Secrétariat, Nations Unies, Conseil économique et social, Commission économique pour l'Europe, 26e Session, 28 novembre-2 décembre 1977, Genève, TRADE/R.349, 31 octobre 1977.

Revue des tendances, des politiques et des problèmes récents du commerce international, note du Secrétariat, Nations Unies, Conseil économique et social, Commission économique pour l'Europe, Comité pour le dévelopement du commerce, Genève, TRADE/R.366, 9 octobre 1978.

Saunders, C. T., (ed.), *East-West Co-operation in Business Inter-Firm Studies,* East-West Co-operation Workshop Papers, Volume 2, Springer Verlag, Wien-New York, 1977.

Saunders, C. T., *Concentration and Specialisation in Western Industrial Countries,* Workshop on East-West European Economic Interaction, Vienna Institute for Comparative Economic Studies, Baden near Vienna, Austria, 3rd-7th April, 1977.

Schmidt, Max, "East-West Economic Relations against the Background of New Trends and Developments in World Economy and in International Distribution of Activities", Paper for the IEA Round Table Conference, Dresden, DDR, June 29 to July 3, 1976.

Sdobnikov, Ju. A., *Ekonomicheskaya integraciya v usloviyakh dvukh sistem* (Economic Integration Under the Condition of Two Systems). Ed. Nauka, Moscow, 1976.

Shlaim, Avi and Yannopoulos, G. N., editors, *The EEC and Eastern Europe,* Cambridge University Press, 1978.

Smith, Glen Alden, *Soviet Foreign Trade: Organisation, Operations and Policy — 1918-1971,* Praeger Publishers, New York, Washington, London, 1973.

Soviet Commercial Operations in the West, Central Intelligence Agency, Directorate of Intelligence, ER 77-10486, September 1977.

Soviet Economic Problems and Prospects – A Study, prepared for the Sub-Committee on Priorities and Economy in Government of the Joint Economic Committee, Congress of the United States, 95th Congress, 1st Session, US Government Printing Office, Washington, D.C., August 8, 1977.

Soviet Economy in a New Perspective – A Compendium of Papers, Joint Economic Committee, Congress of the United States, 94th Congress, 2nd Session, US Government Printing Office, Washington, D.C., October 14, 1976.

Standke, Klaus-Heinrich, *Der Handel mit dem Osten, Die Wirtschaftsbeziehungen mit den Staatshandelsländern,* Nomos Verlagsgesellschaft, Baden-Baden, 1968.

Stankovsky, Jan, "Folgewirkungen der KSZE für den Ost-West Handel und Industrielle Kooperation", Sunderdruck aus J. Delbrück, N. Ropers, G. Zellentin, *Grünbuch zu den Folgewirkungen der KSZE,* Verlag Wisschenschaft und Politik, Köln, 1977.

Summary Statement of Report to the Congress, The Government's Role on East-West Trade – Problems and Issues, Comptroller General of the United States, February 4, 1976.

"The European Community and the Eastern European Countries", *Information,* Commission of the European Communities, Brussels, 163/77.

"The Outlook for 79: Where the Opportunities Lie", *Business Eastern Europe,* January 5, 1979.

The Soviet Union and the Third World: a Watershed in Great Power Policy, report to the Committee on International Relations, House of Representatives, 95th Congress, 1st Session, US Government Printing Office, Washington, D.C., 8th May, 1977.

The USSR in the 1980's: Economic Growth and the Role of Foreign Trade, Colloquium, January 17-19, 1978, Directorate of Economic Affairs, NATO, Brussels, 1978.

Tinguy, Anne de, "Les relations économiques et commerciales soviéto-américaines de 1961 à 1974", *Revue d'études comparatives Est-Ouest,* Editions du Centre national de la recherche scientifique, Paris, Volume 6, No. 4, 1975.

Tiraspolsky, Anita, "Les relations entre la CEE et le CAEM", *Le courrier des pays de l'Est,* La Documentation française et Centre d'étude prospectives et d'informations internationales, Paris, octobre 1978.

Tiraspolsky, Anita, "Le commerce extérieur entre les pays de la Communauté et les pays européens du CAEM en 1977 et 1978", *Le courrier des pays de l'Est,* La Documentation française et Centre d'études prospectives et d'informations internationales, Paris, novembre 1978.

Tiraspolsky, Anita, Despiney, B., "Les échanges des pays européens du CAEM de 1971 à 1975. Perspective 1980", *Le courrier des pays de l'Est*, La Documentation française et Centre d'études prospectives et d'informations internationales, Paris, novembre 1976.

Tomsa, Branko, "CEE-CAEM: Vers la normalisation des relations?", *Revue du Marché commun*, octobre 1977.

Trade of European Non-NATO Countries and Japan with Communist Countries 1971-1974, Department of State, Bureau of Intelligence and Research, Report No. 293, Washington, D.C., 22nd January, 1976.

"Trade of NATO Countries with Communist Countries", *Special Report No. 29*, Bureau of Public Affairs, Office of Media Services, Department of State, Washington, D.C., December 1976.

"Trade of NATO Countries with Communist Countries", *Special Report No. 38*, Bureau of Public Affairs, Office of Media Services, Department of State, Washington, D.C., December 1977.

"Ungelöste Wirtschaftsprobleme mit den sozialistischen Ländern", *Neue Zürcher Zeitung*, 16 März, 1979.

Vaganov, B. S., (Introduction to the Collective Work) *Vneshne ekonomicheskie svjazi Sovetskogo Sojuza na novom etape* (Foreign Economic Relations of the Soviet Union at a New Stage), Moscow, 1977.

Voinov, A. M., *East-West Trade in Europe and Ways of Perfecting It*, Soviet Committee for European Security, USSR Academy of Sciences, Moscow, 1973.

Voinov, A. M.; Iokhin, V. Ja.; Rodina, L. A., *Ekonomicheskie otnoshenija mezhdu socialisticheskimi i razvitymi kapitalisticheskimi stranami*, (Economic Relations Among Socialist and Developed Capitalist Countries), Moscow, 1975.

Wilczynski, J., *The Economics and Politics of East-West Trade – A Study of Trade between Developed Market Economies and Centrally Planned Economies in a Changing World*, London, 1969.

Wiles, Peter, J. D., *Communist International Economics*, Praeger Publishers, New York, 1969.

Wipf, Larry J., and Brada, Josef C., "The Impact of West European Trade Strategies on Exports to Eastern Countries", *European Economic Review*, North-Holland Publishing Co., No. 6, 1975.

Wolf, Thomas A., "East-West European Trade Relations", *East European Economies Post-Helsinki – A Compendium of Papers*, Joint Economic Committee, Congress of the United States, 95th Congress, 1st Session, August 25, 1977, US Government Printing Office, Washington, D.C., 1977.

World Economy and East-West Trade, Vienna Institute for Comparative Economic Studies, Springer Verlag, Wien-New York, 1976.

Wurster, Dirk, *Die Ausweitung der Ost-West-Handelsbeziehungen und ihre Grenzen*, Institut für Aussenhandel und Überseewirtschaft der Universität Hamburg, Forschungsbericht Nr. 5, (Az: 12 2686), n.d.

Zaleski, Eugène, *Les courants commerciaux de l'Europe danubienne au cours de la première moitié du 20ᵉ siècle*, Librairie générale de droit et de jurisprudence, Paris, 1952.

Zevin, L. Z., *Economic Co-operation of Socialist and Developing Countries: New Trends*, "Nauka" Publishing House, Central Department of Oriental Literature, Moscow, 1976.

II. SPECIFIC PROBLEMS OF EAST-WEST TRADE

(Organisation, commercial agreements,
statistics of trade, financial problems,
prices, competition, tripartite industrial cooperation.
Studies for industrial sectors.)

"A European Selling Price System for Chemicals?", *The Economist,* June 24, 1978.

Askanas, Benedykt; Askanas, Halina; Levcik, Friedrich, "East-West Trade and CMEA Indebtedness up to 1980", *International Economics — Comparisons and Interdependences,* edited by Friedrich Levcik, Festschrift für Franz Nemschak, Springer-Verlag, Wien-New York, 1978.

Becker, Abraham S., discussion of the paper by John P. Hardt and George D. Holliday, "East-West Financing by Eximbank and National Interest Criteria", in Paul Marer, (ed.), *US Financing of East-West Trade, The Political Economy of Government Credits and the National Interest,* Studies in East European and Soviet Planning, Development and Trade, International Development Research Center, Indiana University, Bloomington, Indiana, No. 22, August 1975.

"Berlin Trade Unit Issues Guide to Countertrade", *Business Eastern Europe,* Vol. 8, No. 5, February 2, 1979.

Betcher, Oleg, *La balance des paiements de l'Europe occidentale avec l'Europe orientale de 1960 à 1970,* Institut universitaire des hautes études internationales, Genève, 1976.

Brada, Josef C. and Jackson, Marvin R., *Foreign Trade Organisation Under Capitalism and Socialism,* New York University, March 1977.

Brainard Lawrence J., "East-West Financial Relations Prospects for the 1980s", in *Economic and Financial Aspects of East-West Cooperation,* Perspektiven, Reports-Analysis, Zentralsparkasse und Kommerzbank, Wien, 1979.

Brainard, Lawrence J., "Financing Soviet Capital Needs in the Eighties, *The USSR in the 1980's: Economic Growth and the Role of Foreign Trade,* Colloquium, January 17-19, 1978, Directorate of Economic Affairs, NATO Brussels, 1978.

Brainard, Lawrence J., "Eastern European Indebtedness", Conférence sur les problèmes monétaires et financiers à l'Est et à l'Ouest, Budapest, 1977.

Brainard, Lawrence J., "Soviet Foreign Trade Planning", *Soviet Economy in a New Perspective — A Compendium of Papers,* Joint Economic Committee, Congress of the United States, 94th Congress, 2nd Session, US Governement Printing Office, Washington, D.C., October 14, 1976.

Brainard, Lawrence J., "East Europe Improves its Trade", *Euromoney,* May 1978.

"Building Countertrade Into Long-Range Planning", *Business Eastern Europe,* February 11, 1977.

Burt, Richard, "US to Use Oil Expertise as Rights Prod on Russia", *International Herald Tribune,* Paris, July 20, 1978.

Calvier, Alain, "Les biens de consommation est-européens sur les marchés occidentaux", *Problémes économiques,* No. 1540, La Documentation française, Paris, 28 septembre 1977.

Carter, James Richard, *The Net Cost of Soviet Foreign Aid,* Foreword by Raymond F. Mikesell, Praeger Publishers, New York, Washington, London, 1969.

Chervyakov, P., *Organisatsiya i tekhnika vneshney torgovli SSSR* (Organisation and Technique of Soviet Foreign Trade), Moscow, 1958.

Cheval, Jean; Ceze, François; Gutman Patrick, *Les accords de coopération industrielle Est-Ouest,* Rapport au Colloque de la DGRST à Saint Maximin, 19-20 octobre 1978, BIPE-EHESS, Paris, 1978.

"Comecon Borrowing on International Credit Markets", *"Financial Market Trends,* No. 2, OECD, Paris, December 1977.

"Comecon Imports under Scrutiny", *Financial Times,* February 27, 1978.

"Counterpurchase Pressure to Continue in Yugoslavia", *Business Eastern Europe,* March 4, 1977.

"Counterpurchases Trouble West German Machine Builders", *Business Eastern Europe,* July 29, 1977.

Cruse, James C., and Wigg, David G., "The Role of Eximbank in US Exports and East-West Trade", in Paul Marer, (ed.), *US Financing of East-West Trade, The Political Economy of Government Credits and the National Interest,* Studies in East European and Soviet Planning, Development and Trade, International Development Research Center, Indiana University, Bloomington, Indiana, No. 22, August 1975.

"Current Countertrade Policies and Practices in East-West Trade", *Business International*, Geneva, 1976.

Dezsenyi-Gueullette, Agota, "Les calculs d'efficacité du commerce extérieur en Hongrie avant la réforme économique de 1968", *Revue d'études comparatives Est-Ouest*, Editions du Centre national de la recherche scientifique, No. 2, Paris, 1977.

"Discount Rates for Sale of Eastern Production in the West", *Eastern Europe Reports*, No. 19, 1976.

Dohan, Michael; Hewett, Edward, *Two Studies in Soviet Terms of Trade, 1918-1970*, International. Development and Research Center, Indiana University, Bloomington, Indiana, November, 1973.

"Eastern Europe Buy-Back Pacts Secure Larger Trade in West", *International Herald Tribune*, Part I, Paris, December 1977.

East-West Countertrade Practices, An Introduction Guide for Business by Pumpiliu Verzariu; Scott Bozek; Jenelle Matheson, US Department of Commerce, Industry and Trade Administration, US Government Printing Office, Washington, D.C., 1978.

Economic and Financial Aspects of East-West Cooperation, Perspektiven, Reports-Analysis, Zentral-sparkasse und Kommerzbank, Wien, 1979.

"EEC to Offer Trade Pacts to Rumania, Yugoslavia", *International Herald Tribune*, Paris, February 7, 1979.

Ericson, Paul, "Soviet Efforts to Increase Exports of Manufactured Products to the West", *Soviet Economy in a New Perspective — A Compendium of Papers*, Joint Economic Committee, Congress of the United States, 94th Congress, 2nd Session, US Government Printing Office, Washington, D.C., 14th October, 1976.

Expériences de coopération industrielle dans des systèmes socio-économiques différents, Université de Paris-Dauphine, UER Science des organisations, Centre de recherche DMTP, Cahier No. 58, année 1978, (présentation Sylvain Wickham).

"Financing Deals with the Soviet Union", *Business Eastern Europe*, July 28, 1978.

Finanzierungs und Währungsprobleme des West-Wirtschaftsverkehrs, Carl Heymanns Verlag KG, Köln, 1979.

Gèze, François; Cheval, Jean; Gutman Patrick; Finkelstein, Janet, *Le rôle des pays de l'Est dans la division internationale du travail*, Tomes I et II, BIPE et EHESS, Paris, April 1979.

Gutman, Patrick; Arkwright, Francis, "Multinationalisation et pays de l'Est", *Politique étrangère*, Nos. 4-5, Centre d'études de politique étrangère, Paris, 1974.

Gutman, Patrick; Arkwright, Francis, "La coopération industrielle tripartite entre les pays à systèmes économiques et sociaux différents, de l'Ouest, de l'Est et du Sud", *Politique étrangère*, No. 6, Centre d'études de politique étrangère, Paris, 1975.

Gutman, Patrick; Arkwright, Francis, "Coopération industrielle tripartite Est-Ouest-Sud; Evaluation financière et analyse des modalités de paiement et de financement", *Politique étrangère*, No. 5, Centre d'études de politique étrangère, Paris, 1976,

Gutman, Patrick; Romer, Jean Christophe, "Coopération industrielle tripartite Est-Ouest-Sud et dynamique des systèmes", rapport présenté aux journées d'études sur les transferts triangulaires de technologie Est-Ouest-Sud, Centre de recherche sur l'URSS et les pays de l'Est, Université de Strasbourg III, 8 décembre 1978,

Hardt, John P. and Holliday, George D., "East-West Financing by Eximbank and National Interest Criteria", in Paul Marer, (ed.), *US Financing of East-West Trade, The Political Economy of Government Credits and the National Interest*, Studies in East European and Soviet Planning, Development and Trade, International Development Research Center, Indiana University, Bloomington, Indiana, No. 22, August 1975.

Holt, John P., "Strategies of US Pharmaceutical Companies in Eastern Europe", *The ACES Bulletin*, Summer 1977.

"How a Swedish Company Handles Counterpurchase Deals", *Business Eastern Europe*, February 25, 1977.

"How Daimler Benz Handles Countertrade Commitments", *Business Eastern Europe*, January 14, 1977.

"Incentives and Problems in Tripartite Co-operation", *Business Eastern Europe*, April 20, 1979.

"International Debt: Current Issues and Implications", Bureau of Public Affairs, Office of Media Services, Department of State, Washington, D.C., August 1977.

411

Klümper, Bernhard, *Finanzierungsprobleme in Ost-West Handel,* Forschungsbericht No. 8, Institut für Aussenhandel und Überseewirtschaft der Universität Hamburg, Hamburg, 1976.

Klümper, Bernhard, *Der Euro-Dollarmarkt als Finanzierungsquelle des Ost-West Handels,* Forschungsbericht No. 2, n. d., Institut für Aussenhandel und Überseewirtschaft, Universität Hamburg,

Kolarik, William F., Jr., "Statistical Abstract of East-West Trade and Finance", *Issues in East-West Commercial Relations — A Compendium of Papers,* submitted to the Joint Economic Committee, 95th Congress, 2nd Session, US Government Printing Office, Washington, D.C., January 12, 1979.

Kostinsky, Barry L., *Description and Analysis of Soviet Foreign Trade Statistics,* US Department of Commerce, Washington, D.C., July 1974.

Kwiecinski, M. Christopher, "Should Eximbank Finance East-West Trade: A Summary of Issues and the National Debate", in Paul Marer, (ed.), *US Financing of East-West Trade, The Political Economy of Government Credits and the National Interest,* Studies in East European and Soviet Planning, Development and Trade, International Development Research Center, Indiana University, Bloomington, Indiana, No. 22, August 1975.

Lemoine, Françoise, "L'endettement des pays de l'Est en devises convertibles", *Le courrier des pays de l'Est,* La Documentation française and Centre d'études prospectives et d'informations internationales, Paris, octobre 1978.

L'endettement des pays de l'Est en devises convertibles — Problèmes d'évaluation, Centre d'études prospectives et d'informations internationales, Paris, septembre 1978.

Lenz, Allen J., "Potential 1980 and 1985 Hard Currency Debt in Perspective", *Issues in East-West Commercial Relations — A Compendium of Papers,* submitted to the Joint Economic Committee, 95th Congress, 2nd Session, US Government Printing Office, Washington, D.C., January 12, 1979.

Lenz, Allen J., and Theriot, Lawrence H., "The Potential Role of Eximbank Credits and Financing US-Soviet Trade", in *Issues in East-West Commercial Relations — A Compendium of Papers,* submitted to the Joint Economic Committee, Congress of the United States, 95th Congress, 2nd Session, US Government Printing Office, Washington, D.C., January 12, 1979.

"Les pratiques de dumping des pavillons de l'Est", *l'Echo de la Bourse,* Bruxelles, 30 août 1977; given in *Problèmes économiques,* La Documentation française, Paris, 14 décembre 1977.

Levcik, Friedrich, *Ostverschuldung und Ost-West Wirtschafisbeziehungen,* Wiener Institut für Internationale Wirtschaftsvergleiche, No. 27, May 1977, reproduced from West-Ost Journal. Donaueuropaisches Institut, No. 2, Wien, 1977.

Levcik, Friedrich; Stankovsky, Jan, *Kredite des Westens und Österreich an Osteuropa und die UdSSR,* reproduced from the Monatsbericht des Österreichischen Institut für Internationale Wirtschaftvergleiche, 28th June, 1977.

Lotarski, Susanne S., "Institutional Developments and the Joint Commissions in East-West Commercial Relations" *East European Economies Post-Helsinki — A Compendium of Papers,* Joint Economic Committee, Congress of the United States, 95th Congress, 1st Session, August 25, 1977, US Government Printing Office, Washington, D.C., 1977.

Marer, Paul, (ed.), *US Financing of East-West Trade, The Political Economy of Government Credits and the National Interest,* Studies in East European and Soviet Planning, Development and Trade, International Development Research Center, Indiana University, Bloomington, Indiana, No. 22, August 1975.

Masnata, Albert, "Les échanges commerciaux avec l'Est. Dans quelle mesure le financement disproportionné des échanges Est-Ouest contribue-t-il à leur expansion?, *Revue de l'OTAN,* août 1977.

Matheson, Jenelle; McCarthy, Paul; Flanders, Steven, "Countertrade Practices in Eastern Europe", *East European Economies Post-Helsinki — A Compendium of Papers,* Joint Economic Committee, Congress of the United States, 95th Congress, 1st Session, US Government Printing Office, August 1977, Washington, D.C., 1977.

Möglichkeiten einer Kooperation Ost und Westeuropas auf dem Verkehrssektor, Stiftung Wissenschaft und Politik, Ebenhausen, Januar 1975.

Murphy, John, "Plastic Makers Feel the Heat", *New Scientist,* March 8, 1979.

Nagy, Ference, "Hungary and Long-Term CMEA Co-operation", *US Joint Publications Research Service,* No. 729000, US Government Printing Office, Washington, D.C., March 1, 1979.

New Issues Affecting the Energy Economy of the ECE Region In the Medium and Long Term (Preliminary Version), Economic Commission for Europe, United Nations, Geneva, ECE (XXXIII)/2, 18 January, 1978.

412

Nove, Alec; Matko, Dubravko, "L'augmentation du déficit commercial de l'URSS avec les pays occidentaux", *Problèmes économiques,* La Documentation française, Paris, 1ᵉʳ septembre 1976.

Observation on East-West Economic Relations: USSR and Poland, Joint Economic Committee, Congress of the United States, 93rd Congress, 1st Session, US Government Printing Office, Washington, D.C., 16th February, 1973.

Pertuiset, Florence, "La structure et les mécanismes des crédits occidentaux dans les échanges Est-Ouest", *Le courrier des pays de l'Est,* La Documentation française and Centre d'études prospectives et d'informations internationales, Paris, janvier 1979.

Pfeffer, Joachim, "Zunehmende Verschuldung Osteuropas gegenüber dem Westen", *Neue Zürcher Zeitung,* 16 August, 1977.

Poirier d'Ange d'Orsay, Philippe, "Pour notre flotte commerciale une menace venue de l'Est", *Le Figaro,* Paris, 10-11 avril 1976.

"Poland Limits Investments for Self-Repaying Credits", *Business Eastern Europe,* Vol. 6, No. 29, July 22, 1977.

"Poland to Update FT Code: Co-operation No Panacea", *Business Eastern Europe,* Vol. 8, No. 5, February 2, 1979.

Portes, Richard, "East Europe's Debt to the West: Interdependence is a Two-way Street", *Foreign Affairs,* July 1977.

Practical Measures to Remove Obstactles to Intra-Regional Trade and to Promote and Diversify Trade; Long-term Agreements on Economic Co-operation and Trade, Note by the Secretariat, United Nations Economic and Social Council, Economic Commission for Europe, Commitee on the Development of Trade, 26th Session, 28 November-2 December, 1977, TRADE/R/351, Geneva, 18 October, 1977.

Promotion of Trade Through Industrial Co-operation. Review of the Development of Banking and Financial Institutions in East-West Trade, Note by the Secretariat, United Nations Economic and Social Council, Economic Commission for Europe, Committee on the Development of Trade, 26th Session, 28 November-2 December, 1977, Geneva, TRADE/R/355, Add. 1, 18 October, 1977.

Rominskyj, Alexandre, "Experiences of French Businessmen in East-West Trade", *The ACES Bulletin,* Association for Comparative Economic Studies, Bloomington, Indiana, Summer 1977.

"Russia's Merchant Fleet", *The Economist,* London, given in *Problèmes économiques,* La Documentation française, Paris, 14 décembre 1978.

Schoppe, Siegried G., *Funktionwandlungen des Goldes in Rahmen der Sowjetischen Aussenwirtschaft seit 1945/1946,* Forschungsbericht, No. 10, Institut für Aussenhandel und Überseewirschaft der Universität Hamburg, 1977.

"Sicherung der Unternehmensautonomie. Warnung Olivier Longs vor protektionistischen Welthandelspraktiken", *Neue Zürcher Zeitung,* 12 May, 1978.

Sokoloff, Georges, "L'URSS mise sur la stabilité de ses partenaires occidentaux", *Le Figaro,* Paris 9-10 septembre 1978.

Stevenson, Adlai E., III, "Views on Eximbank Credits to the USSR", in Paul Marer, (ed.), *US Financing of East-West Trade, The Political Economy of Government Credits and the National Interest,* Studies in East European and Soviet Planning, Development and Trade, International Development Research Center, Indiana University, Bloomington, Indiana, No. 22, August 1975.

Stolz, Peter, "Machines contre tomates — Les échanges avec les pays de l'Est", Revue hebdomadaire de presse, *Economies des pays communistes,* NATO, 14 March, 1978, reproduced from *Kölner Anzeiger,* 18 February, 1978.

Stoupnitzky, A., *Statut international de l'URSS, Etat commerçant,* Librairie générale de droit et de jurisprudence, Paris, 1952.

Taylor, Karen C., "Import Protection and East-West Trade: A Survey of Industrialised Country Practices", *East European Economies, Post-Helsinki – A Compendium of Papers,* submitted to the Joint Economic Committee, Congress of the United States, 95th Congress, 1st Session, August 25, 1977, US Government Printing Office, Washington, D.C., 1977.

Taylor, Karen, C., "A Summary of US Laws Applying to Imports of Communist Products", *Issues in East-West Commercial Relations – A Compendium of Papers,* submitted to the Joint Economic Committee, 95th Congress, 2nd Session, US Government Printing Office, Washington, D.C., January 12, 1979.

Theriot, Lawrence H., "Communist Country Hard Currency Debt in Perspective", *Issues in East-West Commercial Relations – A Compendium of Papers,* submitted to the Joint Economic Committee, 95th Congress, 2nd Session, US Government Printing Office, Washington, D.C., January 12, 1979.

413

Tiraspolsky, Anita, "Les termes de l'échange des pays de l'Est de 1970 à 1977", *Le courrier des pays de l'Est*, La Documentation française and Centre d'études prospectives et d'informations internationales, Paris, May 1978.

Trade Patterns of the West 1973, Department of State, Bureau of Intelligence and Research, Washington, D.C., November 18th, 1974.

"Trade Patterns of the West, 1976", Bureau of Public Affairs, Office of Media Services, *Special Report No. 35,* Department of State, Washington, D.C., August 1977.

Treaties and Other International Acts, Series 7910, Economic, Industrial and Technical Co-operation, Agreement between the United States of America and the Union of Soviet Socialist Republics.

Trzeciakowski, Witold, "Evolution de la planification et de la gestion du commerce extérieur en Pologne", *Revue de l'Est,* Vol. 2, No. 1, Centre national de la recherche scientifique, Paris, 1971.

"US Trade with the European Community, 1958-1976", *Special Report No. 32,* Bureau of Public Affairs, Office of Media Services, Department of State, Washington, D.C., April 1977.

Wajs, Vincent, *Politique d'exportation des entreprises françaises vers les pays de l'Est,* Résultats d'une enquête, Etude No. 2, 1976, Centre d'étude sur la coopération économique avec les pays de l'Est, Faculté d'économie appliquée d'Aix-Marseille, Aix-en-Provence, 1976.

Welcker, Calsu (interview), "Counterpurchases Trouble West German Machine Builders", *Business Eastern Europe,* Vol. 6, No. 30, July 29, 1977.

"Western Banks Not Worried About East Bloc Borrowing", *International Herald Tribune,* Part I, Paris, December 1977.

Weygand, Robert E. "L'art du troc international", *Harvard Business Review,* November-December, 1977, "Trade Without Money", reproduced from *Problèmes économiques,* La Documentation française, Paris, 14 June, 1978.

Wilczynski, Josef, *Joint Ventures and Right of Ownership,* Institute of Soviet and East European Studies, Carleton University, Ottawa, Canada, Working Paper No. 6, 1975.

Wilczynski, Josef, *The Multinationals in East-West Relations,* Macmillan, London, 1976.

Wilczynski, Josef, *Comparative Monetary Economics. Capitalist and Socialist Monetary Systems and their Interrelations in the Changing Scene,* The Macmillan Press Limited, London and Basingstoke, 1978.

Wolf, Thomas A., "East-West Trade Credit Policy: A Comparative Analysis", in Paul Marer, editor, *US Financing of East-West Trade, The Political Economy of Government Credits and the National Interest,* Studies in East European and Soviet Planning, Development and Trade, International Development Research Center, Indiana University, Bloomington, Indiana, No. 22, August, 1975.

Woroniak, Alexander, "Le problème de la conversion du rouble en dollar", *Revue de l'Est,* Vol. 5, No. 1, Centre national de la recherche scientifique, Paris, janvier 1974.

Zakharov, S. N., *Raschety effektivnosti vneshne ekonomicheshkikh svajazei* (Efficiency Calculations of Foreign Economic Relations), Moscow, 1975.

Zoeter, Joan Parpart, "Eastern Europe: The Growing Hard Currency Debt", *East European Economies Post-Helsinki—A Compendium of Papers,* Joint Economic Committee, Congress of the United States, 95th Congress, 1st Session, US Government Printing Office, Washington, D.C., August 25, 1977.

Zurawicki, Leon, *Multinational Enterprises in the West and East,* Sijthoff and Noordhoff, Alphen aan den Rijn, The Netherlands, and Germantown, Maryland, United States, 1979.

III. TECHNOLOGY TRANSFER BETWEEN EAST AND WEST

Adahl, Andreas; Perlowski, Adam, *Huvudlinjerna i licenshandeln med Sovjetunionen och övriga öststater*, Styrelsen för Teknisk Utveckling Informationssektion, STU-utredning Nr. 56-1976, Stockholm, 1978.

Altmann, Franz-Lothar; Clement, Hermann, *Die Kompensation als Instrument im Ost-West-Handel*, G. Olzog Verlag, München, 1979.

Amann, Ronald; Slama, Jiri; Vogel, Heinrich, *Der Handel mit chemischen Erzeugnissen zwischen Ost und West — eine Analyse der Kilogrammpreise 1960-1972*, Osteuropa Institut München, Nr. 5, Dezember 1975.

Analytical Report on Industrial Co-operation Among ECE Countries, Economic Commission for Europe, United Nations, Geneva, 1973.

Ancker-Johnson, Betsy and Chang, David B., *US Technology Policy — A Draft Study*, US Department of Commerce, National Technical Information Service, Washington, D.C., PB-263806, March 1977.

Aperçu statistique de l'évolution récente de la coopération industrielle, Note by the Secretariat, United Nations Economic and Social Council, Economic Commission for Europe, Committee on the Development of Trade, 26 th Session, 28 November-2 December 1977, Geneva, 14 November, 1977, TRADE/R Add. 2.

Background Materials on US-USSR Co-operative Agreements in Science and Technology, Report prepared for the Sub-Committee on Domestic and International Scientific Planning and Analysis of the Committee on Science and Technology, US House of Representatives, 94th Congress, 1st Session, Science Policy Research Division, Congressional Research Service, Library of Congress, Serial P, US Government Printing Office, Washington, D.C., November 1975.

Baranson, Jack, *International Transfers of Industrial Technology by US Firms and their Implication for the US Economy*, report for the Office of Foreign Economic Research, International Labor Affairs Bureau, US Department of Labor, Washington, D.C., 24th September, 1976.

Beitel, Werner; Nötzold, J., "Les relations économiques entre l'Allemagne et l'URSS au cours de la période 1918-1932 considérées sous l'angle des transferts de technologie", *Revue d'études comparatives Est-Ouest*, Editions du Centre national de la recherche scientifique, Paris, Vol. III, n° 2, juin 1977.

Berliner, Joseph S., "Some International Aspects of Soviet Technological Progress", *The South Atlantic Quarterly*, Summer 1973.

Bolz, Klaus; Plotz, Peter, *Kooperationserfahrungen der Bundesrepublik Deutschland mit den Sozialistischen Ländern Osteuropas*, HWWA Institut Für Wirtschaft Forschung, Hamburg, October 1973.

Boucheny, M., *La coopération scientifique et technique*, rapport présenté au Conseil de l'Europe concernant la mise en œuvre de l'Acte final de la Conférence sur la sécurité et la coopération en Europe, Conseil de l'Europe, Doc. 3954, 20 avril 1977.

Brada, Josef C., "Markets, Property Rights, and the Economics of Joint Ventures in Socialist Countries", *Journal of Comparative Economics 1*, Association for Comparative Economic Studies, Academic Press, New York and London, 1977.

Brzost, W., "Wspolpraca naukowo techniczna" (Scientific and Technical Co-operation), Handel Zagraniczny, No. 8, Izba Handlu Zagranicznego, Warszava, Poland, 1977.

Burgess, Jay A., "An Analysis of the United States-Rumanian Long-Term Agreement on Economic, Industrial, and Technical Co-operation", *East European Economies Post-Helsinki — A Compendium of Papers*, Joint Economic Committee, Congress of the United States, 95th Congress, 1st Session, US Government Printing Office, Washington, D.C., August 25, 1977.

Bykov, Aleksandr N., *Perspectives in East-West Relations in Technology Transfer and Related Problems of Dependence*, Workshop on East-West European Economic Interaction, Vienna Institute for Comparative Economic Studies, Baden near Vienna, Austria, 3rd-7th April, 1977.

Campbell, Robert, *Technology Transfer in the Energy Sector East and West — Global Issues*, Workshop on East-West European Economic Interaction, Vienna Institute for Comparative Economic Studies, Baden near Vienna, Austria, 3rd-7th April, 1977.

Campbell, Robert W., and Marer, Paul (ed.), *East-West Trade and Technology Transfer — An Agenda of Research Needs*, Proceedings of a Conference, Studies in East European and Soviet Planning, Development and Trade, International Development Research Center, Indiana University, Bloomington, Indiana, 1975.

415

Carrick, R. J., *East-West Technology Transfer in Perspective*, Institute of International Studies, University of California, Berkeley, 1978.

Carter, Luther J., "United States and Bulgaria to Co-operate in Research", *Science*, Vol. 199, No. 4333, March 10, 1978.

"Chemical Buy-Back: Where Will it All End?", *Business Eastern Europe*, Vol. 7, No. 23, June 9, 1978.

"Chemicals in the East Explode West", *The Economist*, February 10, 1979.

Clawson, Robert W., and Kolarik, William F., Jr., "Trade Technology, and Soviet R & D: US Corporate Executives' View", A Paper presented at the 1976 AAASS Annual Conference Panel, "The State of Soviet Research and Development", Kent State University, Kent, Ohio, October 8, 1976.

Commercialisation des licences et crédit-bail, Comittee on Development of Trade, United Nations Economic and Social Council, Geneva, 1976.

Czege von, Andreas Wass, *Die Planungs und Entscheidungsstrukturen in der Importorganisation Ungarns und der UdSSR als Einflussfaktor auf die Wahl der Transfermechanismen beim Import Westlicher Technologien*, Institut für Aussenhandel und Überseewirtschaft der Universität Hamburg, Forschungsbericht Nr. 9, (Az: 14 0285), Hamburg, 1977.

Czege von, Andreas Wass, *Mechanismen zum Intersystemaren Technologietransfer – Ihre Klassifizierung und Unterschiedliche Bewertung in Ost und West*, Institut für Aussenhandel und Überseewirtschaft der Universität Hamburg, Forschungsbericht Nr. 11, (Az: 14 0285), Hamburg, 1977.

Czege von, Andreas Wass, *Wirkungen der Intersystemaren Zusammenarbeit auf vor und Nachgelagerte Industriezweige: Die Akzelerationsthese*, Forschungsbericht Nr. 17, Institut für Aussenhandel und Überseewirtschaft der Universität Hamburg, Hamburg, 1978.

Czege von, Andreas Wass, *Strukturelle Folgewirkungen der Ost-West-Kooperation – Dargestellt am Beispiel des Ungarischen Werkzeugmachinenbaus*, Forschungsbericht Nr. 16, Institut für Aussenhandel und Überseewirtschaft der Universität Hamburg, Hamburg, 1978.

De Haven, James C., *Technology Exchanges: Import Possibilities from the USSR*, Rand Corp., Santa Monica, California, R-1414 ARPA, April 1974.

Demidov, V., *Sous licence soviétique*, Ed. Novosti, Moscow, 1976.

Die Bedeutung des Technologietransfers in der Wirtschaftlichen Ost-West Kooperation, Stiftung Wissenschaft und Politik, Ebenhausen, Februar 1974.

Donaghue, Hugh P., *A Unique Manufacturing Joint-Venture in the Computer Industry*, Economic Commission for Europe, Senior Adviser to ECE Governments on Science and Technology, Seminar on the Management of the Transfer of Technology within Industrial Co-operation, 14-17 July, 1977, Geneva, 6 July, 1977.

Dunajewski, Henri; Brami, Michèle; Double, Gérard, *Etude des transfers de la technologie chimique. Contrats France-URSS*, Enquête France-Europe de l'Est, Centre d'études sur la coopération économique avec les pays de l'Est, Faculté d'économie appliquée d'Aix-Marseille, Université de droit, d'économie et des sciences sociales d'Aix-Marseille, décembre 1978.

East-West Technological Co-operation, Main findings of Colloquium held 17-19th March, 1976, in Brussels, NATO, Directorate of Economic Affairs, NATO Information Service, Brussels, 1976.

Engert, Manfred; Reich, Manfred, *Concentration, Specialisation and Co-operation in the CMEA Region*, Workshop on East-West European Economic Interaction, Vienna Institute for Comparative Economic Studies, Baden near Vienna, Austria, 3rd-7th April, 1977.

Fabinc, Ivo, *The International Features of the Transfer of Technology in Yugoslavia*, Workshop on East-West Economic Interaction, Vienna Institute for Comparative Economic Studies, Baden near Vienna, Austria, 3rd-7th April, 1977.

Finer, Harlan, S.; Gobstein, Howard; Holliday, George D., "KamAZ: US Technology Transfer to the Soviet Union", *Technology Transfer and US Foreign Policy*, Praeger Publishers Inc., New York, 1976.

Gaps in Technology – General Report, Organisation for Economic Co-operation and Development, Paris, 1968.

Geze, François, "La coopération Est-Ouest dans l'industrie électronique", *Le courrier des pays de l'Est*, La Documentation française et Centre d'études prospectives et d'informations internationales, Paris, juin 1979.

Goldman, Marshall I., "US Policies on Technology Sales to the USSR", *East-West Technological Co-operation, Colloquium 1976*, NATO, Directorate of Economic Affairs, Brussels, March 17-19, 1976.

Gomulka, Stanislaw, "Investment Imports, Technical Change and Economic Growth: Poland 1971-1980", paper, London School of Economics, n.d.

Gomulka, Stanislaw, "Growth and the Import of Technology: Poland 1971-1980", *Cambridge Journal of Economics*, No. 2, 1978.

Gorodisski, J. M. L., *Licencii po vneshnej torgovle SSSR* (Licences in Soviet Foreign Trade), Moscow, 1972.

Graham, Loren, "How Valuable are Scientific Exchanges with the Soviet Union?", *Science*, Vol. 202, No. 4366, October 27, 1978.

Grands problèmes découlant du transfert des techniques aux pays en voie de développement – Hongrie (manuscript), Study carried out by the Hungarian Chamber of Commerce and the Institute of World Economic Sciences of the Hungarian Academy of Sciences, United Nations' Conference on Trade and Development, Genève, TD/B/AC.II/18, 15 mai 1974.

Green, Donald W. and Levine, Herbert S., "Implications of Technology Transfer for the USSR", *East-West Technological Co-operation, Colloquium 1976*, NATO, Directorate of Economic Affaires, Brussels, March 17-19, 1976.

Hanson, Philip, *Forms and Dimensions of Technology Transfer between East-West*, Workshop on East-West European Economic Interaction, Vienna Institute for Comparative Economic Studies, Baden near Vienna, Austria, 3-7 April, 1977.

Hanson, Philip, "International Technology Transfer from the West to the USSR", *Soviet Economy in a New Perspective – A Compendium of Papers*, Joint Economic Committee, Congress of the United States, 94th Congress, 2nd Session, US Government Printing Office, Washington, D.C., October 14, 1976.

Hanson, Philip, *The Impact of Western Technology: A Case Study of the Soviet Mineral Fertiliser Industry* (revised draft), Conference on Integration in Eastern Europe and East-West Trade, Bloomington, Indiana, 28-31 October, 1976.

Hanson, Philip, "External Influences on the Soviet Economy since the mid-1950's: The Impact of Western Technology", CREES discussion paper, No. 7, Birmingham, n.d.

Hardt, John P.; Holliday, George D.; Kim, Young, C., *Western Investment in Communist Economies, A Selected Survey on Economic Interdependence*, prepared for the Sub-Committee on Multinational Corporations of the Committee on Foreign Relations, United States Senate, 93rd Congress, 2nd Session, US Government Printing Office, Washington, D.C., August 5, 1974.

Hayden, Eric W., "Transferring Technology to the Soviet Bloc: US Corporate Experience", *Research Management*, September 1976.

Hayden Eric W., *The Transfer of Industrial Technology to East Europe: A Study of US Corporate Experience*, Johns Hopkins University, 1975 (dissertation).

Hayden, Eric W. and Nau, Henry R., "East-West Technology Transfer: Theoretical Models and Practical Experiences", *Columbia Journal of World Business*, Fall 1975.

Högberg, B.; Adahl, A., *East-West Industrial Co-operation: A Study of Swedish Firms*, Swedish Board for Technical Development (forthcoming).

Holliday, George D., *Technology Transfer to the USSR, 1928-1937, and 1966-1975: The Role of Western Technology in Soviet Economic Development*, Westview Press, Boulder, Colorado, 1979.

Huntington, Samuel P.; Holzman, Franklyn; Portes, Richard; Kiser, John W.; Mountain, Maurice J.; Klitgaard, Robert E., "Trade Technology and Leverage", *Foreign Policy*, Fall 1978.

Impact of US Commercial Interests on Technology Transfer to the USSR, External Research Study, Department of State, Washington, D.C., 6 August, 1976.

International Transfer of Technology, Report of the President to the Congress together with Assessment of the Report by the Congressional Research Service, Library of Congress, prepared for the Sub-Committee on International Security and Scientific Affairs of the Committee on International Relations, US House of Representatives, 95th Congress, 2nd Session, December 1978, US Government Printing Office, Washington, D.C., 1979.

"Kehrseiten des Technologietransfers: Klagen deutscher Osthandelsfirmen", *Neue Zürcher Zeitung*, 1-2 January, 1978.

Kiser, John W. III, *Report on the Potential for Technology Transfer from the Soviet Union to the United States*, Prepared for the Department of State and the National Science Foundation, Contract No. 1722-620217, July 1977.

Kiser, John W. III, "Technology is Not a One Way Street", *Foreign Policy*, No. 23, Summer 1976.

Kiser, John W. III, *Commercial Technology Transfer from Eastern Europe to the United States and Western Europe*, prepared for the US Department of State, January 1980, Washington, D.C., 1980.

417

Klages, Robert D., *Transfer of Technology and East-West Co-operative Venture*, n.d. (report of the Vice-President and Central Council, Sperry Univer Computer Systems).

Klümper, Bernhard and Leise, Norbert, *Lizensgeschäft mit Staatshandelsländern-Bestimmungsgründe der Verhandlungsposition und der Vertragsgestaltung*, Institut für Aussenhandel und Überseewirtschaft der Universität Hamburg, Forschungsbericht Nr. 6, (AZ: 12 2686) Hamburg, 1976.

Kravalis, Hedija; Lenz, Allen J.; Raffel, Helen and Young, John, "Quantification of Western Exports of High Technology Products to Communist Countries", *Issues in East-West Commercial Relations – A Compendium of Papers*, Submitted to the Joint Economic Committee, Congress of the United States, 95th Congress, 2nd Session, US Government Printing Office, Washington, D.C., January 12, 1979.

Kretschmar, Robert S.; Foor, Robin, Jr., *The Potential for Joint Ventures in Eastern Europe*, Praeger Publishers, New York, 1972.

Leise, Norbert, *Transfermechanismen des technischen Fortschritts in und zwischen verschiedenen Wirtschafissystemen*, Institut für Aussenhandel und Überseewirtschaft der Universität Hamburg, Forschungsbericht Nr. 4, (Az: 12 2686), n.d.

Leise, Norbert, *Die Industrielle Ost-West Kooperation*, Forschungsbericht Nr. 7, Institut für Aussenhandel und Überseewirtschaft der Universität Hamburg, Hamburg, 1976.

"Les ventes d'usines clefs en main: les risques d'un essor dans la confusion", *La Croix*, Paris, 10 février 1978.

Levine, Herbert S.; Movit, Charles H.; Earle, Mark M.; Liebermann, Anne R., *Transfer of US Technology to the Soviet Union: Impact on US Commercial Interests*, Stanford Research Institute, February 1976.

Loeber, Dietrich A., "Capital Investment in Soviet Enterprises? Possibilities and Limits of East-West Trade", *Adelaide Law Review*, September 1978.

Marer, Paul, *Western Multinational Corporations in Eastern Europe and CMEA Integration*, International Conference on "The Choice of Partners in East-West Economic Relations", Montebello, Quebec, April 26-29, 1978.

Marer, Paul; Miller, Joseph C., "US Participation in East-West Industrial Co-operation Agreements", *Journal of International Business Studies*, Fall-Winter 1977.

McMenamin, Robert J., "Western Technology and the Soviets in the 1980s", paper presented at the NATO Colloquium, *The USSR in the 1980s: Economic Growth and the Role of Foreign Trade*, Directorate of Economic Affairs, NATO, Brussels, January 17-19, 1978.

McMillan, Carl H., "Forms and Dimensions of East-West Interfirm Co-operation", *East-West Co-operation in Business: Inter-firm Studies*, edited by C.T. Saunders, the Vienna Institute for Comparative Economic Studies, Springer-Verlag, Wien-New York, 1977.

McMillan, Carl H., *Direct Soviet and East European Investment in the Industrialised Western Economies*, Institute of Soviet and East European Studies, Carleton University, Ottawa, Canada, Working Paper No. 7, February 1977.

McMillan, Carl H., "East-West Industrial Co-operation", *East European Economies Post-Helsinki – A Compendium of Papers*, Joint Economic Committee, Congress of the United States, 95th Congress, 1st Session, US Government Printing Office, Washington, D.C., August 25, 1977.

McMillan, Carl H.; Charles, D.P., *Joint Ventures in Eastern Europe: A Three Country Comparison*, Canadian Economic Policy Committee, C.D. Howe Research Institute, Canada, March 1974.

Monkiewicz, Jan, *Closing the Technological Gap: The Experience of the Socialist Countries*, Wiener Institut für Internationale Wirtschaftsvergleiche, Forschungsbericht No. 39, Wien, April 1977.

Nau, Henry R., *Technology Transfer and US Foreign Policy*, Praeger Special Studies, New York, September 1976.

Nordmann-Zimmermann, Ursula, *Le régime des inventions dans la coopération Est-Ouest*, Institut universitaire des hautes études internationales, Genève, 1977.

Norris, William C., "High Technology Trade with the Communists", *Datamation*, January 1978.

Nötzold, Jürgen, "Wirtschaftwachstum und technischer Fortschritt in der Ostlichen Welt und ihre Auswirkungen auf die Ost-West-Kooperation", extract from *Moderne Welt Jahrbuch für Ost-West-Fragen*, ("Elements des Wandels in der östlichen Welt"), Markus Verlag, Köln, 1976.

Nötzold, Jürgen, "Probleme des Transfers technischen Fortschritts in die RGW Länder", *Osteuropa*, Deutsche Gesellschaft für Osteuropakunde, Deutsche Verlags-Anstalt, Stuttgart, December 1974.

Nötzold, Jürgen; Slama, Jiri, "Der Transfer von Technologie zwischen den Beteiligten Volkswirtschaften", *Osteuropa Wirtschaft*, Deutsche Gesellschaft für Osteuropakunde, Deutsche Verlags-Anstalt, Stuttgart, March 1975.

Nye, Jr.; Joseph S., "US Foreign Policy and Technology", *Les nouvelles,* September 1978.

Oppenlander, Karl Heinrich, *The Role of Business and Government in Promotion of Innovation and Transfer of Technology – Experience in the Federal Republic of Germany,* Workshop on East-West European Economic Interaction, Vienna Institute for Comparative Economic Studies, Baden near Vienna, Austria, 3rd-7th April, 1977.

Otten, Alan L., "Selling Know-How", *Wall Street Journal,* New York, September 15, 1977.

Parrot, Bruce, *Soviet Technological Progress and Western Technology Transfer to the USSR: An Analysis of Soviet Attitudes,* paper prepared for the Office of External Research, Bureau of Intelligence and Research, US Department of State, Washington, D.C., July 1978.

Petrov, D., "Export of Soviet-Made Industrial Equipment", *Foreign Trade,* No. 1, USSR Ministry of Foreign Trade, Moscow, (Foreign translation), 1979.

Podolski, T. M., "Evolution of East-West Technological Transfer Channels", *East-West Technological Co-operation,* Colloquium 1976, Directorate of Economic Affairs, NATO, Brussels, March 17-19, 1976.

Pompiliu, Verzariu, Jr. and Burgess, Jay A., "The Development of Joint Economic and Industrial Co-operation in East-West Trade", *East European Economies Post-Helsinki – A Compendium of Papers,* Joint Economic Committee, Congress of the United States, 95th Congress, 1st Session, August 25, 1977, US Government Printing Office, Washington, D.C., 1977.

Proceedings of the East-West Technological Trade Symposium, sponsored by the US Department of Commerce, Washington, D.C., November 19, 1975.

Proceedings of the UN/ECE Seminar on the Management of the Transfer of Technology within Industrial Co-operation, United Nations Economic and Social Council, Geneva, 14-17 July, 1975.

Promotion of Trade Through Industrial Co-operation – Recent Trends in East-West Industrial Co-operation, Note by the Secretariat, United Nations Economic and Social Council, Economic Commission for Europe, Committee on the Development of Trade, 27th Session, 27 November-1 December, 1978, Geneva, TRADE/R. 373, 31 August, 1978, Annex I; Annex II; Annex III; Annex IV.

Promotion of Trade Through Industrial Co-operation – Yugoslav Enterprises' Experience of Industrial Co-operation: Results of an Inquiry, Note by the Secretariat, United Nations Economic and Social Council, Economic Commission for Europe, Committee on the Development of Trade, 27th Session, 27 November-1 December, 1978, Geneva, TRADE/R. 373/ADD. 2, 18 September, 1978.

Promotion of Trade Through Industrial Co-operation – Case Studies in Industrial Co-operation: Results of a Survey of Five Western Enterprises, Note by the Secretariat, United Nations Economic and Social Council, Economic Commission for Europe, Committee on the Development of Trade, 27th Session, 27 November-1 December, 1978, Geneva, TRADE/R. 373/ADD. 3, 18 September, 1978.

Promotion of Trade Through Industrial Co-operation – Statistical Outline of Recent Trends in Industrial Co-operation, Note by the Secretariat, United Nations Economic and Social Council, Economic Commission for Europe, Committee on the Development of Trade, 27th Session, 27 November-1 December, 1978, Geneva, TRADE/R. 373/ADD. 5, 19 October, 1978.

Promotion of Trade Through Industrial Co-operation – Proposed Creation of an ECE Information Centre on Industrial Co-operation, Note by the Secretariat, United Nations Economic and Social Council, Economic Commission for Europe, Committee on the Development of Trade, 27th Session, 27 November-1 December, 1978, Geneva, TRADE/R. 373, ADD. 6, 30 October, 1978.

Promotion of Trade Through Industrial Co-operation – Tripartite Industrial Co-operation Contracts: Results of an Inquiry, Note by the Secretariat, United Nations Economic and Social Council, Economic Commission for Europe, Committee on the Development of Trade, 27th Session, 27 November-1 December, 1978, Geneva, TRADE/R. 373/Add. 1, 12 October, 1978.

Promotion of Trade Through Industrial Co-operation – The Experience of Selected Western Enterprises Engaging in East-West Industrial Co-operation: Results of a Survey of Fifteen Firms in the Machine-Tool Sector, Note by the Secretariat, United Nations Economic and Social Council, Economic Commission for Europe, Committee on the Development of Trade, 27th Session, 27th November-1st December, 1978, Geneva, TRADE/R. 373/Add. 4, 14 August, 1978.

"Quantification of Western Exports of High Technology Products to the Communist Countries", Office of East-West Policy and Planning, Bureau of East-West Trade, Washington, D.C., October 17, 1977.

Rapport analytique sur la coopération industrielle entre les pays de la CEE, Economic Commission for Europe, United Nations, Geneva, 1973.

419

Rapport de la Commission du transfert des techniques sur sa 1ʳᵉ session, Genève, 25 novembre-5 décembre 1975, Conseil du commerce et·du développement, Session extraordinaire, United Nations, Geneva, 22 December, 1975.

Ray, G. F., "Thoughts on Innovation and the Transfer of Technology", *Workshop on East-West European Economic Interaction,* Baden near Vienna, Austria, April 3-7, 1977.

Review of the US-USSR Co-operative Agreements on Science and Technology, Special Oversight Report No. 6, Sub-Committee on Domestic and International Scientific Planning and Analysis of the Committee on Science and Technology, US House of Representatives, 94th Congress, 2nd Session, Serial VV, US Government Printing Office, Washington, D.C., November 1976.

Review of the US/USSR Agreement on Co-operation in the Fields of Science and Technology, Board on International Scientific Exchange, Commission on International Relations, National Research Council, National Academy of Sciences, May 1977.

Romer, J., Christophe; de Solere, Michel, "Accords conclus par les pays socialistes européens avec les pays en voie de développement", *Transferts de technologie et développement,* Dijon, 1976.

Rushing, F. W., and Lieberman, Anne, *Impact on US Foreign Trade and Investment From Commercial Transfers of Advanced Technology to the Soviet Union and Eastern Europe,* Strategic Studies Center, Stanford Research Institute, January 10, 1975.

Saunders, C. T. (ed.), *Industrial Policies and Technology Transfers between East and West,* East-West European Economic Interaction Workshop Papers, Volume 3, Springer Verlag, Wien-New York, 1977.

Schenk, Karl-Ernst, *Arbeitshypothesen über die Perspektiven und Probleme gesamteuropäischer wirtschaftlicher Kooperation,* Institut für Aussenhandel und Überseewirtschaft der Universität Hamburg, Forschungsbericht Nr. 1, (Az: 12 2686), n.d.

Schenk, K. E.; von Czege, A., Wass *Technologietransfer und Strukturwandel durch Industrielle Ost-West Kooperation: Ergebnisse eines Forschungsprojekts,* Institut für Aussenhandel und Überseewirtschaft der Universität Hamburg, Forschungsbericht Nr. 18, Hamburg, 1979.

Schneiderman, Ron, "High Technology Flow", *Electronics,* January 8, 1976.

Schwartz, Guy, "Les modalités de la coopération industrielle entre la France et les pays de l'Est: quelques exemples significatifs", *Problèmes économiques,* La Documentation française, Paris, 1ᵉʳ septembre 1976.

Science, Technology and American Diplomacy, An Extended Study of the Interactions of Science and Technology with United States Foreign Policy, Volumes I, II and III, Committee on International Relations, US House of Representatives, US Government Printing Office, Washington, D.C., 1977.

Science, Technology and Diplomacy in the Age of Interdependence, Prepared for the Sub-Committee on International Security and Scientific Affairs of the Committee on International Relations, US House of Representatives, by the Congressional Research Service, Library of Congress, US Government Printing Office, Washington, D.C., June 1976.

Scott, Norman, *Technology Specialisation and Foreign Trade,* Workshop on East-West European Economic Interaction, Vienna Institute for Comparative Economic Studies, Baden near Vienna, Austria, 3-7 April, 1977.

"Selling Turnkey Plans to EE: Profitability Questioned", *Business Eastern Europe,* Vol. 6, No. 36, September 9, 1977.

Shmelyon, N. P., *Scope for Industrial, Scientific and Technical Co-operation between East and West,* Paper for the IEA Round Table in Dresden, German Democratic Republic, 29 June-3 July, 1976.

Simpson, G. S.; Jr., Hill, J. D.; Judy, D. R.; Kleiman, H. S., *Final First Year Technical Report on Development of an Analytical Methodology for Assessing Technology Transfer in the USSR,* Battelle, Columbus Laboratories, 31 January, 1976.

Smith, Maureen A., "Industrial Co-operation Agreements: Soviet Experience and Practice", *Soviet Economy in a New Perspective – A Compendium of Papers,* Joint Economic Committee, Congress of the United States, 94th Congress, 2nd Session, US Government Printing Office, Washington, D.C., October 14, 1976.

Soviet Chemical Equipment Purchases From the West: Impact on Production and Foreign Trade, A, Research Paper, National Foreign Assessment Center, Central Intelligence Agency, Washington, D.C., October 1978.

Special Report Technology Transfer: US Government Policy and its Impact on International Business, Washington International Business Report, December 1977.

Sronek, I., *The Experience of Socialist Countries of Eastern Europe in the Transfer of Technology to Developing Countries,* United Nations Conference on Trade and Development, United Nations, 1978.

420

Steindl, Josef, *Import and Production of Know-how in a Small Country – The Case of Austria*, Workshop on East-West European Economic Interaction, Vienna Institute for Comparative Economic Studies, Baden near Vienna, Austria, 3-7 April, 1977.

Sutton, Antony C., *Western Technology and Soviet Economic Development, 1917-1930*, Vol. 1, Hoover Institution Press, Stanford University, Stanford, California, 1968.

Sutton, Antony C., *Western Technology and Soviet Economic Development 1930-1945*, Hoover Institution Press, Stanford University, Stanford, California, 1971.

Sutton, Antony C., *Western Technology and Soviet Economic Development 1945 to 1965*, Hoover Institution Press, Stanford University, Stanford, California, 1973.

Taylor, Karen and Lamb, Deborah, "Communist Exports to the West in Import Sensitive Sectors", *Issues in East-West Commercial Relations – A Compendium of Papers*, submitted to the Joint Economic Committee, Congress of the United States, 95th Congress, 2nd Session, US Government Printing Office, Washington, D.C., January 12, 1979.

Technological Factors Contributing to the Nation's Foreign Trade Position, Draft report, National Academy of Engineering, Washington, D.C., 18th April, 1977.

Technology Transfer and Innovation, Proceedings of a Conference, National Planning Association and National Science Foundation, 15-17 May, 1966, National Science Foundation, Washington, D.C., 1967.

Technology Transfer and Scientific Co-operation Between the United States and the Soviet Union: A Review, prepared for the Sub-Committee on International Security and Scientific Affairs, Committee on International Relations, Congressional Research Service, Library of Congress, US Government Printing Office, Washington, D.C., May 26, 1977.

Technology Transfer – A Review of Economic Issues, US International Trade Commission, US Department of Commerce, US Department of Labor, A Study Pursuant to Section 119 of the Export Administration Amendments of 1977 (Public Law No. 95-52), Washington, D.C., June 1978.

Technology Transfer: US Government Policy and its Impact on International Business, Washington International Business Report, Special Report 77.3, December 1977.

The Challenge of Technology, Annual Conference on Science and the Humanities, A Conference Report from the National Industrial Conference Board, New York, 1967.

The Fiat-Soviet Auto Plant and Communist Economic Reforms, A report pursuant to House Resolution 1043, 89th Congress, 2nd Session for the Sub-Committee on International Trade, Committee on Banking and Currency, House of Representatives, March 1, 1967, Washington, D.C., 1967.

The Government's Role in East-West Trade – Problems and Issues, Comptroller General of the United States, Summary Statement of Report to the Congress, Washington, D.C., 4th February, 1976.

Theriot, Lawrence H., "US Governmental and Private Industry Co-operation with the Soviet Union in the Fields of Science and Technology", *Soviet Economy in a New Perspective – A Compendium of Papers*, Joint Economic Committee, Congress of the United States, 94th Congress, 2nd Session, US Government Printing Office, Washington, D.C., October 14, 1976.

The US Perspective on East-West Industrial Co-operation, (Preliminary). International Development Center, Indiana University, Bloomington, 1975.

Thomas, John R., and Kruse-Vaucienne, Ursula M., (ed.), *Soviet Science and Technology, Domestic and Foreign Perspectives*, National Science Foundation, The George Washington University Press, Washington, D.C., 1977.

Tiraspolsky, Anita, "Les investissements occidentaux dans les pays de l'Est", *Le courrier des pays de l'Est*, La Documentation française et Centre d'études prospectives et d'informations internationales, Paris, avril 1979.

Toda, Yasushi, "Technology Transfer to the USSR: The Marginal-Productivity Differencial and the Elasticity of Intra-Capital Substitution in Soviet Industry", *Journal of Comparative Economics*, Association for Comparative Economic Studies, Academic Press, New York and London, June 1979.

Transfert de techniques – Monographies sur le transfert des techniques, Note du Secrétariat, Nations Unies, Conseil économique et social, Commission économique pour l'Europe, Conseillers des gouvernements des pays de la CEE pour la science et la technique, 7e Session, 16-20 octobre 1978, Genève, SC.TECH./R.73, 18 août 1978.

Transfer of Technology to the Soviet Union and Eastern Europe, Hearings before the Permanent Sub-Committee on Investigations of the Committee on Governmental Affairs, United States Senate, 95th Congress, 1st Session, Part 2, US Government Printing Office, Washington, D.C., May 25, 1977.

Transfer of Technology to the Soviet Union and Eastern Europe – Selected Papers, Permanent Sub-Committee on Investigations, Committee on Governmental Affairs, United States Senate, 95th Congress, 1st Session, US Government Printing Office, Washington, D.C., September 1977.

Transfer of US Technology to the Soviet Union: Impact on US Commercial Interests, Strategic Studies Centre, Stanford Research Institute, prepared for the Department of State, Washington, D.C., February 1976.

Trzeciakowski, W.; Tabaczynski, E., *The Impact of Technology Transfer on Economic Growth,* Workshop on East-West European Economic Interaction, Vienna Institute for Comparative Economic Studies, Baden near Vienna, Austria, 3-7 April, 1977.

US-USSR Co-operative Agreements in Science and Technology – Hearings, Sub-Committee on Domestic and International Scientific Planning and Analysis of the Committee on Science and Technology, US House of Representatives, 94th Congress, 1st Session, November 18-20, 1975, US Government Printing Office, Washington, D.C., 1976.

US-USSR Technology and Patents, Sale and License Prospects, Licensing Executives Society USA Inc., Trip Reports, July 10-28, 1973.

Vernon, Raymond and Goldman, Marshall I., *US Policies in the Sale of Technology to the USSR* prepared for the Department of Commerce, Washington, D.C., (mimeographed) October 30, 1974.

"Warming Trend in US-Soviet Science Co-operation", *Science,* Vol. 202, November 17, 1978.

Wasowski, Stanislaw, (ed.), *East-West Trade and the Technology Gap – A Political and Economic Appraisal,* Praeger Special Studies in International Economics and Development, Praeger Publishers Inc., New York, 1970.

Weitzman, Martin L., "Technology Transfer to the USSR: An Econometric Analysis", *Journal of Comparative Economics,* Association for Comparative Economic Studies, Academic Press, New York and London, June 1979.

Western Investment in Communist Economies, Committee on Foreign Relations, US Senate, 93rd Congress, 2nd Session, US Government Printing Office, Washington, D.C., 5th August, 1974.

White, Edward P., (ed.), *US-USSR Technology and Patents Sale and Licence Prospects,* Trip Report, 10-28 July, 1973, Licensing Executive Society USA Inc., 1974.

Wilczynski, J., "Licences in the West-East-West Transfer of Technology", *Journal of World Trade Law,* No. 2, March-April 1977.

Wilczynski, J., "The East-West Technological Gap and the 'Reverse' Flow of Technology", *Acta Oeconomica,* Vol. 15 (3-4), 1975.

Winter, David, "Legal Developments in Technology Transfer in East-West Technological Co-operation", *East-West Technological Co-operation,* Colloquium 1976, Directorate of Economic Affairs, NATO, Brussels, March 17-19, 1976.

Wolf, Charles, Jr., *US Technology Exchange with the Soviet Union: A Summary Report,* R-1520/1-ARPA, report prepared for Defense Advanced Research Projects Agency, Rand, Santa Monica, California, August 1974.

Young, John P., *Quantification of Western Exports of High Technology Products to the Communist Countries,* Office of East-West Policy and Planning, Industry and Trade Administration, Bureau of East-West Trade, US Department of Commerce, Washington, D.C., 1978.

Zaleski, Eugène, "East-West Trade and the Technology Gap", edited by Stanislaw Wasowski, *Revue d'études comparatives Est-Ouest,* Editions du Centre national de la recherche scientifique, Vol. 7, No. 3, Paris, 1976.

Zoubek, Jan, *Prospect of East-West Joint Ventures,* East-West Research Report No. 7, Brussels, May 1974.

Zur Entwicklung des Patent und Lizenzverkehr mit Ausland, Monatsbericht der Deutschen Bundesbank, April 1976.

IV. POLITICAL AND STRATEGIC PROBLEMS
OF TECHNOLOGY TRANSFER

Adler-Karlsson, Gunnar, *Western Economic Warfare, 1947-1967*, Almquist and Wicksell, Stockholm, 1968.

An Analysis of Export Control of US Technology – A DOD Perspective, A Report of the Defense Science Board Task Force on Export of US Technology, Washington, D.C., February 4, 1976.

A New Look at Trade Policy Towards the Communist Bloc: the Elements of a Common Strategy for the West, Materials Prepared for the Sub-Committee on Foreign Economic Policy of the Joint Economic Committee, US Government Printing Office, Washington, D.C., 1961.

A Summary of US Export Administration Regulations, Revised June 1978, Office of Export Administration, US Department of Commerce, Industry and Trade Administration, Washington, D.C.

«Avis aux importateurs et aux exportateurs relatifs aux produits soumis au contrôle de la destination finale», Ministère de l'Economie et des Finances, *Journal officiel de la République française*, Paris, 23 avril 1967, 18 septembre 1970, 13 mai 1973.

Bauer, Robert A., (ed.), *The Interaction of Economics and Foreign Policy*, The University of Virginia Press, Charlottesville, 1975.

Bekanntmachung der Neufassung des Umschlüsselungs-Verzeichnisses zur Ausführliste, Teil L, Abschnitte A, B und C, vom Dezember 1976, Beilag sum Bundesanzeiger, No. 24, Vol. 4, Februar 1977, No. 3, Eschborn, 1977.

Berliner, Joseph S., "Some International Aspects of Soviet Technological Progress", *The South Atlantic Quarterly*, Summer 1973.

Brada, Josef C., and Wipf, Larry J., "The Impact of US Trade Controls on Exports to the Soviet Bloc", reprint from *Southern Economic Journal*, Vol. 41, No. 1, July 1974.

Bradsher, Henry S., "Do Soviets Turn US Technology to Military Use?", *The Washington Star*, September 22, 1977.

Bresnick, Ronda A., "The Setting: The Congress and East-West Commercial Relations", *Issues in East-West Commercial Relations – A Compendium of Papers*, submitted to the Joint Economic Committee, Congress of the United States, 95th Congress, 2nd Session, US Government Printing Office, Washington, D.C., January 12, 1979.

Bucy, J. Fred., "On Strategic Technology Transfer to the Soviet Union", *Current News, Special Edition*, No. 234, Harvard University, Cambridge, Mass., 11 August, 1977.

Burks, R. V., "The Political Hazards of Economic Reforms", *Reorientation and Commercial Relations of the Economies of Eastern Europe – A Compendium of Papers*, Joint Economic Committee, Congress of the United States, 93rd Congress, 2nd Session, US Government Printing Office, Washington, D.C., August 16, 1974.

Burt, Richard, "Technology and East-West Arms Control", *International Affairs*, April 1977.

Carter, Jimmy, "Special Report on Multilateral Export Controls", pursuant to Section 117 of Public Law 95-52, the Export Administration Amendments of 1977, The White House, Washington, D.C., July 10, 1978.

"Carter: Upgrade Soviet Trade Tie", *International Herald Tribune*, Paris, March 1, 1979.

"Cocom under Attack in US Congress", *Business Eastern Europe*, Vol. 7, No. 48, December 1, 1978.

"Comecon Politics: Impact on Integration Plans", *Business Eastern Europe*, Vol. 7, No. 50, December 15, 1978.

Commission on Security and Co-operation, *Implementation of the Final Act of the Conference on Security and Co-operation in Europe: Findings and Recommendations Two Years after Helsinki*, Report to the Congress of the United States, Washington, D.C., 1 August, 1977.

"Consolidated List of Goods Subject to Security Export Control: Amendment 1", *Trade and Industry*, United Kingdom, London, 5 November, 1976.

Détente, Issues Series Number One, American Bar Association, ABA Press, Washington, D.C., 1977.

Dudzinsky, S. J., Jr.; Digby, James, "New Technology and Control of Conventional Arms: Some Common Ground", *International Security*, Spring, 1977, reprint *Current News*, Harvard University, Cambridge, Mass., 6 September, 1977.

423

Export Administration Act of 1969, As Amended, Public Law 95-223 (H.R. 7738), 91 Stat. 1625.

Export Administration Amendments of 1977, Conference Report, House of Representatives, 95th Congress, 1st Session, Report No. 95-354, Washington, D.C., May 18, 1977.

Export Administration Bulletin, Supplement to Export Administration Regulations, Number 185, US Department of Commerce, Industry and Trade Administration, Bureau of Trade Regulations, Office of Export Administration, Washington, D.C., August 1, 1978.

Export Administration Bulletin, Supplement to Export Administration Regulations, Number 188, US Department of Commerce, Industry and Trade Administration, Bureau of Trade Regulations, Office of Export Administration, Washington, D.C., September 26, 1978.

Export Administration Bulletin, Supplement to Export Administration Regulations, Number 189, US Department of Commerce, Industry and Trade Administration, Bureau of Trade Regulations, Office of Export Administration, Washington, D.C., October 26, 1978.

Export Administration Bulletin, Supplement to Export Administration Regulations, Number 190, US Department of Commerce, Industry and Trade Administration, Bureau of Trade Regulations, Office of Export Administration, Washington, D.C., December 11, 1978.

Export Administration Bulletin, Supplement to Export Administration Regulations, Number 191, US Department of Commerce, Industry and Trade Administration, Bureau of Trade Regulations, Office of Export Administration, Washington, D.C., December 21, 1978.

Export Administration Bulletin, Supplement to Export Administration Regulations, Number 192, US Department of Commerce, Industry and Trade Administration, Bureau of Trade Regulations, Office of Export Administration, Washington, D.C., January 19, 1979.

Export Administration Bulletin, Supplement to Export Administration Regulations, Number 193, US Department of Commerce, Industry and Trade Administration, Bureau of Trade Regulations, Office of Export Administration, Washington, D.C., March 23, 1979.

Export Administration Bulletin, Supplement to Export Administration Regulations, Number 194, US Department of Commerce, Industry and Trade Administration, Bureau of Trade Regulations, Office of Exports Administration, Washington, D.C., April 6, 1979.

Export Administration Regulations, June 1, 1977, US Department of Commerce, Domestic and International Business Administration, Bureau of East-West Trade, Office of Export Administration, Washington, D.C.

Export Administration Report, 115th Report on US Export Controls to the President and the Congress, Semi-annual: October 1976-March 1977, US Department of Commerce, Domestic and International Business Administration, Bureau of East-West Trade, Washington, D.C.

Export Licensing of Advanced Technology: A Review, Hearings before the Sub-Committee on International Trade and Commerce of the Committee on International Regulations, House of Representatives, 94th Congress, Second Session, March 11, 15, 24 and 30, 1976, US Government Printing Office, Washington, D.C., 1976.

Gilpin, Robert, "Technology Development, Technology Export and American Security", paper presented to the Symposium on Science and the Future Navy, Washington, D.C., October 27, 1976.

Hardt, John P., "Soviet Commercial Relations and Political Change", in Robert A. Bauer, (ed.), *The Interaction of Economics and Foreign Policy,* University of Virginia Press, Charlottesville, 1975.

Hardt, John P., *Soviet Economic Capabilities and Defence Resources,* Report on the Conference "The Soviet Threat – Myth or Reality?", The Academy of Political Science and the Program of Continuing Education, Columbia University, New York, October 17, 1977.

Hardt, John P., "Military-Economic Implications of Soviet Regional Policy", *Regional Development in the USSR,* Colloquium 1979, NATO, Directorate of Economic Affairs, Brussels, April 25-27, 1979.

Hardt, John P. and Holliday, George D., *US-Soviet Commercial Relations: The Interplay of Economics, Technology Transfer and Diplomacy,* prepared for the Sub-Committee on National Security Policy and Scientific Developments of the Committee on Foreign Affairs, US House of Representatives, US Government Printing Office, Washington, D.C., June 10, 1973.

Harvey, Mose L.; Goure, Leon and Prokofieff, Vladimir, *Science and Technology as an Instrument of Soviet Policy,* Centre for Advanced International Studies, University of Miami, 1972.

Hass-Hürni, Bettina, "Economic Issues at Belgrade", *Journal of World Trade Law,* July/August 1978.

Hoffman, Erik P., "Technology, Values and Political Power in the Soviet Union: Do Computers Matter?", in Frederic J. Fleron, Jr., (ed.), *Technology and Communist Culture: The Socio-Cultural Impact of Technology Under Socialism,* Praeger Publishers, New York, London, 1977.

Holzman, Franklyn D., *International Trade under Communism, Politics and Economics*, Basic Books, Inc. Publishers, New York, 1976.

International Transfer of Technology: An Agenda of National Security Issues, prepared for the Sub-Committee on International Security and Scientific Affairs of the Committee on International Relations, US House of Representatives, by the Congressional Research Service, Library of Congress, 95th Congress, 2nd Session, February 13, 1978, US Government Printing Office, Washington, D.C., 1978.

Issues in East-West Commercial Relations – A Compendium of Papers, submitted to the Joint Economic Committee, Congress of the United States, 95th Congress, 2nd Session, US Government Printing Office, Washington, D.C., January 12, 1979.

Johnston, Oswald, "US Seeks Ways to Press Russia on Dissent Trials", *International Herald Tribune*, Paris, July 12, 1978.

Kennan, George F., "The United States and the Soviet Union, 1917-1976", *Foreign Affairs*, July 1976.

"Kissinger Defends the Selling of Détente" (Interview with Arnaud de Borchgrave), *International Herald Tribune*, Paris, December 6, 1978.

Klose, Kevin, "US Carrot Stick Starts Soviet Trade Talks", *International Herald Tribune*, Paris, December 5, 1978.

Kramer, John M., "Between Scylla and Charybdis: The Politics of Eastern Europe's Energy Problem", *ORBIS*, Vol. 22, No. 4, Foreign Policy Research Institute, Winter 1979.

Lavigne, Marie, "Une nouvelle phase des relations Est-Ouest: espoirs et difficultés d'une intensification des échanges", *Le Monde diplomatique*, Paris, septembre 1976 (reproduced in *Problèmes économiques*, La Documentation française, Paris, 3 novembre 1976).

Lavigne, Marie, "Un ordinateur pour l'Agence TASS? Le rôle d'un organisme très discret", *Le Monde diplomatique*, Paris, septembre 1978.

Mountain, Maurice J., "Technology Exports and National Security", *Issues in East-West Commercial Relations – A Compendium of Papers*, submitted to the Joint Economic Committee, Congress of the United States, 95th Congress, 2nd Session, US Government Printing Office, Washington, D.C., January 12, 1979.

Norris, William C., "High Technology Trade with the Communists", *Datamation*, January 1978.

Nuclear Non-Proliferation Act – S 897, Congressional Record, Senate, Washington, D.C., July 29, 1977.

Oberdorfer, Don., "Sale of Drill Plant to Russia is Assailed", *International Herald Tribune*, Paris, October 5, 1978.

Ofer, Gur, *The Opportunity Cost of the Nonmonetary Advantages of the Soviet Military R & D Effort*, R-1741-DDRE, Rand, Santa Monica, California, August 1975.

O'Toole, Thomas, "White House to 'Observe' Technology Export Deals", *International Herald Tribune*, Paris, October 28, 1978.

Parrott, Bruce, "Technological Progress and Soviet Politics", in John R. Thomas and Ursula Druse-Vaucienne, (ed.), *Soviet Science and Technology, Domestic and Foreign Perspectives*, National Science Foundation, George Washington University Press, Washington, D.C., 1977.

Pipes, Richard, (ed.), *Soviet Strategy in Europe*, Strategic Studies Center, Stanford Research Institute, Crane, Russak and Company, Inc., New York, 1976.

Pisar, Samuel, "Comment sauver la détente", *Le Monde*, Paris, 1er octobre 1977 (I. La croisade de M. Carter) and *Le Monde*, Paris, 2-3 octobre 1977 (II. L'équation de la coexistence).

Public Law 93-618, 93rd Congress, H.R. 10710, January 3, 1975 (Trade Act of 1974).

Public Law 95-52, 95th Congress, 91 Stat. 235, "Export Administration Amendments of 1977", June 22, 1977, H.R. 5840.

Rabbot, Boris, "Détente: The Struggle within the Kremlin", *Washington Post*, July 10, 1977.

Report to the Congress of the United States on Implementation of the Final Act of the Conference on Security and Co-operation in Europe: *Findings and Recommendations Two Years after Helsinki*, Washington, D.C., August 1, 1977.

Ruehl, L., "Soviet Policy and the Domestic Politics of Western Europe", in Richard Pipes, (ed.), *Soviet Strategy in Europe*, Strategic Studies Center, Stanford Research Institute, Crane, Russak and Company, Inc., New York, 1976.

Schmidt, Max, "Der Entspannungsprozess und Probleme der ökonomischen Zusammenarbeit von Staaten unterschiedlicher Gesellschaftsordnung", *IPW Berichte*, Heft 2, Institut für Internationale Politik und Wirtschaft der DDR, Februar 1979.

425

Schmidt, Max, "East-West Economic Relations against the Background of New Trends and Developments in World Economy and in International Distribution of Activities", Paper for the IEA Round Table Conference, Dresden, GDR, June 29th to July 3rd, 1976.

Schneiderman, Ron, "High Technology Flow", *Electronics*, 8 January, 1976.

"Science Exchanges at Stake in SALT Talks, Says Adviser", *Nature*, Vol. 278, March 15, 1979.

Sicherheitspolitische Aspekte der Ost-West Wirtschaftsbeziehung, Stiftungwissenschaft und Politik, Ebenhausen, März 1977.

Sixth Semi-annual Report by the President to the Commission on Security and Co-operation in Europe on the Implementation of the Helsinki Final Act, December 1, 1978, May 31, 1979.

Spulber, Nicolas, "East-West Trade and the Paradoxes of the Strategic Embargo", *International Trade and Central Planning — An Analysis of Economic Interactions*, edited by Alan A. Brown and Egon Neuberger, University of California Press, Berkeley and Los Angeles, 1968.

Technology and East-West Trade, Office of Technology Assessment, Congress of the United States, Washington, D.C., November, 1979.

The Commission on Security and Co-operation in Europe, *Implementation of the Final Act of the Conference on Security and Co-operation in Europe : Findings and Recommendations Two Years after Helsinki*, Report to the Congress of the United States, Washington, D.C., August 1, 1977.

"US MFN and Exim Credits Proposed for the USSR", *Business Eastern Europe*, Vol. 8, No. 7, February 16, 1979.

"US-Soviet Ties: The Implications of Severence", *Science and Government Report*, Vol. VIII, No. 11, June 15, 1978.

"US-USSR Trade Relations at a Turning Point?", *Business Eastern Europe*, Vol. 7, Nos. 51 & 52, December 22, 1978.

Wiles, Peter, J., "On the Prevention of Technology Transfer", in *East-West Technological Co-operation*, Colloquium 1976. NATO, Directorate of Economic Affairs, Brussels, March 17-19, 1976.

Wiles, Peter, "L'embargo technologique", *Contrepoint*, Paris, No. 24, 1977.

Woroniak, Alexander, "Economic Aspects of Soviet-American Détente", paper prepared for the Third Atlantic Economic Conference, Washington, D.C., September 12-13, 1975. Reprint: *Revue d'études comparatives Est-Ouest*, Vol. 7, No. 3, Centre national de la recherche scientifique, Paris, 1976.

V. INTRODUCTION OF NEW TECHNOLOGIES IN EASTERN EUROPE

Allinson, W. G., "High Voltage Electric Power Transmission Technology in the USSR", CREES discussion paper RC/B No. 9, Centre for East European Studies, University of Birmingham, 1976, n.d.

Amann, Ronald; Cooper, Julien; Davies, R. W., *The Technological Level of Soviet Industry,* Yale University Press, New Haven and London, 1977.

Amann, Ronald; Slama, J., "The Organic Chemicals Industry in the USSR: A Case Study in the Measurement of Comparative Technological Sophistication by Means of Kilogramprices", *Research Policy,* No. 5, 1976.

Berliner, Joseph S., *The Innovation Decision in Soviet Industry,* The MIT Press, Cambridge, Massachusetts, 1976.

Berliner, Joseph S., "Prospects for Technological Progress", *Soviet Economy in A New Perspective – A Compendium of Papers,* Joint Economic Committee, Congress of the United States, 94th Congress, 2nd Session, US Government Printing Office, Washington, D.C., October 14, 1976.

Bodnar, A.; Zahn, B., *Rewolucja naukowo-techniczna i socjalizm* (Scientific and Technological Revolution and Socialism), Warszawa, 1971.

Boretsky, M., "Comparative Progress in Technology, Productivity and Economic Efficiency: USSR vs USA", *New Directions in the Soviet Economy,* Joint Economic Committee, 89th Congress, 2nd Session, Part II-A, US Government Printing Office, Washington, D.C., 1966.

Burks, Richard V., *Technological Innovation and Political Change in Communist Eastern Europe,* Rand Corporation Memorandum RM-6051-PR, Santa Monica, California, August 1969.

Campbell, Robert W., "Technological Levels in the Soviet Energy Sector", *East-West Technological Co-operation,* Colloquium 1976, NATO, Directorate of Economic Affairs, Brussels, 17-19 March, 1976.

Campbell, Robert W., "Recent Reforms in Eastern Europe", *Problems of Technological Progress in the USSR,* University of Missouri, Kansas City, 16-17 July, 1970.

Campbell, Heather, *Controversy in Soviet Research and Development: The Airship Case Study,* R-1001-PR, Rand, Santa Monica, California, October 1972.

Cooper, Julian, *Innovation for Innovation in Soviet Industry,* CREES Discussion Papers, Centre for Russian and East European Studies, University of Birmingham, June 1979.

Davis, N. C.; Goodman, S. E., "The Soviet Bloc's Unified System of Computers", *Computing Surveys,* Vol. 10, No. 2, June 1978.

"EE Investments to Improve Use of Capital and Labor", *Business Eastern Europe,* Vol. 8, No. 14, April 6, 1979.

Fleron, Frederic, J., Jr., (ed.), *Technology and Communist Culture — The Socio-Cultural Impact of Technology Under Socialism,* Praeger Publishers, New York, London, 1977.

Gallagher, C. C., *Manufacturing Technology in Planned and Market Economies,* prepared for the Conference on Technology and Communist Culture, Villa Serbelloni, Bellagio, Italy, 22-28 August, 1975.

Golland, E., *Ekonomika i organizacija promyshlennogo proizvodstva,* (Economics and Organisation of Industrial Production), Moscow, 1976.

Granick, David, *Soviet Metal-Fabricating and Economic Development, Practice versus Policy,* The University of Wisconsin Press, Madison, Milwaukee and London, 1967.

Granick, David, "Soviet Research and Development Implementation in Products: A Comparison with the German Democratic Republic", *International Economics – Comparisons and Interdependences,* edited by Friedrich Levcik, Festschrift für Franz Nemschak, Springer-Verlag, Wien-New York, 1978.

Granick, David, *Soviet Introduction of New Technology: A Depiction of the Process,* SRI Project 2625, Strategic Studies Center, Stanford Research Institute, Menlo Park, California, January 1975.

Grossman, Gregory, "Price Control, Incentives and Innovation in the Soviet Economy", in Alan A. Boucher, *The Socialist Price Mechanism,* University Press, Durham, North Carolina, 1977.

Gustafson, Thane, "Why does the Soviet Union Lag Behind the United States in Basic Science?", Kennedy School of Government, Center for Science and International Affairs, Harvard University, September 1978. (Scheduled for publication in Susan Gross Solomon and Luisa Lubrano, eds, *The Social Context of Soviet Science,* Western Press, forthcoming.)

Hanson, Philip, *The Soviet System as a Recipient of Foreign Technology*, manuscript, 1977.

Hardt, John P., "The Role of Western Technology in Soviet Economic Plans", *East-West Technological Co-operation*, Colloquium 1976, NATO, Directorate of Economic Affairs, Brussels, March 17-19, 1976.

Hardt, John P.; Holliday, George D., "Technology Transfer and Change in the Soviet Economic System", in J. Frederic Fleron, Jr.,editor, *Technology and Communist Culture:The Socio-Cultural Impact of Technology Under Socialism*, Praeger Publishers, New York, London, 1977.

Hewer, Ulrich, *Zentrale Planung und Technischer Fortschritt, Probleme seiner Organisation und Durchsetzung am Beispiel der Sowjetischen Industrie*, Giessener Abhandlungen zu Agrar und Wirtschaftsforschung des Europäischen Ostens, Duncker und Humblot, Berlin, 1977.

Hinkelmann, Hansjoachim, *Fundamental Research in Soviet Science – A Comparative Study*, Zum Leistungsstand der sowjetischen naturwissenschaftlichen Forschung im internationalen Vergleich, Berichte des Bundesinstituts für ostwissenschaftliche und internationale Studien, Köln, 1977.

Hutchings, Raymond, *Soviet Science, Technology, Design – Interaction and Convergence*, published for the Royal Institute of International Affairs by Oxford University Press, London, New York, Toronto, 1976.

Johnson, Betsy Ancker and Chang, David B., *US Technology Policy – A Draft Study*, Office of the Assistant Secretary for Science and Technology, US Department of Commerce, National Technical Information Service, Washington, D.C., March 1977.

Khachaturov, T. S., *Sovetskaya ekonomika na sovremennom etape* (Soviet Economy at the Present Stage), Moscow, 1975.

Kruse, Vaucienne, U. M., *Effectiveness of Soviet Science*, Programme of Policy Studies in Science and Technology, The George Washington University, Washington, D.C., January 1977.

Labedz, Leopold, "Science and the Soviet System", in Thomas, John R., and Kruse-Vaucienne, Ursula M., (eds.), *Soviet Science and Technology, Domestic and Foreign Perspectives*, National Science Foundation, The George Washington University Press, Washington, D.C., 1977.

Madej, Z., *Nauka i rozwoj gospodarczy* (Science and Economic Development), Warszawa, 1970.

Moiseyenko, V., "Specialisation and Co-operation in the Machine-Building Industry – An Important Factor in Promoting CMEA Member Countries' Mutual Trade", *Foreign Trade*, No. 2, USSR Ministry of Foreign Trade (English translation), 1979.

Nolting, Louvan E., *The 1968 Reform of Scientific Research, Development and Innovation in the USSR*, US Department of Commerce, Bureau of the Census, Washington, D.C., September 1976.

Nolting, Louvan E., *The Planning of Research, Development and Innovation in USSR*, US Department of Commerce, Bureau of the Census, Washington, D.C., July 1978.

Nötzold, Jürgen, *Untersuchungen zur Durchsetzung des Technischen Fortschritts in der Sowjetischen Wirtschaft*, Stiftung Wissenschaft und Politik, Eggenberg, December 1972.

Olteanu, Ionita; Rausser, Vasile, *Joint Ventures in the Innovation Process in Rumania*, Workshop on East European Economic Interaction, Vienna Institute for Comparative Economic Studies, Baden near Vienna, Austria, 3rd-7th April, 1977.

Perakh, Mark, "Utilization of Western Technological Advances in Soviet Industry", *East-West Technological Co-operation*, Colloquium 1976, NATO, Directorate of Economic Affairs, Brussels, March 17-19, 1976.

Perry, Robert, *Comparisons of Soviet and US Technology*, Rand, R-827-PR, Santa Monica, California.

Siemiaszko, Z. A., "Industrial Process Control", *The Technological Level of Soviet Industry*, Yale University Press, New Haven and London, 1977.

Slama, Jiri; Vogel, Heinrich, *Comparative Analysis of Research and Innovation Processes in East and West*, Workshop on East-West European Economic Interaction, Vienna Institute for Comparative Economic Studies, Baden near Vienna, Austria, 3-7 April, 1977.

Slama, Jiri; Vogel, Heinrich, *Technology Advances in CMEA Countries* (final version), Osteuropa-Institut, München, 1977.

Soviet Space Programs 1971-1975, Vol. I and Vol. II, Committee on Aeronautical and Space Sciences, US Senate, US Government Printing Office, Washington, D.C., 30 August, 1976.

Spechler, Martin C., "The Pattern of Technological Achievement in the Soviet Enterprise", *The Association for Comparative Studies Bulletin*, Summer 1975.

USSR-GDR: 25 Years of Co-operation in Science and Technology, Novosti Press Agency Publishing House, Moscow, 1976.

428

Wilczynski, J., *Technology in Comecon: Acceleration of Technological Progress through Economic Planning and the Market*, MacMillan, London and Basingstoke, 1974.

Young, John P., *Impact of Soviet Ministry Management Practices on the Assimilation of Imported Process Technology* (with examples from the motor vehicle sector), paper presented at the Joint Annual Meeting of the Southwestern and Rocky Mountain Associations of Slavic Studies, Houston, Texas, April 13, 1978.

Zaleski, Eugène, "Planning and Financing of Research and Development in the USSR", in John R. Thomas and Ursula M. Kruse-Vaucienne, *Soviet Science and Technology, Domestic and Foreign Perspectives,* National Science Foundation, The George Washington University Press, Washington, D.C., 1977.

Zaleski, Eugène; Kozlowski, J. P.; Wienert, H.; Davies, R. W.; Berry, M. J.; Amann, R., *Science Policy in the USSR,* OECD, Paris, 1969.

VI. GENERAL STUDIES ON EASTERN ECONOMIES

Allocation of Resources in the Soviet Union and China – 1977, Hearings, Sub-Committee on Priorities and Economy in Government, Joint Economic Committee, Congress of the United States, 95th Congress, 1st Session, Part 3, June 23 and 30 (Executive Sessions), and July 6, 1977, US Government Printing Office, Washington, D.C., 1977.

Alton, Thad P., "Comparative Structure and Growth of Economic Activity in Eastern Europe", *East European Economies Post-Helsinki – A Compendium of Papers,* Joint Economic Committee, Congress of the United States, 95th Congress, 1st Session, US Government Printing Office, Washington, D.C., August 25, 1977.

Bajbakov, N. K., "O gosudarstvennom plane ekonomicheskogo i socialnogo razvitija SSSR na 1979 god", (Comments on the USSR Economic and Social Development Plan for 1979), *Ekonomicheskaya Gazeta,* No. 50, Moscow, December 1978.

Baykov, Alexander, *The Development of the Soviet Economic System,* Cambridge, United Kingdom, 1950.

Bergson, Abram, "Comparative Productivity and Efficiency in the USA and the USSR", *Comparison of Economic Systems: Theoretical and Methodological Approaches,* edited by Alexander Eckstein, University of California Press, Berkeley, Los Angeles, London, 1971.

Bornstein, Morris, "Economic Reforms in Eastern Europe", *East European Economies Post-Helsinki – A Compendium of Papers,* Joint Economic Committee, Congress of the United States, 95th Congress, 1st Session, US Government Printing Office, Washington, D.C., August 25, 1977.

Bornstein, Morris, "Unemployment in Capitalist Market Economies and Socialist Centrally Planned Economies", *The American Economic Review,* May 1978.

Bush, Keith, "Les prix de détail: Moscou et quatre villes occidentales", *Revue de l'Est,* No. 1, Centre national de la recherche scientifique, Paris, janvier 1974.

Collected Reports on Various Activities of Bodies of the CMEA in 1977, Council for Mutual Economic Assistance, Secretariat, Moscow, 1977.

Comecon: Progress and Prospects, Colloquium 1977, NATO, Directorate of Economic Affairs, Brussels, March 16-18, 1977.

"Die Aktionslinien des Comecon", *Neue Zürcher Zeitung,* Fernausgabe, No. 66, 21 ten März, 1978.

"Die 32te Tagung des Rates für Gegenseitige Wirtschaftshilfe", *Wiener Institut für Internationale Wirtschaftsforschung,* Mitgliederinformation, 6/1978, Oktober 1978.

"Die 'sozialistischen multinationalen Unternehmungen' der Comecon-Länder", *Neue Zürcher Zeitung,* Fernausgabe No. 276, 25 ten November, 1977.

Faddeyev, N. V., "CMEA's Role in Strengthening the Community of the Socialist Nations", *Foreign Trade,* No. 1, USSR Ministry of Foreign Trade (English translation), 1979.

Fallenbuchl, Zbigniew M., *Planning, Market and Integration in Eastern Europe,* Discussion Paper Series, Serial, No. 50, Department of Economics, University of Windsor, Windsor, Ontario, March 31, 1978.

Fallenbuchl, Zbigniew M., "The Polish Economy in the 1970's", in *East European Economies Post-Helsinki – A Compendium of Papers,* Joint Economic Committee, Congress of the United States, 95th Congress, 1st Session, US Government Printing Office, Washington, D.C., August 25, 1977.

Feshbach, Murray and Rapawy, Stephen, "Soviet Population and Manpower Trends and Policies", *Soviet Economy in a New Perspective – A Compendium of Papers,* Joint Economic Committee, Congress of the United States, 94th Congress, 2nd Session, US Government Printing Office, Washington, D.C., October 14, 1976.

"Gebremster Westhandel und verstärkte Autarkie im Comecon", *Neue Zürcher Zeitung,* Fernausgabe No. 199, 30. August, 1978.

Haberstroh, John R., "Eastern Europe: Growing Energy Problems", *East European Economies Post-Helsinki – A Compendium of Papers,* Joint Economic Committee, Congress of the United States, 95th Congress, 1st Session, US Government Printing Office, Washington, D.C., August 25, 1977.

Hardt, John P.; Bresnik, Ronda A. and Levine, David, "Soviet Oil and Gas in the Global Perspective", *Project Interdependence: US and World Energy Outlook Through 1990,* Congressional Research Service, Library of Congress, 95th Congress, 1st Session, US Government Printing Office, Washington, D.C., November 1977.

"Higher Growth Planned for this Year, Nikolai Baibakov Reports on State Plan for 1979", *Soviet News,* No. 5956, published by the Press Department of the Soviet Embassy in London, January 23, 1979.

"Indicators of Comparative East-West Economic Strength – 1976", Bureau of Public Affairs, Office of Media Services, *Special Report No. 36,* Department of State, Washington, D.C., November 1977.

Jack, Emily E.; Lee, Richard; Lent, Harold H., "Outlook for Soviet Energy", *Soviet Economy in a New Perspective – A Compendium of Papers,* Joint Economic Committee, Congress of the United States, 94th Congress, 2nd Session, US Government Printing Office, Washington, D.C., October 14th, 1976.

Jacobs, Everett M., "The Global Impact of Foreign Trade on Soviet Growrh", *The USSR in the 1980's: Economic Growth and the Role of Foreign Trade,* Colloquium 1978, Directorate of Economic Affairs, NATO, Brussels, January 17-19th, 1978.

Lister, James P., "Siberia and the Soviet Far East: Development of Policies and the Yakutia Gas Project", *Regional Development in the USSR,* Colloquium 1979, Directorate of Economic Affairs, NATO, Brussels, April 25-27th, 1979.

Kosygin, A. N., "Main Directions of the Development of the National Economy of the USSR for the Years 1976-1980", *XXV S'jezd Kommunisticheskoi Partii Sovetskogo Sojuza* (25th Congress of the Communist Party of the Soviet Union), Vol. 2, Moscow, 1976.

Marczewski, Jan, *Planification et croissance des démocraties populaires – Analyse historique,* Presses universitaires de France, Paris, 1956.

Nutter, G. Warren, "The Structure and Growth of Soviet Industry: Its Comparison with the United States", *Comparisons of the United States and Soviet Economies,* 86th Congress, 1st Session, Part I, Washington, D.C., 1959.

Nutter, G. Warren, *The Growth of Industrial Production in the Soviet Union,* Princeton University Press, Princeton, N.J., 1962.

"Organisation and Management in the Soviet Economy: The Ceaseless Search for Panaceas", Central Intelligence Agency, National Foreign Assessment Center, ER 77-10769, December 1977.

"1978 Plan Failure Hit Bulgarian Export Items", *Business Eastern Europe,* Vol. 8, No. 11, March 16th, 1979.

"Production et utilisation de l'or soviétique", *Report of Chamber of Mines of South Africa,* September 1978, cited in *Problèmes économiques,* La Documentation française, Paris, 18 juillet 1979.

Prokopovicz, Serge N., *Histoire économique de l'URSS,* Flammarion, Paris, 1952.

Prospects for Soviet Oil Production, Central Intelligence Agency, ER 77-10270, Washington, D.C., April 1977.

Prospects for Soviet Oil Production – A Supplemental Analysis, Central Intelligence Agency, Washington, D.C., April 1977.

Seeger, Murray, "Comecon's Growth Below 1978 and 5 year Targets", *International Herald Tribune,* Paris, February 8th, 1979.

Smith, Alan, "Soviet Economic Influence in Comecon", *Comecon: Progress and Prospects,* Colloquium 1977, NATO, Directorate of Economic Affairs, Brussels, March 16-18th, 1977.

Smith, Arthur J., "The Council of Mutual Economic Assistance in 1977: New Economic Power, New Political Perspectives and Some Old and New Problems", *East European Economies Post-Helsinki – A Compendium of Papers,* Joint Economic Committee, Congress of the United States, 95th Congress, 1st Session, US Government Printing Office, Washington, D.C., August 25th, 1977.

Soviet Economic Outlook, Hearing before the Joint Economic Committee, Congress of the United States, 93rd Congress, 1st Session, US Government Printing Office, Washington, D.C., July 1973.

"Soviet Economy Loses Momentum", *Business Eastern Europe,* Vol. 7, No. 49, December 8th, 1978.

"Soviet Economy Worsening, CIA Says", *International Herald Tribune,* Paris, October 5th, 1978.

"Soviet Oil Industry Eager to Get US Technology", *International Herald Tribune,* Paris, February 19th, 1979.

Stoin, Ion, "Principles Guiding Activity in CMEA Discussed", *US Joint Publications Research Service,* No. 72204, US Government Printing Office, Washington, D.C., November 8th, 1978.

"The Business Outlook – Czechoslovakia", *Business Eastern Europe,* Vol. 8, No. 11, March 16th, 1979.

431

"The Business Outlook – Hungary", *Business Eastern Europe,* Vol. 8, No. 8, February 23rd, 1979.

"The Business Outlook – Poland", *Business Eastern Europe,* Vol. 8, No. 13, March 30th, 1979.

"The Business Outlook – Soviet Union", *Business Eastern Europe,* Vol. 8, No. 6, February 9th, 1979.

"The Planetary Product at Near Zero Growth in 1975 (and a Preview for 1976)", External Research Study, INR/XRS-9, Department of State, Washington, D.C., March 1977.

"The Planetary Product in 1975 and a Preview for 1976", *Special Report No. 33,* Bureau of Public Affairs, Office of Media Services, Department of State, Washington, D.C., May 1977.

The Soviet Oil Situation: An Evaluation of CIA Analyses of Soviet Oil Production, Staff Report of the Senate Select Committee on Intelligence, United States Senate, 95th Congress, 2nd Session, US Government Printing Office, Washington, D.C., May 1978.

The USSR in the 1980s, Economic Growth and the Role of Foreign Trade, Colloquium 17th-19th January, 1978, Directorate of Economic Affairs, NATO, Brussels, 1978.

Tiraspolsky, A., "Le pouvoir d'achat du rouble en 1972", *Revue de l'Est,* No. 1, Centre national de la recherche scientifique, Paris, janvier 1974.

XXV S'ezd Kommunisticheskoj Partii Sovetskogo Sojuza (25th Congress, Communist Party of the Soviet Union), Vol. 2, Moscow, 1976.

USSR: Development of the Gas Industry, National Foreign Assessment Center, ER 78-10393, Central Intelligence Agency, Washington, D.C., July, 1978.

"USSR 1978 Plan Results Pinpoint Problem Areas", *Business Eastern Europe,* Vol. 8, No. 5, February 2nd, 1979.

Vernet, Daniel, "Le rythme de développement de l'industrie s'est ralenti au cours des derniers mois de 1978", *Le Monde,* Paris, 24 janvier 1979.

"Vorsichtigere Planung in der Ostblockwirtschaft", *Neue Zürcher Zeitung,* Fernausgabe, Nr. 67, 22 März, 1978.

Weiss, Gerhard, "Resolutely on the Road to Socialist Economic Integration", *US Joint Publications Research Service,* No. 72301, US Government Printing Office, Washington, D.C., November 24th, 1978.

Western Perceptions of Soviet Economic Trends, A Staff Study, Sub-committee on Priorities and Economy in Government, Joint Economic Committee, Congress of the United States, 95th Congress, 2nd Session, US Government Printing Office, Washington, D.C., March 6th, 1978.

"Wird Moskau die Erdölproduktion weiter erhöhen?", *Neue Zürcher Zeitung,* February 1979.

Zaleski, Eugène, *Planning for Economic Growth in the Soviet Union, 1918-1932,* The University of North Carolina Press, Chapel Hill, 1971.

Zaleski, Eugène, *Stalinist Planning for Economic Growth 1933-1952,* Edited by Marie-Christine MacAndrew and John H. Moore, The University of North Carolina Press, Chapel Hill, 1980.

Zoubek, Jan, "Prospects of East-West Joint Ventures", *East-West,* Research Report No. 7, Brussels, May 1974.

VII. ECONOMY AND TECHNOLOGY TRANSFER IN WESTERN COUNTRIES

A Factbook Concerning the Relationship Between Technology and Trade, Vol. II, *Legal/Institutional Data,* Center for Policy Alternatives, MIT, Cambridge, Massachusetts, August 1976.

Analyse des institutions et des procédures relatives à la gestion et à l'organisation de la recherche internationale concertée, Economic Commission for Europe, United Nations, Geneva, SC.TECH/ R.41, 10 May, 1976.

Besoins statistiques des responsables de la politique en matière de transfert des techniques, Note by the Secretariat, Economic Commission for Europe, United Nations, 4th Session, 15-19 September, 1975, Geneva, SC.TECH/R. 27/Rev. 1, 4 December, 1975.

Carter, Anne, "The Economics of Technological Change", *The Scientific American,* April 1966.

Gaps in Technology – Analytical Report, Comparisons between Member Countries, OECD, Paris, 1970.

"Economic Growth of OECD Countries, 1966-1976", *Special Report No. 31,* Bureau of Public Affairs, Office of Media Services, Department of State, Washington, D.C., March 1977.

Guide on Drawing up International Contracts on Industrial Co-operation, United Nations, New York, 1976.

Lave, Lester B., *Technological Change: Its Conception and Measurement,* Prentice Hall Inc., Englewood Cliffs, New Jersey, 1966.

Le rôle et la place des industries mécaniques et électriques dans les économies nationales et dans l'économie mondiale, Vol. I et II, Economic Commission for Europe, United Nations, New York, 1974.

Le transfert inverse des techniques, son ampleur, ses conséquences économiques et ses incidences en matière de politique générale, Etude du Secrétariat de la CNUCED, TD/B/C.6/7, United Nations, Geneva, 13 October, 1975.

"Les transferts de technologie", *Mondes en développement,* No. 14, *Mondes en développement,* No. 15, Institut des sciences mathématiques et économiques appliquées, Paris, 1976.

Marczewski, Jean, *Inflation et chômage en France,* Editions Economica, Paris, 1977.

Nabseth, L.; Ray, G. F., (ed.), *The Diffusion of New Industrial Processes,* Cambridge, 1974.

Rapport du groupe d'experts officieux pour la création de centres de transfert et de développement de techniques, Conseil du Commerce et du Développement, Conférence des Nations Unies sur le Commerce et le Développement, United Nations, Geneva, TD/B/595, 6 January, 1976.

Tendances et perspectives du marché des produits des industries mécaniques et électriques utilisés dans le secteur de l'énergie et de l'équipement des télécommunications, Nations Unies, Conseil économique et social, Commission économique pour l'Europe, Geneva, December 1976.

Transfert des techniques – La dépendance technique, sa nature, ses conséquences et ses incidences en matière de politique générale, Rapport du Secrétariat de la CNUCED, TD/190, 31 décembre 1975, Quatrième session, Nairobi, 5 May, 1976.

Un code international de conduite pour le transfert des techniques, Rapport du Secrétariat de la CNUCED, United Nations, New York, 1975.

VIII. YEARBOOKS, PERIODICALS AND JOURNALS CITED

A. Western

Economic Bulletin for Europe, Vol. 20, No. 1, Prepublication, Trade (XXVII), United Nations, Geneva.

Economic Bulletin for Europe, Vol. 25, United Nations, New York, 1974.

Economic Bulletin for Europe, Vol. 28, United Nations, New York, 1976.

Economic Bulletin for Europe, Vol. 29, United Nations, New York, 1977.

Economic Bulletin for Europe, Vol. 30, United Nations, Geneva, 1978.

Economic Bulletin for Europe, Vol. 31, United Nations, Geneva, 1979.

Bulletin mensuel de statistique (Monthly Statistical Bulletin), United Nations, New York, 1978.

Bulletin of Statistics on World Trade in Engineering Products, 1974, 1975, United Nations, New York, 1976, 1977.

Business Eastern Europe, Business International, Chemin Riou, Geneva, 1976-1979.

Doing Business with Eastern Europe, Business International, Geneva, October 1975-November 1977.

Eastern Europe, Editor: Kurt Weisskopf, Production Editor: Roger Craik, London Chamber of Commerce and Industry (fortnightly).

East-West Markets, Stockholm, 1976, 1977, 1978.

Economic Survey of Europe in 1978, Part I. *The European Economy in 1978*, (Prepublication, 20th March, 1979).

International Financial Statistics, IMF, Washington, D.C., 1977.

Marer, Paul, *Soviet and East European Foreign Trade, 1946-1969*, Statistical Compendium and Guide, Indiana University Press, Bloomington and London, 1972.

Ost-Wirtschaftsreport Digest, Handelsblatt, Dusseldorf, Business International, Genève, 1977.

Revue d'études comparatives Est-Ouest, Editions du Centre national de la recherche scientifique, Paris, 1972-1979.

Science Resources/Newsletter, OECD/DSTI, "Science Resources" Unit, No. 2, OECD, Paris, Spring 1977.

Selected Trade and Economic Data of the Centrally Planned Economies, US Department of Commerce, Industry and Trade Administration, Bureau of East-West Trade, Washington, D.C., December 1977-January 1978.

Statistical Yearbook of the League of Nations, Years 1937-1938.

Statistics of the Foreign Trade, Series A, OECD, Paris, 1977.

Statistics of Foreign Trade, Trade by Commodities, Market Summaries, Series C., OECD, Paris, 1965.

Statistics of Foreign Trade, Trade by Commodities, Market Summaries, Series C., OECD, Paris, 1970.

Statistics of Foreign Trade, Trade by Commodities, Market Summaries, Series C., OECD, Paris, 1974, 1975, 1976 and 1977.

Wochenbericht, DIW (Deutsches Institut für Wirtschaftsforschung), Berlin (West), 1977, 1978.

B. Eastern

Annuaire du commerce extérieur tchécoslovaque, Editions de la Chambre de commerce tchécoslovaque, Prague, 1977, 1978 and 1979.

Commerce extérieur (French edition of the monthly review *Vneshnyaya Torgovlya*), Moscow, Supplément to No. 3, 1979.

Czechoslovakian Foreign Trade Yearbook, Czechoslovakian Chamber of Commerce Publications, Prague, 1977 and 1978.

Ekonomicheskaya Gazeta, Weekly, Moscow, 1976-1980.

Ekonomicheskie Otnoshenjja SSR s Spravochnik (Economic Relations Between the USSR and Foreign Countries, 1917-1967), Moscow, 1967.

Külkereskedelmi Statistikai Evkönyv (Statistical Yearbook of Foreign Trade), Budapest, years 1976, 1977.

Marketing in Hungary (monthly), Budapest, 1975-1979.

Maly Rocznik Statystyczny (The Small Statistical Yearbook), Warsaw, 1939.

Maly Rocznik Statystyczny, 1979 (The Small Statistical Yearbook), Warsaw, 1979.

Mishustin D., Ed., *Vneshnyaya Torgovlya Sovetskogo Suyuza* (Foreign Trade of the Soviet Union), Moscow, 1938.

Moscow, Narodnyj Bank Press Bulletin, London, 1975-1979.

Narodnogo Khosjajstvo SSSr za 60 let, (National Economy of the USSR for 60 Years), Moscow, 1977.

50 let Sovetskoy Vneshnoj torgovli, (Fifty Years of Soviet Foreign Trade), Moscow, 1967.

Polityka, Revue, Warsaw, 3 March 1978.

Pravda, 1965-1979.

SSSR v cifrakh 1977 godu (The USSR in Figures in 1977), Moscow, 1978.

Statisticheski Godisnik na Carstvo Bolgarija, Godina XXX, (Statistical Yearbook of the Kingdom of Bulgaria, XXX year), Sofia, 1938.

Statisticheskij Ezhegodnik Strans Chlenov Soveta Ekonomicheskoij Vzaimopomoschi – 1978 (Statistical Yearbook for CMEA countries for 1978), Moscow, 1978.

Statisticka Rocenka CSSR (Statistical Yearbook of Czechoslovakia), Prague, 1977 and 1978.

Spravochnik po Veshnei Torgovle (Handbook of Foreign Trade), Moscow, 1958.

Vneshnyaya Torgovlya (Foreign Trade), Moscow, May 1966.

Vneshnyaya Torgovlya SSSR v 1973 godu (USSR Foreign Trade in 1973), Moscow, 1974.

Vneshnyaya Torgovlya SSSR v 1974 godu (USSR Foreign Trade in 1974), Moscow, 1975.

Vneshnyaya Torgovlya SSSR v 1975 godu (USSR Foreign Trade in 1975), Moscow, 1976.

Vneshnyaya Torgovlya SSSR v 1976 godu (USSR Foreign Trade in 1976), Moscow, 1977.

Vneshnyaya Torgovlya SSSR v 1977 godu (USSR Foreign Trade in 1977), Moscow, 1978.

Vneshnyaya Torgovlya SSSR v 1978 godu (USSR Foreign Trade in 1978), Moscow, 1979.

Vneshnyaya Torgovlya SSSR za 1918-1940 gg., Statisticheskij Obzor (Foreign Trade of the USSR for 1918-1940, Statistical Review), Moscow, 1960.

OECD SALES AGENTS
DÉPOSITAIRES DES PUBLICATIONS DE L'OCDE

ARGENTINA – ARGENTINE
Carlos Hirsch S.R.L., Florida 165, 4° Piso (Galería Guemes)
1333 BUENOS-AIRES, Tel. 33-1787-2391 Y 30-7122

AUSTRALIA – AUSTRALIE
Australia & New Zealand Book Company Pty Ltd.,
23 Cross Street, (P.O.B. 459)
BROOKVALE NSW 2100 Tel. 938-2244

AUSTRIA – AUTRICHE
OECD Publications and Information Center
4 Simrockstrasse 5300 BONN Tel. (0228) 21 60 45
Local Agent:
Gerold and Co., Graben 31, WIEN 1. Tel. 52.22.35

BELGIUM – BELGIQUE
LCLS
44 rue Otlet, B1070 BRUXELLES .Tel. 02-521 28 13

BRAZIL – BRÉSIL
Mestre Jou S.A., Rua Guaipà 518,
Caixa Postal 24090, 05089 SAO PAULO 10. Tel. 261-1920
Rua Senador Dantas 19 s/205-6, RIO DE JANEIRO GB.
Tel. 232-07. 32

CANADA
Renouf Publishing Company Limited,
2182 St. Catherine Street West,
MONTREAL, Quebec H3H 1M7 Tel. (514) 937-3519

DENMARK – DANEMARK
Munksgaards Boghandel,
Nørregade 6, 1165 KØBENHAVN K. Tel. (01) 12 85 70

FINLAND – FINLANDE
Akateeminen Kirjakauppa
Keskuskatu 1, 00100 HELSINKI 10. Tel. 65-11-22

FRANCE
Bureau dés Publications de l'OCDE,
2 rue André-Pascal, 75775 PARIS CEDEX 16. Tel. (1) 524.81.67
Principal correspondant :
13602 AIX-EN-PROVENCE : Librairie de l'Université.
Tel. 26.18.08

GERMANY – ALLEMAGNE
OECD Publications and Information Center
4 Simrockstrasse 5300 BONN Tel. (0228) 21 60 45

GREECE – GRÈCE
Librairie Kauffmann, 28 rue du Stade,
ATHÈNES 132. Tel. 322.21.60

HONG-KONG
Government Information Services,
Sales and Publications Office, Baskerville House, 2nd floor,
13 Duddell Street, Central. Tel. 5-214375

ICELAND – ISLANDE
Snaebjörn Jónsson and Co., h.f.,
Hafnarstraeti 4 and 9, P.O.B. 1131, REYKJAVIK.
Tel. 13133/14281/11936

INDIA – INDE
Oxford Book and Stationery Co.:
NEW DELHI, Scindia House. Tel. 45896
CALCUTTA, 17 Park Street. Tel. 240832

INDONESIA – INDONÉSIE
PDIN-LIPI, P.O. Box 3065/JKT., JAKARTA, Tel. 583467

IRELAND – IRLANDE
TDC Publishers – Library Suppliers
12 North Frederick Street, Dublin 1 Tel. 744835-749677

ITALY – ITALIE
Libreria Commissionaria Sansoni:
Via Lamarmora 45, 50121 FIRENZE. Tel. 579751
Via Bartolini 29, 20155 MILANO. Tel. 365083
Sub-depositari:
Editrice e Libreria Herder,
Piazza Montecitorio 120, 00 186 ROMA. Tel. 6794628
Libreria Hoepli, Via Hoepli 5, 20121 MILANO. Tel. 865446
Libreria Lattes, Via Garibaldi 3, 10122 TORINO. Tel. 519274
La diffusione delle edizioni OCSE è inoltre assicurata dalle migliori
librerie nelle città più importanti.

JAPAN – JAPON
OECD Publications and Information Center,
Landic Akasaka Bldg., 2-3-4 Akasaka,
Minato-ku, TOKYO 107 Tel. 586-2016

KOREA · CORÉE
Pan Korea Book Corporation,
P.O.Box n° 101 Kwangwhamun, SÉOUL. Tel. 72-7369

LEBANON – LIBAN
Documenta Scientifica/Redico,
Edison Building, Bliss Street, P.O.Box 5641, BEIRUT.
Tel. 354429–344425

MALAYSIA – MALAISIE
and/et SINGAPORE-SINGAPOUR
University of Malaya Co-operative Bookshop Ltd.
P.O. Box 1127, Jalan Pantai Baru
KUALA LUMPUR Tel. 51425, 54058, 54361

THE NETHERLANDS – PAYS-BAS
Staatsuitgeverij
Verzendboekhandel Chr. Plantijnstraat
S-GRAVENHAGE Tel. nr. 070-789911
Voor bestellingen: Tel. 070-789208

NEW ZEALAND – NOUVELLE-ZÉLANDE
The Publications Manager,
Government Printing Office,
WELLINGTON: Mulgrave Street (Private Bag),
World Trade Centre, Cubacade, Cuba Street,
Rutherford House, Lambton Quay, Tel. 737-320
AUCKLAND: Rutland Street (P.O.Box 5344), Tel. 32.919
CHRISTCHURCH: 130 Oxford Tce (Private Bag), Tel. 50.331
HAMILTON: Barton Street (P.O.Box 857), Tel. 80.103
DUNEDIN: T & G Building, Princes Street (P.O.Box 1104),
Tel. 78.294

NORWAY – NORVÈGE
J.G. TANUM A/S Karl Johansgate 43
P.O. Box 1177 Sentrum OSLO 1 Tel (02) 80 12 60

PAKISTAN
Mirza Book Agency, 65 Shahrah Quaid-E-Azam, LAHORE 3.
Tel. 66839

PHILIPPINES
National Book Store, Inc.
Library Services Division, P.O.Box 1934, Manila,
Tel. Nos. 49-43-06 to 09 40-53-45 49-45-12

PORTUGAL
Livraria Portugal, Rua do Carmo 70-74,
1117 LISBOA CODEX. Tel. 360582/3

SPAIN – ESPAGNE
Mundi-Prensa Libros, S.A.
Castelló 37, Apartado 1223, MADRID-1. Tel. 275.46.55
Libreria Bastinos, Pelayo, 52, BARCELONA 1. Tel. 222.06.00

SWEDEN – SUÈDE
AB CE Fritzes Kungl Hovbokhandel,
Box 16 356, S 103 27 STH, Regeringsgatan 12,
DS STOCKHOLM. Tel. 08/23 89 00

SWITZERLAND – SUISSE
OECD Publications and Information Center
4 Simrockstrasse 5300 BONN Tel. (0228) 21 60 45
Agents locaux :
Librairie Payot, 6 rue Grenus, 1211 GENÈVE 11. Tel: 022.31.89.50
Freihofer A.G., Weinbergstr. 109, CH-8006 Zürich Tel: 01-3624282

TAIWAN – FORMOSE
National Book Company,
84-5 Sing Sung South Rd., Sec. 3, TAIPEI 107. Tel. 321.0698

THAILAND – THAILANDE
Suksit Siam Co., Ltd., 1715 Rama IV Rd.
Samyan, BANGKOK 5 Tel. 2511630

UNITED KINGDOM – ROYAUME-UNI
H.M. Stationery Office, P.O.B. 569,
LONDON SEI 9 NH. Tel. 01-928-6977, Ext. 410 or
49 High Holborn, LONDON WC1V 6 HB (personal callers)
Branches at: EDINBURGH, BIRMINGHAM, BRISTOL,
MANCHESTER, CARDIFF, BELFAST.

UNITED STATES OF AMERICA – ÉTATS-UNIS
OECD Publications and Information Center, Suite 1207,
1750 Pennsylvania Ave., N.W. WASHINGTON. D.C.20006.
Tel. (202)724 1857

VENEZUELA
Libreria del Este, Avda. F. Miranda 52, Edificio Galipàn,
CARACAS 106. Tel. 32 23 01/33 26 04/33 24 73

YUGOSLAVIA – YOUGOSLAVIE
Jugoslovenska Knjiga, Terazije 27, P.O.B. 36, BEOGRAD.
Tel. 621-992

Les commandes provenant de pays où l'OCDE n'a pas encore désigné de dépositaire peuvent être adressées à :
OCDE, Bureau des Publications, 2 rue André-Pascal, 75775 PARIS CEDEX 16.
Orders and inquiries from countries where sales agents have not yet been appointed may be sent to:
OECD, Publications Office, 2 rue André-Pascal, 75775 PARIS CEDEX 16.

OECD PUBLICATIONS, 2 rue André-Pascal, 75775 Paris Cedex 16 - N° 41 379 1980
PRINTED IN FRANCE
(5000 D-- 92 80 02 1) ISBN 92-64-12125-0